Nonparametric Statistical Methods and Related Topics

A Festschrift in Honor of Professor P K Bhattacharya on the Occasion of His 80th Birthday

Nonparametric Statistical Methods and Related Topics

A Festschrift in Honor of Professor P K Bhattacharya on the Occasion of His 80th Birthday

editors

J Jiang ♦ G G Roussas ♦ F J Samaniego

University of California, Davis, USA

World Scientific

NEW JERSEY • LONDON • SINGAPORE • BEIJING • SHANGHAI • HONG KONG • TAIPEI • CHENNAI

Published by

World Scientific Publishing Co. Pte. Ltd.
5 Toh Tuck Link, Singapore 596224
USA office: 27 Warren Street, Suite 401-402, Hackensack, NJ 07601
UK office: 57 Shelton Street, Covent Garden, London WC2H 9HE

British Library Cataloguing-in-Publication Data
A catalogue record for this book is available from the British Library.

NONPARAMETRIC STATISTICAL METHODS AND RELATED TOPICS
A Festschrift in Honor of Professor P K Bhattacharya on the Occasion of His 80th Birthday

ISBN-13 978-981-4366-56-4
ISBN-10 981-4366-56-0

Printed in Singapore by B & Jo Enterprise Pte Ltd

Professor Bhattacharya in his office at the University of California, Davis, in the early 1980s.

Preface

This Festschrift is dedicated to our dear friend and esteemed colleague, Dr. Prodyot K. Bhattacharya, Professor Emeritus of the Department of Statistics at the University of California, Davis. It is being published in celebration of his 80th birthday. It is difficult to think of another individual whom the phrase "gentleman and scholar" more aptly fits. As a colleague, P. K. was a consummate professional whose analytical proclivity and good intentions were applied to all problems brought to him, large and small. His gentle nature and his desire to be helpful made him both a delight to work with and an inspiration. Over his long professional career, he accumulated a collection of contributions to Probability and Statistics that were admirably broad as well as deep. The present volume represents new work by his students, friends and colleagues that touches upon themes that P. K. himself dedicated much of his time investigating and writing about. Most of the articles treat topics in nonparametric inference, P. K.'s favored area of study.

The Festschrift begins with an article which reviews P. K.'s scholarly work and highlights the contributions he made that were both important at the time and have had a lasting impact on the field. The first section contains two other review papers covering topics in causal inference and nonparametric tests for symmetry. Section 2 features six papers on general nonparametric inference, a mixture of papers investigating nonparametric estimation and nonparametric testing. The third section is dedicated to six investigations positing linear or generalized linear models and, for the most part, developing nonparametric inference procedures in such frameworks. The four papers in Section 4 treat problems in the general area of time series analysis. The fifth and final section consists of three papers on asymptotic analysis, another domain in which P. K. was deeply immersed.

Compiling this festschrift has been an honor and a pleasure. The faculty of the Department of Statistics at UC Davis benefited, both individually and collectively, from P. K.'s presence and from his tireless dedication to the growth and welfare of the fledgling program that he joined in 1980. He was the backbone of a new graduate program in Statistics that was both rigorous and modern, and he is remembered and admired by his students for his approachability, his helpfulness and his remarkably broad knowledge of the field. We have all been influenced by his example.

Special thanks to Pat Aguilera and to Elizabeth Dudley for their kind assistance in various aspects of this project. We are especially indebted to Michael McAssey, who took a highly heterogeneous collection of research papers and converted them into the entries you see in this elegant volume. Finally, thanks to Wolfgang Polonik, Chair of the Department of Statistics, for his encouragement and support of this project.

Jiming Jiang, George G. Roussas, Francisco J. Samaniego
Editors

Contents

PART 1
Review Papers

Chapter 1

On the Scholarly Work of Prodyot K. Bhattacharya

Peter Hall and Francisco J. Samaniego

University of Melbourne and University of California, Davis

P. K. Bhattacharya's research over an academic career spanning some 50 years covered a wide spectrum of topics. It utilized, and often invented, state-of-the-art mathematical and stochastic tools for studying the subtle features of statistical procedures. Much of his work was nonparametric in nature. He maintained a special interest, throughout his career, in topics such as regression analysis, both parametric and nonparametric, change point problems, asymptotic theory, stochastic processes and sequential analysis. An interesting fact, not as widely known among his statistical colleagues, is that he also participated in a variety of productive collaborations with scientists from other disciplines. He used his expertise in statistical modeling and inference to make meaningful contributions to quantitative problems in mathematical biology, economics, social processes, physics and cosmology. In this brief review, we discuss some of the highlights of Bhattacharya's most influential papers and call attention to the many creative ideas he contributed to the theory and practice of statistical science.

Contents

1. Introduction

P. K. Bhattacharya has made remarkably rich and eclectic contributions to statistical science, during a career that has spanned more than half a century. He has turned his considerable talent, particularly his theoretical skills, to problems involving hypothesis testing, sequential methods and optimal stopping, nonparametric curve estimation, order statistics, a variety of topics in regression, properties of stochastic processes, statistical and stochastic control, and other decision procedures. In this all-too-brief survey of his contributions, we shall try to give a flavor of his many research achievements and to honor the breadth and depth of his contributions to statistical science.

2. Early work and foundations for the future

Bhattacharya's first paper, published in the Calcutta Statistical Association Bulletin in 1955, addressed the problem of testing the null hypothesis that the mean and variance of a normal distribution are equal to specified values, against the complementary alternative. He suggested three approaches to implementing the test, noting that in large samples they were all rather similar. This work, of interest in its own right, was to play a major role as a precursor to Bhattacharya's better known research in areas such as hypothesis testing and properties of Gaussian processes and distributions.

Six years later, in a paper with K. R. Parthasarathy in 1961, Bhattacharya established extreme-value properties of means of concomitants of order statistics, related to results for extremes in more conventional settings. This intriguing and highly original work made use of Mahalanobis' (1960) technique of fractile graphical analysis, originally devised to compare specific socio-economic circumstances of different strata within a population. Indeed, the contributions made in this paper provided some of the first theoretical underpinnings for Mahalanobis' ideas, which were to prove influential in economics. Therefore, the importance of Parthasarathy and Bhattacharya's work went well beyond its generally accepted interpretation as an early contribution to contemporary nonparametric methodology.

Both the methodology and the theoretical work of Parthasarathy and Bhattacharya (1961) were precursors to ideas introduced in papers that we shall discuss below. The theoretical arguments established consistency of function estimators and hypothesis tests, and involved decomposing scalar explanatory variables into k fractile classes of equal size (or approximately equal size, if the sample size was not divisible by k). Bhattacharya's deft analysis involved allowing the sample size n and the number of classes k to diverge together.

To those schooled on textbooks written by English authors, regression analysis is conventionally understood to be conditional on the explanatory variables. In particular, optimality properties are established in that setting. However, Bhattacharya (1962) showed that if the explanatory variables are treated as random, then, under certain conditions, those optimality results continue to hold. In particular, as Bhattacharya demonstrated in his Theorem 2, least squares estimators are minimax optimal for the vector of slope parameters.

In related work, Bhattacharya (1963) introduced estimators of conditional quantile functions for the distribution of the response variable, conditional on the explanatory variable, and suggested hypothesis tests relating to the shape of conditional quantile functions. The results established uniform consistency of estimators, asymptotic level accuracy for hypothesis tests, and consistency of those tests.

A particularly interesting aspect of the methodology introduced in this paper is the way in which the number, k, of "fractile groups" could diverge with sample size. In particular, the scalar explanatory variables were distributed evenly into k classes, with the first class containing the k largest values of the explanatory variables, the next class containing the variables between the $(k+1)$st and $2k$th largest, and so forth. Observations within a quantile group were used to calculate the quantile function estimator at the appropriate quantile. The asymptotic theory involved both sample size and k diverging together, and required careful analysis in order to control the various effects that can arise. Indeed, Bhattacharya's technical arguments used sophisticated methods drawn from probability theory, and involved applying the inequalities of Bernstein and Esseen. However, the regularity conditions were remarkably mild.

3. Forays into decision theory

As is well recognized today, the papers by Stein (1956) and James and Stein (1961) constituted major breakthroughs in the development of mathematical statistics, shocking the statistical community of that era with the news that the mean $\bar{\mathbf{X}}$ of a sample from a multivariate normal population, the time-honored estimator of the population mean $\boldsymbol{\mu}$, was in fact an inadmissible estimator of $\boldsymbol{\mu}$ when the dimension of the observed data strictly exceeded 2. These papers introduced the notion of shrinkage in estimating high dimensional parameters and demonstrated its salutary effects. In the years that followed, much research was done seeking to understand the Stein effect more comprehensively and trying to determine the extent to which shrinkage was a useful estimation strategy.

In the course of a postdoctoral appointment he held at Stanford in 1962-63, Bhattacharya joined the ranks of the many mathematical statisticians involved in such investigations. His findings were published in the Annals of Mathematical Statistics in 1966. In that paper, he extended the work of James and Stein by treating their estimation problem utilizing the general quadratic loss function that was advocated by James and Stein (that is, the loss $L(\boldsymbol{\theta}, \mathbf{a}) = (\boldsymbol{\theta} - \mathbf{a})'\mathbf{D}(\boldsymbol{\theta} - \mathbf{a})$, with $\mathbf{D}$ a nonnegative diagonal matrix), but relaxing their assumption that the common variance σ^2 of the components of an observed $\mathbf{X}$ is known. He broke new ground in several ways, establishing the inadmissibility of $\bar{\mathbf{X}}$ in dimension $k \geq 3$ in a more general framework and giving a new and explicit representation for the estimator that uniformly improves upon $\bar{\mathbf{X}}$. He also demonstrated that this estimator is uniformly superior to the least squares estimator of the parameter vector of the standard linear model with independent and identically distributed normal errors.

Bhattacharya and Dasgupta (1964) introduced a class of admissible Bayes solutions to classification problems when inference was based on data from exponential family populations. Particular attention was paid to cases where the data had a Gaussian distribution. The authors showed that in this setting, the usual procedure is minimax optimal. One of several attractive features of this work is that, unlike some earlier contributions, it does not require the populations to be completely specified; rather, it requires only that they come from a parametric family, for example, the multivariate normal family with a common covariance matrix.

4. Work on density estimation and related problems

Bhattacharya's 1967 Sankhyā paper was a pioneering contribution to the study of uniform consistency properties of kernel estimators of density functions and their derivatives. Earlier work had demonstrated that, under appropriate conditions, kernel estimators were consistent, but of course those results said little about the rate of convergence. In the context of density estimation, Bhattacharya (1967) showed that the convergence rate, uniformly on an interval that expanded to the real line in a polynomial manner, was almost as good as the rate of convergence at a point. Indeed, under the assumption that the density has a bounded derivative, the optimal pointwise rate is $n^{-1/(2r+3)}$, and Bhattacharya established a uniform convergence rate of $n^{-1/(2r+4+\varepsilon)}$ for any $\varepsilon > 0$.

Related work of Woodroofe (1967) pertained only to the density, not derivatives of the density. More "standard" versions of Bhattacharya's result, now well known, did not appear until about a decade later, and in many cases used the elaborate machinery of the "Hungarian embedding" or other stochastic approximation schemes. By way of contrast, Bhattacharya's approach was simple and elegantly transparent, being founded in part on Hoeffding's inequality.

More generally, Bhattacharya has maintained a long-term interest in the nonparametric estimation of functions, including empirical approaches to tuning the estimators so as to obtain optimal performance. In this direction, Bhattacharya and Mack (1987) derived key theoretical results underpinning the use of so-called plug-in methods, and related techniques, for smoothing parameter choice. They developed a general theory addressing properties of estimators when a nearest-neighbor order, or a bandwidth, is replaced by a function of the data, and showed how to use the resulting methodology to construct approximate best linear unbiased density estimators.

5. Special explorations

When developing theoretical results for discrete random variables, it is not uncommon to transform them so that their distributions are Gaussian, even if only approximately. For example, a variety of embedding theorems and stochastic approximations proceed in this manner. The natural analogue of this approach, when working with a stochastic process, is to transform the process to Brownian motion. Doob (1949), when developing an intuitive argument describing the relationship between empirical distribution approximations and the true distributions, noted the potential usefulness of methodology of this type. Bhattacharya (1971) introduced just such a transformation, and also discussed properties of an alternative approach suggested by Doob. An advantage of Bhattacharya's method is that it is more direct.

Bhattacharya and Smith (1972) discussed sequential probability ratio tests for testing simple hypotheses, in the context of Gaussian processes with known co-variance function. Using earlier work of Parzen (1959), Bhattacharya and Smith addressed properties of the Radon–Nikodym derivative of one Gaussian process measure with respect to another. When the derivative is evaluated at the first of these two processes, its logarithm is a Gaussian process in its own right. Bhattacharya and Smith addressed properties of that process as well as of the derivative itself, this time interpreted as a random variable rather than a process. They used the results obtained in this way to describe properties of sequential probability likelihood ratio tests, for example to show that they are well-defined (that is, terminate with probability 1), and to describe the error probabilities of tests and detailed properties of stopping rules.

Given a bivariate random sample $(X_1, Y_1), \ldots, (X_n, Y_n)$, the concomitants of the order statistics $X_{1:n} < X_{2:n} < \cdots < X_{n:n}$ are called the induced order statistics of the Y values. Specifically, $Y_{k:n} = Y_j$ if $X_{k:n} = X_j$. Induced order statistics arise naturally in regression problems, since estimating the regression function $m(x) = E(Y|X = x)$ generally involves operating with the ordered Xs and working with the Y values corresponding to them. Motivated by the problem of testing the null hypothesis that the regression $m(x)$ was equal to a fixed, known function $m_0(x)$, Bhattacharya (1974) studied the distributional properties of partial sums $S_k^{(n)}$ of centered induced order statistics. Using the conditional Skorokhod embedding of $\{S_k^{(n)}\}$, given $\mathbf{X}$, on Brownian motion paths, he demonstrated, under specific regularity conditions, that the sample paths of the normalized sums converge, in an appropriate stochastic sense, to a process obtained from the standard Brownian motion process by a transformation of the time axis. This yields an invariance principle similar to Donsker's. The limit theory established in the paper leads naturally to a valid asymptotic test, based on $\max_k\{|S_k^{(n)}|, k = 1, 2, \ldots, n\}$, of the hypothesis that $m(x)$ is a specified function.

Bhattacharya and Brockwell (1976) focused on the detection of very faint signals, with applications to image processing, storage processes and integrals of Markov chains. The context of their work was that of additive processes having no negative jumps and enjoying right-continuous sample paths with left-hand limits. They developed a representation for, first, the joint distribution of the process's infimum on the interval $[0, t]$, and, secondly, the first time that the process achieved that infimum in $[0, t]$. Particular attention was paid to the case where the process was stable. The results generalized earlier work of Chernoff and Zacks (1964) and Hinkley (1970), on change point methods, and foreshadowed a vast contemporary literature on the detection of weak signals.

Bhattacharya's 1981 Annals of Statistics paper stands alone, topically, in his bibliography. Bhattacharya's insightful and creative contribution to nonparametric Bayesian inference in a biometric setting well exemplifies the impressive breadth of his interests and technical skills. Suppose that X is the tolerance level of an individual to a drug and F represents the distribution of X in the population as a whole. Bhattacharya studied the problem of estimating F based on quantal response data that is based only on whether or not individual subjects had tolerance for the drug when administered at a particular dose level. The experimental design for gathering such data generally stipulates that separate random samples of subjects be assigned to participate at each of several dose levels. When Bhattacharya considered a nonparametric Bayesian treatment of this problem, it was known, from Antoniak (1974) work, that if F was distributed according to a Dirichlet process prior, the posterior distribution of F, given quantal response data, was a mixture of Dirichlet processes. However, no representation or feasible computational algorithm was available for obtaining the posterior's finite dimensional distributions (fdds).

Bhattacharya (1981) developed a systematic procedure for obtaining these fdds, showing that they could be represented as multivariate beta distributions with random parameter vectors. A linkage was established between the sequence of random parameters associated with a set of r increasing dose levels and an ensemble of r separate Markov chains. From this association, Bhattacharya derived the asymptotic posterior distribution. The random parameter vector of the posterior beta distribution was shown to be asymptotically normal with the mean and the covariance matrix satisfying particular recursion formulae. The novel approach and techniques in this paper have proven useful in a variety of later investigations. For example, Neath and Samaniego (1966a; 1966b) used Bhattacharya's construction in executing a Bayesian treatment of the estimation of the (nonidentifiable) multiple decrement function within a nonparametric competing risks framework.

6. Work on statistical quality control

Walter Shewhart is widely considered the "father" of the area of industrial statistics known as statistical quality control. His development of the control chart (see Shewhart (1931)) as a tool for reducing variation in manufacturing processes is arguably one of his most significant and lasting contributions. This tool continues to be widely used in industrial and other applications. Among the important variations of Shewhart's control chart is the CUSUM chart introduced by Page (1954). As noted by Bhattacharya and Frierson in their 1981 Annals of Statistics article, "the existing literature mainly deals with a change in the mean of normal random variables or a change in the probability of Bernoulli trials."

Bhattacharya and Frierson (1981) broke important new ground by providing the necessary tools for the construction and application of nonparametric control charts. Their framework postulated the availability of sequentially sampled observations $X_1, X_2, \ldots, X_N$, whose distribution changes from F to G after the first $[N\theta]$ variables, with F, G and $\theta \in [0,1]$ unknown, where $[x]$ is the greatest integer no larger than x. After reviewing relevant work on change point detection and on nonparametric methods based on ranks in the nonsequential case, Bhattacharya and Frierson proceeded to develop a nonparametric control chart based on sequential ranks and, under suitable assumptions, they characterized the procedure's asymptotic behavior. The generality of their results is reflected by the fact that the type of change to be detected is quite broadly defined, with changes in location or in scale or via contamination all being special cases of their treatment. They proposed the use of the sums $\{S_k\}$ of standardized sequential ranks as a means of detecting "disorders" in the sequential observations. Theorem 1, their main result, established weak convergence of the linearly interpolated version of $\{0, S_1, S_2, \ldots, S_N\}$ to the process $\{X(t),\ 0 \le t \le 1\}$ given by

$$X(t) = B(t) + \sqrt{12}\delta\theta \log(t/\theta) I_{[\theta,1]}(t),$$

where $B(t)$ is standard Brownian motion and δ is a constant associated with magnitude of the (suitably small) change after the $[N\theta]$th observation. This result, which indicates that, in the framework studied, a process that is "out of control" will behave like Brownian motion with a logarithmic drift, gave rise to the proposed "logarithmic stopping rule" advocated in the paper. The authors examined, via simulation, the comparative performance of the traditional parametric control chart and the nonparametric counterpart and identified a variety of situations in which the latter provides better performance.

Bhattacharya and Zhou (1994) developed a rank-sign analogue of Page's (1954) CUSUM chart. This paper is motivated from the same concern that led to Bhattacharya and Frierson's (1981) paper, namely, the lack of robustness of the traditional procedure when its parametric assumptions are violated. Bhattacharya and Zhou (1994) treat the situation in which one wishes to detect a small change in location of an unknown symmetric density f. Both CUSUM procedures are based on a density g used as an (assumed reasonable) approximation of f. The asymptotic behaviors of both control charts are shown to be equivalent when the "score functions" used in both procedures are appropriate for the underlying true density f. Under violations of the latter condition, the nonparametric CUSUM procedure is seen to maintain its prescribed false alarm rate and achieve a detection rate which is superior to that of the Page CUSUM chart. In a separate paper, Bhattacharya and Zhou (1996) provided a rigorous treatment of the general asymptotic theory of the nonparametric CUSUM procedure.

7. Work of cosmic significance

Motivated by a cosmological controversy concerning Hubble's law in astronomy, Bhattacharya, et al. (1983) suggested a nonparametric slope estimator in linear regression problems where the response variable is subject to truncation. Hubble's Law, which responds to the observation (through red shifts due to a Doppler effect) that the velocities at which galaxies are moving away from the Earth are proportional to their distance from us, is at variance with Irving Segal's chronometric theory, which introduces a "law of conservation of curved energy" to explain the red shift without requiring the universe to be expanding. A key regression problem that arises in this setting involves errors that potentially have infinite variance, and where the response variable is truncated because objects of high magnitude are not visible. Bhattacharya, Chernoff and Yang introduced ingenious Mann–Whitney type nonparametric methods for undertaking inference in this problem.

In closely related work, Bhattacharya (1983) provided theoretical justification for an alternative approach to addressing the same problem, this time developed by the astrophysicist Turner (1979). Turner had introduced an approach entirely different from that of Bhattacharya, et al. (1983), this time based on a Kolmogorov–Smirnov test applied to exponential ranks. However, Turner's approach lacked adequate justification. Heuristic support for it was relatively easy to give in the non-truncated case, but, as we noted above, truncation is an essential feature of any realistic model in this area. It had been observed, through numerical work in a special case, that Turner's technique was valid, but more general support for the approach was lacking. Bhattacharya (1983) provided that support, extending the heuristic argument to the truncation case and providing valid theoretical arguments. More recently the chronometric cosmology interpretation of red shifts and relative velocities has experienced difficulties accommodating the cosmic microwave background, and has declined in popularity.

8. Inference about change points

Bhattacharya has had a strong interest in change point problems, and studied such problems in a variety of different scenarios. His joint work with Brockwell on additive processes has a clear connection to the change point literature, and his interest in quality control may be viewed as dealing with a special formulation of the change point framework. But Bhattacharya's interests also included what might be called mainstream formulations of problems in the area. We mention two specific examples here.

Bhattacharya (1987) considered the large sample behavior of the maximum likelihood estimator of an unknown change point $\lambda \in [0,1]$ under the assumption that the probability density function (pdf) of the first $[\lambda n]$ observations in a sample of size n is $f(x,\theta)$ and the remaining observations are drawn from the pdf $f(x,\theta+\delta v_n^{-1})$, where v_n grows to ∞ at a rate slower than $n^{1/2}$. Under regularity conditions, it is shown that if the random variables $Y_n = n v_n^{-2}(\hat{\lambda} - \lambda)$ are bounded in probability as $n \to \infty$, where $\hat{\lambda}$ is the MLE of λ, then Y_n converges in distribution to the time $T_{(\delta' J \delta)^{1/2}}$ at which a two-sided Brownian motion with drift $\delta' J \delta^{1/2}|t/2|$ on the entire real line attains its almost sure unique maximum (where J is the Fisher Information matrix). This generalizes findings described in Bhattacharya and Brockwell (1976). In his 1987 paper, Bhattacharya also considered the general case in which the condition that Y_n is bounded in probability is not required. In this latter case, he proposed a three-stage estimation procedure yielding an estimator $\hat{\lambda}^*$ with the same asymptotic behavior as the estimator $\hat{\lambda}$ above. For the special case of multiparameter exponential families satisfying simple regularity conditions, these results take a particularly manageable form. Several applications of the paper's theoretical results are treated, including a) changes in the mean and variance of a univariate normal distribution, b) changes in regression of a bivariate normal distribution and c) changes in the median of a Cauchy distribution.

Bhattacharya's published work shows a longstanding interest in regression problems. Included in this body of work is his early study of shrinkage estimators, varied contributions to k-nearest neighbor type methods and work on the asymptotic theory of nonparametric regression. He also studied change point problems in a regression setting. Bhattacharya (1990) considered a two-phase linear regression model with normal errors, where the regression equation, given fixed independent variables $x_1 < \cdots < x_n$, is postulated to be

$$Y_i|\mathbf{x} = \theta_{11} + \theta_{12}x_i + \theta_3\varepsilon_i \quad \text{for } x_i \leq \tau$$

and

$$Y_i|\mathbf{x} = \theta_{21} + \theta_{22}x_i + \theta_3\varepsilon_i \quad \text{for } x_i > \tau,$$

where $\varepsilon_1, \varepsilon_2, \ldots, \varepsilon_n \overset{\text{iid}}{\sim} N(0,1)$ and the θ's and the parameter τ are unknown. Two specific change point problems are addressed, first the restricted problem in which it is assumed that the two regression lines intersect at the change point τ and then the unrestricted problem in which there is no hypothesized relationship between the two phases. In the first case, Bhattacharya characterized the asymptotic behavior of the log likelihood (LL) process, establishing its weak convergence, with respect to uniform convergence on compact sets, to an explicitly displayed weak limit. Under additional regularity conditions, he identified the weak limit of LL in the unrestricted case. In the paper's final section he provided under suitable conditions, the asymptotic distribution of the maximum likelihood estimator of the parameter vector $(\theta_{11}, \theta_{12}, \theta_{21}, \theta_{22}, \tau)$.

9. Discussion

There is much to be admired in P. K. Bhattacharya's research work. In the review we have presented here, we have attempted to give the reader a feeling for many of the important ideas contained in his major papers. Because of time and space restrictions on the present work, we have limited our remarks to some twenty of his publications. While this survey does convey the quite correct impression that the scope of his interests, and his elegant use of a broad array of powerful technical tools, are both impressive and laudable, a more careful review of Bhattacharya's curriculum vitae reveals that the topics and tools we have mentioned leave many of his interests, techniques and contributions unmentioned. We close here with a brief listing of other problem areas to which Bhattacharya's research made notable contributions: stochastic process theory, mathematical biology (focusing on the study of human cells, with papers published in *Acta Cytologica* and in the *Annals of Statistics*), sequential analysis, stochastic modeling (with applications, for example, to chain letter schemes), semiparametric modeling and inference, and statistical inference based on observational studies.

It has been a distinct pleasure to have had this opportunity to reexamine Bhattacharya's many contributions to Probability and Statistics. After engaging in such a project, one can reach no other conclusion than that our community is in Bhattacharya's debt.

References

Antoniak, C. E. (1974). Mixtures of Dirichlet processes with applications to Bayesian nonparametric problems. *Annals of Statistics* **2**, 641–47.

Bhattacharya, P. K. (1955). Joint test for the mean and variance of a normal population. *Calcutta Statistical Association Bulletin* **6**, 7390.

Bhattacharya, P. K. (1962). Some properties of the least squares estimator in regression Analysis when the predictor variables are stochastic. *Annals of Mathematical Statistics* **33**, 1365–74.

Bhattacharya, P. K. (1963). On an analog of regression analysis. *Annals of Mathematical Statistics* **34**, 1459–73.

Bhattacharya, P. K. and Dasgupta, S. (1964). Classification between univariate exponential populations. *Sankhyā*, Ser. A, **26**, 17–24.

Bhattacharya, P. K. (1966). Estimating the mean of a multivariate normal population with general quadratic loss function. *Annals of Mathematical Statistics* **37**, 1819–24.

Bhattacharya, P. K. (1967). Estimation of a probability density function and its derivatives. *Sankhyā*, Ser. A, **29**, 373–82.

Bhattacharya, P. K. (1971). Transformations of Gaussian processes to the Brownian motion. *Annals of Mathematical Statistics* **42**, 2008–17.

Bhattacharya, P. K. (1974). Convergence of sample paths of normalized sums of induced order statistics, *Annals of Statistics* **2**, 803–11.

Bhattacharya, P. K. (1981). Posterior distribution of a Dirichlet process from quantal response data. *Annals of Statistics* **9**, 641–47.

Bhattacharya, P. K. (1983). Justification for a K–S type test for the slope of a truncated regression. *Annals of Statistics* **11**, 697–7.

Bhattacharya, P. K. (1987) Maximum likelihood estimation of a change point in the distribution of independent Random variables: general multiparameter case, *Journal of Multivariate Analysis* **23**, 183–207.

Bhattacharya, P. K. (1990) Weak Convergence of the log likelihood process in the two-phase linear regression problem. In R.R. Bahadur (Editor) *Proceedings of the R. C. Bose Symposium on Probability, Statistics and Design of Experiments*, New Dehli: Wiley Eastern.

Bhattacharya, P. K. and Brockwell, P. J. (1976). The minimum of an additive process with applications to signal estimation and storage theory. *Z. Wahrscheinlichkeitstheorie und Verw. Gebiete* **37**, 51–75.

Bhattacharya, P. K., Chernoff, H. and Yang, S.S. (1983). Nonparametric estimation of the slope of a truncated regression. *Annals of Statistics* **11**, 505–14.

Bhattacharya, P. K. and Frierson, D. (1981) A nonparametric control chart for detecting small disorders. *Annals of Statistics* **14**, 544–54.

Bhattacharya, P. K. and Mack, Y. P. (1987). Weak convergence of k-NN density and regression estimators with varying k and applications. *Annals of Statistics* **15**, 976–94.

Bhattacharya, P. K. and Smith, R. P. (1972). Sequential probability ratio test for the mean value function of a Gaussian process. *Annals of Mathematical Statistics* **43**, 1861–73.

Bhattacharya, P. K. and Zhou, H. (1994) A rank CUSUM procedure for detecting small changes in a symmetric distribution. In Carlstein, E., Mueller, H. G. and Siegmund, D. (Editors), *Change Point Problems*, Hayward, CA: IMS Lecture Notes – Monograph Series **23**, 57–65.

Bhattacharya, P. K. and Zhao, H. (1996) A generalized CUSUM procedure for sequential detection of change-point in a parametric family when the initial parameter is unknown. *Sequential Analysis* **15**, 4, 311–325.

Chernoff, H. and Zacks, S. (1964). Estimating the current mean of a normal distribution which is subjected to changes in time. *Annals of Mathematical Statistics* **35**, 999–1018.

Doob, J. L. (1949). Heuristic approach to the Kolmogorov–Smirnov theorems. *Annals of Mathematical Statistics* **20**, 393–403.

Hinkley, D. V. (1970). Inference about the change-point in a sequence of random variables. *Biometrika* **57**, 1–17.

James, W. and Stein, C. (1961) Estimation with quadratic loss. In J. Neyman (Editor) *Proceedings of the Fourth Berkeley Symposium on Mathematical Statistics and Probability* **1**, 361–79, Berkeley: University of California Press.

Mahalanobis, P. C. (1960). A method of fractal graphical analysis. *Econometrica* **28**, 325–51.

Neath, A. A. and Samaniego, F. J. (1996a). On Bayesian estimation of the multiple decrement function in the competing risks problem. *Statistics and Probability Letters* **31**, 75–83.

Neath, A. A. and Samaniego, F. J. (1996b). On Bayesian estimation in the competing risks problem: An example of the distinguished role played by the multivariate exponential distribution. *Statistics and Probability Letters* **31**, (1996), 69–74.

Page, E. S. (1954). Continuous inspection schemes. *Biometrika* **41**, 100–15.

Parthasarathy, K. R. and Bhattacharya, P. K. (1961). Some limit theorems in regression theory. *Sankhyā*, Ser. A **23**, 91–102.

Parzen, E. (1959). Statistical analysis on time series by Hilbert space methods, I. Technical Report No. 23, Applied Mathematics and Statistical Laboratory, Stanford University.

Shewhart, W. A. (1931). *Economic Control of Quality in Manufactured Product*, New York, D. Van Nostrand.

Stein, C. (1956). Inadmissibility of the usual estimator for the mean of a multivariate normal distribution. In J. Neyman (Editor) *Proceedings of the Third Berkeley Symposium on Mathematical Statistics and Probability* **1**, 197–206, Berkeley: University of California Press

Turner, E. L. (1979). Statistics of the Hubble diagram, I. Determination of $q0$

and luminosity evolution with application to quasars. *Astrophysics Journal* **230**, 291–303.

Woodroofe, M. (1967). On the maximum deviation of the sample density. *Annals of Mathematical Statistics* **38**, 475–81.

Chapter 2

The Propensity Score and Its Role in Causal Inference

Christiana Drake and Travis Loux

University of California, Davis

Observational studies compare treatment groups without the benefit of randomization and are often done when randomized studies are not feasible. Association does not imply causation and causal inference from such studies is often dismissed. Epidemiologic studies and evidence based medicine rely on observational studies to investigate harmful exposures and study treatments in the clinical setting. Randomized studies estimate average treatment effects which are unbiased. Regression based estimates depend on other variables in the model and may not equal average effects in non-linear models.

The propensity score was developed as a method of adjusting for confounders. It leads to simple estimators of average exposure effects. The focus is on identification of subpopulations where the distribution of potential confounders is balanced between exposure groups. The propensity score is a consequence of the Rubin Causal Model (RCM) and has gained increasing acceptance as a tool to reduce confounding. Several aspects of the propensity score are the focus of debate and investigation: 1) how to adjust by the propensity score, 2) how to estimate the propensity score and 3) conceptual differences among the propensity score approach and a parametric model. The focus of this paper is a discussion of causal inference and the role of the propensity score in conceptual aspects of causal inference as well as in applications in observational studies.

Contents

1. Introduction

Causation has been debated by philosophers for centuries. With the development of quantitative methods such as statistics, causation also employed the techniques developed by these disciplines in addressing issues of cause and effect. Freedman (1999) gives an account of the historical development of causation and the role it has played in the history and development of statistics. He provides philosophical definitions of "cause and effect" which could form a basis for statistical models. Hume (1739) defines "... a cause is an object followed by another, ..., where if the first object had not been the second had never existed." MacMahon and Pugh (1960) gave this definition: "An association may be classed as presumptively causal when it is believed that, had the cause been altered, the effect would have changed." Statistics proceeds from the particular (the data) to the general (the population) and thus proceeds by inductive reasoning. In a broad sense, one can interpret causal inference as the process by which sample information in the form of statistics is used to estimate population parameters. The roots of such inference go back to De Moivre (1738), Bayes (1764), LaPlace (1774) and De Morgan (1838). In a comparative study where the objective is to establish a cause A, preceding an effect B, two samples are compared, one in which A is present, and one in which A is absent. This comparison can be in the form of the probability of the occurrence of B if preceded by A to the probability of B when A did not precede B or in the form of parameters that determine the probability of B given A and given A^c (the complement of A). One implicit assumption in these studies: the populations from which the samples are obtained have the same probability for outcome B in the absence of an effect of A.

The study of causal inference proceeded along with the development of modern statistics. Neyman (1923) and Fisher (1925; 1935) were among the statisticians in the early twentieth century who were instrumental in advancing statistical theory and methods. Their research was often motivated by applications, particularly in agriculture. Fisher was the first statistician to propose randomization as a way to study and compare treatments. Fisher argued that establishment of cause and effect by inductive reasoning is associated with some degree of uncertainty. However, such inference is validated by the rigor with which it is possible to quantify the nature and degree of this uncertainty. This validity, Fisher argued, comes from randomization.

Neyman (1923) explores randomization in the comparison of yields of different crop varieties when planted in different plots in the same field. He defines for each plot i and variety j a "true" yield U_{ij} and an average $\mu_j = (1/N)\sum_i U_{ij}$ where N is the number of plots in the field, estimated by the average of the plots planted with variety j. In his analysis Neyman introduces what could be argued is a counterfactual model (Rubin (1990a)).

Kempthorne (1955) and Cox (1958) explore the randomization argument further and discuss randomization analysis vs. the use of a linear model with normally distributed errors. Kempthorne (1977) criticizes the practice of textbooks in statistics to present solutions to the analysis of comparative studies that make no reference to the way in which the data were obtained. Cox (1958) and Kempthorne (1980) point out that randomization validates the analysis using normally distributed errors.

While the case for randomization is made convincingly by many statisticians, there is no uniform agreement. Basu (1980) questions the need for it. Furthermore, in many situations treatment or exposure effects have to be estimated without the benefit of randomization, sometimes for ethical or for practical reasons. Such studies and the causal conclusions to be drawn from them have received increasing attention in recent years.

Observational studies are the primary type of study conducted in fields as diverse as epidemiology, health services research, psychology, anthropology, sociology, economics and many others. Environmental sciences and criminology rely on a mixture of experimental and observational studies for the assessment of causes and effects. The point is made that experimental studies in the form of randomized clinical trials are carried out to establish the efficacy of a new intervention or drug whereas observational studies in the clinical setting are necessary to assess the effectiveness of a drug or intervention.

Cochran (1965) defines a comparative observational study as an empirical investigation in which: "the objective is to elucidate cause and effect relationships ... [in which it] is not feasible to use controlled experimentation, in the sense of being able to impose the procedures or treatments whose effects it is desired to discover or assign subjects at random to different procedures." As a consequence of this lack of randomization the populations to be compared may be quite different with regard to experiencing the effect that is to be studied. Comparative observational studies, as pointed out by Cochran, are studies of association. To interpret such an association as causal requires further assumptions. Rubin (1974; 1977; 1978; 1980) develops a model that provides a framework for causality.

The Rubin Causal Model (RCM)

Each observational unit in the Rubin Causal Model has two potential outcomes, (y_{1i}, y_{0i}). The first response y_{1i} is observed when the condition, which we will call treatment, is acting on the unit, and y_{0i} when the cause is not acting on the unit. Treatment effect is then defined as

$$\Delta_i = y_{1i} - y_{0i} \qquad \text{and} \qquad \Delta = E(\Delta_i) = E(y_{1i} - y_{0i})$$

for the ith unit and the average over all units, respectively. Only one of y_{1i}, y_{0i} is observed, the other quantity being counterfactual, thus Δ_i is not observable and Δ is not estimable by direct observation. Much of the statistical literature on estimating causal effects centers on substitute values for the unobserved counterfactual values.

Holland (1986) uses the Rubin Causal Model to state a set of conditions, which if true justify interpreting an association as causative. Stone (1993) uses the counterfactual approach to discuss and outline a set of assumptions, covariate sufficiency, in causal inference. Covariate sufficiency is closely related to a condition called strongly ignorable treatment assignment introduced by Rosenbaum and Rubin (1983; 1984) and Rosenbaum (1984a).

Definition 1. Treatment assignment is said to be *strongly ignorable* or *X-adjustable* if

$$(y_1, y_0) \perp\!\!\!\perp t \mid x \qquad \text{and} \qquad 0 < p(t \mid x) < 1 \quad \text{for all } t, x$$

In this case

$$E(y_0 \mid t = 1, x) = E(y_0 \mid t = 0, x)$$

and

$$\Delta = E(y_1 \mid t = 1, x) - E(y_0 \mid t = 0, x).$$

It requires that the distribution of the vector of responses (y_1, y_0) conditional on a set of sufficient covariates x is independent of the actual treatment received. The need for strongly ignorable treatment assignment, regardless of the mode of statistical inference is further underscored by Rubin (1986; 1990b; 1991) where he considers model based, Bayesian and randomization based inference. For observational studies, strongly ignorable treatment assignment is the basis for causal inference. Randomization achieves that while any observational studies must identify groups for which strongly ignorable treatment assignment holds. Traditional methods of adjustment to balance covariates are stratification, post-hoc subclassification, matching and regression adjustment. Stratification is a model free method and works well for one or two categorical variables. Matching is also a model free method suitable for one or two continuous variables. These methods tend to break down in the presence of many confounders. Regression accommodates several quantitative and categorical variables. However, the parameter associated with the causal effect has an interpretation that is dependent on the particular model chosen. Furthermore, the validity of the parameter depends on the correct formulation of the regression model. An incorrect model can create serious bias even if all confounders are adjusted for. Gail, et al. (1984) show that regression models can result in biased estimates of treatment effect even in randomized studies. A linear predictor fit incorrectly with only linear terms in the confounders (Drake (1993)) will produce a bias similar to that of an omitted confounder. Most observational studies, particularly in subjects such as epidemiology, have to adjust for several confounders, at least some of them are continuous. Thus, a method that identifies subgroups in which treatment assignment is strongly ignorable but which summarizes many covariates into a single score is highly desirable. Such a score is called a balancing score and the propensity score is the coarsest.

2. The propensity score

Let $t = 1, 0$ be a binary treatment indicator and x a set of covariates. Rosenbaum and Rubin (1983) define the propensity score as $e(x) = p(t = 1 \mid x)$ and showed that $p(x \mid t, e(x)) = p(x \mid e(x))$. The propensity score is a balancing score in the sense that for any outcome y, set of covariates x and binary variable t, the distribution of (y, x) conditional on $e(x)$ is independent of t. For any fixed value of the propensity score, treated and control subjects have the same response distribution. Thus, conditional on the propensity score, treatment assignment is strongly ignorable and we have

$$(y_1, y_0) \perp\!\!\!\perp t \mid e(x) \quad \text{and} \quad \Delta = E(y_1 \mid t = 1, e(x)) - E(y_0 \mid t = 0, e(x)).$$

In practice, the propensity score is unknown and must be estimated. What if any advantage does the propensity score possess over regression models? Rosenbaum and Rubin (1983) showed that any consistent estimate of the propensity score leads to unbiased estimates of treatment effects. It has also been shown (Drake (1993)) that bias due to misspecification of the propensity score is considerably less than is observed in a misspecified regression model. Furthermore, the propensity score approach estimates average treatment effects. Regression models estimate effects that are conditional on the model and the variables adjusted for. Thus, interpretation of results must be made within the context of the model. Finally, conditional on the propensity score, a test of a treatment effect requires no further assumptions, similar to a randomized experiment.

In recent years, uses of the propensity score have been extended both in the method of adjustment and the studies where they are used. It is used routinely for non-additive treatment effects. Examples are estimates of odds ratios or relative risks. Propensity scores have also found uses when the randomization used in clinical trials fails to balance all covariates. Properties of adjustment using propensity scores have been further investigated, particularly with regard to the type of estimator and the efficiency of the estimators derived.

An in-depth look at propensity score stratification and weighting was given by Lunceford and Davidian (2004). In their paper, the focus was on estimating the marginal difference $\Delta = E(y_1) - E(y_0)$. While stratification on the propensity score can reduce the bias by a large amount, there likely remains bias resulting from non-constant propensity scores within each stratum. The residual bias, however, can be accounted for by the use of a regression model within each stratum, so that the within-stratum estimate becomes

$$\hat{\Delta}^{(j)} = \frac{1}{n_j} \sum_{i=1}^{n} I(\hat{e}_j \in \tilde{Q}_j)\Big(m^{(j)}(1, x_i, \hat{\alpha}^{(j)}) - m^{(j)}(0, x_i, \hat{\alpha}^{(j)})\Big)$$

where n_j is the number of observations in stratum $\tilde{Q}_j$ and $m^{(j)}$ is the same function of t and x for all j. In the case of a linear model, $m^{(j)}(t, \mathbf{X}, \hat{\alpha}^{(\mathbf{j})}) = \alpha_{\mathbf{0}}^{(\mathbf{j})} + \alpha_{\mathbf{1}}^{(\mathbf{j})}\mathbf{t} + \mathbf{x}^{\mathbf{T}}\alpha_{\mathbf{2}}^{(\mathbf{j})}$ so that $\hat{\Delta}^{(j)} = \hat{\alpha}_1^{(j)}$. The estimate of the marginal difference Δ is then the weighted sum of the estimates from the K strata: $\hat{\Delta}_{SR} = \frac{1}{K}\sum_{i=1}^{K} \hat{\Delta}^{(j)}$. Lunceford and Davidian (2004) show that $\hat{\Delta}_{SR}$ is consistent for Δ so long as the models $m^{(j)}$ are correctly specified.

The simplest of the weighted estimates Lunceford and Davidian consider is

$$\hat{\Delta}_{IPW1} = n^{-1} \sum_{i=1}^{n} \frac{t_i y_i}{\hat{e}_i} - n^{-1} \sum_{i=1}^{n} \frac{(1 - t_i) y_i}{1 - \hat{e}_i}$$

which they show is unbiased for Δ assuming strongly ignorable treatment assignment and the correct specification of the propensity score model. Though $\hat{\Delta}_{IPW1}$ is unbiased for Δ, gains in efficiency can be obtained by adjusting the terms to form $\hat{\Delta}_{SR}$

$$= \frac{1}{n}\sum_{i=1}^{n}\frac{t_i y_i - (t_i - \hat{e}_i)m_1(x, \hat{\alpha}_1)}{\hat{e}_i} - \frac{1}{n}\sum_{i=1}^{n}\frac{(1-t_i)y_i + (t_i - \hat{e}_i)m_1(x, \hat{\alpha}_1)}{1-\hat{e}_i}$$

where $m_t(x, \alpha_{\mathbf{t}})$ is the regression of y on x within the treatment group defined by $t = 0, 1$. Not only is $\hat{\Delta}_{SR}$ the asymptotically efficient estimator in the class of semiparametric estimators, it is also known to be "doubly-robust." The double-robustness property means that the estimator is consistent if (1) the propensity score is correctly specified but the regression model is not, or (2) the regression models are correctly specified but the propensity score is not, though the estimator need no longer be most efficient.

Senn, et al. (2007) compare results from propensity score stratification to linear regression in the case of discrete covariates. When covariates are discrete there will only be a finite number of propensity scores, so one can reduce all bias by creating strata of individual propensity score values. Using the model $E(y \mid t, x) = \beta_0 + \beta_1 t + x'\tilde{\beta}$ with $Var(y \mid t, x) = \sigma^2$, Senn, et al., show that $\beta_1 = \Delta$:

$$\begin{aligned}\beta_1 &= E(\beta_1) = E[E(y \mid t = 1, x) - E(y \mid t = 0, x)] \\ &= E[E(y_1 \mid x) - E(y_0 \mid x)] = E(y_1) - E(y_0) = \Delta\end{aligned}$$

assuming strongly ignorable treatment assignment. Thus, the comparison between regression and propensity score stratification is a reasonable one to make. It is well-known that $\hat{\beta}_{1,LS}$ is the best linear unbiased estimator. However, $\hat{\Delta}$, the weighted average of the within-propensity score stratum estimates, is not conditionally unbiased. Senn, et al., in a study from the German Breast Cancer Study Group obtain estimates of $\hat{\beta}_1$ and $\hat{\Delta}$ that are very similar and have nearly identical standard errors.

Loux and Drake (2011) study adjustment methods by the propensity in estimating an odds ratio in the case of binary treatment and outcome. They investigate the properties of the Mantel–Haenszel estimator, using stratification, matching and weighting by an estimated propensity score. The Mantel–Haenszel estimator is known to be consistent under both regular and sparse asymptotics, therefore, can be used for stratifying on a categorical covariate as well as matching on a continuous confounder. Unlike an additive treatment effect for a continuous outcome, the odds ratio is not collapsible and the marginal and conditional odds ratio differ. The Mantel–Haenszel estimator converges to the conditional odds ratio when stratifying on a categorical confounder or matching on a continuous covariate. Loux and Drake show that matching by the propensity score does not lead to a consistent estimate, however, weighting by the propensity score does.

3. Propensity score estimation

In order for the propensity score to balance a variable, that variable must be included in the model used to estimate the propensity score. A simple rule, and one that is commonly advocated, is to include any variable which is suspected of being related to the exposure probability (Rubin and Thomas (1996), Shah, et al. (2005), Zanutto, et al. (2005)). The model may also include higher-order and interaction terms. Because the interest is on estimation of the propensity score and not the model itself, there is no need to simplify the model by excluding insignificant predictors. This suggestion becomes problematic, however, when the number of covariates is large relative to the number of observations. Including too many predictors may over-saturate the model. In this case it may be prudent to include only covariates which are predictive of the outcome of interest (Judkins, et al. (2007)) and are confounders in the classic sense. Furthermore, several studies (Rosenbaum (1987), McCaffrey, et al. (2004), Senn, et al. (2007)) have shown the estimated propensity score to be superior in its ability to reduce bias. While the true propensity scores will reduce imbalances in true confounders, estimated propensity scores will also allow one to adjust for random imbalances which may appear in any given sample.

Propensity score estimation is basically a classification problem and so is treated as such. The most common method to estimate the propensity score is through binary regression, usually a logistic model. Recently, other non-parametric computational methods have gained interest. McCaffrey, et al. (2004) suggest that generalized boosted modeling can yield better propensity score estimates when the functional form of the propensity score model is unknown and may contain non-linear terms. Along similar lines, Berk (2005) suggests propensity scores as an application of any one of a number of "ensemble methods," which take the average of many fits obtained through a given stochastic algorithm. While boosting is one special ensemble method, Berk also discusses bagging and random forests. Setoguchi, et al. (2008) compare logistic regression to regression trees and neural networks. Though, they find that no single method obtained optimal covariate balance across all situations, neural networks lead to the least bias in many scenarios. Woo, et al. (2008) compare logistic regression to generalized additive models (GAMs). Here, Woo, et al., find that GAMs tend to give better balance when there is a large overlap of covariates between exposed and unexposed units and more clearly shows poor overlap if that is the case. Since poor overlap suggests groups which are not comparable, this feature of GAMs is as important as precise propensity score estimates. It is recommended that analysts fit both a logistic and GAM to the data, and choose the one which yields the best overall balance. While logistic regression was one of the first methods for modeling the propensity score and continues to remain common, methods like regression tree boosting and bagging, neural networks, and GAMs lead to less restrictive models that also lessen the need to explicitly fit higher order terms and interactions. When the propensity score model is unknown, as in most applied cases, these non-parametric methods may yield better balance and more accurate estimates of the causal effects of interest.

4. Applications of propensity scores

When the propensity score methodology was first introduced, typical applications estimated the propensity score and formed about 5 to 6 strata on the basis of the propensity score. Rosenbaum and Rubin (1984) apply stratification on the estimated propensity score to a study comparing two treatments for coronary artery disease and 74 test results that are likely confounders. They use logistic regression to estimate the propensity score and then assess the balance achieved in the variables used to estimate the propensity score. They show further, that the propensity score can also be used to achieve balance in the missingness patterns of incompletely observed covariates. The study involved the comparison of several outcome variables. The advantage of the propensity score in this case is that one needs to establish the subclasses and assess the balance achieved in the covariates only once. A regression approach would require fitting a separate model for each outcome. Rosenbaum (1984b) derives a permutation test conditional on the propensity score that is an extension of the Fisher randomization test to observational studies under the assumption of strongly ignorable treatment assignment and a logistic regression model for the propensity score.

In another application Rosenbaum and Rubin (1985) use the Mahalanobis metric to establish a matched control group with the propensity score as one of the variables in the algorithm. Rosenbaum and Rubin show that the propensity score is equal percent bias reducing if used as the sole matching variable. Such a matching method is desirable as it reduces the bias in all variables used in calculating the propensity score proportionately. Generally, multivariate matching algorithms can decrease the bias due to some variables while increasing it due to others. Rubin and Thomas (1992) further study the properties of matching on the propensity score when the joint distribution of the covariates is ellipsoidal. The multivariate normal distribution has this property. They show that matching on the estimated propensity score is affinely invariant and, in addition to controlling the bias, can lead to a large reduction in variance.

Though the application of propensity scores began primarily in the epidemiologic literature, it has since spread to any field in which researchers must deal with observational data. Keating, et al. (2001) used propensity scores in their article looking to assess whether having a discussion with a medical oncologist affected type of breast cancer surgery (breast-conserving surgery or mastectomy) and satisfaction. They used a sample of 2426 women in Massachusetts and Minnesota and recorded a number of patient characteristics, whether they discussed possible surgical procedures with a medical oncologist and others, their final surgical choice, and the satisfaction with the choice they made. Analyses were carried out separately for each of the two states. Patients were stratified by propensity for discussion with a medical oncologist (the number of strata was unreported), within-stratum estimates of both outcomes were made, estimates were combined across strata to form population estimates, then the difference between overall estimates for those who did and did not discuss surgery with a medical oncologist were calculated. Keating, et al., found that in Massachusetts, where breast-conserving surgery is more common, discussion with a medical oncologist led to similar rates in choice of surgery but higher satisfaction, and in Minnesota, where breast-conserving surgery is much less common, discussion with a medical oncologist led to higher rates of breast-conserving surgery but similar satisfaction scores.

Dehejia and Wahba (2002) looked to determine whether an on-the-job training program improves earning prospects for individuals with current social or economic problems. They compared their results, obtained by various propensity score matching techniques, to those obtained by the National Supported Work (NSW) Demonstration, an experimental study with the same purpose. As their treatment group, they used the treated individuals from the NSW. For their control group, they used the untreated individuals from two observational studies in two different analyses, the Current

Population Survey (CPS) and the Population Survey of Income Dynamics (PSID). The control group from the CPI provides many comparable units to the NSW treated units, while the PSID is much less comparable. For a comparison of matching techniques, Dehejia and Wahba used matching with replacement using single nearest neighbor and caliper matching, and without replacement using low-to-high, high-to-low, and random ordering. Dehejia and Wahba found that for both control groups, propensity score matching mimics the results of a randomized study by selecting only comparable pairs, creating samples with very similar characteristics. The estimate of the final outcome of interest, increase in earning due to the training program, is also very similar to that found in the NSW experimental study. The authors noted that choice of matching (with or without replacement) should be influenced by the number of comparable units: when there is significant overlap of the propensity score, matching without replacement reduces the standard errors of the estimates, but when there is minimal overlap matching with replacement reduces the bias from poorly matched pairs.

Saitto, et al. (2004) used subclassification on propensity score quintiles to adjust for confounding in their analysis of effectiveness of Coronary Care Units (CCUs) in Lazaio, Italy. Their study used retrospective discharge data of 181,049 patients, 8620 of which were admitted to CCUs. Saitto, et al. created a hierarchical ranking of thirteen disease classes, wishing to determine whether CCU treatment was related to the relative risk of death for various heart diseases. They found that propensity scores (estimated probability of being admitted to a CCU) were distributed similarly in CCU and non-CCU patients, though shifted slightly higher for CCU patients. The odds ratio estimated within each propensity score quintile suggested that CCU admittance decreased the risk of mortality. These results confirmed the relationship found between CCU admittance and mortality in standard logistic regression in all disease groups.

Schneeweiss headed two studies (Schneeweiss, et al. (2007; 2010)) which focused on propensity score-adjusted analyses. The 2007 study looked to compare the risk of death associated with conventional and atypical antipsychotic drugs in elderly patients. Using data from 37,241 new antidepressant users from 1996 to 2004, Schneeweiss estimated propensities for atypical drug prescription by logistic regression. These scores were then used for adjustment in a Cox regression model of patient survival. After propensity score adjustment, the conventional drug was associated with an approximately 50% increase in risk of death within 180 days of start.

These results were similar to those of a meta-analysis of randomized studies. Schneeweiss' second study sought to explain the variability in risk of suicide attempts and antidepressant drugs. Patient characteristics were assessed at the initiation of treatment based on medical claims and used to estimate a propensity score for one of two types of antidepressant medications. After adjusting for propensity score decile and doing a second analysis using a high-dimensional propensity model (including 500 explanatory variables), there were no significant differences in attempted or completed suicides between the two classes of antidepressants.

In an analysis to determine whether community college students are less likely to complete a baccalaureate degree than students who enter directly into four-year universities, Long and Kurlaender (2009) used propensity score stratification to adjust for imbalances between the two groups. Using census data for all students who entered Ohio public colleges and universities in Fall 1998, Long and Kurlaender had access to student characteristics such as family background, high school preparation, and postsecondary intentions. Thus, they were able to include only community college students who listed a four-year degree as an academic goal at the outset of the community college career. Community college and university students' baccalaureate degree completion rates were assessed at 4, 6, and 9 years after their post secondary education began. Students were stratified by propensity score into ten blocks and a model was created using community college as a fixed effect on the propensity score strata. The analysis found that students starting at community colleges had a significantly lower chance of completing a four-year degree.

Though standard propensity score methods are used to account for dichotomous treatments, Imbens (2000) expanded the use of propensity scores to multiple-level treatments. This generalization made propensity score analysis more widely applicable. Foster (2003) used propensity score weighting in a study of children's mental health using data from the Fort Bragg Demonstration. Children ages 5 to 18 were followed for twelve months, with the number of counseling sessions and their improvement in two standard assessment tests recorded, along with entry-time characteristics. With an interest in determining the relationship between volume of services and mental health outcomes, Foster categorized treatment volume into seven levels. He then used methods for multiple-level treatments to estimate propensity scores for each of the seven categories of treatment. The inverse of the propensity scores were used as weights in calculating an average outcome for each treatment level. Through this analysis, Foster found that higher number of visits increased children's functioning to a certain point, but that any more than 12 to 18 visits in the year resulted in diminishing returns.

Multiple-level treatments were also necessary in the case of Zanutto, et al. (2005). At question here was the effectiveness of a national anti-drug media campaign on children aged 9 to 18. Since the campaign was nation-wide, there was no control group to reference, so the authors divided treatment (campaign) exposure into five groups. They then used McCullagh's (1980) ordinal logit model $\log(P(z_k \geq d)/P(z_k < d)) = \alpha_d + \beta^T x_k$ for $d = 2, 3, 4, 5$. The data were subclassified into five strata based on quintiles of $\beta^T x_k$, with analysis limited to the region of overlap for all five treatment levels ($0.4728 \leq \beta^T x_k \leq 2.190$), including 466 of the 521 sampled individuals. Each individual was asked a series of questions about the likelihood of using marijuana or inhalants over the next twelve months, with responses scaled from 1 (definitely no) to 4 (definitely yes). Responses were combined into a single outcome variable. By combining within-quintile estimates into a single estimate of the mean response for each exposure level, Zanutto, et al., found little evidence of a convincing downward trend with increased exposure.

5. Summary

The propensity score arose out of the discussion surrounding causal inference. Randomization is now widely accepted as providing a basis for valid inference about the treatments effects by guaranteeing strongly ignorable treatment assignment. This condition tends not be true in observational studies. If a set of variables can be identified such that conditional on these variables, treatment assignment is strongly ignorable, then valid causal inferences are possible in observational studies. Typically, the variables on which one must condition to ensure strongly ignorable treatment assignment are high dimensional and one cannot find treated and non-treated units with equal values in these variables. The propensity score is a summary measure, such that in sub-populations with the same value of the propensity score, strongly ignorable treatment assignment holds. Thus, inference conditional on the propensity score allows for valid comparisons of treated and non-treated groups. The propensity score is conceptually simple and less dependent on model specifications. Its uses are expanding and its properties are being studied in diverse applications. The question whether or not it provides a reasonable alternative to regression models and similar approaches to inference is still being debated. However, the condition of strongly ignorable treatment assignment cannot be confirmed with observational studies, only assumed to be true.

References

Basu, D. (1980). Randomization analysis of experimental data: the Fisher randomization test, *Journal of the American Statistical Association* **75**, 575–582.

Bayes, T. (1764). An essay towards solving a problem in the doctrine of chances, *Philosophical Transactions of the Royal Society* **53**, 370–418. (reprinted in Studies in the history of probability and statistics IX: Thomas Bayes' essay towards solving a problem in the doctrine of chances, *Biometrika* **45**, 293–315, 1958).

Berk, R. (2005). An Introduction to Ensemble Methods for Data Analysis. *Department of Statistics, UCLA*, Paper 2005032701.

Cochran, W. G. (1965). The planning of observational studies of human populations (with Discussion), *Journal of the Royal Statistical Society Series A* **128**, 134–155.

Cox, D. R. (1958). *Planning of Experiments.* John Wiley & Sons.

Dehejia, R. H. and Wahba, S. (2002). Propensity score-matching methods for nonexperimental causal studies, *The Review of Economics and Statistics* **84**, 151–161.

De Moivre, A. (1738). *The Doctrine of Chances*, 2nd ed., Woodfall, London.

De Morgan, A. (1838). *An essay on probabilities and their application to life contingencies and insurance offices*, Longman, Orme, Brown, Green & Longmans, London.

Drake, C. M. (1993). Effects of misspecification of the propensity score on estimators of treatment effect, *Biometrics* **49**, 1231–1236.

Fisher, R. A. (1925). *Statistical methods for research workers*, Oliver& Boyd, Edingburgh.

Fisher, R. A. (1935). *Design of Experiments*, Oliver& Boyd, Edingburgh.

Foster, E. M. (2003). Propensity score matching: an illustrative analysis of dose response. *Medical Care* **41**, 1183–1192.

Freedman, D. (1999). From association to causation: some remarks on the history of statistics, *Statistical Science* **14**, 243–258.

Gail, M. H., Wieand, S., and Piantadosi, S. (1984). Biased estimates of treatment effect in randomized experiments with nonlinear regressions and omitted covariates, *Biometrika* **71**, 431–444.

Holland, P. W. (1986). Statistics and causal inference, *Journal of the American Statistical Association* (with Discussion and Reply) **81**, 945–970.

Hume, D. (1739). *A Treatise of Human Nature* (LA Selby–Bigge ed. 1888 and later printings), Clarendon, Oxford.

Imbens, G. W. (2000). The role of the propensity score in estimating dose-response functions, *Biometrika* **87**, 706–710.

Judkins, D. R., Morganstein, D., Zador, P., Piesse, A., Barrett, B. and Mukhopadhyay, P. (2007). Variable selection and raking in propensity scoring. *Statistics in Medicine* **26**, 1022–1033.

Keating, et al. (2001). Discussion of treatment options for early-stage breast cancer: effect of provider specialty on type of surgery and satisfaction, *Medical Care* **39**, 681–691.

Kempthorne, O. (1955). The randomization theory of experimental inference, *Journal of the American Statistical Association* **50**, 946–967.

Kempthorne, O. (1977). Why randomize, *Journal of Statistical Planning and Inference* **1**, 1–25.

Kempthorne, O. (1980). Comment on Basu: Randomization analysis of experimental data: the Fisher randomization test, *Journal of the American Statistical Association* **75**, 584–587.

LaPlace, P. S. de (1774). Mémoire sur la probabilité des causes par les évènemens [Translated in Stigler 1986].

Long, B. T. and Kurlaender, M. (2009). Do community colleges provide a viable pathway to a baccalaureate Degree? *Educational Evaluation and Policy Analysis* **31**, 30–53.

Loux, T. and Drake, C. (2011). Estimating the marginal odds ratio in an observational study; unpublished manuscript.

Lunceford, J. K. and Davidian, M. (2004). Stratification and weighting via the propensity score in estimation of causal treatment effects: a comparative study, *Statistics in Medicine* **23**, 2937–2960.

MacMahon, B. and Pugh, T. F. (1960). *Epidemiologic Methods*, Little, Brown and Company, Boston.

McCaffrey, D. F., Ridgeway, G. and Morral, A. R. (2004). Propensity Score Estimation With Boosted Regression for Evaluating Causal Effects in Observational Studies. *Psychological Methods* **9**, 403–425.

McCullagh, P. (1980). Regression models for ordinal data, *Journal of the Royal Statistical Society Series B* **42**, 109–142.

Neyman, J. (1923). Sur les applications de la théoriedes probabilities aux experinecs agricoles: essai des principes, Roczniki Nauk Rolniczki **10**, 1–51. (In Polish. English translation by D. Dabrowska and T. Speed, *Statistical Science* **5**, 463–472, (1990).

Rosenbaum, P. R. and Rubin, D. B. (1983a). The central role of the propensity score in observational studies for causal effects, *Biometrika* **70**, 41–55.

Rosenbaum, P. R. and Rubin, D. B. (1984). Reducing bias in observational studies using subclassification on the propensity score, *Journal of the American Statistical Association* **79**, 516–524.

Rosenbaum, P. R. and Rubin, D. B. (1985b). Constructing a control group using multivariate matched sampling methods that incorporate the propensity score, *American Statistician* **39**, 33–38.

Rosenbaum, P. R. (1984a). From association to causation: the role of tests of strongly ignorable treatment assignment. *Journal of the American Statistical Association* **79**, 41–48.

Rosenbaum, P. R. (1984b). Conditional permutation tests and the propensity score in observational studies. *Journal of the American Statistical Association* **79**, 565–574.

Rosenbaum, P. R. (1987). Model-Based Direct Adjustment. *Journal of the American Statistical Association* **82**, 387–394.

Rubin, D. B. and Thomas, N. (1992). Affinely invariant matching methods with ellipsoidal distributions. The Annals of Statistics, **20**, 2, 1079–1093

Rubin, D. B. and Thomas, N. (1996). Matching using estimated propensity scores: relating theory to practice, *Biometrics* **52**, 249–264.

Rubin, D. B. (1974). Estimating causal effects in randomized and non-randomized studies, *Journal of Educational Psychology* **66**, 688–701.

Rubin, D. B. (1977). Assignment to treatment on the basis of a covariate, *Journal of Educational Statistics* **2**, 1–26.

Rubin, D. B. (1978). Bayesian inference for causal effects: the role of randomization, *Annals of Statistics* **6**, 34–58.

Rubin, D. B. (1980). Comment: Basu, D. (1980). Randomization analysis of experimental data: the Fisher randomization test, *Journal of the American Statistical Association* **75**, 591–593.

Rubin, D. B. (1986). Which ifs have causal answers, *Journal of the American Statistical Association* **81**, 961–962.

Rubin, D. B. (1990a). Comment: Neyman (1923) Causal inference in experiments and observational studies, *Statistical Science* **5**, 472–480.

Rubin, D. B. (1990b). Formal modes of statistical inference for causal effects, *Journal of Statistical Planning and Inference* **25**, 279–292.

Rubin, D. B. (1991). Practical implications of modes of statistical inference for causal effects and the critical role of the assignment mechanism, *Biometrics* **47**, 1213–1234.

Saitto, et al. (2004). Outcome of patients with cardiac diseases admitted to coronary care units: a report from Lazio, Italy, *Medical Care* **42**, 147–154.

Schneeweiss, et al. (2007). Risk of death associated with the use of conventional versus atypical antipsychotic drugs among elderly patients, *Canadian Medical Association Journal* **176**, 627–632.

Schneeweiss, et al. (2010). Variation in the risk of suicide attempts and completed suicides by antidepressant agent in adults, *Archives of General Psychiatry* **67**, 497–505.

Senn, S., Graf, E. and Caputo, A. (2007). Stratification for the propensity score compared with linear regression techniques to assess the effect of treatment or exposure, *Statistics in Medicine* **26**, 5529–5544.

Setoguchi, S., Schneeweiss, S., Brookhart, M. A., Glynn, R. J. and Cook, E. F. (2008). Evaluating uses of data mining techniques in propensity score estimation: a simulation study. *Pharmacoepidemiology and Drug Safety* **17**, 546–555.

Shah, B. R., Laupacis, A., Hux, J. and Austin, P. (2005) Propensity score methods gave similar results to traditional regression modeling in observational studies: a systematic review. *Journal of Clinical Epidemiology* **58**, 550–559.

Stone, R. (1993). The assumptions on which causal inference rests, *Journal of the Royal Statistical Society Series B* **55**, 455–466.

Woo, M. J., Reiter, J.P. and Karr, A. F. (2008). Estimation of propensity scores using generalized additive models. *Statistics in Medicine* **27**, 3805–3816.

Zanutto, E., Lu, B. and Hornik, R. (2005). Using propensity score subclassification for multiple treatment doses to evaluate a national antidrug media campaign, *Journal of Educational and Behavioral Statistics* **30**, 59–73.

Chapter 3

Recent Tests for Symmetry with Multivariate and Structured Data: A Review

Simos G. Meintanis and Joseph Ngatchou-Wandji

National and Kapodistrian University of Athens, Athens, Greece, and EHESP, Département INFOBIOSTAT Rennes and Université Henri Poincaré, INSERM 954 Nancy, France

We review some recent tests of symmetry (i) for the distribution of i.i.d. multivariate observations, and (ii) for the unobserved errors in structured models. The structured models considered fall into one of the following three categories: (a) Parametric regression models, (b) Non-parametric regression models, and (c) Dynamic models. Adaptation of certain procedures to random effects models is also considered.

Contents

1. Introduction

The problem of testing symmetry of an unknown distribution has been considered under various data-generating mechanisms. In the univariate i.i.d. context, with known and estimated center there is a large number of papers; see, for instance, Arcones and Giné (1991), Hollander (1988), Husková (1984), and references therein. The current review is confined to more recently developed procedures for multivariate symmetry with i.i.d. data, and testing for symmetric error distribution under some common data-generating mechanisms such as regression and time series.

We start by emphasizing the critical importance of symmetry. This importance became evident in the statistics literature under different settings. With i.i.d. data, for instance, identifiability of location and consistency of M–estimators requires the assumption of symmetry of the underlying distribution; see Beran (1974) and Freedman and Diaconis (1982). Also in the two-sample design with paired observations, Munzel (1999) warns against the use of the popular Wilcoxon–signed-rank test unless we accept a symmetry hypothesis for the distribution of the differences of the pairs.

Turning to the linear regression setting, Bickel (1982) showed (refer also to Newey (1988)) that under symmetry of errors, adaptive efficient estimators of the parameters can be constructed. Bickel's results were subsequently generalized by Klaassen and Putter (2005). In the related context of analysis of covariance, Ceylan and Goad (2009) show by simulations the effect that the assumption of symmetry of errors has on the performance of some parametric and non-parametric methods. The assumption of symmetry of errors plays also a key role in extending certain optimality results for regression estimators to the case of non-linear, dynamic and non-parametric regression models; see for instance, Koul (2002), and Hillebrand and Müller (2006). Passing on to multivariate rank-based procedures we note that a natural context in this case is that of elliptical (or spherical) symmetry, and in this context optimal tests have been proposed by Choi and Marden (1977) and Hallin and Paindaveine (2002a; 2002b), for both independent observations and under ARMA type dependence. Finally, the assumption of symmetry appears to be also relevant in heteroscedastic and conditionally heteroscedastic models. In the former case, Im (2000) has shown that the celebrated Glejser test is asymptotically invalid unless the error distribution is symmetric, while in the latter case, Newey and Steigerwald (1997) and Gonzalez–Rivera and Drost (1999) show in detail the significant impact that the assumption of symmetry of innovations has on the consistency and efficiency of the so-called quasi maximum likelihood (QML) estimator in GARCH models. In the multivariate context, Boudt and Croux (2010), echoing the views of several researchers that Gaussian QML is very innacurate in the presence of outliers, propose robust M-estimators for GARCH models, but only under the framework of elliptically symmetric innovations.

The rest of the paper unfolds as follows. In Section 2, we mention some notions of symmetry in the multivariate i.i.d. setting and review the corresponding test procedures. Section 3 is devoted to tests of symmetry in linear, non-parametric regression, and in certain time series models. In Section 4 some of the methods already mentioned above are modified in order to test symmetry of the error distribution in random effects models. The paper is concluded with a discussion in Section 5.

2. Notions of and testing for multivariate symmetry

2.1. *Diagonal symmetry*

The notion of symmetry of a univariate distribution is extended to the multivariate case in various ways; for a nice exposition the reader is referred to Zuo and Serfling (2000). The more natural extension, known as the diagonal or reflective symmetry, is defined as follows: a random vector X is said to have a symmetric distribution around a center point μ if and only if $(X - \mu)$ has the same distribution as $-(X - \mu)$. It may also be proved that reflective symmetry is equivalent to the property that $e'(X - \mu)$ has the same distribution as $-e'(X - \mu)$ for all vectors e in the unit sphere, where e' stands for the transpose of e. The advantage in using this property of univariate projections of vector-valued observations is that it can allow for use of results established in the univariate case.

Most of the tests constructed for diagonal symmetry are based on the idea that a random vector X (or its distribution) is reflectively symmetric around μ if and only if the imaginary part of the characteristic function of $X - \mu$ is identically zero. These tests are then constructed on the basis of a stochastic process derived from the imaginary part of the empirical characteristic function. Such a process is typically of the form

$$S_n(t) = n^{-1/2} \sum_{j=1}^{n} \sin[t' g_{nj}(X_j)], \quad t \in \mathbb{R}^d, \tag{1}$$

where the function $g_{nj}(\cdot)$ changes according to the situation being tested and the specific test statistic considered. For example, if the center of symmetry is known, and without loss of generality, we set this center equal to zero, and consider the process $S_n(t)$ in (1) with $g_{nj}(X_j) = X_j$, $j = 1, \ldots, n$. For this case where the center of symmetry is the origin, Ghosh and Ruymgaart (1992) propose a Cramér–von Mises test

$$GR = \int_{e \in \mathcal{U}} \int_0^{\rho} S_n^2(\rho e) d\eta_1(\rho) d\eta_2(e), \tag{2}$$

(where $\eta_j(\cdot)$, $j = 1, 2$, denote arbitrary measures), which combines the empirical characteristic function approach with that of univariate projections, by computing $S_n(\cdot)$ at all possible univariate projections of direction $c \subset \mathcal{U}$, in the unit sphere $\mathcal{U}$ of varying length ρ. The authors derive the spectral decomposition of the limit null distribution of the test statistic in (2), propose an approximation thereof incorporating a finite number of the corresponding eigenvalues, but do not investigate the performance of this approximation by Monte Carlo. It should be noted that the paper by Ghosh and Ruymgaart (1992) rejuvenated the characteristic function approach initially proposed by Feuerverger and Mureika (1977). Regarding the derivation of the spectral decomposition of symmetry statistics with known center, we mention that Baringhaus (1996) considers the test statistic

$$B_\Phi = \int_{t \in \mathbb{R}^d} S_n^2(t) d\Phi(t), \tag{3}$$

where $\Phi(x)$ denotes the distribution function of the d-variate standard normal. The author was able to derive closed formulas for the eigenvalues of the corresponding operator and thus provide a theoretical solution to the limit null distribution of B_Φ in (3). However, the same problem with estimated center remains open.

Turning now to the case of unknown center we consider (1) with $g_{nj}(X_j) = X_j - \widehat{\mu}_n$, $j = 1, \ldots, n$, where $\widehat{\mu}_n$ denotes a consistent and asymptotically normally distributed estimator of the center of symmetry μ. For diagonally symmetric distributions, $S_n(t)$ is generally proved to converge, in compact t intervals, to a centered Gaussian random process with covariance kernel, $\Sigma(s,t)$ depending on the characteristic function of X, the value of μ and the type of estimator $\widehat{\mu}_n$. This result is utilized by Heathcote, et al. (1995) who define a Kolmogorov–Smirnov type statistic

$$H_n(t) = \sup_{t \in D_n} \left| \frac{S_n(t)}{\sigma_n(t)} \right|, \tag{4}$$

where $\sigma_n^2(t)$ is an estimator of the variance function $\sigma^2(t) = \Sigma(t,t)$. In this estimator, the characteristic function of X is replaced by the corresponding empirical characteristic function, and μ is replaced by $\widehat{\mu}_n$. Also in (4), D_n is a working region around the origin over which the supremum could be taken, confined in size by the first zero of the real part of the empirical characteristic function. Since the tails of the limiting distribution of $H_n(t)$ are difficult to compute, the authors propose a test with critical region given by

$$\left|\frac{S_n(t_n)}{\sigma_n(t_n)}\right| > z_{1-\alpha/2},$$

where t_n is obtained as the point that maximizes $\sigma_n^2(t)$ over D_n, and $z_{1-\alpha/2}$ is the $(1-\alpha/2)$-quantile of the standard normal distribution. In view of the possibility of heavy tails a trimmed mean is used as an estimator of μ, and the resulting test is illustrated by a real-data example.

Given the difficulty of approximating the limit null distribution in the statistics in (2), and (4) when the center is estimated, Neuhaus and Zhu (1998) proposed permutation versions of these tests and examined their limit properties. These new statistics are proved to be asymptotically equivalent to their correponding 'non-permutation' ones, and their quantiles are then used for determining the critical regions of the tests. The behavior of these tests are next studied under a sequence of local alternatives and a simulation experiment is conducted to evaluate the performance of the tests for moderate values of the sample sizes. To define the permutation statistics, consider independent random variables $e_1, \ldots, e_n$ independent of the random vectors $X_1, \ldots, X_n$ such that

$$P(e_j = 1) = P(e_j = -1) = 1/2, \quad j = 1, \ldots, n. \tag{5}$$

Then conditionally on $X_j = x_j$, $j = 1, \ldots, n$, consider test statistics based on $S_n(t)$, with $g_{nj}(x_j) = e_j(x_j - \widehat{\mu}_n)$, $j = 1, \ldots, n$, in (1), where $\widehat{\mu}_n = \bar{x}_n$, the sample mean. With this change, Neuhaus and Zhu (1998) define slightly modified versions of the test statistics in (2) and (4), and show the asymptotic validity of these 'permutation statistics.' Note that the permutation procedure borrows from the wild bootstrap method first proposed by Wu (1986).

Since the property of reflective symmetry is affine invariant, that is, invariant under any transformation of the type $X \mapsto MX + b$, for any non-singular matrix M and any $b \in \mathbb{R}^d$, it would be desirable for any test statistic T to also be affine invariant, i.e., to satisfy $T(X_1, \ldots, X_n) = T(MX_1 + b, \ldots, MX_n + b)$. This is the case of the test statistic of Henze, et al. (2003), which corresponds to the statistic in (3) with $g_{nj}(X_j) = \widehat{A}^{-1}(X_j - \bar{X}_n)$ in (1), where $\widehat{A}$ is the square root of the sample covariance matrix $\mathcal{S}_n = n^{-1}\sum_{j=1}^{n}(X_j - \bar{X}_n)(X_j - \bar{X}_n)'$. The properties of the test are studied both under the null hypothesis and a sequence of local alternatives converging to the null hypothesis from a fixed direction. The authors also define a permutation statistic as a slight variation of the original statistic, and show its asymptotic equivalence to the original 'non-permutation' test. They also show that the test which rejects the null hypothesis for large values of the 'non-permutation' statistic, but with critical value obtained from the permutation statistic is consistent against non-symmetric alternatives, and conduct intensive simulations showing that, even for small sample sizes, the test performs well.

Another line of research was followed by Ngatchou–Wandji (2009a) in an effort to numerically approximate the limit null distribution of test statistics utilizing $S_n(t)$, and thus avoid the use of the time-consuming bootstrap and permutation tests. Extending the earlier approach of Koutrouvelis (1985) to the multivariate setting, the author first proposes a robust estimator of the center of symmetry. This estimator utilizes the equation $\tan^{-1}[S(t)/C(t)] = t'\mu,\ t : S(t) \neq 0$, which holds true under reflective symmetry around μ, where $C(t)$ and $S(t)$ denote the real and the imaginary part, respectively, of the characteristic function of X. The test statistic incorporates this estimator as $\widehat{\mu}_n$ and may be written as

$$NW_{\omega,\nu} = \int_{t\in\mathbb{R}^d} S_n^2(t)\omega^2(t)d\nu(t),$$

with $S_n(t)$ defined by $g_{nj}(X_j) = X_j - \widehat{\mu}_n$ in (1), where ω denotes a weight function, and the measure $\nu(t)$ is such that $d\nu(t) = \omega(t)d\eta(t)$, where $\eta(t)$ is a probability measure. Under the null hypothesis of reflective symmetry, the process $S_n(t)\omega(t)$ is proved to converge in distribution to a centered Gaussian process with covariance kernel depending on the underlying distribution of X, the specific type of location estimator and the true value of the location parameter μ. The behavior of $S_n(t)\omega(t)$ is also studied under a general alternative, and also under a sequence of local aternatives converging to the null hypothesis from a fixed direction. Using the Karhunen–Loève expansion (or spectral decomposition) of the limit process, the author shows that the limiting null distribution of his test statistic is the same as that of a weighted sum of independent chi-squared random variables with one degree of freedom, whose cumulative distribution function can be approximated by the results of Imhof (1961). The limit null distribution of a discrete version of the test statistic, which results by replacing $\nu(t)$ by the empirical distribution function of $X_j,\ j = 1,\ldots,n$, is approximated by the sum of a finite number of chi-squared random variables with eigenvalues numerically computed. This version of the test shows good behavior in simulation experiments.

We close our review on diagonal symmetry by referring to some alternative procedures which do not use the Fourier approach. For instance, tests of symmetry may be constructed which utilize different notions of multivariate skewness and corresponding sample estimates; see Gutjahr, et al. (1999), Klar (2002), Balakrishnan, et al. (2007), and Kollo (2008). This approach may be easily implemented and generally leads to test statistics with a distribution-free limit behavior, but has the drawback that, apart from assuming finite moments, it also leads to procedures which are not consistent against arbitrary deviations from the symmetry hypothesis. These shortcomings of Mardia's skewness statistic have been carefully studied by Gutjahr, et al. (1999) and Baringhaus (1992) in the context of testing for multivariate normality. Another idea is to employ the aforementioned notion of univariate projections. Zhu, et al. (1997) employed this idea of projections of vector-valued observations and proposed tests based on the Wilcoxon type statistic

$$ZFL_n(e) = \frac{1}{n^{1/2}(n-1)} \sum_{i \neq j} \left[I\left(e'X_i \leq -e'X_j\right) - \frac{1}{2} \right],$$

where $I(A)$ denotes the indicator function of condition A. Since the asymptotic null distributions of the Kolmogorov–Smirnov and Cramér–von Mises tests based on $ZFL_n(e)$ are difficult to obtain, the authors suggest the use of the chi-squared type statistic $J_{ns} R_{ns}^{-1} J'_{ns}$ where $J_{ns} = (ZFL_n(e_1), \ldots, ZFL_n(e_s))$ and R_{ns} is a consistent estimate of the covariance matrix of J_{ns}, and $e_1, \ldots, e_s$, are suitable chosen points in the unit sphere.

2.2. *Spherical symmetry*

Spherical symmetry is another generalization of the notion of univariate symmetry for distributions. This type of symmetry is discussed for instance in Chmielewski (1981). A random vector X is said to have a spherically symmetric distribution about μ if $X-\mu$ has the same distribution as $H(X-\mu)$ for each orthogonal matrix H. It is proved that a random vector X with a spherically symmetric distribution about μ is characterized by the property that the Euclidian norm $\|X-\mu\|$ of $X-\mu$ and its normalization $\|X-\mu\|^{-1}(X-\mu)$ are independent, and $\|X-\mu\|^{-1}(X-\mu)$ is uniformly distributed on the unit sphere. Henceforth, we shall denote this property by **(P)**.

Spherical symmetries about known center points have received extensive attention. The particular case where the center point is the origin ($\mu = 0$) has been studied amongst others by Ghosh and Ruymgaart (1992), Kariya and Eaton (1977), Baringhaus (1992).

Ghosh and Ruymgaart (1992) prove that under the hypothesis of spherical symmetry, the limiting distribution of the test statistic (2) can be specified. They therefore propose this statistic for testing the null hypothesis of spherical symmetry around the origin. This null hypothesis has also been tested against three types of alternatives by Kariya and Eaton (1977). More formally, they denote by $\mathcal{O}(d)$ the group of orthogonal matrices on $\mathbb{R}^d$. Given Σ a $d \times d$ positive definite matrix, $\mu \in \mathbb{R}$ and $a_0 \in \mathbb{R}^d$ with $\|a_0\| = 1$, they consider the following classes of functions

$$\mathcal{F}_0 = \{f | f \text{ is a density function on } \mathbb{R}^d, f(x) = f(gx),\ x \in \mathbb{R}^d,\ g \in \mathcal{O}(n)\}$$

$$\mathcal{Q} = \{q | q : [0, \infty) \to [0, \infty), q \text{ is increasing, and } \int_{\mathbb{R}^d} q(||x||^2) dx = 1\}$$

$$\mathcal{F}_1(\Sigma) = \{f | f(x) = (\det\Sigma)^{-1/2} q(x'\Sigma^{-1}x),\ q \in \mathcal{Q}\}$$

$$\mathcal{F}_2(\mu) = \{f | f(x) = q(||x - \mu a_0||^2),\ q \in \mathcal{Q}\}$$

$$\mathcal{F}_3(\mu) = \{f | f(x) = q(||x - \mu a_0||^2),\ q \in \mathcal{Q},\ q \text{ convex}\}.$$

Then, they propose a UMP (Uniformly Most Powerful) test for testing $\mathcal{F}_0$ against $\mathcal{F}_1(\Sigma)$ for fixed Σ, and extend their results to cover some situations where Σ is not fixed any longer. Modifying a technique used in Lehmann and Stein (1949), they derive a UMP test for testing $\mathcal{F}_0$ against $\{\mathcal{F}_2(\mu),\ \mu > 0\}$. Next, they apply the above two tests to a regression model. Turning to testing $\mathcal{F}_0$ against $\{\mathcal{F}_3(\mu),\ \mu \neq 0\}$, they use the generalized Neyman–Pearson lemma to derive a UMPU (Uniformly Most Powerful Unbiaised) test. Finally, they discuss a conjecture concerning the existence of a UMPU test for testing $\mathcal{F}_0$ against $\{\mathcal{F}_2(\mu), \mu \neq 0\}$. Their work also includes the study of the distributions of $X/\|X\|$, $e'X/\|X\|$ and $X'\Lambda X/\|X\|$, for a unit vector e, a symmetric matrix Λ and a random vector X with spherically symmetric distribution. These results are applied to the derivation of the distribution of the sample correlation coefficient under a weaker assumption than the usual normal assumption.

Based on **(P)**, Baringhaus (1991) proposes a class of test statistics indexed by infinite linear combinations of either Chebyschev or Gegenbauer polynomials. These statistics result from a modification of the Cramér–von Mises type test of Smith (1977) for testing angular symmetry in the case of bidimensional distributions. Baringhaus (1991) shows that, under the null hypothesis of spherical symmetry, the limiting distribution of each test statistic is the same as that of a sum of some weighted iid chi-squared random variables. He also shows his tests are consistent against the various alternatives he considers.

A more general null hypothesis is investigated by Diks and Tong (1999) who, for a density function $f(x)$ defined on $\mathbb{R}^d$, consider testing

$$f(x) = f(gx), \; g \in G, \tag{6}$$

where G is a given compact group of isometries on $\mathbb{R}^d$. Extending the ideas of Besag and Diggle (1977) and Eigen and Lillegard (1997), they propose a conditional Monte Carlo test. Such tests are based on comparing an observed dataset, using any test criterion of choice, with a set of reference datasets obtained by simulations under the null hypothesis. Diks and Tong derive their test as follows. Let $X_1, \ldots, X_n$ be a sample from a distribution with probability density function $f(x)$ and let $\mathcal{T}(X_1, \ldots, X_n)$ be a given test statistic. The reference set of values of the test statistic required by Monte Carlo procedures is obtained by sampling from $\mathcal{T}(X_1', \ldots, X_n')$, where $X_1', \ldots, X_n'$ are distributed according to the conditional distribution of $X_1, \ldots, X_n$ given $\{G(X_1), \ldots, G(X_n)\}$, with $G(x) = \{gx; g \in G\}$ standing for the G-orbit of $x \in \mathbb{R}^d$. The authors show that this conditional sampling procedure is equivalent to a conditional scheme in which exactly one random observation is taken from each G-orbit. Now, let $\mathcal{T}_0 = \mathcal{T}(X_1, \ldots, X_n)$ and $\mathcal{T}_1, \ldots, \mathcal{T}_B$ be obtained from the Monte Carlo procedure. The p-values of conditional Monte Carlo tests are generally estimated by

$$\widehat{p} = \frac{k}{B+1}, \tag{7}$$

where k is the number of the $\mathcal{T}_j$, $j = 0, 1, \ldots, B$, that are larger or equal to $\mathcal{T}_0$. Denoting by π the invariant probability measure on G, Diks and Tong define the symmetrised version of $f(x)$ by $f_s(x) = \int_G f(gx)\, d\pi(g)$ and observe that testing (6) is tantamount to testing $f(x) = f_s(x)$. Then they define a measure of distance between $f(x)$ and $f_s(x)$ by $Q = \langle f - f_s,\, f - f_s \rangle$, where

$$\langle f_1,\, f_2 \rangle = \iint f_1(r) f_2(s) K(r-s)\, dr\, ds,$$

with $K(r-s) = \exp(-\|r-s\|^2/4d^2)$. Using the fact that $Q = \langle f, f\rangle - \langle f_s, f_s\rangle$, the authors estimate Q by $\widehat{Q} = DT - DT_s$ where DT_s is an unbiased and consistent U-estimator of $\langle f_s, f_s\rangle$ and DT is the U-estimator of $\langle f, f\rangle$ defined by

$$DT = \frac{2}{n(n-1)} \sum_{i=1}^{n-1} \sum_{j=i+1}^{n} K(X_i - X_j).$$

Since T_s is difficult to compute and since it does not affect the estimation of the p-values of the test based on $\widehat{Q}$, Diks and Tong consider DT as the test statistic for their testing problem rather than $\widehat{Q}$. Then they apply their approach to testing spherically symmetric and reflectively symmetric distributions in the case the center of symmetry is the origin. They also illustrate their method by a numerical simulation, and apply it to a real data set.

The case of unknown μ is considered in Koltchinsky and Li (1998). The test proposed is based on the extension of the distribution and the quantile function to the multivariate case. This test is a multivariate version of those suggested by Schuster and Barker (1987) and by Arcones and Giné (1991). Specifically, denote by ν the uniform distribution function on the unit sphere $\mathcal{U}$ (with dimension $d-1$). The approach here is to use, as a measure of symmetry of P, the functional

$$\gamma(P) = \sup_{s \in \mathbb{R}^d} \left|\left| F_P(s + F_P^{-1}(0)) - \psi_P(||s||)\frac{s}{||s||} \right|\right|,$$

where

$$F_P(s) = \int_{x\in\mathbb{R}^d} \frac{s-x}{||s-x||} P(dx), \;\; s \in \mathbb{R}^d$$

$$\psi_P(\lambda) = \int_{x\in\mathbb{R}^d} \int_{e\in\mathcal{U}} \frac{(\lambda e - x)'e}{||\lambda e - x||} \nu(de) P_{-\mu}(dx),$$

with $P_{-\mu}$ standing for the $-\mu$-shift of P, $F_P^{-1}(s)$ the inverse of $F_P(s)$ (the multivariate quantile function) and $F_P^{-1}(0) = \mu$. Indeed, $\gamma(P)$ is proved to be invariant with respect to all translations and all orthogonal transformations on $\mathbb{R}^d$. Therefore it seems natural to base a test for spherical symmetry on the statistic $\gamma(P_n)$ where P_n is the empirical distribution of $X_1, \ldots, X_n$. The test statistic considered is

$$KL = n^{1/2}\gamma(P_n).$$

Its limiting behavior is investigated under the null hypothesis and under a sequence of asymmetric alternatives converging to a spherically symmetric distribution. The asymptotic power of the test is also evaluated under those hypotheses. Since the limiting distribution of KL under the null hypothesis depends on P which is unknown, its quantiles are approximated using a symmetrized bootstrap. Finally, the performance of the test is investigated by numerical simulations.

2.3. *Elliptical symmetry*

Another main type of multivariate symmetry dealt with in the literature is ellipsoidal (or elliptical) symmetry. A random vector X is said to have an ellipsoidal symmetric distribution with parameters μ and Σ if it has the same distribution as that of $A'Y + \mu$, where Y is a spherically symmetric random vector around the origin, and A is $k \times d$ matrix satisfying $A'A = \Sigma$ with $\text{rank}(\Sigma) = k \leq d$. If the density of such a random vector exists, it would have the form $[\det\Sigma]^{-1/2} g[(x-\mu)'\Sigma^{-1}(x-\mu)]$ for some nonnegative real-valued function $g(z)$. In the case of $d \times d$ nonsingular A, the form of the density reduces to $[\det A]^{-1} g_0[A^{-1}(x-\mu)]$, for a density $g_0(z)$ spherically symmetric around the origin.

Beran (1979) considers the case where the matrix A is lower triangular with positive diagonal elements and determinant one. Based on the property **(P)** and the link between ellipsoidal and elliptically symmetric random vectors, he derives a test for ellipsoidal symmetry. His method runs as follows. Let ν be the distribution of the angular polar coordinates $\Theta = (\theta_1, \theta_2, \ldots, \theta_{p-1})$ of $\|Y\|^{-1}Y$, where Y has a spherically symmetric distribution around the origin. Recall that if X has an elliptically symmetric distribution with parameters μ and Σ then for A as above, $A^{-1}(X-\mu)$ is spherically symmetric around the origin. Denote by $\{R_i(\mu, A)\}$ the ranks divided by $n+1$ of the $\{\|A^{-1}(X_i - \mu)\|\}$ and $\{\Theta_i(\mu, A)\}$ the angular polar coordinates of the direction vectors $\{\|A^{-1}(X_i-\mu)\|A^{-1}(X_i-\mu)\}$. Let $\{a_k; k \geq 1\}$ be a family of functions orthonormal with respect to Lebesgue measure on $[0,1]$ and orthogonal to the constant function on $[0,1]$. If $p \geq 2$, let $\{b_k; k \geq 1\}$ be a family of functions orthonormal with respect to the uniform measure ν on $\mathbf{s}=[0,\pi]^{p-2} \times [0, 2\pi)$ and orthogonal to the constant function on $\mathbf{s}$. Let $(\widehat{\mu}_n, \widehat{A}_n)$ be an estimator of (μ, A) based on $X_1, \ldots, X_n$.

Beran proposes as test statistic $Be(\widehat{\mu}_n, \widehat{A}_n)$, where

$$Be(\mu, A) = \sum_{k=1}^{M_n}\sum_{\ell=1}^{L_n}\left\{n^{-\frac{1}{2}}\sum_{i=1}^{n} a_k[R_i(\mu, A)]b_\ell[\Theta_i(\mu, A)]\right\}^2,$$

with K_n and L_n denoting sequences of integers allowed to increase with n. Indeed, letting $F(\cdot)$ be the cumulative distribution function of $\|A^{-1}(X-\mu)\|$ under the hypothesis of spherical symmetry, and putting $U_i = F(\|A^{-1}(X_i-\mu)\|)$, Beran observes that under the null hypothesis, for suitable choosen K_n and L_n, the function

$$\widehat{g}_n(u,\theta) = 1 + \sum_{k=1}^{M_n}\sum_{\ell=1}^{L_n}\left\{n^{-\frac{1}{2}}\sum_{i=1}^{n} a_k[R_i(\mu, A)]b_\ell[\Theta_i(\mu, A)]\right\} a_k(u)b_\ell(\theta),$$
$$u \in [0,1],\ \theta \in \mathbf{s}$$

should consistently estimate the density of $(U_i, \Theta_i(\mu, A))$. He deduces from this observation that the statistic $n\int[\widehat{g}_n(u,\theta) - 1]^2 du d\nu(\theta)$, which equals $Be(\mu, A)$, can be used for this testing problem. He studies the behavior of this statistic under the null hypothesis and under a broad class of alternatives and analyses the effects of replacing μ and A by their estimates $\widehat{\mu}_n$ and $\widehat{A}_n$.

A different approach is studied in Zhu and Neuhaus (2000) who propose a conditional Monte Carlo test for testing multivariate distributions which are independently decomposable. This class of distributions includes amongst others elliptically symmetric, reflectively symmetric, Liouville-Dirichlet and symmetric scale mixture distributions. A random vector X is said to be independently decomposable if $X = Y \cdot Z$, where Y has an analytically tractable distribution, Y and Z are independent and $Y \cdot Z$ denotes the dot product of Y and Z.

The Monte Carlo testing procedure of Zhu and Neuhaus (2000) runs as follows. Let $X_1, \ldots, X_n$ be copies of X. If the X_i's are independently decomposable with $X_i = Y_i \cdot Z_i$, then any test statistic $\mathcal{T}(X_1, \ldots, X_n)$ has the same distribution as $\mathcal{T}(Y_1 \cdot Z_1, \ldots, Y_n \cdot Z_n)$. The reference set of values of the test statistic required by conditional Monte Carlo procedures is constructed by sampling from $\mathcal{T}(Y_1' \cdot Z_1, \ldots, Y_n' \cdot Z_n)$, where the Y_i''s have the same distribution as the Y_i's. The p-values of the test are estimated along the same lines, as in Diks and Tong (1999), by $\widehat{p}$ given by (7).

Since X and $Y \cdot Z$ have the same characteristic function under the null hypothesis, Zhu and Neuhaus (2000) base their test statistic on a measure of deviation between these functions. They propose as test statistic

$$ZN_1 = \int \left\{ \left[\frac{1}{n} \sum_{i=1}^{n} \cos(t'X_i) \right]^2 + \left[\frac{1}{n} \sum_{i=1}^{n} \sin(t'X_i) \right]^2 \right\} \omega_a(t)\, dt,$$

where $\omega_a(t)$ is a continuous weight function, with parameter a, on its support set such that the integral is finite. ZN_1 is obtained in the same spirit, and using the same arguments as those in Diks and Tong (1999). It is computed for $\omega_a(t) = (2\pi a)^{-d/2} \exp\{-\|t\|^2/(2a^2)\}$, $t \in \mathbb{R}^d$ and $\omega_a(t) = (2a)^{-d}$, $t \in [-a, a]^d$. Zhu and Neuhaus (2000) extend their test procedure to handle cases with a nuisance location parameter. They conclude their work by an illustration of their method with a real data set and by a simulation experiment in which elliptical symmetry is investigated. They also compare the performance of their test with bootstrap tests.

Using once more a Monte Carlo test procedure, Zhu and Neuhaus (2003) consider again testing elliptical symmetry with center μ and shape Σ. Let $\widehat{\mathcal{A}}$ stand for $\Sigma^{1/2}$ or $\widehat{A}$ the square-root of the sample covariance matrix, and $\widehat{\mu} = \mu$ or $\bar{X}$ be the sample mean depending on whether the parameters are known or not. Extending the statistic GR of Ghosh and Ruymgaart (1992), they propose as test statistic

$$ZN_2 = \int_{e \in \mathcal{U}} \int_{\rho \in I} S_n^2(\rho e) d\omega(\rho) d\nu(e),$$

where $S_n(t)$ is defined by (1) with $g_{ni}(X_j) = \widehat{\mathcal{A}}^{-1}(X_i - \widehat{\mu})$, $\omega(t)$ being a weight function with compact support, ν being the uniform distribution on $\mathcal{U}$ and I being a working region. They study the behavior of ZN_2 under the null hypothesis of elliptical symmetry, under general alternatives and under a sequence of local alternatives. Since the limiting distribution of the test statistic under the null hypothesis is untractable, they use a conditional Monte Carlo test procedure based on the test statistic ZN_2. Their tests hinge on the key property that X is elliptically symmetric with center μ and shape matrix Σ if and only if

$$A^{-1}(X - \mu) = U \|A^{-1}(X - \mu)\|,$$

where A^{-1} is the inverse of the nonsingular matrix defined at the begining of this subsection, $U = A^{-1}(X - \mu)/\|A^{-1}(X - \mu)\|$ is uniformly distributed on $\mathcal{U}$ and independent of $\|A^{-1}(X - \mu)\|$. Indeed, from this property, it is easy to see that for a given test statistic $\mathcal{T}$, $\mathcal{T}(A^{-1}(X_1 - \mu), \ldots, A^{-1}(X_n - \mu))$ has the same distribution as $\mathcal{T}(U_1\|A^{-1}(X_1 - \mu)\|, \ldots, U_n\|A^{-1}(X_n - \mu)\|)$, for uniformly distributed U_i's on $\mathcal{U}$, and, given the $\|A^{-1}(X_i - \mu)\|$'s, the conditional distributions of $\mathcal{T}(A^{-1}(X_1 - \mu), \ldots, A^{-1}(X_n - \mu))$ and $\mathcal{T}(U_1\|A^{-1}(X_1 - \mu)\|, \ldots, U_n\|A^{-1}(X_n - \mu)\|)$ are identical. For $\mathcal{T} = NZ_2$, the authors establish the validity of their tests, and prove that the conditional distribution based on the generated data given the $\|A^{-1}(X_i - \mu)\|$'s has the same limiting distribution as NZ_2 under the null hypothesis. This important result justifies the approximation of the p-values of the test based on NZ_2 by those obtained by the conditional Monte Carlo procedure. Here, the authors obtain the reference set of values of the test statistic required by conditional Monte Carlo tests by sampling from $\mathcal{T}(U_1\|A^{-1}(X_1-\mu)\|, \ldots, U_n\|A^{-1}(X_n-\mu)\|)$, with the U_i's being uniformly distributed on $\mathcal{U}$. The p-values are estimated in the same way as in Zhu and Neuhaus (2000), and the case of unkown μ and A is treated in the same spirit.

We now present some approaches which have the common feature of being based on a *necessary condition* statistic for elliptical symmetry. Such tests are generally not omnibus tests since they may not have asymptotic power equal to one against all alternatives. We start with Manzotti, et al. (2002). The authors consider the random vectors $\dot{Y}_i = Y_i/\|Y_i\|$, where the Y_i's are the scaled residuals $Y_i = \widehat{A}^{-1}(X_i - \bar{X})$. They observe that if the X_i's have an elliptical symmetric distribution, then the $\dot{Y}_i$'s should be asymptotically uniformly distributed on the unit sphere $\mathcal{U}$. The idea of their test is then to test for nonuniformity using a test statistic based on averaging spherical harmonics over the $\dot{Y}_i$'s. Recall that a spherical harmonic of degree j is the restriction to $\mathcal{U}$ of a homogeneous polynomial $p(x)$ on $\mathbb{R}^d$ of degree j such that $\sum_{j=1}^{d} \partial^2 p(x)/\partial x_j^2$ is zero on $\mathbb{R}^d$. Denote $\tilde{F}_n(y)$ the empirical cumulative distribution of $\|Y_1\|, \ldots, \|Y_n\|$ and $\mathcal{J}_{jl} = \cup_{j \le i \le l} \mathcal{H}_i$, where $\mathcal{H}_j$ stands for the set of spherical harmonics of degree j in the orthonormal basis described by Müller (1966). For a fixed ϵ and for a function h defined on $\mathcal{U}$, let

$$\mathcal{R}_n(h) = \frac{1}{n} \sum_{i=1}^{n} h(\dot{Y}_i) I\left(\|Y_i\| > \tilde{F}_n^{-1}(\epsilon)\right).$$

The authors consider as test statistic

$$MPQ = n \sum_{h \in \mathcal{J}_{jl}} \mathcal{R}_n^2(h), \quad 3 \le j \le l.$$

Denoting by $N(j,l)$ the number of elements in $\mathcal{J}_{jl}$, they prove that MPQ is affine invariant, and that under the null hypothesis of elliptical symmetry, it converges in distribution to $(1-\epsilon)\chi^2_{N(j,l)}$, where χ^2_{df} stands for a chi-square random variable with *df* degrees of freedom. From the results of a bivariate simulation experiment conducted, the authors find that the finite sample quantiles of their test statistic converge to the theoretical limiting quantiles, and that their test statistic has good power properties.

Another *necessary condition* test is the one presented in Schott (2002). This test results from analyzing the sample covariance matrix whose limiting distribution depends on the fourth-order moments of the distribution under study. Recall that given two matrices $A = (a_{ij} : 1 \leq i \leq p, 1 \leq j \leq q)$ and $B = (b_{ij} : 1 \leq i \leq p', 1 \leq j \leq q')$, $A \otimes B = (a_{ij}B : 1 \leq i \leq p, 1 \leq j \leq q)$ and $\text{vec}(A) = (a_{11}, a_{12}, \ldots, a_{1q}, \ldots, a_{p1}, a_{p2}, \ldots, a_{pq})$. Schott (2002) reduces the problem of testing elliptical symmetry to the *necessary* problem of testing whether the fourth-order moment $M_4 = E\{(X-\mu)(X-\mu)' \otimes (X-\mu)(X-\mu)\}$ has the form

$$M_4 = (1+\kappa)\{(I_{d^4} + K_{dd})(\Sigma \otimes \Sigma) + \text{vec}(\Sigma)\text{vec}(\Sigma)'\}$$

where κ is a number depending on the distribution under study and K_{dd} is a *commutation* matrix (see Magnus (1988), Chapter 3). For deriving a feasible test, the author defines the random matrices

$$\widehat{M_4} = \frac{1}{n}\sum_{i=1}^{n}(X_i - \bar{X})(X_i - \bar{X})' \otimes (X_i - \bar{X})(X_i - \bar{X})'$$

$$\widehat{M_{4*}} = (S^{-1/2'} \otimes S^{-1/2'})\widehat{M_4}(S^{-1/2} \otimes S^{-1/2})$$

$$N_4 = I_{d^2} + K_{dd} + \text{vec}(I_d)\text{vec}(I_d)', \quad R = \frac{1}{3d(d+2)}\text{vec}(N_4)\text{vec}(N_4)'$$

and proves that the vector $n^{1/2}v = n^{1/2}(I_{d^4} - R)\text{vec}(\widehat{M_{4*}})$ converges in distribution to a centered Gaussian random vector with covariance matrix $\Omega = (I_{d^4} - R)\Xi(I_{d^4} - R)$, where Ξ is the limiting covariance of $n^{1/2}\text{vec}(\widehat{M_{4*}})$. Estimating Ξ, he finds a consistent estimator of Ω denoted by $\widehat{\Omega}$, and proposes the Wald statistic

$$SC = nv'\widehat{\Omega}^{-1}v,$$

where $\widehat{\Omega}^{-1}$ stands for any generalized inverse of $\widehat{\Omega}$. It is proved that under the null hypothesis of elliptical symmetry with eighth-order moment, SC converges in distribution to a chi-square random variable with $df = d^2 - 1 + d(d-1)(d^2+7d-6)/24$ degrees of freedom. Under the null hypothesis of Gaussian distribution, $v_* = \text{vec}(\widehat{M_{4*}} - N_4)$ is substituted for v in SC, and it is proved that the resulting statistic converges in distribution to a chi-square random variable with $df + 1$ degrees of freedom. The results of a simulation study for assessing the finite sample properties of the tests are also reported.

We close our review on elliptical symmetry with the Pearson-type chi-square test proposed in Huffer and Park (2005). The idea of this test is to divide the space $\mathbb{R}^d$ into cells which have approximatively equal expected cell counts under the hypothesis of elliptical symmetry, and determine the observed counts for these cells. The chi-square statistic for comparing observed and expected cell counts can then be derived. The authors find it easy to describe and compute this test statistic in terms of the scaled residuals Y_i defined in a preceding paragraph. They proceed as follows. Denote by $\overline{C}$ the closure of a set C. Let $\mathcal{G} = \{\Gamma_1, \ldots, \Gamma_r\}$ be a finite group of orthogonal $d \times d$ matrices with $\Gamma_1 = I$. Let $O \in \mathbb{R}^p$ be an open set such that $O \cap \Gamma_i O = \emptyset$, $i \neq 1$, $\cup_{i=1}^r \overline{\Gamma_i O} = \mathbb{R}^p$ and there exist vectors $a_1, a_2, \ldots, a_k \in \mathbb{R}^d$ for which $O = \{x \in \mathbb{R}^d : a_i' x > 0, i = 1, \ldots, k\}$. Putting $O_i = \Gamma_i O$, $i = 1, \ldots, r$, and denoting by $\widehat{F}_n(y)$ the empirical cumulative distribution of $\|Y_1\|^2, \ldots, \|Y_n\|^2$, the authors first construct the s shells

$$SH_{nj} = \left\{ y : \widehat{F}_n^{-1}\left(\frac{j-1}{s}\right) \leq \|y\|^2 < \widehat{F}_n^{-1}\left(\frac{j}{s}\right) \right\}, \; j = 1, \ldots, s,$$

with $\widehat{F}_n^{-1}(0) = 0$ and $\widehat{F}_n^{-1}(s) = \infty$, and divide the space of the Y_i's into the rs cells $\Lambda_{n\tau} = O_i \cap SH_{nj}$, $\tau = (i, j) \in \{(k, l) : 1 \leq k \leq r, 1 \leq l \leq s\}$. Next, considering the cell counts $U_{n\tau} = \sum_{k=1}^n I\left(Y_k \in \Lambda_{n\tau}\right)$ and letting $p_0 = 1/rs$, they propose as test statistic

$$HP = \sum_\tau \frac{(U_{n\tau} - np_0)^2}{np_0}.$$

For Gaussian X_i's, they prove that HP converges in distribution to a sum of weighted independent chi-square random variables. For the X_i's having an elliptical symmetric distribution, the limiting distribution of HP is not easy to obtain. For this reason, the authors use boostrap for approximating its quantiles. They finally evaluate the performance of their test by means of a simulation experiment, and compare it to that of some other existing tests.

3. Testing symmetry with structured data

The general representation of the models considered in this section is:

$$Y_j = m\left(Z_j\right) + \sigma(Z_j)\varepsilon_j, \;\; j \in \mathbb{Z}, \tag{8}$$

where $Y_j \in \mathbb{R}$, $X_j = (X_j^{(1)}, \ldots, X_j^{(d)}) \in \mathbb{R}^d$, $Z_j = (Y_{j-1}, \ldots, Y_{j-p}, X_j, X_{j-1}, \ldots, X_{j-q})$, with (p,q) standing for possibly infinite nonnegative integers, and (ε_j) a sequence of i.i.d. centered random variables with common density $f(\varepsilon|Z)$, and cumulative distribution $F(\varepsilon|Z)$. Here ε is a generic notation for ε_j. Also in what follows we will often supress Z in the quantities of interest which however should be understood conditionally on the regressor Z. There is an increasing amount of literature on testing whether the distribution of ε is symmetric or not. Such a test can be expressed in the form of densities $H_0 : f(\varepsilon) = f(-\varepsilon)$, for every ε against the alternative that there exist ε, such that $H_1 : f(\varepsilon) \neq f(-\varepsilon)$, or equivalently, as $H_0 : F(\varepsilon) = 1 - F(-\varepsilon)$ versus $H_1 : F(\varepsilon) \neq 1 - F(-\varepsilon)$. The alternative Fourier approach is formulated in terms of the imaginary part $S(t)$ of the characteristic function of ε as $H_0 : S \equiv 0$ against the alternative that there exist $t \in \mathbb{R}$, such that $H_1 : S(t) \neq 0$. Each of these formulations generally leads to the construction of a suitable test statistic which utilizes the residuals

$$\widehat{\varepsilon}_j = \frac{Y_j - \widehat{m}(Z_j)}{\widehat{\sigma}(Z_j)}, \; j = 1, \ldots, n, \tag{9}$$

obtained from suitable estimators $\widehat{m}(Z_j)$ and $\widehat{\sigma}(Z_j)$ of the conditional mean function $m(\cdot)$ and the variance function $\sigma(\cdot)$.

3.1. *Linear regression*

We begin with the case of linear regression which corresponds to (8) with $p = q = 0$, in which case Z_j is reduced to a vector of fixed design points X_j, $m(X_j) = m(\rho; X_j) = \rho_0 + \sum_{j=1}^d \rho_j X_i^{(j)}$ and $\sigma(z) = 1$. For this model Fan and Gencay (1995) propose a test based on

$$\lambda_\gamma := \frac{\delta_\gamma(\widehat{\varepsilon})}{\Delta_{\widehat{\varepsilon}}(\widehat{f})}, \text{ where } \Delta_\varepsilon(f) = \int f^2(\varepsilon) d\varepsilon, \tag{10}$$

with $\widehat{f}(\cdot)$ denoting the kernel estimator of $f(\cdot)$ based on the kernel $K(\cdot)$ with bandwidth denoted by h_n

$$\widehat{f}(\varepsilon) = \frac{1}{nh_n} \sum_{j=1}^n K((\varepsilon - \varepsilon_j)/h_n). \tag{11}$$

The method utilizes the symmetric version $\int f(\varepsilon) f(-\varepsilon) d\varepsilon / \int f^2(\varepsilon) d\varepsilon$ of the discrepancy measure of Ahmad and van Belle (1974), but in the numerator of λ_γ in (10), the weighted form $\delta_\gamma(\varepsilon) = n_\gamma^{-1} \sum_{j=1}^n C_j(\gamma) \widehat{f}(-\varepsilon_j)$

appears, where $n_\gamma = n$, if n is even, and $n_\gamma = n+\gamma$, $\gamma > 0$, if n is odd, and $C_j(\gamma) = 1-\gamma$, if j is even, and $C_j(\gamma) = 1+\gamma$, if j is odd. This formulation, unlike the non-weighted case corresponding to $\gamma = 0$ which renders the test statistic asymptotically degenerate, allows for an asymptotically normally distributed test statistic with the usual $n^{-1/2}$ rate. In a simulation study, the test based on least squares residuals with a proper choice of bandwidth h_n in (11) shows good level and power characteristics, but the power results are based on grossly asymmetric errors. The choice of γ inherent in (10) may be avoided by utilizing the simple discrepancy measure $D := (1/2)\int[f(\varepsilon)-f(-\varepsilon)]^2 d\varepsilon = \int[f(\varepsilon) - f(-\varepsilon)]dF(\varepsilon)$. A corresponding test statistic can be constructed by replacing in the discrepancy measure D, $f(\varepsilon)$ by $\widehat{f}(\widehat{\varepsilon})$, and the distribution function $F(\cdot)$ by the empirical distribution function of the residuals. This test was proposed by Ahmad and Li (1997) and was shown to be asymptotically normal, but with a convergence rate depending on the value of the bandwidth. Also, the authors did not investigate the finite-sample performance of the method by Monte Carlo.

In order to avoid dependence on the bandwidth and at the same time establish the usual $n^{-1/2}$ rate of convergence, Holger, Dette and co-workers initiated a study of classical symmetry statistics which utilize, instead of the density, the properties of the corresponding empirical distribution function of residuals. In the context of linear regression, Neumeyer, et al. (2005) proposed the Kolmogorov–Smirnov and Cramér–von Mises statistics,

$$\sup_{t\in\mathbb{R}} |\widehat{Z}_n(t)| \text{ and } \int \widehat{Z}_n^2(t)d\widehat{H}_n(t), \tag{12}$$

where

$$\widehat{Z}_n(x) = \sqrt{n}\left[\widehat{F}_n(x) - \widehat{F}_n(-x^-) - F(x) - F(-x),\right] x \in \mathbb{R}$$

with $\widehat{F}_n(x)$ and $\widehat{H}_n(x)$ denoting the empirical distribution function of $\widehat{\varepsilon}$ and $|\widehat{\varepsilon}|$, respectively. The authors show that under the null hypothesis of symmetric errors, $\widehat{Z}_n(t)$, converges weakly to a centered Gaussian process $Z(t)$ with covariance kernel depending on the distribution of ε, the design points X_j and the type of estimator involved in computing the residuals. Consequently the test statistics converge, by the continuous mapping theorem, to the corresponding quantities $\sup_{t\in\mathbb{R}} |Z(t)|$ and $\int S^2(t)dH(t)$, where $H(x) = F(x) - F(-x-)$. Since the test statistics are not asymptotically distribution-free under the null hypothesis, bootstrap versions are proposed which then employed in the computation of critical values of the tests. In particular, a version of the wild bootstrap, introduced earlier in the context of testing symmetry with i.i.d. observations, is employed, adapted to the linear regression model. Specifically, for given sample size n, and conditionally on the original data (X_j, Y_j), the corresponding estimators $(\widehat{\rho}_0, \ldots, \widehat{\rho}_d)$, and the resulting residuals $\widehat{\varepsilon}_j$, $j = 1, \ldots, n$, these quantities are used as follows:

(i) Independently of the data (X_j, Y_j), generate random variables e_j, $j = 1, \ldots, n$ as in (5).
(ii) Compute the bootstrap residuals $\varepsilon_j^* = e_j \widehat{\varepsilon}_j$, $j = 1, \ldots, n$.
(iii) Compute bootstrap observations $Y_j^* := m(\widehat{\rho}; X_j) + \varepsilon_j^* = \widehat{\rho}_0 + \sum_{m=1}^{d} \widehat{\rho}_m X_j^{(m)} + \varepsilon_j^*$.
(iv) Based on the new data (X_j, Y_j^*), $j = 1, \ldots, n$, calculate the bootstrap estimator $(\widehat{\rho}_0^*, \ldots, \widehat{\rho}_p^*)$, and the corresponding estimated residuals $\widehat{\varepsilon}_j^* = Y_j^* - m(\widehat{\rho}^*; X_j)$.
(v) Calculate the value of the test statistic, say T^*, on the basis of the estimated residuals.

Then, if the procedure (i)–(v) is repeated a number of times, say B, the critical points are computed by (7). The authors prove that this version of the wild bootstrap leads to a consistent test, and compare their test with a version of the λ_γ test in (10) by Monte Carlo. The finite-sample results show that the classical tests dominate the test of Ahmad and Li (1997) in the simulated situations.

3.2. *Nonparametric regression*

Turning to the problem of testing symmetry of errors in nonparametric regression, we begin with Dette, et al. (2002) who generalize the test of Ahmad and Li (1997) to this situation by replacing the parametrically obtained residuals by nonparametric counterparts (see below). The authors consider the homoscedastic case with fixed design points and prove that the test detects local alternatives at a rate $n^{-1} h_n^{-1/2}$. On the other hand, Neumeyer and Dette (2007) considered modified versions of the classical statistics in (12) appropriate under heteroscedasticity, thereby reestablishing the usual $n^{-1/2}$ of consistency, at the cost however of introducing complicated asymptotics. These modifications correspond to the way that the residuals $\widehat{\varepsilon}_j$ are defined in the process $\widehat{Z}_n(t)$. In particular let in (8), $p = q = 0$ and $d = 1$, with $m(\cdot)$ and $\sigma(\cdot)$ left unspecified, and consider the residuals in (8) computed from the quantities

$$\widehat{m}(x) = \frac{1}{nh_n\widehat{f}_X(x)} \sum_{j=1}^{n} K((x - X_j)/h_n)Y_j, \tag{13}$$

$$\widehat{\sigma}^2(x) = \frac{1}{nh_n\widehat{f}_X(x)} \sum_{j=1}^{n} K((x - X_j)/h_n)(Y_j - \widehat{m}(x))^2,$$

where the density estimate $\widehat{f}_X(x)$ of the regressor is obtained by using (11) with the errors replaced by the regressors. Under general assumptions it is proved that the resulting process $\widehat{Z}_n(t)$ should be bias-corrected (unless $nh_n^4 = o_P(1)$) with the bias factor depending on the error and regressor densities, and the regression and variance functions. Then the bias corrected process converges weakly to a centered Gaussian process with covariance which involves the error density. With the same reasons as in the case of linear regression, a bootstrap procedure is proposed in order to actually perform the tests, and it is shown to lead to consistent tests. This procedure corresponds to the aforementioned steps (i)–(v) with $\widehat{\varepsilon}_j$ replaced by $\widehat{\sigma}(X_j)\widehat{\varepsilon}_j$ and $m(\widehat{\rho}; X_j)$ replaced by $\widehat{m}(X_j)$, $j = 1, \ldots, n$, where $\widehat{m}(\cdot)$ and $\widehat{\sigma}^2(\cdot)$ are defined by (13). The authors compare by simulation their Cramér–von Mises test to the corresponding test proposed by Dette, et al. (2002) under homoscedasticity, and observe notable improvements. A Monte Carlo study with heteroscedastic errors is also included.

The Fourier approach for symmetry in linear as well as in nonparametric regression is presented in Husková and Meintanis (2010). This approach utilizes the test statistic

$$HM_\nu = \int_{-\infty}^{\infty} S_n^2(t)d\nu(t),$$

where $S_n(t)$ is obtained from (1) by defining $g_{nj}(\widehat{\varepsilon}_j) := \widehat{\varepsilon}_j$ and $\nu(\cdot)$ denoting a (possibly not normalized) symmetric probability measure satisfying $\int t^4 d\nu(t) < \infty$. The residuals are obtained either by M–estimation (in linear regression) or by kernel estimation (in nonparametric regression), as in Neumeyer and Dette (2007). Accordingly asymptotic results run parallel to those in Neumeyer, et al. (2005) and Neumeyer and Dette (2007), thereby establishing a limit null distribution which is not distribution-free and the usual consistency rate at $n^{-1/2}$. Special features of the test are studied under $\nu(t) = \int_{-\infty}^{t} e^{-au^2}du$, $a > 0$. The corresponding test statistic properly normalized is shown to approach a limit value as $a \to \infty$, which boils down to a symmetry test based on the empirical skewness of the residuals.

Moreover in the nonparametric regression case, the test statistic is shown to be identical (apart from irrelevant constants) to that of Ahmad and Li (1997) obtained from the Gaussian kernel, but with fixed bandwidth equal to $a^{-1/2}$. Bootstrap versions of the procedures analogous to those already described are employed in a simulation comparison. This comparison shows that in the linear regression case the proposed tests outperform the classical tests of Neumeyer, et al. (2005), while in nonparametric regression case the new tests are the most powerful along with the Cramér–von Mises test of Neumeyer and Dette (2007).

3.3. *Conditional symmetry in time series*

We now review some tests for conditional symmetry in time series models. Additional motivations for such tests can be found in Bai and Ng (2001), Delgado and Escanciano (2007) or Pérez–Alonso (2007).

We start with the problem of Bai and Ng (2001). This paper deals with testing the conditional symmetry of Y_i given Z_i in the setting of model (8) with $m(z) = m(\rho; z)$ and $\sigma(z) = \sigma(\theta; z)$ for parameter vectors ρ and θ belonging to suitable sets. For each $i \in \mathbb{Z}$, ε_i is assumed to have a unit variance and is independent of Z_i, and the process $\{(Y_i, X_i)\}$ is not necessarily stationary. In this paper, the test based on the coefficient of skewness of a residual series from a linear regression fit is investigated. A theoretical result about this test is established and its advantages and its limitations are discussed. But the main focus of the authors is the derivation of a nonparametric test based on comparing the empirical distributions of ε_i, $i = 1, \ldots, n$ and $-\varepsilon_i$, $i = 1, \ldots, n$. From this idea, they define the empirical process

$$W_n(x) = \frac{1}{\sqrt{n}} \sum_{i=1}^{n} \left[I\left(\widehat{\varepsilon}_i \le x\right) - I\left(-\widehat{\varepsilon}_i \le x\right) \right], \tag{14}$$

where $\widehat{\varepsilon}_i$ is given by (9) with $\widehat{m}(z) = m(\widehat{\rho}_n; z)$, $\widehat{\sigma}(z) = \sigma(\widehat{\theta}_n; z)$ and $(\widehat{\rho}_n, \widehat{\theta}_n)$ standing for a consistent estimator of the true parameter (ρ_0, θ_0). Since $W_n(x)$ is not asymptotically distribution-free, for obtaining a tractable test statistic, the authors construct its martingale transformation (see Khmaladze (1981)). They proceed in the following way. Let $g = \dot{f}/f$, where $\dot{f}$ denotes the derivative of f. Let f_n and g_n be estimates of f and g respectively. Define the functions

$$h_n^-(y) = g_n(y) f_n(y) \left[\int_{-\infty}^{y} g_n^2(z) f_n(z) dz \right]^{-1} \int_{-\infty}^{y} g_n(z) dW_n(z)$$

$$h_n^+(y) = g_n(y) f_n(y) \left[\int_y^\infty g_n^2(z) f_n(z) dz \right]^{-1} \int_y^\infty g_n(z) dW_n(z).$$

The martingale transformation of $W_n(x)$ denoted by $MT_n(x)$ is defined by

$$MT_n(x) = \begin{cases} W_n(x) - W_n(0) + \int_x^0 h_n^-(y)dy, & x \leq 0 \\ W_n(x) - W_n(0) - \int_0^x h_n^+(y)dy, & x > 0. \end{cases}$$

For testing conditional symmetry, Bai and Ng (2001) propose as test statistics $BN^- = \max_{x \leq 0} |MT_n(x)|$ and $BN^+ = \max_{x \geq 0} |MT_n(x)|$ and prove they converge in distribution to the sup-norm of a standard brownian motion on $[0, 1]$. Observing that these statistics are equivalent, the authors define the more advantageous statistic

$$BN = \max\{BN^-, BN^+\} = \max_x |MT_n(x)|$$

which has the same distribution as BN^- and BN^+. Consequently, the test based on BN is obviously asymptotically distribution free under the hypothesis of conditional symmetry. It is shown to be consistent and to have a non-trivial power under a sequence of local alternatives. Its performance is studied by means of numerical simulations and is applied to some financial and nonfinancial real time series.

Delgado and Escanciano (2007) work in a different context, and use a different approach. They consider a $\mathbb{R}^{1+d}$-valued strictly stationary process $(Y_i, X_i)_{i \in \mathbb{Z}}$ with an information set $Z_i = \{(Y_{s-1}, X_s),\ i - q + 1 \leq s \leq i\} \in \mathbb{R}^{q(1+d)}$ at time i. Given a suitable parameter space $\Theta \subset \mathbb{R}^p$ and a function $m : \mathbb{R}^{q(1+d)} \times \Theta \to \mathbb{R}$, the authors aim at testing the symmetry of the conditional distribution function of Y_i given $Z_i = \cdot$ around $m(\theta_0; \cdot)$ for some $\theta_0 \in \Theta$. The null hypothesis considered is

$$H_0 : \text{There exists } \theta_0 \in \Theta \text{ such that}$$

$$F_{\theta_0}(v|u) \in \{F \text{ cdf} : F(v) = 1 - F(v)\} \ a.s.,$$

where $F_\theta(v|u)$ is the *conditional* cumulative distribution function (cdf) of $\varepsilon_i(\theta) = Y_i - m(\theta; Z_i)$ given $Z_i = u$ evaluated at $\varepsilon_i = v$. Thus, under H_0, Y_i, Z_i and ε_i are related by a parametric model of the form (9) with $\sigma(z) = 1$ and the $\varepsilon_i = \varepsilon_i(\theta_0)$'s being the innovations of unknown serial dependance structure, with a conditional symmetry center equal to zero. Observing that testing H_0 is tantamount to testing $DE_{\theta_0}(u, v) = 0$ for some $\theta_0 \in \Theta$, where

$$DE_\theta(u, v) = K_\theta(u, v) - K_\theta(u, \infty) + K_\theta(u, -v),$$

and

$$K_\theta(u,v) = E\left[I(Z_1 \le u)I(\varepsilon_1(\theta) \le v)\right], \; u \in \mathbb{R}^{q(1+d)}, \; v \in \mathbb{R},$$

the authors derive test statistics on the basis of the process

$$DE_{n,\theta_n}(u,v) = \sqrt{n}\left[K_{n,\theta_n}(u,v) - K_{n,\theta_n}(u,\infty) + K_{n,\theta_n}(u,-v)\right],$$

where θ_n is some suitable $\sqrt{n}$-consistent estimator of θ_0 and

$$K_{n,\theta}(u,v) = \frac{1}{n}\sum_{i=1}^{n} I(Z_i \le u)I(\varepsilon_i(\theta) \le v), \; u \in \mathbb{R}^{q(1+d)}, \; v \in \mathbb{R}.$$

Observing again that a symmetric distribution around zero is characterized by the fact that the imaginary part of its characteristic function is zero, they derive test statistics based on the process

$$R_{n,\theta_n}(u,v) = \frac{1}{\sqrt{n}}\sum_{i=1}^{n} \sin(\varepsilon_i(\theta_n)v)I(Z_i \le u), \; u \in \mathbb{R}^{q(1+d)}, \; v \in \mathbb{R}$$

which, under H_0, can be written as

$$R_{n,\theta_n}(u,v) = \frac{1}{2}\int_{-\infty}^{+\infty} \sin(\overline{v}v)DE_{n,\theta_n}(u,d\overline{v}), \; u \in \mathbb{R}^{q(1+d)}, \; v \in \mathbb{R}.$$

The authors prove that the process $DE_{n,\theta_n}(u,v)$ converges in distribution to a centered Gaussian process $DE_\infty(u,v)$ with covariance kernel depending on the distribution of $(\varepsilon_1(\theta_0), Z_1)$, and obtain from the convergence mapping theorem the limiting distributions of their test statistics. They also study the behavior of these statistics under a sequence of local alternatives contiguous to the null hypothesis. Since they are not asymptotically pivotal, the boostrap is used for the computation of the critical values. The authors conclude their work with intensive simulations in which they study two Cramér–von Mises type tests based on $DE_{n,\theta_n}(u,v)$ and $R_{n,\theta_n}(u,v)$, and compare these tests to the one proposed in Bai and Ng (2001).

Pérez–Alonso (2007) studies several tests that can be used for testing symmetry of the error distribution in time series models of the form (8) with : $q = 0$, $m(z) = m(\theta; z)$, $\sigma(z) = \sigma(\theta; z)$ for $\theta \in \Theta \subset \mathbb{R}^k$ and ε_i with unit variance, independent of Z_i for all $i \in \mathbb{Z}$. Let $\{V_i\}$ be a stationary process. Define $Y_i^{(\ell)} = \Psi_{Y,\ell}(V_i, V_{i-1}, \ldots, V_{i-\ell}) \in \mathbb{R}$ and $Z_i^{(\ell)} = \Psi_{Z,\ell}(V_i, V_{i-1}, \ldots, V_{i-\ell}) \in \mathbb{R}^d$, for some measurable functions $\Psi_{Y,\ell}(z)$ and $\Psi_{Z,\ell}(z)$. Here, it is required that $Y_i = Y_i^{(\infty)}$ and $Z_i = Z_i^{(\infty)}$ and that for some positive δ, $E|Y_i - Y_i^{(\ell)}|^{(2+\delta)} + E\|Z_i - Z_i^{(\ell)}\|^{(2+\delta)}$ tends to zero as ℓ tends to ∞. For testing the symmetry of the conditional distribution of Y_i given Z_i, the following tests for unconditional symmetry are applied to regression residuals : the test of skewness reviewed in Gupta (1967), the Wilcoxon signed-rank test studied in Gibbons and Chakrabort (1992), the runs test of McWilliams (1990) and the triplet test of Randles, et al. (1980). Since the asymptotic null distributions of the resulting test statistics are unknown, the author uses a symmetric boostrap for computing their quantiles. These tests, as well as the one of Bai and Ng (2001) are compared by means of numerical simulations. It comes out from this study that the Wilcoxon test seems to be better than the others.

With regard to the tests for conditional symmetry, the test in Ngatchou–Wandji (2009b) and that of Klar, et al. (2010) are based on the empirical characteristic function. The first test is derived in the same model context as in Bai and Ng (2001) but with the process $\{(Y_i, X_i)\}$ assumed to be stationary and ergodic. This test is constructed in the same spirit as in Ngatchou–Wandji (2009a), with the observations replaced by a residual series from an appropriate fit of the model under study. More precisely, the test statistic proposed is of the Cramér–von Mises type, based on the statistic

$$JN = \int S_n^2(t) d\widehat{F}_n(t) = \frac{1}{n} \sum_{j=1}^{n} S_n^2(\widehat{\varepsilon}_j)\omega^2(\widehat{\varepsilon}_j),$$

where $\widehat{F}_n(t)$ stands for the empirical distribution function of the $\widehat{\varepsilon}_1, \ldots, \widehat{\varepsilon}_n$ given by (14), $S_n(t)$ is defined by (1) with $g_{nj}(\widehat{\varepsilon}_j) = \widehat{\varepsilon}_j$ and $\omega(t)$ is a weight function. Note that here, $S_n(t)$ is a normalized empirical version of the imaginary part $S(t)$ of the characteristic function of ε_i which should be zero under the hypothesis of symmetry aroud 0. The author proves the convergence in distribution of $S_n(t)$ to a centered Gaussian process with covariance kernel depending on the functions $m(\rho_0; z), \sigma(\theta_0; z)$ and the distribution of ε_1. Proceeding as in Ngatchou–Wandji (2009a), he shows that the cumulative distribution function of the limit of JN can be approximated by that of a weighted sum of iid chi-squared random variables with one degree of freedom. He next uses the results of Imhof (1961) for approximating the cumulative distribution function of this limiting distribution. However, some notation problems are encountered in this paper and we take the opportunity to correct some of them. On page 1467, line 9, rather denote $DU = UD = (D_1U_1, \ldots D_\kappa U_\kappa)'$, leaving the notation $U'V = UV'$ for the Euclidian scalar product of the vectors U, V with the same length. With this, in assumption (A4), one would instead have $\Pi(\psi_0; Z_i)\Gamma[\varepsilon_i(\psi_0)]$. For a good understanding of the proof of Theorem 3.2, begining from page 1478, one would write $\Pi'(\psi_0; Z_i)\{\Gamma[\varepsilon_i(\psi_0)]\Lambda(t)t\}$ for $\Pi'(\psi_0; Z_i)\Gamma[\varepsilon_i(\psi_0)]t\Lambda(t)$. It is important to note that these notation-related problems do not affect the simulation results.

The test of Klar, et al. (2010) aims at testing symmetry of the innovation distribution in the context of GARCH models. These models result from (8) with $d = 1$ and $m \equiv 0$, by replacing X_j by σ_j in Z_j, if we let

$$\sigma_j^2 := \sigma^2(Z_j) = \sum_{k=1}^{p} \rho_k Y_{j-k}^2 + \sum_{k=1}^{q} \tilde{\rho}_k \sigma_{j-k}^2.$$

The test statistics extends the procedure of Feuerverger and Mureika (1977) and takes the form

$$\int_{-\infty}^{\infty} \left((n - \max(p,q))^{-1/2} \sum_{j=\max(p,q)+1}^{n} \sin(t\varepsilon_j) \right)^2 w(t)dt,$$

where the unobserved innovations ε_j are replaced by $\widehat{\varepsilon}_j$ of (9) which in turn result from the quasi maximum likelihood estimator (QMLE) of σ_j^2. Critical points are obtained by the resampling procedure previously described in the context of linear regression, suitably modified to the present set–up. When tested by Monte Carlo, the method is shown to be consonant with the nominal size of the test under Gaussian and other popular symmetric GARCH models, and to have satisfactory power against GARCH models with skewed innovations.

We close this section with Fan and Ullah (1999). This paper proposes some goodness-of-fit tests for weakly dependent processes. These tests include a two sample goodness-of-fit test, a symmetry test and a test for the goodness-of-fit of a parametric density function. More precisely, let $\{V_i\}$ and $\{\dot{V}_i\}$ be two strictly stationary univariate processes not necessarily independent of each other, such that $\{V_i, \dot{V}_i\}$ is absolutely regular with geometric decay rate. For a positive integer d, let $\tilde{X}_i = (V_i, V_{i+1}, \ldots, V_{i+d-1})$ and $\tilde{Y}_i = (\dot{V}_i, \dot{V}_{i+1}, \ldots, \dot{V}_{i+d-1})$ be two random vectors with probability density functions $f_{\tilde{X}}(u)$ and $f_{\tilde{Y}}(u)$ respectively. One of the aims of the authors is to test $H_0 : f_{\tilde{X}}(u) = f_{\tilde{Y}}(u)$ against $H_1 : f_{\tilde{X}}(u) \neq f_{\tilde{Y}}(u)$, given that $\tilde{X}_i$, $i = 1, \ldots, n_1$ and $\tilde{Y}_i$, $i = 1, \ldots, n_2$ are observed. For deriving a two sample goodness-of-fit test in the case $f_{\tilde{X}}(u)$ and $f_{\tilde{Y}}(u)$ are unknown, they consider $n_1 = n_2 = n$ and use as test statistic a suitable normalized version of the statistic

$$FU = \frac{1}{n^2 h^d} \sum_i \sum_{j \neq i} \left[K\left(\frac{\tilde{X}_i - \tilde{X}_j}{h}\right) + K\left(\frac{\tilde{Y}_i - \tilde{Y}_j}{h}\right) - K\left(\frac{\tilde{X}_i - \tilde{Y}_j}{h}\right) - K\left(\frac{\tilde{Y}_i - \tilde{X}_j}{h}\right) \right],$$

where $K(u)$ is the kernel and h is the smoothing parameter, both used for estimating $f_{\tilde{X}}(u)$ and $f_{\tilde{Y}}(u)$. In fact, FU is derived from an estimator of an integreted squared difference between these functions. Using techniques of U-statistics, the authors study the convergence of FU under H_0 and under H_1. In the case $\tilde{Y}_i = -\tilde{X}_i$, the above two sample goodness-of-fit test for $n_1 = n_2$ reduces to a symmetry test for the distribution of $\tilde{X}_i$. The authors investigate this problem for $d = 1$. The results obtained in this case can be applied to testing unconditional or conditional symmetries in time series. Finally, given a density function of known form f_0, the authors also consider testing $H_0 : f_{\tilde{X}}(u) = f_0(u, \theta_0)$ for some parameter $\theta_0 \in \Theta \subset \mathbb{R}^p$ against $H_1 : f_{\tilde{X}}(u) \neq f_0(u, \theta)$ for all $\theta \in \Theta$. They propose a test based on a suitable normalization of the bias-corrected statistic

$$T_n = \int \left[\widehat{f}(u) - K_h * f_0(u, \widehat{\theta})\right]^2 du,$$

where $\widehat{\theta}$ is the quasi maximum likelihood estimator of θ_0, $\widehat{f}(u)$ is the kernel estimator of $f_{\tilde{X}}(u)$ and

$$K_h * f_0(u, \widehat{\theta}) = \frac{1}{h^d} \int K\left(\frac{u - v}{h}\right) f_0(v, \widehat{\theta}) dv.$$

Using again techniques of U-statistics, they study the behavior of T_n under the null hypothesis and prove the consistency of the test based on this statistic.

4. Testing for symmetry in random effect models

4.1. *Model and tests*

In random effects models an observation Y is driven by the model

$$Y = X'\rho + u + \varepsilon, \tag{15}$$

with X, ρ, u and ε denoting a vector of covariates, the parameter vector, the random effect, and the random error, respectively. Under standard assumptions, the random effects u are independent of the random errors ε and both have common mean zero, and constant variances σ_u^2 and σ_ε^2, respectively. On the basis of data $(X_{j,t}, Y_{j,t})$, with $X'_{j,t} = (X_{j,t}^{(1)}, X_{j,t}^{(2)}, \ldots, X_{j,t}^{(d)}), (j = 1, \ldots, n,\ t = 1, \ldots, T)$, we wish to test the following null hypothesis against general alternatives

H_0: The law of ε is symmetric around zero.

It should be noted in this connection that these models are extremely popular but also highly parametrized as the assumption is often made that errors and effects follow normal distributions. It has been argued, however, that by relaxing this assumption to symmetry will not significantly affect statistical procedures.

There are various general approaches for testing H_0, most of them depending on ranks or on the empirical distribution function of suitable observables. In the present context, as in subsection 4.2, all of these approaches would depend instead on the errors which are unobserved, on the corresponding standardized residuals

$$\widehat{\varepsilon}_{j,t} = \frac{Y_{j,t} - X'_{j,t}\widehat{\rho} - \widehat{u}_j}{\widehat{\sigma}_\varepsilon},\ (j = 1, \ldots, n,\ t = 1, \ldots, T), \tag{16}$$

where in obvious notation $(\widehat{\rho}, \widehat{u}_j, \widehat{\sigma}_\varepsilon)$ denote estimates of the corresponding theoretical quantities. Recalling that $I(A)$ denotes the indicator function of entity A, the signed rank tests are on the form

$$\mathrm{SR} = \frac{1}{nT} \sum_{j=1}^{n} \sum_{t=1}^{T} s\left(\frac{R_{j,t}^{+}}{nT+1}\right) I(\widehat{\varepsilon}_{j,t} > 0),$$

where $R_{1,1}^{+}, \ldots, R_{n,T}^{+}$, denote the ranks of $|\widehat{\varepsilon}_{1,1}|, \ldots, |\widehat{\varepsilon}_{n,T}|$, and $s(\cdot)$ are score functions.

The classical omnibus test statistics depend on the empirical distribution function

$$F_{n,T}(v) = \frac{1}{nT} \sum_{j=1}^{n} \sum_{t=1}^{T} I(\widehat{\varepsilon}_{j,t} \le v).$$

In particular, the Cramér–von Mises (CM) statistic proposed by Srinivasan and Godio (1974) may be written as

$$\mathrm{CM} = \int (F_{n,T}(v) - F_{n,T}^{*}(v))^2 dF_{n,T}(v),$$

where $F_{n,T}^{*}(v) = (1/2)[F_{n,T}(v) + \{1 - F_{n,T}(-v)\}]$. A corresponding Kolmogorov–Smirnov (KS) test statistic is given by

$$\mathrm{KS} = \sup_v |F_{n,T}(v) - \widehat{F}_{n,T}^{*}(v)|.$$

Finally, the empirical characteristic function (CF) test is given by

$$\mathrm{CF} = \int_{-\infty}^{\infty} S_{n,T}^2(\tau) w(\tau) d\tau, \tag{17}$$

where

$$S_{n,T}(\tau) := \int \sin(\tau v) dF_{n,T}(v) = \frac{1}{nT} \sum_{j=1}^{n} \sum_{t=1}^{T} \sin(\tau \widehat{\varepsilon}_{j,t}), \ \tau \in \mathbb{R}, \ i = \sqrt{-1}$$

is the imaginary part of the empirical characteristic function, and $w(\tau)$ denotes a weight function.

4.2. *Specification of estimation and test statistics*

The test statistic utilizes the standardized data given by (16). The estimate $\widehat{\rho}$ of ρ is computed by regressing $\widehat{Y}_{j,t}$ on $\widehat{X}'_{j,t}$, with $\widehat{X}_{j,t} := (\widehat{X}^{(1)}_{j,t}, \ldots, \widehat{X}^{(d)}_{j,t})'$, $(j = 1, \ldots, n,\ t = 1, \ldots, T)$, where

$$\widehat{Y}_{j,t} = Y_{j,t} - \widehat{\theta} \bar{Y}_j, \quad \widehat{X}^{(k)}_{j,t} = X^{(k)}_{j,t} - \widehat{\theta} \bar{X}_{k,j}, \quad k = 1, \ldots, d,$$

and

$$\bar{Y}_j = T^{-1} \sum_{t=1}^{T} Y_{j,t}, \quad \bar{X}_{k,j} = T^{-1} \sum_{t=1}^{T} X^{(k)}_{j,t}, \quad k = 1, \ldots, d.$$

This estimate of ρ is called a feasible (or estimated) generalized LS estimator, since the parameter $\theta := 1 - \left(\frac{\sigma^2_\varepsilon}{T\sigma^2_u + \sigma^2_\varepsilon} \right)^{1/2}$ required in order to transform the data is estimated from these data. Its estimator is $\widehat{\theta} = 1 - \left(\frac{\widehat{\sigma}^2_\varepsilon}{T\widehat{\sigma}^2_u + \widehat{\sigma}^2_\varepsilon} \right)^{1/2}$, where $\widehat{\sigma}^2_\varepsilon$ is computed as the estimator of the error variance based on the within groups estimator of the parameter β, while

$$\widehat{\sigma}^2_u = \frac{1}{n} \sum_{j=1}^{n} \widehat{u}^2_j - \frac{\widehat{\sigma}^2_\varepsilon}{T}.$$

Finally, $\widehat{u}_j$ is set equal to the residuals resulting from the between groups regression model. For more details on the estimation step the reader is referred to Baltagi (2002), Arellano (2003).

Let $M = nT$, and write $\widehat{\epsilon}_{(1)} \le \widehat{\epsilon}_{(2)} \le \ldots \le \widehat{\epsilon}_{(M)}$ for the order statistics of $\{\widehat{\varepsilon}_{j,t}\}$, $(j,t) = (1,1), \ldots, (n,T)$. For the rank-based statistics we employ in SR the score function $s \equiv 1$ which gives the sign test

$$\mathrm{SN} = \frac{1}{M} \sum_{m=1}^{M} I(\widehat{\epsilon}_{(m)} > 0),$$

and the score function

$$s(u) = \sum_{m=1}^{M} \frac{m}{M} I\left(\frac{(m-1)}{M} < u \leq \frac{m}{M}\right),$$

which gives the Wilcoxon (WX) test.

For the statistics based on the empirical distribution function, write $|\widehat{\epsilon}|^{(1)} \leq |\widehat{\epsilon}|^{(2)} \leq \ldots \leq |\widehat{\epsilon}|^{(M)}$ for the order statistics of the absolute residuals $|\widehat{\epsilon}_m|$, $m = 1, \ldots, M$. Following Srinivasan and Godio (1974), if N_m denotes the number of negative $\widehat{\epsilon}_j$'s such that $|\widehat{\epsilon}_j| \leq |\widehat{\epsilon}|^{(m)}$ and P_m denotes the number of positive $\widehat{\epsilon}_j$'s such that $|\widehat{\epsilon}_j| \leq |\widehat{\epsilon}|^{(m)}$, then

$$\mathrm{CM} = \sum_{m=1}^{M} (N_m - P_m)^2,$$

and

$$\mathrm{KS} = \sup_{1 \leq m \leq M} |N_m - P_m|.$$

For the CF-based statistic we have by (17) and direct computation of integrals,

$$\mathrm{CF} = \frac{1}{2M^2} \sum_{\ell,m=1}^{M} \left\{\mathcal{I}(\widehat{\epsilon}_\ell - \widehat{\epsilon}_m) - \mathcal{I}(\widehat{\epsilon}_\ell + \widehat{\epsilon}_m)\right\},$$

where

$$\mathcal{I}(z) = \int_{-\infty}^{\infty} \cos(\tau z) w(\tau) d\tau.$$

Standard choices for the weight function are $w(\tau) = e^{-a|\tau|}$ and $w(\tau) = e^{-a\tau^2}$, $a > 0$, which render the test statistic in closed form.

4.3. *Simulations*

In this section simulation results are reported for all tests based on 2,000 replications with sample size $(n, T) = (10, 20), (10, 40)$. For computing critical points, we apply a resampling procedure similar to that of subsection 1. Specifically, let $e_{j,t}$ be i.i.d. as in (5) ($j = 1, \ldots, n$, $t = 1, \ldots, T$):

- Compute the bootstrap residuals $\varepsilon^*_{j,t} = e_{j,t}\widehat{\sigma}_\varepsilon \widehat{\varepsilon}_{j,t}$.
- Build new observations $Y^*_{j,t} = X'_{j,t}\widehat{\rho} + \widehat{u}_j + \varepsilon^*_{j,t}$.
- Estimate the residuals $\widehat{\varepsilon}^*_{j,t} = (Y^*_{j,t} - X'_{j,t}\widehat{\rho}^* - \widehat{u}^*_j)/\widehat{\sigma}^*_\varepsilon$, where all quantities are defined analogously to those in Sections 4.1–4.2, but with $Y_{j,t}$ replaced by $Y^*_{j,t}$.

For the results below we used $B = 200$ bootstrap samples, and the modified percentage points $\tilde{p}_{0.05} = S^*_{(190)} + 0.95[S^*_{(191)} - S^*_{(190)}]$ and $\tilde{p}_{0.10} = S^*_{(180)} + 0.90[S^*_{(181)} - S^*_{(180)}]$.

Observations from the model $Y = \rho_1 X^{(1)} + \rho_2 X^{(2)} + u + \varepsilon$, ($d = 2$ covariates), were simulated, with design values $X^{(1)}_{j,t} = [(j-1)T + t]/nT$ and $X^{(2)}_{j,t} = (X^{(1)}_{j,t})^2$, ($j = 1, \ldots, n$, $t = 1, \ldots, T$), and $\rho_1 = \rho_2 = 0$. The distribution of random effects u was taken to be normal, while the distribution of random errors ε was taken to be normal (N), Laplace (LP), logistic (LG), uniform (UN), triangular (TR) and Student's t_θ, under the null hypothesis H_0. Under the alternative the law of ε was simulated as chi-squared, skew-normal or Tukey's. Note that the last two distributions generalize the notion of the (symmetric) normal distribution by allowing for skewness. The skew-normal (SN) is defined as $\varepsilon = (1+\theta^2)^{-1/2}(\theta|z_1| + z_2)$ with z_1, z_2 independent standard normal variates, while the Tukey variate (TU) is $\varepsilon = (\exp(\theta z_1) - 1)/\theta$, with z_1 as before. Notice that both the SN_θ and TU_θ reduce to the normal distribution as $\theta \to 0$.

In Table 1, Table 2 and Table 3 we report the percentage of rejection rounded to the nearest integer for the test statistics. An asterisk denotes power 100%. The CF statistic with $e^{-a|\tau|}$ (respectively, $e^{-a\tau^2}$) as weight function is denoted by $\mathrm{CF}^{(1)}_a$ (resp. $\mathrm{CF}^{(2)}_a$). All test statistics are often seen to over-reject or under-reject under the null hypothesis H_0, but most of the times the situation improves with larger sample size. Among the standard tests, the Wilcoxon test is clearly the most powerful, despite the fact that it is often seen to be quite conservative or liberal in rejecting the null hypothesis H_0, when H_0 is true. The two CF statistics attain roughly the same rates of rejection (though at different values of the weight parameter a), with a slight edge for $\mathrm{CF}^{(2)}_a$. Moreover, the last statistic, with a proper value of a, say $\mathrm{CF}^{(2)}_2$ or $\mathrm{CF}^{(2)}_{2.5}$, is seen to outperform all its competitors, uniformly for all sample sizes and alternatives distributions considered.

5. Other procedures for testing symmetry and conclusion

This paper is a review covering general notions of symmetry and methods to test corresponding hypotheses under different set–ups. It is shown that the symmetry hypothesis appears in meaningful context under various data generating mechanisms, from those that generate multivariate i.i.d. observations, to other more complicated mechanisms which include explanatory variables and dependence. We mainly stress methods which utilize the empirical distribution function and the empirical characteristic function, but several other competing procedures have also been reviewed. Some of the methods reviewed are subsequently adapted to the case of random effects models in order to test for symmetry of the error distribution in these models.

We close by mentioning that a substantial portion of the literature has been covered in the present review. Nevertheless no completeness is claimed. In fact, there are some new notions of symmetry and symmetry measures, such as those in Zuo and Serfling (2000) and Rousseeuw and Struyf (2000), that are not included in this review. Most of the skewness measures not included also lead to test statistics for the hypothesis of symmetry. In particular, Q-Q plots are considered by Li, et al. (1997) for testing spherical symmetry, tests for spherical and elliptical symmetry can be constructed using certain notions of vector-valued runs defined by Marden (1999) and Paindaveine (2009), rank based methods are applied by Hettmansperger, et al. (2002) in order to test symmetry in the regression set-up , tests for diagonal or shperical symmetry based on the notion of 'data depth' appear in Rousseeuw and Struyf (2002) and Hallin, et al. (2010), tests for spherical symmetry by projection methods are discussed in Zhu, et al. (1995), while Kankainen, et al. (2004) employ robust measures of skewness in the context of testing multinormality.

References

Ahmad, I. A. and Li, Q. (1997). Testing symmetry of an unknown density function by kernel method, *J. Nonparametr. Statist.* **7**, 279–293.

Ahmad, I. A. and van Belle, G. (1974). Measuring affinity of distributions, In: *Reliability and Biometry–Statistical Analysis of Lifelength*, 651–688. SIAM, Philadelphia.

Arcones, M. A. and Giné, E. (1991). Some bootstrap tests of symmetry for univariate continuous distributions, *Ann. Statist.* 19, 1496–1511.

Arellano, M. (2003). *Panel Data Econometrics.* Oxford University Press, New York.

Bai, J. and Ng, S. (2001). A consistent test for conditional symmetry in time series models, *J. Econometr.* **103**, 225–258.

Balakrishnan, N., Brito, M. R. and Quiroz, A. J. (2007). A vectorial notion of skewness and its use in testing multivariate symmetry, *Commun. Statist.–Theor. Meth.* **36**, 1757–1767.

Baltagi, B. H. (2002). *Econometric Analysis of Panel Data.* John Wiley & Sons , New York.

Baringhaus, L. (1991). Testing for spherical symmetry of a multivariate distribution, *Ann. Statist.*, **19**, 899–917.

Baringhaus, L. and Henze, N. (1992). Limit distributions for Mardia's measure of multivariate skewness, *Ann. Statist.* **20**, 1889–1902.

Baringhaus, L. (1996). Fibonacci numbers, Lucas numbers, and integrals of certain Gaussian processes, *Proc. Amer. Math. Soc.* **124**, 3875–3884.

Beran, R. (1974). Asymptotically efficient adaptive rank estimates in location models, *Ann. Statist.* **2** , 63–74.

Beran, R. (1979). Testing for ellipsoidal symmetry of a multivariate density, *Ann. Statist.* **7**, 150–162.

Besag, J. and Diggle, P. J. (1977). Simple Monte Carlo tests for spatial pattern, *Appl. Statist.* **26**, 327–333.

Bickel, P. (1982). On adaptive estimation, *Ann. Statist.* **10**, 647–671.

Boos, D. D. (1982). A test for symmetry associated with Hodges-Lehmann estimator, *J. Amer. Statist. Assoc.* **77**, 647–651.

Boudt, K. and Croux, C. (2010). Robust M–estimation of multivariate GARCH models, *Comput. Statist. Dat. Anal.*, to appear.

Ceyhan, E. and Goad, C. L. (2009). A comparison of analysis of covariate–adjusted residuals and analysis of covariance, *Commun. Statist.–Simul. Comput.* **38**, 2019–2038.

Chmielewski, M. A. (1981). Elliptical symmetric distributions: a review and bibliography, *Int. Stat. Rev.* **49**, 67–74.

Choi, K. and Marden, J. (1997). An approach to multivariate rank tests in multivariate analysis of variance, *J. Amer. Statist. Assoc.* **92**, 1581–1590.

Csörgő, S. and Heathcote, C. R. (1987). Testing for symmetry, *Biometrika* **74**, 177–184.

Delgado, M. A. and Escanciano, J. C. (2007). Nonparametric tests for conditional symmetry in dynamic models, *J. Econometrics* **141**, 652–682

Dette, H., Kusi–Appiah, S. and Neumeyer, N. (2002). Testing symmetry in nonparametric regression models, *J. Nonparametr. Statist.* **14**, 477–494.

Diks, C. and Tong, H. (1999). A test for symmetries of multivariate probability distributions, *Biometrika* **86**, 605–614.

Eigen, S. and Lillegard, M. (1997). Stochastic simulations conditioned on sufficient statistics, *Biometrika* **84**, 235–240.

Eubank, R. L., LaRiccia, V. and Rosenstein, R. B. (1992). Testing symmetry about an unknown median, via linear rank procedure, *J. Nonparametr. Statist.* **1**, 301–311.

Fan, Y. and Gencay, R. (1995). A consistent nonparametric test of symmetry in linear regression models, *J. Amer. Statist. Assoc.* **90**, 551–557.

Fan, Y. and Ullah, A. (1999). On goodness-of-fit tests for weakly dependent processes using kernel method, *J. Nonparametr. Statist.* **11**, 337–360.

Feuerverger, A. and Mureika, R. (1977). The empirical characteristic function and its applications, *Ann. Statist.* **5**, 88–97.

Freedman, D. A. and Diaconis, P. (1982). On inconsistent M–estimators, *Ann. Statist.* **10**, 454–461.

Gastwirth, J. L. (1971). On the sign test for symmetry, *J. Amer. Statist. Assoc.* **66**, 821–823.

Ghosh, S. and Ruymgaart, F. H. (1992). Application of empirical characteristic functions in some multivariate problems, *Canad. J. Statist.* **20**, 429–440.

Gibbons, J. D. and Chakraborti, S. (1992). *Nonparametric statistical inference*, Third edition. Marcel Dekker, Inc., New York.

Gonzalez–Rivera, G. and Drost, F. C. (1999). Efficiency comparisons of maximum-likelihood-based estimators in GARCH models, *J. Econometr.* **93**, 93–111.

Gupta M. K. (1967). A asymptotically nonparametric test of symmetry, *Ann. Math. Statist.* **38**, 849–866.

Gutjahr, S., Henze, N. and Folkers, M. (1999). Shortcomings of generalized affine invariant skewness measures, *J. Multivar. Anal.* **71**, 1–23.

Hallin, M. and Paindaveine, D. (2002a). Optimal tests for multivariate location based on interdirections and pseudo-Mahalanobis ranks, *Ann. Statist.* **30**, 1103–1133.

Hallin, M. and Paindaveine, D. (2002b). Optimal procedures based on interdirections and pseudo–Mahalanobis ranks for testing multivariate elliptic white noise against ARMA dependence, *Bernoulli* **8**, 787–816.

Hallin, M., Paindaveine, D. and Siman, M. (2010). Multivariate quantiles and multiple–output regression quantiles: From L_1 to halfspace depth, *Ann. Statist.* **38**, 635–669.

Härdle, W. and Mammen, E. (1993). Comparing nonparametric versus parametric regression fits, *Ann. Statist.* **21**, 1926–1947.

Heathcote, C. R., Rachev, S. T. and Cheng, B. (1995). Testing multivariate symmetry, *J. Multivar. Anal.* **54**, 91–112.

Henze, N., Klar, B. and Meintanis, S. G. (2003). Invariant tests for symmetry about an unspecified point based on the empirical characteristic function, *J. Multivar. Anal.* **87**, 275–397.

Hettmansperger, T. P., McKean, J. W. and Sheather, S. J. (2002). Finite sample performance of tests for symmetry of the errors in a linear model, *J. Statist. Comput. Simul.* **72**, 863–879.

Hillebrand, M. and Müller, C. H. (2006). On consistency of redescending M–kernel smoothers, *Metrika* **63**, 71–90.

Hollander, M. (1988). Testing for symmetry, In: *Encyclopedia of Statistical Sciences* **9**, 211–216. Wiley, New York.

Huffer, F. W. and Park, C. (2005). A test for elliptical symmetry, *J. Multivar. Anal.* **98**, 256–281.

Husková, M. (1984). Hypothesis of symmetry, In: *Handbook of Statistics* **4**, Nonparametric Methods, 63–78. North Holland, Amsterdam.

Husková, M., Meintanis, S. G. (2010). Tests for symmetric error distribution in linear and non–parametric regression models, *J. Statist. Plann. Infer.*, to appear.

Im, K.-S. (2000). Robustifying Glejser test for heteroskedasticity, *J. Econometr.* **97**, 179–188.

Imhof, J. P. (1961). Computing the distribution of quadratic forms in normal variables, *Biometrika* **48**, 419–426.

Kankainen, A., Taskinen, S. and Oja, H. (2004). On Mardia's test of multinormality, In: *Theory and Applications of Recent Robust Methods*, 153–164, Birkhäuser, Basel.

Kariya, T. and Eaton, M. L. (1977). Robust tests for spherical symmetry, *Ann. Statist.* **5**, 206–215.

Khmaladze, E. V. (1981). Martingale approach in the theory of goodness-of-fit tests, *Theor. Probab. Appl.*, XXVI, 240–257.

Klaassen, C. A. J. and Putter, H. (2005). Efficient estimation of Banach parameters in semiparametric models, *Ann. Statist.* **33**, 307–346.

Klar, B. (2002) A treatment of multivariate skewness, kurtosis, and related statistics, *J. Multivar. Anal.* **83**, 141–165.

Klar, B., Lindner, F. and Meintanis, S. G. (2010). Specification tests for the error distribution in GARCH models, *Comput. Statist. Dat. Anal.* In press.

Kollo, T. (2008). Multivariate skewness and kurtosis measures with an application to ICA, *J. Multivar. Anal.* **99**, 2328–2338.

Koltchinsky, V. I. and Li, L. (1998). Testing spherical symmetry of a multivariate distribution, *J. Multivar. Anal.* **65**, 228–244.

Koul, H. (2002). Asymptotic distributions of some scale estimators in nonlinear models, *Metrika* **55**, 75–90.

Koutrouvelis, I. A. (1985). Distribution-free procedures for location and symmetry inference problems based on the empirical characteristic function, *Scand. J. Statist.* **12**, 257–269.

Lehmann, E. L. and Stein, C. (1949). On the theory of nonparametric hypothesis, *Ann. Math. Statist.* **20**, 28–45.

Li, R-Z, Fang, K-T and Zhu, L-X (1997). Some Q–Q probability plots to test spherical and elliptical symmetry, *J. Comput. Graph. Statist.* **6**, 435–450.

Marden, J. I. (1999). Multivariate rank tests, in: S. Ghosh (Ed.), *Multivariate Analysis, Design of Experiments, and Survey Sampling*, Marcel Dekker,

New York, 401–432.

Magnus, J. R. (1988). *Linear Structures*, Oxford University Press, New York.

Manzotti, A., Pérez, F. J. and Quiroz, A. J. (2002). A statistic for testing the null hypothesis of elliptical symmetry, *J. Multivar. Anal.* **81**, 274–285.

McWilliams, T. P. (1990). A distribution-free test for symmetry based on a run statistic, *J. Amer. Statist. Assoc.* **85**, 1130–1133.

Müller, C. (1966). *"Spherical Harmonics"*, *Lecture Notes in Mathematics*, vol 17, Springer–Verlag, New York.

Munzel, U. (1999). Nonparametric methods for paired samples, *Statist. Neerlan.* **53**, 277–286.

Neuhaus, G. and Zhu, L-X (1998). Permutation tests for reflected symmetry, *J. Multivar. Anal.* **67**, 129–153.

Neumeyer, N. and Dette, H. (2007). Testing for symmetric error distribution in nonparametric regression models, *Statist. Sinica* **17**, 775–795.

Neumeyer, N., Dette, H. and Nagel, E-R (2005). A note on testing symmetry of the error distribution in linear regression models, *J. Nonparametr. Statist.* **17**, 697–715.

Newey, W. K. (1988). Adaptive estimation of regression models via moment restrictions, *J. Econometr.* **38**, 301–339.

Newey, W. K. and Steigerwald, D. G. (1997). Asymptotic bias for quasi-maximum-likelihood estimators in conditional heteroskedasticity models, *Econometrica* **65**, 587–599.

Ngatchou–Wandji, J. (2009a). Testing for symmetry in multivariate distributions, *Statist. Methodol. 6*, 230–250.

Ngatchou–Wandji, J. (2009b). Testing symmetry of the error distribution in nonlinear heteroscedastic models, *Commun. Statist.–Theor. Meth.* **38**, 1465–1485.

Paindaveine, D. (2009). On multivariate runs tests for randomness, *J. Amer. Statist. Assoc.* **104**, 1525–1538.

Pérez–Alonso, A. (2007). A bootstrap approacgh to test the conditional symmetry in time series models. *Comput. Statist. Dat. Anal.* **51**, 3484–3504.

Randles, R. H., Flinger, M. A., Policello, , G. E. and Wolfe, D. A. (1980). An asymptotically distribution-free test for symmetry versus asymmetry, *J. Amer. Statist. Assoc.* **75**, 168–172.

Rousseeuw, P. J. and Struyf, A. (2000). Characterizing angular symmetry and regression symmetry, *J. Statist. Plann. Infer.* **122**, 161–173.

Rousseeuw, P. J. and Struyf, A. (2002). A depth test for symmetry. In: Huber–Carol, et al. (Eds.), *Goodness–of–fit Tests and Model Validity*, 401–412. Birkhäuser, Basel.

Schott, J. R. (2002). Testing for elliptical symmetry in covariance-matrix-based analyses, *Statist. Probab. Letters* **60**, 395-404.

Schuster, E. F. and Barker, R. C. (1987). Using boostrap in testing symmetry versus asymmetry, *Comm. Statist.–Simul. Comput.* **16**, 69–84.

Smith, P. J. (1977). A nonparametric test for bivariate circular symmetry based on the empirical cdf, *Comm. Statist.–Theor. Meth.* **6**, 209–220.

Srinivasan, R. and Godio, L. B. (1974). A Cramér–von Mises type statistic for testing symmetry. *Biometrika* **61**, 196–198.

Wu, C-F-J (1986). Jackknife, bootstrap and other resampling methods in regression analysis. With discussion and a rejoinder by the author, *Ann. Statist.* **14**, 1261–1350.

Zhu, L-X, Fang, K-T and Li, R-Z (1997). A new approach for testing symmetry of a high-dimensional distribution. *BHKMS* **1**, 35–46.

Zhu, L-X, Fang, K-T and Zhang, J-T (1995). A projection NT-type test for spherical symmetry of a multivariate distribution. In: Tiit, et al. (Eds.), *Multivariate Statistics and Matrices in Statistics*, Proc. of the 5th Tartu Conference, 109–122, VSP, Utrecht.

Zhu, L-X and Neuhaus, G. (2000). Nonparametric Monte Carlo tests for multivariate distributions. *Biometrika* **87**, 919–928.

Zhu, L-X and Neuhaus, G. (2003). Conditional tests for elliptical symmetry, *J. Multivar. Anal.* **84**, 284–298.

Zuo, Y. and Serfling, R. (2000). On the performance of some robust nonparametric location measures relative to a general notion of multivariate symmetry, *J. Statist. Plann. Infer.* **84**, 55–79.

Table 1. 5% (left entry) and 10% (right entry) percentage of rejection with sample size $(n,T) = (10,20)$ (left part) and $(n,T) = (10,40)$ (right part).

Density	$CF^{(1)}_{0.5}$	$CF^{(1)}_{1.0}$	$CF^{(1)}_{2.0}$	$CF^{(1)}_{3.0}$	$CF^{(1)}_{5.0}$	$CF^{(1)}_{0.5}$	$CF^{(1)}_{1.0}$	$CF^{(1)}_{2.0}$	$CF^{(1)}_{3.0}$	$CF^{(1)}_{5.0}$
N	6 11	6 11	5 10	5 10	5 10	6 11	5 11	5 10	5 10	5 11
LG	6 11	6 11	6 11	6 11	6 11	5 10	5 11	5 11	5 11	5 11
LP	7 13	7 13	7 13	7 13	7 12	7 11	6 10	6 10	6 10	5 11
UN	6 11	5 10	5 10	4 10	4 9	7 13	7 12	7 12	6 12	6 12
TR	5 10	5 10	5 9	4 9	4 8	6 11	6 11	5 11	5 11	5 10
t_4	6 12	7 12	7 13	7 13	8 14	6 11	6 11	6 11	6 11	6 11
t_6	5 10	5 10	6 11	7 11	7 12	6 10	6 10	6 10	5 11	6 11
t_{10}	5 10	5 10	6 11	6 11	6 11	6 11	6 10	6 10	6 11	6 12
χ^2_{10}	91 95	95 98	98 99	99 99	99 *	* *	* *	* *	* *	* *
χ^2_{15}	75 84	84 91	91 95	92 96	94 97	98 99	99 *	* *	* *	* *
χ^2_{20}	60 71	71 82	80 88	83 91	85 92	92 96	96 99	99 99	99 *	99 *
χ^2_{25}	50 63	62 74	71 81	75 84	77 86	86 92	92 96	96 98	97 98	98 99
SN_1	8 14	9 16	10 17	10 18	10 18	9 18	12 21	14 23	15 24	16 25
$SN_{1.5}$	20 31	24 36	29 41	31 43	32 44	36 49	45 59	53 67	57 70	60 71
SN_2	41 54	50 64	59 70	62 72	63 74	75 85	84 91	90 94	91 96	93 96
SN_3	80 87	87 93	92 96	93 96	93 97	99 *	* *	* *	* *	* *
$TU_{0.05}$	7 15	9 16	10 18	11 18	11 19	13 20	15 23	18 28	19 30	21 31
$TU_{0.10}$	18 27	22 33	27 40	30 43	31 45	34 47	43 56	52 66	56 69	59 72
$TU_{0.15}$	35 47	44 57	53 65	56 69	57 71	66 77	77 86	85 91	88 93	90 94
$TU_{0.20}$	56 69	67 79	77 85	80 88	82 90	89 95	95 97	97 99	98 99	99 99

Table 2. 5% (left entry) and 10% (right entry) percentage of rejection with sample size $(n, T) = (10, 20)$ (left part) and $(n, T) = (10, 40)$ (right part).

Density	$CF^{(2)}_{0.5}$	$CF^{(2)}_{1.0}$	$CF^{(2)}_{1.5}$	$CF^{(2)}_{2.0}$	$CF^{(2)}_{2.5}$	$CF^{(2)}_{0.5}$	$CF^{(2)}_{1.0}$	$CF^{(2)}_{1.5}$	$CF^{(2)}_{2.0}$	$CF^{(2)}_{2.5}$
N	6 10	5 10	5 10	5 10	5 10	5 10	5 10	5 10	5 10	5 10
LG	5 11	6 11	6 11	6 11	6 11	5 11	5 11	5 10	5 10	5 11
LP	7 13	7 13	7 12	7 12	7 12	6 10	6 11	6 11	5 11	5 11
UN	5 10	4 10	4 9	4 9	4 9	7 12	6 12	6 12	6 12	6 11
TR	5 9	5 9	4 8	4 8	4 8	6 11	5 10	5 10	5 10	5 10
t_4	7 13	7 13	8 13	8 13	8 13	5 11	6 11	6 11	6 11	6 11
t_6	6 11	7 11	7 12	7 12	7 12	5 10	5 10	6 11	6 11	6 11
t_{10}	5 11	6 11	6 11	6 11	6 11	6 10	6 11	6 11	5 11	6 12
χ^2_{10}	97 99	99 99	99 99	99 *	99 *	* *	* *	* *	* *	* *
χ^2_{15}	89 95	92 96	93 97	94 97	94 97	* *	* *	* *	* *	* *
χ^2_{20}	78 87	82 91	84 92	85 92	86 92	98 99	99 *	99 *	99 *	99 *
χ^2_{25}	70 80	75 84	77 85	77 86	78 86	96 97	97 98	97 99	98 99	98 99
SN_1	10 17	10 18	10 18	10 18	10 18	14 22	15 24	16 25	16 25	16 25
$SN_{1.5}$	29 41	31 43	32 44	32 44	32 45	52 65	57 70	59 70	60 71	60 72
SN_2	58 70	62 72	63 74	63 74	63 74	89 94	91 96	92 97	93 97	93 97
SN_3	91 95	93 96	93 97	93 97	94 97	* *	* *	* *	* *	* *
$TU_{0.05}$	10 18	11 18	12 19	12 19	11 19	19 20	19 20	20 31	21 31	21 31
$TU_{0.10}$	26 38	30 43	31 45	31 45	31 45	50 64	56 69	58 70	59 71	60 72
$TU_{0.15}$	51 64	56 68	57 70	57 71	58 71	83 90	88 93	89 94	90 94	90 94
$TU_{0.20}$	75 84	80 88	81 89	82 90	82 90	97 98	98 99	98 99	99 99	99 99

Table 3. 5% (left entry) and 10% (right entry) percentage of rejection with sample size $(n, T) = (10, 20)$ (left part) and $(n, T) = (10, 40)$ (right part).

Density	KS	CM	SN	WX	KS	CM	SN	WX
N	6 11	6 11	5 10	5 9	5 10	5 10	5 8	5 10
LG	6 11	6 12	5 10	6 11	5 10	5 10	5 9	6 11
LP	7 12	7 12	6 11	8 14	6 11	6 11	6 10	7 12
UN	6 10	6 11	5 8	2 6	6 12	6 11	5 9	3 7
TR	5 10	5 10	4 8	3 8	5 9	5 10	4 9	4 10
t_4	6 10	6 11	6 10	8 14	6 12	6 11	6 10	6 12
t_6	5 10	6 11	5 9	7 12	6 10	6 11	5 10	6 11
t_{10}	5 10	5 10	4 9	6 12	5 10	6 11	6 10	6 11
χ^2_{10}	76 84	61 74	68 78	98 99	98 99	92 97	95 97	* *
χ^2_{15}	57 68	43 57	49 62	91 96	90 94	77 86	84 90	* *
χ^2_{20}	46 56	36 48	41 52	81 89	79 87	63 76	72 81	99 *
χ^2_{25}	36 47	27 39	31 43	72 82	68 77	52 66	61 71	97 98
SN_1	6 11	6 12	5 11	10 18	9 16	8 15	9 14	15 24
$\mathrm{SN}_{1.5}$	15 24	13 22	14 22	31 42	26 37	21 32	25 36	58 69
SN_2	30 40	24 35	26 37	60 72	59 70	47 59	54 66	91 96
SN_3	64 75	49 63	55 68	92 96	94 97	86 92	89 94	* *
$\mathrm{TU}_{0.05}$	7 12	7 12	6 11	11 18	10 17	9 16	10 16	19 30
$\mathrm{TU}_{0.10}$	12 20	11 18	11 18	29 42	24 35	19 29	23 33	57 69
$\mathrm{TU}_{0.15}$	24 34	19 28	21 31	55 68	50 61	38 50	45 57	88 93
$\mathrm{TU}_{0.20}$	39 53	30 43	36 48	79 88	76 85	60 73	69 79	98 99

PART 2

Papers on General Nonparametric Inference

Chapter 4

On Robust Versions of Classical Tests with Dependent Data

Jiming Jiang

University of California, Davis, USA

We provide a unified treatment of robust versions of Wald, score, and likelihood-ratio tests for complex hypotheses under M-estimation and with dependent observations. Asymptotic null distributions are obtained for the proposed tests, which are χ^2 for the robust versions of Wald and score tests and a "weighted" χ^2 for that of the likelihood-ratio test. The general theory is applied to mixed linear models to obtain robust tests regarding the fixed effects and variance components without the normality assumption.

Contents

1. Introduction

Robust testing procedures have been studied extensively in the literature. In particular, robust versions of the classical tests, that is, the Wald, score, and likelihood-ratio tests (e.g., Lehmann (1999), §7) have been considered. In the case of i.i.d.observations, see Foutz and Srivastava (1977), Kent (1982), Hampel, et al. (1986), Heritier and Ronchetti (1994), among others. In the case of independent but not identically-distributed observations, see, e.g., Schrader and Hettmansperger (1980), Chen (1985), Silvapulle (1992), Kim and Cai (1993). In many cases, such as in biomedical research and survey sampling, the observations are dependent. One important class of models for dependent observations are mixed linear models (e.g., Searle, et al. (1992)). In a mixed linear model, the observations are correlated because of the presence of the random effects. Tests with dependent data have been developed in the case of mixed linear models (e.g., Khuri et al. (1998)). Other areas in which dependent data are frequently encountered include time series and stochastic processes, in which tests with dependent data have also been developed (e.g., Basawa and Rao (1980), Dzhaparidze (1986). Note that, in these tests with dependent data, it is assumed that the distribution of the data is known up to a set of parameters (therefore the likelihood function is available). In contrast to the independent cases, the literature on robust testing with dependent observations is not extensive. For example, in the case of mixed linear models, Richardson and Welsh (1996) proposed a robust likelihood-ratio test based on restricted maximum likelihood (REML) estimation. The purpose of this paper is to provide a unified and rigorous treatment of these robust versions of classical tests for the case of dependent observations under a full or partial parametrization. It should be noted that not all, or nearly all, models for dependent data fall into the parametrization framework. Nevertheless, the results established in this paper are applicable to a number of areas. In particular, we apply the results to mixed linear models to obtain robust dispersion tests without the normality assumption.

In many cases, people are interested in testing a hypothesis regarding some parameters associated with the population, but the entire population distribution is unknown given these (and possibly other) parameters. In such cases, the likelihood function, by which the classical tests are defined, is not available, because it relies on full specification of the distribution of the observations, given the parameters. However, one may replace the likelihood function by a known objective function, and thus obtain robust versions of the classical tests. In this paper, the objective function is assumed to be a function for M-estimation (Huber (1981)).

It should be pointed out that there has been a quite unified extension of the classical tests in terms of quasi-likelihood (e.g., Heyde (1997)). The quasi-likelihood method is a special case of M-estimation, in which the estimating functions are closer to a (log-)likelihood in the sense that the quasi-score function maintains the properties of the first two moments of a score function. However, it seems that the class of quasi-likelihood functions is not rich enough to include some useful methods. For example, in mixed linear models the Gaussian restricted likelihood function has been used for estimating the variance components in the non-normal case (e.g., Richardson and Welsh (1994), Jiang (1996; 1997)). It is clear that, when normality does not hold, such a method may be regarded as a method of M-estimation, but not that of quasi-likelihood. In fact, Heyde (1994) has shown that the Gaussian restricted likelihood corresponds to a quasi-likelihood only when some restricted moment conditions (e.g., about the kurtosis) are satisfied. Also, it is not always easy to construct a quasi-likelihood function, especially when the parameter is multivariate, and the observations are dependent (e.g., McCullagh and Nelder (1989) §9.3.2).

The results obtained in this paper have the following features: (i) they are for dependent data; (ii) they are based on a general objective function (rather than the likelihood); and (iii) the hypotheses to be tested are of a quite general form. The robustness of these tests is in the sense of (ii), that is, the tests require weak distributional assumptions, and therefore are robust to violation of a specific distributional assumption such as normality. Although most researchers would apply such tests for dependent data and assume the asymptotic distribution to hold, at least under certain conditions, the exact conditions seem to be unclear for a testing problem that has the characteristics of (i)–(iii).

The rest of the paper is organized as follows. In Section 2 we introduce robust versions of the classical tests. Asymptotic null distributions of the proposed tests and other theoretical results are given in Section 3. In Section 4 we apply the general theory to mixed linear models, and obtain robust tests for (possibly complex) hypotheses regarding the fixed effects and variance components. The proofs are given in an Appendix.

2. Robust Tests

2.1. *Basic idea, assumptions, and examples*

The likelihood-ratio test is known to be applicable to a broad range of testing problems. Let $L(\theta, y)$ be the likelihood function, where y is a vector of observations and θ a vector of parameters. Then, the likelihood-ratio test of the hypothesis

$$H_0: \quad \theta \in \Theta_0 \quad \text{versus} \quad H_1: \quad \theta \notin \Theta_0 , \tag{1}$$

where $\Theta_0 \subset \Theta$ and Θ is the parameter space, is based on the likelihood-ratio statistic

$$\frac{\sup_{\theta \in \Theta_0} L(\theta, y)}{\sup_{\theta \in \Theta} L(\theta, y)} = \frac{L(\hat{\theta}_0, y)}{L(\hat{\theta}, y)} , \tag{2}$$

where $\hat{\theta}$ is the maximum likelihood estimator (MLE) of θ, and $\hat{\theta}_0$ the MLE of θ under H_0. The Wald test (e.g., Lehmann (1999), p.525), on the other hand, is known to apply to some more restricted testing problems. The original Wald test is for the hypothesis H_0: $\theta = \theta_0$ versus H_1: $\theta \neq \theta_0$, where θ_0 is a known parameter vector. The test statistic can be expressed as $(\hat{\theta} - \theta_0)' I(\theta_0)(\hat{\theta} - \theta_0)$, where $I(\theta)$ is the Fisher information matrix. The method can be extended to test (1), provided that H_0 can be expressed as $R(\theta) = 0$, where $R(\cdot)$ is a known (vector-valued) function. In such a case, the test statistic has the form

$$R(\hat{\theta})' \left[\left(\left. \frac{\partial R}{\partial \theta} \right|_{\hat{\theta}} \right)' I(\hat{\theta})^{-1} \left(\left. \frac{\partial R}{\partial \theta} \right|_{\hat{\theta}} \right) \right]^{-1} R(\hat{\theta}) . \tag{3}$$

However, the extension still seems a somewhat restrictive, because it rules out hypotheses that, for example, involve inequalities. Therefore, we consider a Wald test of (1) based on the following test statistic

$$(\hat{\theta} - \hat{\theta}_0)' \hat{W}^{-1} (\hat{\theta} - \hat{\theta}_0) , \tag{4}$$

where $\hat{\theta}$ is an M-estimator of θ, $\hat{\theta}_0$ is a M-estimator of θ under the null hypothesis, and $\hat{W}$ is a "consistent estimator" of the asymptotic covariance matrix of $\hat{\theta} - \hat{\theta}_0$ under the null hypothesis. Such an idea has been used by, e.g., Gourieroux and Monfort (1995) to construct tests that take into account inequality constraints. There is another interesting feature of the Wald test. It is known that, while the likelihood-ratio test involves both $\hat{\theta}$ and $\hat{\theta}_0$, the Wald test based on (3) only involves $\hat{\theta}$. The score test, on the other hand, involves only $\hat{\theta}_0$, because its test statistic is given by

$$\left(\left. \frac{\partial l}{\partial \theta} \right|_{\hat{\theta}_0} \right)' I(\hat{\theta}_0)^{-1} \left(\left. \frac{\partial l}{\partial \theta} \right|_{\hat{\theta}_0} \right) , \tag{5}$$

where $l = l(\theta, y)$ is the log-likelihood. Now, the point is, under (4), all three tests may be thought as involving both $\hat{\theta}$ and $\hat{\theta}_0$. Note that (5) can be written as

$$\left(\left. \frac{\partial l}{\partial \theta} \right|_{\hat{\theta}} - \left. \frac{\partial l}{\partial \theta} \right|_{\hat{\theta}_0} \right)' I(\hat{\theta}_0)^{-1} \left(\left. \frac{\partial l}{\partial \theta} \right|_{\hat{\theta}} - \left. \frac{\partial l}{\partial \theta} \right|_{\hat{\theta}_0} \right) ,$$

because $(\partial l/\partial\theta)|_{\hat{\theta}} = 0$.

We now consider robust versions of Wald, score, and likelhood-ratio tests, which we call W-, S-, and L-tests, respectively. Let $y = (y_k)_{1\le k\le n}$ be a vector of observations not necessarily independent. Let θ be a vector of unknown parameters that are associated with the joint distribution of y, but the entire distribution of y may not be known given θ (and possibly other parameters). We are interested in testing the hypothesis:

$$H_0: \quad \theta \in \Theta_0 \tag{6}$$

versus H_1: $\theta \notin \Theta_0$, where $\Theta_0 \subset \Theta$, and Θ is the parameter space.

Suppose that there is a new parametrization ϕ such that, under the null hypothesis (6), $\theta = \theta(\phi)$ for some ϕ. Here $\theta(\cdot)$ is a map from Φ, the parameter space of ϕ, to Θ. Note that such a reparametrization is almost always possible, but the key is to try to make ϕ unrestricted (unless completely specified, such as in Example 1 below). The following are some examples.

Example 1. Suppose that, under the null hypothesis, θ is completely specified, i.e.,

$$H_0: \quad \theta = \theta_0 \ .$$

Then, under H_0, $\theta = \phi = \theta_0$.

Example 2. Let $\theta = (\theta_1, \dots, \theta_p, \theta_{p+1}, \dots, \theta_q)'$, and suppose that one wishes to test the hypothesis

$$H_0: \quad \theta_j = \theta_{0j}\ , \quad p+1 \le j \le q\ ,$$

where θ_{0j}, $p+1 \le j \le q$ are known constants. Then, under the null hypothesis, $\theta_j = \phi_j$, $1 \le j \le p$, and $\theta_j = \theta_{0j}$, $p+1 \le j \le q$ for some (unrestricted) $\phi = (\phi_j)_{1\le j\le p}$.

Example 3. Let $y_1, \dots, y_n$ be independent with distribution $N(\mu, \sigma^2)$, so $\theta = (\mu, \sigma^2)'$. Suppose that one wishes to test the hypothesis H_0: $\sigma^2 = \mu^2$. Then, under the null hypothesis, $\theta_1 = \phi$, $\theta_2 = \phi^2$ for some unrestricted ϕ.

Example 4. Suppose that the null hypothesis includes inequality constraints:

$$H_0: \quad \theta_j \ > \ \theta_{0j}\ , \quad p_1+1 \le j \le p\ , \quad \text{and} \quad \theta_j \ = \ \theta_{0j}\ , \quad p+1 \le j \le q\ ,$$

where $p_1 < p < q$. Then, under the null hypothesis, $\theta_j = \phi_j$, $1 \le j \le p_1$, $\theta_j = \theta_{0j} + e^{\phi_j}$, $p_1+1 \le j \le p$, and $\theta_j = \theta_{0j}$, $p+1 \le j \le q$ for some (unrestricted) $\phi = (\phi_j)_{1\le j\le p}$.

Let $L(\theta, y)$ be a function of θ and y that takes positive values, and $l(\theta, y) = \log L(\theta, y)$. Also, Let $L_0(\phi, y) = L(\theta(\phi), y)$, and $l_0(\phi, y) = \log L_0(\phi, y)$. Let q and p be the dimensions of θ and ϕ, respectively. Then, we have

$$\frac{\partial l_0}{\partial \phi} = \left(\frac{\partial \theta}{\partial \phi}\right)' \frac{\partial l}{\partial \theta}, \tag{7}$$

$$\frac{\partial^2 l_0}{\partial \phi^2} = \left(\frac{\partial \theta}{\partial \phi}\right)' \frac{\partial^2 l}{\partial \theta^2} \left(\frac{\partial \theta}{\partial \phi}\right) + \sum_{i=1}^{q} \frac{\partial l}{\partial \theta_i} \cdot \frac{\partial^2 \theta_i}{\partial \phi^2}, \tag{8}$$

where

$$\frac{\partial l_0}{\partial \phi} = \left(\frac{\partial l_0}{\partial \phi_j}\right)_{1 \le j \le p}, \quad \frac{\partial l}{\partial \theta} = \left(\frac{\partial l}{\partial \theta_i}\right)_{1 \le i \le q},$$

$$\frac{\partial \theta}{\partial \phi} = \left(\frac{\partial \theta_i}{\partial \phi_j}\right)_{1 \le i \le q, 1 \le j \le p};$$

$$\frac{\partial l_0}{\partial \phi^2} = \left(\frac{\partial^2 l_0}{\partial \phi_j \partial \phi_{j'}}\right)_{1 \le j, j' \le p}, \quad \frac{\partial^2 l}{\partial \theta^2} = \left(\frac{\partial^2 l}{\partial \theta_i \partial \theta_{i'}}\right)_{1 \le i, i' \le q},$$

$$\frac{\partial^2 \theta_i}{\partial \phi^2} = \left(\frac{\partial^2 \theta_i}{\partial \phi_j \partial \phi_{j'}}\right)_{1 \le j, j' \le p}.$$

Let $\hat{\theta}$ be an estimator of θ, and $\hat{\phi}$ an estimator of ϕ. Note that here we do not require that $\hat{\theta}$ and $\hat{\phi}$ are the (global) maximizers of $l(\theta, y)$ and $l_0(\phi, y)$, respectively. But we shall require that $\hat{\theta}$ is a solution to $\partial l / \partial \theta = 0$, and $\hat{\phi}$ a solution to $\partial l_0 / \partial \phi = 0$. We use θ_0 to represent the true θ, and ϕ_0 the true ϕ under (6).

2.2. *The W-, S-, and L- test statistics*

We first introduce the following quantity, which is closely related to the W-test:

$$\tilde{\chi}_w^2 = [\hat{\theta} - \theta(\hat{\phi})]' G Q_w^- G [\hat{\theta} - \theta(\hat{\phi})], \tag{9}$$

where Q_w^- is the unique Moore-Penrose inverse (e.g., Searle (1971), §1.3) of

$$Q_w = [A^{-1} - C(C'AC)^{-1}C'] \Sigma [A^{-1} - C(C'AC)^{-1}C']. \tag{10}$$

Note that the expression inside $[\cdots]$ in (10) can be written in a familiar form: Write $C = A^{-1}D$. Then, $A^{-1} - C(C'AC)^{-1}C' = A^{-1} -$

$A^{-1}D(D'A^{-1}D)^{-1}D'A^{-1} = K(K'AK)^{-1}K' \equiv P$, where K is any matrix of maximum column rank such that $K'D = 0$ (e.g., Searle, et al. (1992), p.451). The matrix P is used extensively in linear statistical inference, especially in the context of mixed linear models (e.g., Searle, et al. (1992), §6). The definitions of the matrices G, A, C, and Σ will be given in Theorem 1 in the next section, but here we first give some interpretations: A is the limit of the matrix of second derivatives of l with respect to θ; B is the limit of the matrix of second derivatives of l_0 with respect to ϕ; C is the limit of the matrix of first derivatives of θ with respect to ϕ; and Σ is the asymptotic covariance matrix of $\partial l/\partial\theta$, all after suitable normalizations. Theorem 1 in the sequel shows that, as the symbol indicates, $\tilde{\chi}^2_w$ has an asymptotic χ^2 distribution. Typically, the normalizing matrix G is completely specified. In fact, in many cases G is a diagonal matrix, where the diagonal elements are normalizing constants corresponding to the components of $(\partial l/\partial\theta)|_{\theta_0}$. On the other hand, the matrix Q_w may involve a vector of parameters, say, ϑ. In some cases such as generalized linear models (e.g., McCullagh and Nelder (1989)), ϑ depends only on θ. In other cases, ϑ may involve some additional parameters, which may also need to be estimated. In Section 4 we consider a case of estimating the additional parameters in the context of mixed linear models. Let $\hat{Q}^-_w$ be a consistent estimator of Q^-_w in the sense that $\|\hat{Q}^-_w - Q^-_w\| \to 0$ in probability, where for any matrix M, $\|M\| = [\lambda_{\max}(M'M)]^{1/2}$ with $\lambda_{\max}$ representing the largest eigenvalue (and similarly $\lambda_{\min}$ the smallest). Note that this also covers the case in which Q^-_w may represent a sequence of matrices (see in Theorem 1' in Section 3). We define the W-test statistic as

$$\hat{\chi}^2_w = [\hat{\theta} - \theta(\hat{\phi})]' G \hat{Q}^-_w G [\hat{\theta} - \theta(\hat{\phi})] \,. \tag{11}$$

Similarly, we define the following:

$$\tilde{\chi}^2_s = \left(\left. \frac{\partial l}{\partial \theta} \right|_{\theta(\hat{\phi})} \right)' G^{-1} A^{-1/2} Q^-_s A^{-1/2} G^{-1} \left(\left. \frac{\partial l}{\partial \theta} \right|_{\theta(\hat{\phi})} \right) , \tag{12}$$

where Q^-_s is the unique Moore-Penrose inverse (e.g., Searle (1971), §1.3) of

$$Q_s = (I - P) A^{-1/2} \Sigma A^{-1/2} (I - P) \,, \tag{13}$$

$P = A^{1/2}C(C'AC)^{-1}C'A^{1/2}$, and the matrices G, A, C are the same as in Theorem 1. Let $\hat{A}$ and $\hat{Q}^-_s$ be consistent estimators of A and Q^-_s, respectively. Note that, quite often, A only depends on θ, so that a consistent estimator of it is already available (based on $\hat{\theta}$). We define the S-test statistic as

$$\hat{\chi}_s^2 = \left(\left. \frac{\partial l}{\partial \theta} \right|_{\theta(\hat{\phi})} \right)' G^{-1} \hat{A}^{-1/2} \hat{Q}_s^- \hat{A}^{-1/2} G^{-1} \left(\left. \frac{\partial l}{\partial \theta} \right|_{\theta(\hat{\phi})} \right) . \tag{14}$$

Finally, the L-ratio for testing the hypothesis (6) is defined as

$$R = \frac{L_0(\hat{\phi}, y)}{L(\hat{\theta}, y)} . \tag{15}$$

As noted before, here we do not require that $\hat{\theta}$ and $\hat{\phi}$ are, respectively, (global) maximizers of $l(\theta, y)$ and $l_0(\phi, y)$. In particular, if $\hat{\theta}$ and $\hat{\phi}$ are, indeed, the global maximizers, R is equal to $\sup_{\Theta_0} L(\theta, y)/\sup_{\Theta} L(\theta, y)$, which is the original definition of the likelihood-ratio when $L(\theta, y)$ is a likelihood function [see (2)]. The L-test statistic is then $-2 \log R$.

3. Asymptotic theory

The main results of this section are the asymptotic distributions of the W-, S-, and L- test statistics. The following notation will be used here: if $a = (a_1, \dots, a_k)'$ is a vector, then $|a|_v = (|a_1|, \dots, |a_k|)'$. Here the index v refers to "vector". Also, if $b = (b_1, \dots, b_k)$, then $a \le b$ iff $a_i \le b_i$, $1 \le i \le k$; and $a \vee b = (a_1 \vee b_1, \dots, a_k \vee b_k)$, where $a \vee b = \max(a, b)$ [and similarly $a \wedge b = \min(a, b)$]. For any matrix M, $M > 0$ means that M is positive definite; and $\mathrm{diag}(x_i, 1 \le i \le k)$ represents a diagonal matrix with $x_i, 1 \le i \le k$ on its diagonal.

Theorem 1. *Suppose that the following hold:*

i) $l(\cdot, y)$ is twice continuously differentiable for fixed y, and $\theta(\cdot)$ is twice continuously differentiable;
ii) with probability $\to 1$, $\hat{\theta}$, $\hat{\phi}$ satisfy $\partial l/\partial \theta = 0$, $\partial l_0/\partial \phi = 0$, respectively;
iii) there are sequences of nonsingular symmetric matrices $\{G\}$ and $\{H\}$ and matrices A, B, C with A, $B > 0$ such that the following $\to 0$ in

probability:

$$\sup_{|\theta^{(i)}-\theta_0|_v \le |\hat{\theta}-\theta_0|_v \vee |\theta(\hat{\phi})-\theta(\phi_0)|_v,\ 1\le i\le q} \left\| G^{-1} \left(\left.\frac{\partial^2 l}{\partial\theta_i \partial\theta_j}\right|_{\theta^{(i)}} \right)_{1\le i,j\le q} G^{-1} + A \right\| ,$$

$$\sup_{|\phi^{(i)}-\phi_0|_v \le |\hat{\phi}-\phi_0|_v,\ 1\le i\le p} \left\| H^{-1} \left(\left.\frac{\partial^2 l_0}{\partial\phi_i \partial\phi_j}\right|_{\phi^{(i)}} \right)_{1\le i,j\le p} H^{-1} + B \right\| ,$$

$$\sup_{|\phi^{(i)}-\phi_0|_v \le |\hat{\phi}-\phi_0|_v,\ 1\le i\le q} \left\| G \left(\left.\frac{\partial\theta_i}{\partial\phi_j}\right|_{\phi^{(i)}} \right)_{1\le i\le q, 1\le j\le p} H^{-1} - C \right\| ;$$

iv) $D(\partial l/\partial\theta)|_{\theta_0} \to 0$ *in probability, where* $D = \mathrm{diag}(d_i, 1 \le i \le s)$ *with* $d_i = \|H^{-1}(\partial^2\theta_i/\partial\phi^2)|_{\phi_0} H^{-1}\|$, *and*

$$G^{-1} \left.\frac{\partial l}{\partial\theta}\right|_{\theta_0} \longrightarrow N(0,\Sigma) \quad \text{in distribution} . \tag{16}$$

Then, under the null hypothesis, the asymptotic distribution of $\tilde{\chi}^2_w$ *is* χ^2_r, *where* $r = \mathrm{rank}[\Sigma^{1/2}A^{-1/2}(I-P)]$ *with* P *given in the line below (13). In particular, if* Σ *is nonsingular, then* $r = q-p$.

The theorem may be extended to allow the matrices A, B, etc., to be replaced by sequences of matrices. Such an extension may be useful. For example, suppose G is a diagonal normalizing matrix, then, in many cases, A can be chosen as $-G^{-1}[E(\partial^2 l/\partial\theta^2)|_{\theta_0}]G^{-1}$, but the latter may not have a limit as $n \to \infty$.

Theorem 1.1' Suppose that, in Theorem 1, A, B, C are replaced by sequences of matrices $\{A\}$, $\{B\}$, and $\{C\}$, such that A, B are symmetric,

$$0 \ < \ \liminf[\lambda_{\min}(A) \wedge \lambda_{\min}(B)] \le \limsup[\lambda_{\max}(A) \vee \lambda_{\max}(B)] \ < \ \infty ,$$

and $\limsup \|C\| < \infty$. Furthermore, suppose that (16) is replaced by

$$\Sigma^{-1/2} G^{-1} \left.\frac{\partial l}{\partial\theta}\right|_{\theta_0} \longrightarrow N(0,I) \quad \text{in distribution} , \tag{17}$$

where $\{\Sigma\}$ is a sequence of positive definite matrices such that $0 < \liminf \lambda_{\min}(\Sigma) \le \limsup \lambda_{\max}(\Sigma) < \infty$, and I is the p-dimensional identity matrix. Then, the asymptotic distribution of $\tilde{\chi}^2_w$ is χ^2_{q-p}.

Since, by the proof of Theorem 1 in the Appendix, $G[\hat{\theta}-\theta(\hat{\phi})] = O_P(1)$, we have

$$\begin{aligned}\hat{\chi}_w^2 &= [\hat{\theta}-\theta(\hat{\phi})]'G[Q_w^- + o_P(1)]G[\hat{\theta}-\theta(\hat{\phi})]\\ &= \tilde{\chi}_w^2 + o_P(1)\ .\end{aligned}$$

Thus, by Theorem 1, we have the following.

Corollary 4.1. *Under the conditions of Theorem 1, the asymptotic distribution of $\hat{\chi}_w^2$ is χ_r^2, where r is the same as in Theorem 1. Thus, in particular, if Σ is nonsingular, $r = q - p$. Under the conditions of Theorem 1.1', the asymptotic distribution of $\hat{\chi}_w^2$ is χ_{q-p}^2.*

Theorem 2. *Suppose that the conditions of Theorem 1 are satisfied with the following changes: 1) in ii), that $\hat{\theta}$ satisfies $\partial l/\partial\theta = 0$ with probability $\to 1$ is not required; and 2) in iii), the supremum for the first quantity (involving A) is now over $|\theta^{(i)} - \theta_0|_v \leq |\theta(\hat{\phi}) - \theta(\phi_0)|_v$, $1 \leq i \leq q$. Then, under the null hypothesis, the asymptotic distribution of $\tilde{\chi}_s^2$ is χ_r^2, where r is the same as in Theorem 1. In particular, if Σ is nonsingular, then $r = q - p$.*

In exactly the same way, we have the following.

Theorem 1.2' Suppose that, in Theorem 2, A, B, and C are replaced by $\{A\}$, $\{B\}$, and $\{C\}$, and (16) by (17), where the sequences of matrices $\{A\}$, $\{B\}$, $\{C\}$, and $\{\Sigma\}$ satisfy the conditions of Theorem 1.1'. Then, the asymptotic distribution of $\tilde{\chi}_s^2$ is χ_{q-p}^2.

By a similar argument, we have

Corollary 4.2. *Under the conditions of Theorem 2, the asymptotic distribution of $\hat{\chi}_s^2$ is χ_r^2, where r is the same as in Theorem 1. Thus, in particular, if Σ is nonsingular, $r = q - p$. Under the conditions of Theorem 1.2', the asymptotic distribution of $\hat{\chi}_s^2$ is χ_{q-p}^2.*

It is seen that the asymptotic distributions for W- and S- tests are both χ^2. However, the following theorem states that the asymptotic distribution for the L-test is not χ^2 but a "weighted" χ^2 (e.g., Chernoff and Lehmann (1954)). Let

$$Q_l = [A^{-1} - C(C'AC)^{-1}C']^{1/2}\Sigma[A^{-1} - C(C'AC)^{-1}C']^{1/2}\ . \tag{18}$$

Theorem 3. *Suppose that the conditions of Theorem 1 are satisfied except that the third quantity in iii) (involving C) $\to 0$ in probability is replaced by*

$G[(\partial\theta/\partial\phi)|_{\phi_0}]H^{-1} \to C$. *Then, under the null hypothesis, the asymptotic distribution of* $-2\log R$ *is the same as* $\lambda_1\xi_1^2 + \cdots + \lambda_r\xi_r^2$, *where* r *is the same as in Theorem 1;* $\lambda_1, \ldots, \lambda_r$ *are the positive eigenvalues of* Q_l; *and* $\xi_1, \ldots, \xi_r$ *are independent* $N(0,1)$ *random variables. In particular, if* Σ *is nonsingular, then* $r = q - p$.

It should be pointed out that if $L(\theta, y)$ is, indeed, the likelihood function, in which case L-test is the likelihood-ratio test, the asymptotic distribution of $-2\log R$ reduces to χ^2 [see Weiss (1975)].

Let $\hat{Q}_l$ be a consistent estimator of Q_l. Then, by Weyl's eigenvalue perturbation theorem [e.g., Bhatia (1997, page 63)], the eigenvalues of $\hat{Q}_l$ are consistent estimators of those of Q_l, and therefore can be used to obtain the asymptotic critical values for the L-test.

In Theorem 1, the consistency of $\hat{\theta}$ and $\hat{\phi}$ is not explicitly required. However, condition iii) does implicate some consistent properties of the estimators. In the following, we give sufficient conditions under which $\hat{\theta}$, $\hat{\phi}$ are consistent, and conditions ii), iii) of Theorem 1 are satisfied. In the theorem below, the assumption of G and H being diagonal is not essential, but it much simplifies the result.

Theorem 4. *Suppose that* $G = \text{diag}(g_j, 1 \le j \le q)$ *and* $H = \text{diag}(h_j, 1 \le j \le p)$, *where* g_j*'s and* h_j*'s are sequences of positive numbers that* $\to \infty$ *as sample size increases. Furthermore, suppose that the following conditions are satisfied:*

i) $l(\cdot, y)$ *is three-times continuously differentiable for fixed* y, *and* $\theta(\cdot)$ *is three-times continuously differentiable;*

ii) $G^{-1}(\partial l/\partial\theta)_{\theta_0} = O_P(1)$;

iii) there are matrices A, $B > 0$ *and* C *such that*

$$G^{-1}\left(\left.\frac{\partial^2 l}{\partial\theta^2}\right|_{\theta_0}\right)G^{-1} + A \longrightarrow 0\,, \quad H^{-1}\left(\left.\frac{\partial^2 l_0}{\partial\phi^2}\right|_{\phi_0}\right)H^{-1} + B \longrightarrow 0$$

in probability, and $G[(\partial\theta/\partial\phi)|_{\phi_0}]H^{-1} \to C$;

iv) there is $\delta > 0$ *such that*

$$\frac{1}{g_i g_j}\sup_{|\theta-\theta_0|\le\delta}\left|\frac{\partial^3 l}{\partial\theta_i\partial\theta_j\partial\theta_k}\right|\,, \quad 1 \le i,j,k \le q\,,$$

$$\frac{1}{h_i h_j}\sup_{|\phi-\phi_0|\le\delta}\left|\frac{\partial^3 l_0}{\partial\phi_i\partial\phi_j\partial\phi_k}\right|\,, \quad 1 \le i,j,k \le p$$

are bounded in probability, and $g_i h_j^{-1} \sup_{|\phi-\phi_0|\le\delta} |\partial^2\theta_i/\partial\phi_j\partial\phi_k|$, $1 \le i \le q$, $1 \le j, k \le p$ *are bounded.*

Then, there exist $\hat\theta$ *and* $\hat\phi$ *such that 1) ii) of Theorem 1 holds; 2)* $G(\hat\theta-\theta_0) = O_P(1)$, $H(\hat\phi-\phi_0) = O_P(1)$, *therefore* $\hat\theta$ *and* $\hat\phi$ *are consistent; and 3) iii) of Theorem 1 holds for the matrices given above.*

Example 1.2 (continued) For any positive constants g_j, $1 \le j \le q$, let $h_j = g_j$, $1 \le j \le p$. Then,

$$\frac{\partial\theta}{\partial\phi} = \begin{pmatrix} I \\ 0 \end{pmatrix}, \quad \text{hence} \quad G\left(\frac{\partial\theta}{\partial\phi}\right)H^{-1} = \begin{pmatrix} I \\ 0 \end{pmatrix},$$

where I is the $p \times p$ identity matrix. Also, $g_i h_j^{-1} \sup_\phi |\partial^2\theta_i/\partial\phi_j\partial\phi_k| = 0$, $\forall i, j, k$. Furthermore, let $\theta_{(1)} = (\theta_1, \dots, \theta_p)'$. Then, by (8),

$$\frac{\partial^2 l_0}{\partial\phi^2} = \left(\frac{\partial\theta}{\partial\phi}\right)' \frac{\partial^2 l}{\partial\theta^2}\left(\frac{\partial\theta}{\partial\phi}\right) = \frac{\partial^2 l}{\partial\theta_{(1)}^2}.$$

Thus,

$$-G^{-1}\left(\frac{\partial^2 l}{\partial\theta^2}\right)G^{-1} \longrightarrow A = \begin{pmatrix} B & * \\ * & * \end{pmatrix} > 0 \text{ in probability}$$

implies that

$$-H^{-1}\left(\frac{\partial^2 l_0}{\partial\phi^2}\right)H^{-1} = -H^{-1}\left(\frac{\partial^2 l}{\partial\theta_{(1)}^2}\right)H^{-1} \longrightarrow B > 0 \text{ in probability.}$$

Theorem 4 can be easily extended to allow A, B, C to be replaced by sequences of matrices, as in Theorem 1.1' and Theorem 2. The statement of the extension is omitted.

4. Application to mixed linear models

In this section, we consider the following mixed linear model:

$$y = X\beta + Z_1\alpha_1 + \cdots + Z_s\alpha_s + \epsilon, \tag{19}$$

where y is an $N \times 1$ vector of observations; X is an $N \times p$ known matrix of covariates, and, without loss of generality, is assumed to have full rank p; β is a $p \times 1$ vector of unknown constants (the fixed effects); Z_i is an $N \times m_i$ known matrix; α_i is an $m_i \times 1$ vector of unobservable i.i.d.random variables (the random effects); and ϵ is an $N \times 1$ vector of i.i.d.errors. Furthermore, it is assumed that the random effects and errors have mean 0, and variances $\sigma_0^2, \sigma_1^2, \ldots, \sigma_s^2$ corresponding to the components of $\epsilon, \alpha_1, \ldots, \alpha_s$, respectively; and $\alpha_1, \ldots, \alpha_s, \epsilon$ are independent. Note that it is not assumed that the random effects and errors are normally distributed, nor that their distributions are specified up to a set of parameters.

Following Hartley and Rao (1967), we consider the following parameters of variance components: $\lambda = \sigma_0^2$, $\gamma_i = \sigma_i^2/\sigma_0^2$, $1 \le i \le s$. Let $\gamma = (\gamma_i)_{1 \le i \le s}$, and $\theta = (\beta' \ \lambda \ \gamma')'$. Then, without normality, θ is a vector of parameters, which alone may not completely determine the distribution of y. Nevertheless, in many cases, people are still interested in testing the following hypothesis:

$$H_0: \quad \theta \in \Theta_0 , \tag{20}$$

where $\Theta_0 \subset \Theta = \{\theta : \lambda > 0, \gamma_i \ge 0, 1 \le i \le s\}$, versus H_1: $\theta \notin \Theta_0$. When normality is assumed, the use of likelihood-ratio test for potentially complex hypotheses and unbalanced data was first proposed by Hartley and Rao (1967), although rigorous justification was not given. Welham and Thompson (1997) showed the equivalence of the likelihood ratio, score, and Wald tests under normality. For discussions about other testing precedures in mixed linear models under normality, see Khuri et al. (1998). On the other hand, Richardson and Welsh (1996) considered likelihood ratio test without assuming normality, whose approach is similar to our L-test, but their goal was to select the (fixed) covariates. Under the normality assumption, the log-likelihood function for estimating θ is

$$\begin{aligned} l(\theta, y) &= \text{constant} \\ -\frac{1}{2}&\left[N \log\lambda + \log(|V|) + \frac{1}{\lambda}(y - X\beta)'V^{-1}(y - X\beta)\right], \end{aligned} \tag{21}$$

where $V = V_\gamma = I + \sum_{i=1}^s \gamma_i V_i$ with I being the N-dimensional identity matrix, $V_i = Z_i Z_i'$, $1 \le i \le s$, and $|V|$ the determinant of V. The restricted log-likelihood for estimating λ, γ is

$$\begin{aligned} l_R(\lambda, \gamma, y) &= \text{constant} \\ -\frac{1}{2}&\left[(N-p)\log\lambda + \log(|K'VK|) + \frac{y'Py}{\lambda}\right], \end{aligned} \tag{22}$$

where K is any $N \times (N-p)$ matrix such that $\text{rank}(K) = N-p$ and $K'X = 0$, and $P = P_\gamma = K(K'VK)^{-1}K' = V^{-1} - V^{-1}X(X'V^{-1}X)^{-1}X'V^{-1}$ (e.g., Searle, et al. (1992), page 451). The restricted log-likelihood is only for estimating the variance components. It is then customary to estimate β by the empirical best linear unbiased estimator:

$$\hat{\beta} = (X'\hat{V}^{-1}X)^{-1}X'\hat{V}^{-1}y\,, \tag{23}$$

where $\hat{V} = V_{\hat{\gamma}}$, and $\hat{\gamma} = (\hat{\gamma}_i)_{1\le i\le s}$ is the REML estimator of γ (e.g., Jiang (1998)). Alternatively, one may define the following "restricted log-likelihood" for θ:

$$\begin{aligned} l_R(\theta, y) &= \text{constant} \\ -\frac{1}{2}&\left[(N-p)\log\lambda + \log|K'VK| + \frac{1}{\lambda}(y - X\beta)'V^{-1}(y - X\beta)\right]. \end{aligned} \tag{24}$$

It is easy to show that the maximizer of (24) is $\hat{\theta} = (\hat{\beta}'\ \hat{\lambda}\ \hat{\gamma}')'$, where $\hat{\lambda}$ and $\hat{\gamma}$ are the REML estimators, and $\hat{\beta}$ is given by (23) with $\hat{V} = V_{\hat{\gamma}}$. The difference is that, unlike $l(\theta, y)$, $l_R(\theta, y)$ is not a log-likelihood even if normality holds. Nevertheless, we now show that both $l(\theta, y)$ and $l_R(\theta, y)$ can be used to test (20) without the normality assumption.

In mixed linear models, conditions of Theorems 1–3 and their extensions correspond to mild restrictions. However, the asymptotic covariance matrix of the REML estimator involves higher (i.e., third or fourth) moments of the random effects and errors. More specifically, if the test is only about the fixed effects, no higher moments are involved in Σ; if the test is only about the variance components, the fourth moments, or kurtoses, will appear in Σ; if the test is about both the fixed effects and the variance components, the third moments will be involved, in addition to the kurtoses. The matrices A, B and C do not involve higher moments. To simplify the results, we make the following additional assumption:

$$E(\epsilon_1^3) = 0\,, \quad E(\alpha_{i1}^3) = 0\,, \quad 1 \le i \le s\,. \tag{25}$$

Such conditions hold if, in particular, the distributions of the random effects and errors are symmetric. Another action of simplification is that we shall consider testing a special class of hypotheses (20). Let $\theta(\cdot) = (\theta_i(\cdot))_{1\le i\le q}$, where $q = p + s + 1$, be the map such that, under (20), $\theta = \theta(\phi)$ for some $\phi = (\phi_k)_{1\le k\le a}$, where $a \le q$ (see Section 2; note that p therein is replaced by a). We assume that there is a subset of indexes $1 \le i_1 < \cdots < i_a \le q$ such that

$$\begin{cases} \theta_{i_k}(\phi) \text{ is a function of } \phi_k\,, \quad 1 \le k \le a\,, \quad \text{and} \\ \theta_i(\phi) \text{ is a constant}\,, \quad i \in \{1, \ldots, q\} \setminus \{i_1, \ldots, i_a\}\,. \end{cases} \tag{26}$$

Among the examples in Section 2, Examples 1, 2, and 4 satisfy (26), while Example 3 does not. Also, for notational simplicity, in the rest of this section, θ, ϕ, etc., represent the true vectors of parameters. We need the following notation: Let $A_1 = (\mathrm{tr}(V^{-1}V_i)/2\lambda\sqrt{Nm_i})_{1\le i\le s}$, $A_2 = (\mathrm{tr}(V^{-1}V_iV^{-1}V_j)/2\sqrt{m_im_j})_{1\le i,j\le s}$, and

$$A = \begin{pmatrix} X'V^{-1}X/\lambda N & 0 & 0 \\ 0 & 1/2\lambda^2 & A_1' \\ 0 & A_1 & A_2 \end{pmatrix}. \tag{27}$$

Let $b = (I \ \sqrt{\gamma_1}Z_1 \ \cdots \ \sqrt{\gamma_s}Z_s)$, $B_0 = b'V^{-1}b$, $B_i = b'V^{-1}V_iV^{-1}b$, $1 \le i \le s$. We define

$$D_{0,ij} = \sum_{l=1}^{N} B_{i,ll}B_{j,ll}, \quad D_{1,ij} = \sum_{l=N+1}^{N+m_1} B_{i,ll}B_{j,ll}, \quad \dots$$

$$D_{s,ij} = \sum_{l=N+m_1+\cdots+m_{s-1}+1}^{N+m_1+\cdots+m_s} B_{i,ll}B_{j,ll},$$

where $B_{i,kl}$ is the (k,l) element of B_i, $0 \le i,j \le s$. The kurtoses of the errors and random effects are defined by $\kappa_0 = (E\epsilon_1^4/\sigma_0^4) - 3$, and $\kappa_i = (E\alpha_{i1}^4/\sigma_i^4) - 3$, $1 \le i \le s$. Let $\Delta_1 = (\Delta_{0i}/\sqrt{Nm_i})_{1\le i\le s}$, $\Delta_2 = (\Delta_{ij}/\sqrt{m_im_j})_{1\le i,j\le s}$, and

$$\Delta = \begin{pmatrix} 0 & 0 & 0 \\ 0 & \Delta_{00}/N & \Delta_1' \\ 0 & \Delta_1 & \Delta_2 \end{pmatrix}, \tag{28}$$

where $\Delta_{ij} = [4\lambda^{1_{(i=0)}+1_{(j=0)}}]^{-1}\sum_{t=0}^{s}\kappa_tD_{t,ij}$, $0 \le i,j \le s$. Let $W = b'V^{-1}X(X'V^{-1}X)^{-1/2}$, and W_l' be the lth row of W, $1 \le l \le N+m$, where $m = m_1 + \cdots + m_s$.

Theorem 5. *Suppose that*

i) $\theta(\cdot)$ *is three-times continuously differentiable and satisfies* (26), *and* $\partial\theta_{i_k}/\partial\phi_k \ne 0$, $1 \le k \le a$;

ii) $E\epsilon_1^4 < \infty$, $\mathrm{var}(\epsilon_1^2) > 0$, $E\alpha_{i1}^4 < \infty$, $\mathrm{var}(\alpha_{i1}^2) > 0$, $1 \le i \le s$, *and (25) holds; and*

iii) $N \to \infty$, $m_i \to \infty$, $1 \le i \le s$, $0 < \liminf \lambda_{\min}(A) \le \limsup \lambda_{\max}(A) < \infty$, *and* $\max_{1\le l\le N+m}|W_l| \to 0$;

Then, for $l(\theta, y)$ *there exist* $\hat\theta$ *and* $\hat\phi$ *such that the conditions of Theorem 1.1' (and hence Theorem 1.2') are satisfied with* $G = \mathrm{diag}(\sqrt{N}, \dots, \sqrt{N}, \sqrt{m_1}, \dots, \sqrt{m_s}) = \mathrm{diag}(g_i, 1 \le i \le q)$, $H = \mathrm{diag}(g_{i_k}, 1 \le k \le a)$, *A given by* (27), $C = \partial\theta/\partial\phi$, $B = C'AC$, *and*

$\Sigma = A + \Delta$, where Δ is given by (28). The same conclusion holds for $l_R(\theta, y)$ as well.

Note that the ith row of $\partial\theta/\partial\phi$ is $\partial\theta_i/\partial\phi'$, which, under (26), is $(0, \dots, 0)$ if $i \notin \{i_1, \dots, i_a\}$, and is $(0, \dots, 0, \partial\theta_{i_k}/\partial\phi_k, 0, \dots, 0)$ (kth component nonzero) if $i = i_k$, $1 \le k \le a$.

Theorem 6. *Suppose that the conditions of Theorem 5 are satisfied except that, in iii), the condition about A is strengthened to that $A \to A_0$, where $A_0 > 0$, and $\Sigma \to \Sigma_0$. Then, the conditions of Theorem 3 are satisfied with $A = A_0$, $\Sigma = \Sigma_0$, and everything else given by Theorem 5.*

It is seen from (28) that Δ, and hence Σ, depends on the kurtoses κ_i, $0 \le i \le s$, in addition to the variance components σ_i^2, $0 \le i \le s$. One already has consistent estimators of σ_i^2, $0 \le i \le s$ (e.g., the ML or REML estimators). As for κ_i, $0 \le i \le s$, they can be estimated by the empirical method of moments of Jiang (2003). Alternatively, Σ can be estimated by the partially observed information matrix of Jiang (2005), which does require the condition (25).

Acknowledgments

The research in this paper was supported, in part, by National Security Agency Grant MDA904-9810038 and National Science Foundation Grant SES-9978101.

Appendix A. Proofs

Proof of Theorem 1. First, we note that, by the assumption below (6), we have

$$\theta_0 = \theta(\phi_0)\,. \tag{A.1}$$

By Taylor series expansion, and condition ii), we have, with probability $\to 1$,

$$\begin{aligned} 0 &= \left.\frac{\partial l}{\partial \theta_i}\right|_{\hat\theta} \\ &= \left.\frac{\partial l}{\partial \theta_i}\right|_{\theta(\hat\phi)} + \left[\left.\frac{\partial^2 l}{\partial \theta' \partial \theta_i}\right|_{\theta^{(i)}}\right](\hat\theta - \theta(\hat\phi))\,, \quad i = 1, \dots, q\,, \end{aligned}$$

where $\theta^{(i)}$ lies between $\theta(\hat{\phi})$ and $\hat{\theta}$, hence

$$0 = \left.\frac{\partial l}{\partial \theta}\right|_{\theta(\hat{\phi})} + \left(\left.\frac{\partial^2 l}{\partial \theta_i \partial \theta_j}\right|_{\theta^{(i)}} \right)_{1 \le i,j \le q} (\hat{\theta} - \theta(\hat{\phi})) \,. \tag{A.2}$$

Again, by Taylor expansion, we have

$$\left.\frac{\partial l}{\partial \theta}\right|_{\theta(\hat{\phi})} = \left.\frac{\partial l}{\partial \theta}\right|_{\theta_0} + \left(\left.\frac{\partial^2 l}{\partial \theta_i \partial \theta_j}\right|_{\theta_{(i)}} \right)_{1 \le i,j \le q} (\theta(\hat{\phi}) - \theta_0) \,, \tag{A.3}$$

where $\theta_{(i)}$ lies between θ_0 and $\theta(\hat{\phi})$. Combining (A.1)–(A.3), and by condition iii), we have

$$0 = G^{-1} \left.\frac{\partial l}{\partial \theta}\right|_{\theta_0} - [A + o_P(1)]G[\theta(\hat{\phi}) - \theta_0] - [A + o_P(1)]G[\hat{\theta} - \theta(\hat{\phi})] \,. \tag{A.4}$$

On the other hand, by Taylor expansion and condition ii), we have, with probability $\to 1$,

$$\begin{aligned} 0 &= \left.\frac{\partial l_0}{\partial \phi_i}\right|_{\hat{\phi}} \\ &= \left.\frac{\partial l_0}{\partial \phi_i}\right|_{\phi_0} + \left[\left.\frac{\partial^2 l_0}{\partial \phi' \partial \phi_i}\right|_{\phi^{(i)}} \right] (\hat{\phi} - \phi_0) \,, \quad i = 1, \ldots, p \,, \end{aligned}$$

where $\phi^{(i)}$ lies between ϕ_0 and $\hat{\phi}$. It follows, by condition iii), that

$$\begin{aligned} 0 &= H^{-1} \left.\frac{\partial l_0}{\partial \phi}\right|_{\phi_0} + H^{-1} \left(\left.\frac{\partial^2 l_0}{\partial \phi_i \partial \phi_j}\right|_{\phi^{(i)}} \right)_{1 \le i,j \le p} H^{-1} H(\hat{\phi} - \phi_0) \\ &= H^{-1} \left.\frac{\partial l_0}{\partial \phi}\right|_{\phi_0} - [B + o_P(1)]H(\hat{\phi} - \phi_0) \,, \end{aligned}$$

and hence

$$H(\hat{\phi} - \phi_0) = [B + o_P(1)]^{-1} H^{-1} \left.\frac{\partial l_0}{\partial \phi}\right|_{\phi_0} \,. \tag{A.5}$$

Once again, by Taylor expansion, and (A.1), we have

$$\theta(\hat{\phi}) - \theta_0 = \left(\left.\frac{\partial \theta_i}{\partial \phi_j}\right|_{\phi_{(i)}} \right)_{1 \le i \le q, 1 \le j \le p} (\hat{\phi} - \phi_0) \,,$$

and hence, by condition iii),

$$G[\theta(\hat{\phi}) - \theta_0] = [C + o_P(1)]H(\hat{\phi} - \phi_0) \,. \tag{A.6}$$

Combining (A.4)–(A.6), and using (A.1), (7), we have

$$\begin{aligned}
&[A+o_P(1)]G[\hat{\theta}-\theta(\hat{\phi})]\\
&=G^{-1}\left.\frac{\partial l}{\partial \theta}\right|_{\theta_0}-[A+o_P(1)][C+o_P(1)][B+o_P(1)]^{-1}\left(H^{-1}\left.\frac{\partial l_0}{\partial \phi}\right|_{\phi_0}\right)\\
&=G^{-1}\left.\frac{\partial l}{\partial \theta}\right|_{\theta_0}-[ACB^{-1}+o_P(1)]\left[H^{-1}\left(\left.\frac{\partial \theta}{\partial \phi}\right|_{\phi_0}\right)'G\right]\left(G^{-1}\left.\frac{\partial l}{\partial \theta}\right|_{\theta_0}\right)\\
&=\{I-[ACB^{-1}+o_P(1)][C'+o_P(1)]\}\left(G^{-1}\left.\frac{\partial l}{\partial \theta}\right|_{\theta_0}\right).
\end{aligned}$$

It follows that

$$\begin{aligned}
G[\hat{\theta}-\theta(\hat{\phi})]&=[A+o_P(1)]^{-1}[I-ACB^{-1}C'+o_P(1)]\left(G^{-1}\left.\frac{\partial l}{\partial \theta}\right|_{\theta_0}\right)\\
&=(A^{-1}-CB^{-1}C')\left(G^{-1}\left.\frac{\partial l}{\partial \theta}\right|_{\theta_0}\right)+o_P(1)\\
&\longrightarrow\ N(0,Q_w) \qquad\qquad (\text{A.7})
\end{aligned}$$

in distribution. Note that, by (8) and condition iii), iv), it is easy to show that, in fact,

$$B=C'AC\ , \qquad (\text{A.8})$$

and hence the right side of (A.8) is nonsingular, therefore Q_w in (10) is well-defined. Equation (A.7) implies that $\tilde{\chi}_w^2 \to \xi' Q_w^- \xi$ in distribution, where $\xi \sim N(0, Q_w)$. Because $Q_w^- Q_w$ is idempotent, the distribution of $\xi' Q_w^- \xi$ is χ_r^2, where $r = \text{rank}(Q_w^- Q_w) = \text{rank}(Q_w) = \text{rank}[\Sigma^{1/2}A^{-1/2}(I-P)]$ (e.g., Searle (1971), page 58). If Σ is nonsingular, then $r = \text{rank}(I-P) = \text{tr}(I-P) = q-p$. □

Proof of Theorem 1.1'. First, we note that, strictly speaking, the sequences $\{A\}$, $\{B\}$, etc. should be denoted by $\{A_n\}$, $\{B_n\}$, etc., but for notation simplicity the index n is suppressed. Under the conditions, for any subsequence $\{n'\}$ of $\{n\}$, there is a further subsequence $\{n''\}$ of $\{n'\}$, such that, for the subsequence $\{n''\}$, the limits of $\{A\}$, $\{B\}$, $\{C\}$, and $\{\Sigma^{1/2}\}$ exist. Denote these limits by $\tilde{A}$, $\tilde{B}$, $\tilde{C}$, and $\tilde{\Sigma}^{1/2}$, respectively. Then, $\tilde{A}, \tilde{B} > 0$, and it is easy to see that condition iii) of Theorem 1 holds with A, B, C replaced by $\tilde{A}$, $\tilde{B}$, and $\tilde{C}$. Furthermore, (16) holds with Σ replaced by $\tilde{\Sigma}$, which is nonsingular. It follows, by Theorem 1, that the asymptotic distribution of $\tilde{\chi}^2_w$ for the subsequence $\{n''\}$ is χ^2_{q-p}. Because the limit distribution does not depend on the subsequence, the result follows. □

The proofs of Theorem 2, Theorem 1.2', and Theorem 3 are similar and therefore omitted.

Proof of Theorem 4. Write $\tilde{a} = (\partial l/\partial\theta)|_{\theta_0}$, and $\tilde{A} = (\partial^2 l/\partial\theta^2)|_{\theta_0}$. Let $\tilde{\theta} = \theta_0 - \tilde{A}^{-1}\tilde{a}$. By condition iii), the inverse of $\tilde{A}$ exists with probability $\to 1$. By Taylor expansion, and the definition of $\tilde{\theta}$, it is easy to show that

$$\begin{aligned} l(\theta, y) &= l(\theta_0, y) + \tilde{a}'(\theta - \theta_0) + (1/2)(\theta - \theta_0)'\tilde{A}(\theta - \theta_0) + r(\theta) \\ &= \tilde{c} + (1/2)(\theta - \tilde{\theta})'\tilde{A}(\theta - \tilde{\theta}) + r(\theta) , \end{aligned}$$

where $\tilde{c}$ does not depend on θ, and

$$r(\theta) = \frac{1}{6}\sum_{i,j,k}\left(\left.\frac{\partial^3 l}{\partial\theta_i\partial\theta_j\partial\theta_k}\right|_{\theta_*}\right)(\theta_i - \theta_{0i})(\theta_j - \theta_{0j})(\theta_k - \theta_{0k})$$

with θ_* lying between θ_0 and θ. Thus, in particular, $l(\tilde{\theta}, y) = \tilde{c} + r(\tilde{\theta})$, and hence

$$l(\theta, y) - l(\tilde{\theta}, y) = (1/2)(\theta - \tilde{\theta})'\tilde{A}(\theta - \tilde{\theta}) + r(\theta) - r(\tilde{\theta}) . \tag{A.9}$$

Now, consider $\partial S = \{\theta : |G(\theta - \tilde{\theta})| = 1\}$. If $\theta \in \partial S$, we have

$$\begin{aligned} |G(\theta - \theta_0)| &\le |G(\theta - \tilde{\theta})| + |G(\tilde{\theta} - \theta_0)| \\ &= 1 + |G\tilde{A}^{-1}\tilde{a}| \\ &\le 1 + \|(G^{-1}\tilde{A}G^{-1})^{-1}\| \cdot |G^{-1}\tilde{a}| \\ &\equiv \eta \;=\; O_P(1) . \end{aligned} \tag{A.10}$$

It follows that, $\sup_{\theta\in\partial S}|\theta - \theta_0| \le \|G^{-1}\|O_P(1) = o_P(1)$, and hence $\le \delta$ with probability $\to 1$. Since

$$r(\theta) = \frac{1}{6}\left((\theta - \theta_0)'G\left[G^{-1}\left(\left.\frac{\partial^3 l}{\partial\theta^2\partial\theta_k}\right|_{\theta_*}\right)G^{-1}\right]G(\theta - \theta_0)\right)'_{1\le k\le q} G^{-1}G(\theta - \theta_0) ,$$

we have, with probability $\to 1$,

$$\sup_{\theta\in\partial S}|r(\theta)| \le \frac{1}{6}\left(\sum_{i,j,k}\frac{1}{g_i g_j}\sup_{|\theta-\theta_0|\le\delta}\left|\frac{\partial^3 l}{\partial\theta_i\partial\theta_j\partial\theta_k}\right|\right)\frac{\eta^3}{\min_{1\le i\le q} g_i} \;=\; o_P(1) .$$

Similarly, it can be shown that $|r(\tilde{\theta})| = o_P(1)$. Thus, we have, by (A.9), that with probability $\to 1$,

$$\sup_{\theta \in \partial S} [l(\theta, y) - l(\tilde{\theta}, y)] \leq -\frac{1}{2}\lambda_{\min}(A) + \frac{1}{2}\|G^{-1}\tilde{A}G^{-1} + A\| + o_P(1)$$
$$= -\frac{1}{2}\lambda_{\min}(A) + o_P(1) \, .$$

It follows that, with probability tending to one, there is a solution to $\partial l/\partial\theta = 0$ in $S^o = \{\theta : |G(\theta - \tilde{\theta})| < 1\}$. Define this solution as $\hat{\theta}$. Then, by the argument of (A.10), $|G(\hat{\theta} - \theta_0)| = O_P(1)$, hence $\hat{\theta}$ is consistent.

By exactly the same arguments, one can show the conclusions about $\hat{\phi}$. Note that, by (7) and (A.1),

$$H^{-1}\left(\left.\frac{\partial l_0}{\partial \phi}\right|_{\phi_0}\right) = \left[G\left(\left.\frac{\partial \theta}{\partial \phi}\right|_{\phi_0}\right)H^{-1}\right]'\left(G^{-1}\left.\frac{\partial l}{\partial \theta}\right|_{\theta_0}\right) ,$$

which is $O_P(1)$, given the conditions. Furthermore, we have

$$\theta(\hat{\phi}) - \theta(\phi_0) = \left(\left.\frac{\partial \theta}{\partial \phi}\right|_{\phi_0}\right)(\hat{\phi} - \phi_0)$$
$$+\frac{1}{2}\left((\hat{\phi} - \phi_0)'\left[\left.\frac{\partial^2 \theta_i}{\partial \phi^2}\right|_{\phi_{*,i}}\right](\hat{\phi} - \phi_0)\right)_{1 \leq i \leq q} ,$$

where $\phi_{*,i}$ lies between ϕ_0 and $\hat{\phi}$, hence, if $|\hat{\phi} - \phi_0| \leq \delta$, we have

$$|G[\theta(\hat{\phi}) - \theta(\phi_0)]| \leq \left\|G\left(\left.\frac{\partial \theta}{\partial \phi}\right|_{\phi_0}\right)H^{-1}\right\| |H(\hat{\phi} - \phi_0)|$$
$$+\frac{1}{2}\sum_{i,j,k}\frac{g_i}{h_j h_k}\sup_{|\hat{\phi} - \phi_0| \leq \delta}\left|\frac{\partial^2 \theta_i}{\partial \phi_j \partial \phi_k}\right| |H(\hat{\phi} - \phi_0)|^2$$
$$= O_P(1) \, ,$$

hence $|\theta(\hat{\phi}) - \theta(\phi_0)| = o_P(1)$.

We now show that iii) of Theorem 1 holds. Let I_l, $l = 1, 2, 3$ represent the three quantities there that are supposed to $\to 0$ in probability. If $|\hat{\theta} - \theta_0| \vee |\theta(\hat{\phi}) - \theta(\phi_0)| \leq \delta/2$, then, for I_1 we have, by Taylor expansion,

$$\left|\frac{1}{g_i g_j}\left.\frac{\partial^2 l}{\partial \theta_i \partial \theta_j}\right|_{\theta^{(i)}} - \frac{1}{g_i g_j}\left.\frac{\partial^2 l}{\partial \theta_i \partial \theta_j}\right|_{\theta_0}\right|$$
$$\leq |\theta^{(i)} - \theta_0| \sum_k \frac{1}{g_i g_j}\sup_{|\theta - \theta_0| \leq \delta}\left|\frac{\partial^3 l}{\partial \theta_i \partial \theta_j \partial \theta_k}\right|$$
$$\leq O_P(1)|\hat{\theta} - \theta_0| \vee |\theta(\hat{\phi}) - \theta(\phi_0)| \, ,$$

hence $I_1 \le O_P(1)|\hat\theta - \theta_0| \vee |\theta(\hat\phi) - \theta(\phi_0)| + \|G^{-1}\tilde{A}G^{-1} + A\| = o_P(1)$. Similarly, if $|\hat\phi - \phi_0| \le \delta$, we have $I_2 \le O_P(1)|\hat\phi - \phi_0| + \|H^{-1}\tilde{B}H^{-1} + B\| = o_P(1)$, where $\tilde{B} = (\partial^2 l_0/\partial\phi^2)|_{\phi_0}$. Finally, if $|\hat\phi - \phi_0| \le \delta$, by Taylor expansion, we have, for I_3,

$$\begin{aligned}
&\left| \left(\frac{g_i}{h_j}\right) \frac{\partial\theta_i}{\partial\phi_j}\bigg|_{\phi^{(i)}} - \left(\frac{g_i}{h_j}\right) \frac{\partial\theta_i}{\partial\phi_j}\bigg|_{\phi_0} \right| \\
&\le \sum_k \left(\frac{g_i}{h_j}\right) \sup_{|\phi-\phi_0|\le\delta} \left|\frac{\partial^2\theta_i}{\partial\phi_j\partial\phi_k}\right| |\hat\phi - \phi_0| \\
&= O_P(1)|\hat\phi - \phi_0| \,.
\end{aligned}$$

Thus, $I_3 \le O_P(1)|\hat\phi - \phi_0| + \|G\tilde{C}H^{-1} - C\| = o_P(1)$, where $\tilde{C} = (\partial\theta/\partial\phi)|_{\phi_0}$. This completes the proof. □

Proof of Theorem 5. By the proof of Theorem 4.3 in Jiang (1996), it can be shown that there is $\hat\theta$ which satisfies $\partial l/\partial\theta = 0$ with probability $\to 1$, such that the first quantity in iii) of Theorem 1 involving A (which now represents a sequence of matrices) $\to 0$ in probability. Similar conclusions about $\hat\phi$ (and B) can be shown by virtually the same arguments. Note that, under (26), there are simple expressions of the derivatives of $l_0(\cdot, y)$ in terms of those of $l(\cdot, y)$ and $\theta(\cdot)$, that is,

$$\begin{aligned}
\frac{\partial l_0}{\partial\phi_u} &= \frac{\partial l}{\partial\theta_{i_u}} \cdot \frac{\partial\theta_{i_u}}{\partial\phi_u} \\
\frac{\partial^2 l_0}{\partial\phi_u\partial\phi_v} &= \frac{\partial^2 l}{\partial\theta_{i_u}\partial\theta_{i_v}} \cdot \frac{\partial\theta_{i_u}}{\partial\phi_u} \cdot \frac{\partial\theta_{i_v}}{\partial\phi_v} + \frac{\partial l}{\partial\theta_{i_u}} \cdot \frac{\partial^2\theta_{i_u}}{\partial\phi_u^2} 1_{(u=v)} \,,
\end{aligned}$$

$$\begin{aligned}
&\frac{\partial^3 l_0}{\partial\phi_u\partial\phi_v\partial\phi_w} \\
&= \frac{\partial^3 l}{\partial\theta_{i_u}\partial\theta_{i_v}\partial\theta_{i_w}} \cdot \frac{\partial\theta_{i_u}}{\partial\phi_u} \cdot \frac{\partial\theta_{i_v}}{\partial\phi_v} \cdot \frac{\partial\theta_{i_w}}{\partial\phi_w} + \frac{\partial^2 l}{\partial\theta_{i_u}\partial\theta_{i_v}} \cdot \frac{\partial\theta_{i_u}}{\partial\phi_u} \cdot \frac{\partial^2\theta_{i_v}}{\partial\phi_v^2} 1_{(v=w)} \\
&\quad + \frac{\partial^2 l}{\partial\theta_{i_v}\partial\theta_{i_w}} \cdot \frac{\partial\theta_{i_v}}{\partial\phi_v} \cdot \frac{\partial^2\theta_{i_w}}{\partial\phi_w^2} 1_{(w=u)} + \frac{\partial^2 l}{\partial\theta_{i_w}\partial\theta_{i_u}} \cdot \frac{\partial\theta_{i_w}}{\partial\phi_w} \cdot \frac{\partial^2\theta_{i_u}}{\partial\phi_u^2} 1_{(u=v)} \\
&\quad + \frac{\partial l}{\partial\theta_{i_u}} \cdot \frac{\partial^3\theta_{i_u}}{\partial\phi_u^3} 1_{(u=v=w)} \,.
\end{aligned}$$

It is also easy to see that, under (26) (and with G, H given in the theorem), we have

$$G\left(\frac{\partial\theta_i}{\partial\phi'}\bigg|_{\phi^{(i)}}\right)_{1\le i\le q} H^{-1} = \left(\frac{\partial\theta_i}{\partial\phi'}\bigg|_{\phi^{(i)}}\right)_{1\le i\le q} .$$

Thus, the third quantity in iii) of Theorem 1 $\to 0$ in probability, because $\hat{\phi}$ is consistent. Note that, in this case, C is, in fact, a fixed matrix (rather than a sequence of matrices). Also, by the proof of Theorem 3.1 in Jiang (1998), (17) holds. Note that, by Lemma 1 in the following, $0 < \liminf \lambda_{\min}(A) \le \limsup \lambda_{\max}(A) < \infty$ and condition ii) [here we only need that $\text{var}(\epsilon_1^2) > 0$ and $\text{var}(\alpha_{i1}^2) > 0$, $1 \le i \le s$] imply $0 < \liminf \lambda_{\min}(\Sigma) \le \limsup \lambda_{\max}(\Sigma) < \infty$. Thus, the first condition in iv) of Theorem 1 (involving D) is obvious, because $\partial^2\theta_i/\partial\phi^2 = 0$, if $i \notin \{i_1, \dots, i_a\}$.

The same results hold for $l_R(\theta, y)$ as well, because it is easy to show that the difference between $l_R(\theta, y)$ and $l(\theta, y)$ or their any up to third derivative is nonrandom and is bounded by p times a function of the variance components, and the latter is bounded if the variance components are bounded from above and away from zero, which is ensured by condition ii). □

Lemma 1. *Let $B_1, \dots, B_s$ be $n \times n$ symmetric matrices. Define*

$$A = (\text{tr}(B_i B_j))_{1 \le i,j \le s}\,, \quad B = \left(\sum_{k=1}^n B_{i,kk} B_{j,kk}\right)_{1 \le i,j \le s},$$

where $B_{i,kl}$ is the (k, l) element of B_i. Then $A \ge B$.

Proof. Note that $\forall x \in R^s$, we have $x'Ax = \|\sum_{i=1}^s x_i B_i\|_2^2$ and $x'Bx = \sum_{k=1}^n \left(\sum_{i=1}^s x_i B_{i,kk}\right)^2$. □

References

Basawa, I. V., and Rao, B. L. S. P. (1980), *Statistical Inference for Stochastic Processes*, Academic Press, London.

Bhatia, R. (1997), *Matrix Analysis*, Springer, New York.

Chen, C. F. (1985), Robustness aspects of score tests for generalized linear and partially linear regression models, *Technometrics* **27**, 277–283.

Chernoff, H., and Lehmann, E. L. (1954), The use of maximum-likelihood estimates in χ^2 tests for goodness of fit, *Ann. Math. Statist.* **25**, 579–586.

Dzhaparidze, K. (1986), *Parameter Estimation and Hypothesis Testing in Spectral Analysis of Stationary Time Series*, Springer, New York.

Foutz, R. V., and Srivastava, R. C. (1977), The performance of the likelihood ratio test when the model is incorrect, *Ann. Statist.* **5**, 1183–1194.

Gourieroux, C., and Monfort, A. (1995), *Statistics and Econometric Models*, Vol. 2, Cambridge Univ. Press.

Hampel, F. R., Ronchetti, E. M., Rousseeuw, P. J., and Stahe, W. A. (1986), *Robust Statistics: The Approach Based on Influence Functions*, Wiley, New York.

Hartley, H. O., and Rao, J. N. K. (1967), Maximum likelihood estimation for the mixed analysis of variance model, *Biometrika* **54**, 93–108.

Heritier, S., and Ronchetti, E. (1994), Robust bounded-influence tests in general parametric models, *J. Amer. Statist. Assoc.* **89**, 897–904.

Heyde, C. C. (1994), A quasi-likelihood approach to the REML estimating equations, *Statist. & Probab. Letters* **21**, 381–384.

Heyde, C. C. (1997), *Quasi-Likelihood and Its Application*, Springer, New York.

Huber, P. J. (1981), *Robust Statistics*, John Wiley & Sons, New York.

Jiang, J. (1996), REML estimation: Asymptotic behavior and related topics, *Ann. Statist.* **24**, 255–286.

Jiang, J. (1997), Wald consistency and the method of sieves in REML estimation, *Ann. Statist.* **25**, 1781–1803.

Jiang, J. (1998), Asymptotic properties of the empirical BLUP and BLUE in mixed linear models, *Statist. Sinica* **8**, 861–885.

Jiang, J. (2003), Empirical method of moments and its applications, *J. Statist. Plann. Inference* **115**, 69–84.

Jiang, J. (2005), Partially observed information and inference about non-Gaussian mixed linear models, *Ann. Statist.* **33**, 2695–2731.

Kent, J. T. (1982), Robustness properties of likelihood ratio tests, *Biometrika* **69**, 19–27.

Khuri, A. I., Mathew, T., and Sinha, B. K. (1998), *Statistical tests for mixed linear models*, Wiley, New York.

Kim, H. J., and Cai, L. (1993), Robustness of the likelihood ratio test for a change in simple linear regression, *J. Amer. Statist. Assoc.* **88**, 864–871.

Lehmann, E. L. (1999), *Elements of Large-Sample Theory*, Springer, New York.

McCullagh, P., and Nelder, J. A. (1989), *Generalized Linear Models*, 2nd Ed., Chapman & Hall.

Richardson, A. M., and Welsh, A. H. (1994), Asymptotic properties of restricted maximum likelihood (REML) estimates for hierarchical mixed linear models, *Austral. J. Statist.* **36**, 31–43.

Richardson, A. M., and Welsh, A. H. (1996), Covariate screening in mixed linear models, J. Multivariate Anal. **58**, 27–54.

Schrader, R. M., Hettmansperger, T. P. (1980), Robust analysis of variance based upon a likelihood ratio criterion, *Biometrika* **67**, 93–101.

Searle, S. R. (1971), *Linear Models*, John Wiley & Sons, New York.

Searle, S. R., Casella, G., and McCulloch, C. E. (1992), *Variance Components*, John Wiley & Sons, New York.

Silvapulle, M. J. (1992), Robust Wald-type tests of one-sided hypotheses in the linear model, *J. Amer. Statist. Assoc.* **87**, 156–161.

Weiss, L. (1975), The asymptotic distribution of the likelihood ratio in some nonstandard cases, *J. Amer. Statist. Assoc.* **70**, 204–208.

Welham, S. J., and Thompson, R. (1997), Likelihood ratio tests for fixed model terms using residual maximum likelihood, *J. Roy. Statist. Soc. B* **59**, 701–714.

Chapter 5

Density Estimation by Sampling from Stationary Continuous Time Parameter Associated Processes

George G. Roussas and Debasis Bhattacharya

University of California, Davis, and

Visva-Bharati University, Santiniketan, India

On the basis of periodic observations, taken from a continuous time-parameter positively associated stochastic process according to a certain sampling scheme, the usual kernel estimate of a probability density function is constructed, and several of its optimality properties are established. These properties include asymptotic unbiasedness, representation of the bias, asymptotic behavior of the variance and the correlation of the estimate evaluated at distinct points, and strong uniform consistency.

Contents

1. Introduction

The family $\mathbf{X} = \{X_t;\ t \in T \subseteq \mathbb{R}\}$ of (real-valued) random variables (r.v.'s) is said to be *positively associated* (PA) (or just *associated*, according to the original reference Esary, Proschan, and Walkup (1967)) if for any finite $I \subset \mathbb{R}$ and any real-valued coordinate-wise nondecreasing Borel functions g and h, it holds

$$\mathrm{Cov}\,[g(X_t;\ t \in I),\ h(X_t;\ t \in I)] \geq 0\,, \tag{1}$$

provided the r.v.'s $g(X_t;\ t \in I)$ and $h(X_t;\ t \in I)$ have finite second moments. (See also Definitions 1.1, 1.2 and 1.3, page 3, in Bulinski and

Shashkin (2007)). For various applications of PA r.v.'s the reader is referred to Esary, Proschan, and Walkup (1967) or, for example, Roussas (1999). Several results involving PA r.v.'s can be found, for example, in Roussas (1991; 1993; 2000), Cai and Roussas (1992; 1998), Ioannides and Roussas (1998), and Oliveira (2005).

In all that follows $\mathbf{X} = \{X_t;\ t \in [0, \infty)\}$ is a strictly stationary PA stochastic process of real-valued r.v.'s. Although the process may be observed in the continuum, for most practical applications it is a segment of the processes observed over a discrete set of time points which is used. Thus, in the sequel, we assume that we have at our disposal the r.v.'s X_{t_i}, $i = 0, 1, \ldots, n$ with $0 = t_0 < t_1 < \cdots < t_n$; the points t_i, $i = 0, 1, \ldots, n$ need not be equidistant. It is assumed that the probability distribution of X_0 is absolutely continuous with respect to Lebesgue measure, and let f be a specified version of the Radon–Nikodym derivative involved (probability density function, p.d.f.). The purpose of this paper is to estimate $f(x)$, $x \in \mathbb{R}$, by the kernel method and establish several optimality properties of the proposed estimate

$$\hat{f}_n(x) = \frac{1}{nh_n} \sum_{i=1}^{n} K\left(\frac{x - X_{t_i}}{h_n}\right), \tag{2}$$

where the kernel K and the bandwidth h_n satisfy certain regularity conditions to be stated below. Although the observation times may be arbitrary, it is assumed nevertheless that $t_i - t_{i-1} \geq \tau (> 0)$, the *sampling rate*, for $i = 1, \ldots, n$. Actually, this restriction is used only in the derivations of results in Sections 3 and 4, but not Section 5. For notational convenience, set

$$\hat{f}_n(x) = \frac{1}{n} \sum_{i=1}^{n} K_n\left(x - X_{t_i}\right), \quad K_n\left(x - X_{t_i}\right) = \frac{1}{h_n} K\left(\frac{x - X_{t_i}}{h_n}\right). \tag{3}$$

The objective of this paper is to show that the estimate defined by (3) shares many of the optimality properties with such an estimate established in the standard i.i.d. setting. Some of the results discussed here have been dealt with in Masry (1983) in a strong mixing framework, and under various sampling schemes. For the case of equidistant time-parameter observations, there is a large literature. Indicatively we mention Parzen (1962), and Roussas (1991). For non-stationary processes, several results have been obtained in Lenain, Harel and Puri (2011) and elsewhere.

This paper consists of five sections. In Section 2, the object of the present study is to establish asymptotic unbiasedness of the estimate in

(3), and an explicit representation of its bias. The main result in Section 3 is the investigation of the asymptotic behavior of the variance of the estimate, whereas in Section 4, it is shown that the covariance of the estimate, evaluated at distinct points, tends to zero as the sample size tends to infinity. Finally, in Section 5, it is shown that the estimate is strongly and uniformly consistent over intervals tending to the real line as $n \to \infty$.

Assumptions are listed as needed, and limits are always taken as $n \to \infty$. It is worth observing that a restriction imposed on the sampling scheme employed here is used only in the derivation of the main results in Sections 3 and 4. Alternative sampling schemes are under consideration, as well as the case of negatively associated processes.

2. Asymptotic unbiasedness and representation of the bias

Assumptions will be listed as needed. So, it is assumed here that the kernel K satisfies the following conditions.

(A.2.1) $\displaystyle\int_{-\infty}^{\infty} |K(x)|\,\mathrm{d}x < \infty.$

(A.2.2) $\displaystyle\int_{-\infty}^{\infty} K(x)\,\mathrm{d}x = 1.$

(A.2.3) $|K(x)| \leq C(< \infty)$ for all $x \in \mathbb{R}$, where here and in the sequel C is a generic positive constant which may vary from situation to situation,

(A.2.4) $\displaystyle\lim_{|x|\to\infty} |xK(x)| = 0.$

Then we have

Theorem 1. *Let $\boldsymbol{X} = \{X_t;\ t \in [0,\infty)\}$ be any strictly stationary process, let f be the p.d.f. of X_0, and let $0 = t_0 < t_1 < \cdots < t_n$ be any periodic sampling scheme. Then, under assumptions* (A.2.1)–(A.2.4) *and for $0 < h_n \to 0$, the estimate $\hat{f}_n(x)$ given in* (2) *is asymptotically unbiased at all continuity points x of f; i.e.,*

$$\mathcal{E}\hat{f}_n(x) \longrightarrow f(x), \quad x \in C(f). \tag{4}$$

Proof. Indeed, writing h instead of h_n, we have

$$\begin{aligned}
\mathcal{E}\hat{f}_n(x) &= \frac{1}{h}\mathcal{E}K\left(\frac{x-X_0}{h}\right) \qquad \text{(by stationarity)}\\
&= \int_{-\infty}^{\infty} \frac{1}{h}K\left(\frac{x-u}{h}\right) f(u)\,\mathrm{d}u\\
&= \int_{-\infty}^{\infty} K(v)f(x-hv)\,\mathrm{d}v \qquad \left(\text{by setting } \frac{x-u}{h}=v\right)\\
&\longrightarrow f(x)\int_{-\infty}^{\infty} K(v)\,\mathrm{d}v \qquad (5)
\end{aligned}$$

by Bochner's theorem (see Parzen (1962)) for $x \in C(f)$, and the last expression equals $f(x)$ by assumption (A.2.2). □

The next result is an expression of the bias of the estimate. For this purpose, the following additional assumptions are needed.

(A.2.5) $\int_{-\infty}^{\infty} |x^i K(x)|\,\mathrm{d}x < \infty$, $i = 1, \ldots, r+1$, for some positive integer r.

(A.2.6) The p.d.f. f has r-th order derivative in $\mathbb{R}$ which satisfies the Lipschitz condition

$$\left|f^{(r)}(x) - f^{(r)}(y)\right| \leq C|x-y|.$$

Then we have

Theorem 2. *Let the underlying stochastic process, the sampling scheme, and h_n be as in Theorem 1. Then, under assumptions* (A.2.1)–(A.2.6), *it holds,*

$$\text{bias}[\hat{f}_n(x)] = \mathcal{E}\hat{f}_n(x) - f(x) = \sum_{i=1}^{r}\left[\frac{(-h_n)^i}{i!}f^{(i)}(x)\right]a_i + o(h_n^r),$$

where $a_i = \int_{-\infty}^{\infty} x^i K(x)\,dx$, $i = 1, \ldots, r+1$.

Proof. Consider the following expansion of f around x according to Taylor's formula (see, e.g., (8.14.3), page 186 in Diedonné (1960)) to obtain

$$f(x+t) = f(x) + \sum_{i=1}^{r-1}\frac{t^i}{i!}f^{(i)}(x) + \frac{t^r}{(r-1)!}\int_0^1 (1-\lambda)^{r-1} f^{(r)}(x+\lambda t)\,\mathrm{d}\lambda.$$

Replacing t by $-h_n u$ and writing h instead of h_n, we have then

$$\begin{aligned}
&f(x-hu)\\
&= f(x) + \sum_{i=1}^{r-1} \frac{(-hu)^i}{i!} f^{(i)}(x) + \frac{(-hu)^r}{(r-1)!} \int_0^1 (1-\lambda)^{r-1} f^{(r)}(x-\lambda hu)\, \mathrm{d}\lambda\\
&= f(x) + \sum_{i=1}^{r-1} \frac{(-h)^i}{i!} f^{(i)}(x) u^i + \frac{(-h)^r}{(r-1)!} u^r \int_0^1 (1-\lambda)^{r-1} f^{(r)}(x-\lambda hu)\, \mathrm{d}\lambda.
\end{aligned}$$

Multiplying both sides of the last relation by $K(u)$, taking expectations, and using the fact that $\mathcal{E}\hat{f}_n(x) = \int_{-\infty}^{\infty} K(u) f(x-hu)\, \mathrm{d}u$ (as was seen in the proof of Theorem 1), we have

$$\begin{aligned}
\mathcal{E}\hat{f}_n(x) &= \int_{-\infty}^{\infty} K(u) f(x-hu)\, \mathrm{d}u\\
&= f(x) \int_{-\infty}^{\infty} K(u)\mathrm{d}u + \sum_{i=1}^{r-1} \frac{(-h)^i}{i!} f^{(i)}(x) \int_{-\infty}^{\infty} u^i K(u)\, \mathrm{d}u\\
&\quad + \frac{(-h)^r}{(r-1)!} \int_{-\infty}^{\infty} u^r K(u) \left[\int_0^1 (1-\lambda)^{r-1} f^r(x-\lambda hu)\, \mathrm{d}\lambda \right] \mathrm{d}u\,.
\end{aligned} \tag{6}$$

But $f^{(r)}(x-\lambda hu) \to f^{(r)}(x)$ (by continuity of $f^{(r)}$), and

$$|f^{(r)}(x-\lambda hu) - f^{(r)}(x)| \le C|-\lambda hu| = C\lambda h|u| \le C|u|$$

(by assumption (A.2.6), and the facts that $0 \le \lambda \le 1$ and h is bounded since $0 < h \to 0$). Hence $|f^{(r)}(x-\lambda hu)| \le |f^{(r)}(x)| + C|u|$ independent of n and integrable over $[0,1]$. Also,

$$(1-\lambda)^{r-1} f^{(r)}(x-\lambda hu) \to (1-\lambda)^{r-1} f^{(r)}(x) \quad \text{and}$$

$$|(1-\lambda)^{r-1} f^{(r)}(x-\lambda hu)| \le |f^{(r)}(x-\lambda hu)| \le |f^{(r)}(x)| + C|u|.$$

Therefore, by the Dominated Convergence Theorem (DCT),

$$\int_0^1 (1-\lambda)^{r-1} f^{(r)}(x-\lambda hu)\, \mathrm{d}\lambda \longrightarrow \int_0^1 (1-\lambda)^{r-1} f^{(r)}(x)\, \mathrm{d}\lambda = \frac{1}{r} f^{(r)}(x)\,.$$

Then

$$u^r K(u) \int_0^1 (1-\lambda)^{r-1} f^{(r)}(x-\lambda hu)\, \mathrm{d}\lambda \longrightarrow \frac{1}{r} f^{(r)}(x) u^r K(u)\,,$$

whereas

$$\left|u^r K(u) \int_0^1 (1-\lambda)^{r-1} f^{(r)}(x - \lambda h u)\, \mathrm{d}\lambda\right|$$
$$\leq \left|f^{(r)}(x)\right| |u^r K(u)| + C \left|u^{r+1} K(u)\right|$$

independent of n and integrable (by assumption (A.2.5)). Therefore, by the DCT again,

$$\int_{-\infty}^{\infty} u^r K(u) \left[\int_0^1 ((1-\lambda)^{r-1} f^{(r)}(x - \lambda h u)\, \mathrm{d}\lambda\right] \mathrm{d}u$$
$$\longrightarrow \frac{1}{r} f^{(r)}(x) \int_{-\infty}^{\infty} u^r K(u)\, \mathrm{d}u\,,$$

so that

$$\int_{-\infty}^{\infty} u^r K(u) \left[\int_0^1 ((1-\lambda)^{r-1} f^{(r)}(x - \lambda h u)\, \mathrm{d}\lambda\right] \mathrm{d}u$$
$$= \frac{1}{r} f^{(r)}(x) \int_{-\infty}^{\infty} u^r K(u) \mathrm{d}u + o(1)$$

(where $o(1) \to 0$ as $n \to \infty$). It follows that

$$\frac{(-h)^r}{(r-1)!} \int_{-\infty}^{\infty} u^r K(u) \left[\int_0^1 ((1-\lambda)^{r-1} f^{(r)}(x - \lambda h u)\, \mathrm{d}\lambda\right] \mathrm{d}u$$
$$= \frac{(-h)^r}{(r-1)!} f^{(r)}(x) \int_{-\infty}^{\infty} u^r K(u)\, \mathrm{d}u + \frac{(-h)^r}{(r-1)!} o(1)\,,$$

and $\frac{(-h)^r}{(r-1)!} o(1) = o(h^r)$. This, along with relation (6), complete the proof. □

3. Asymptotic behavior of the variance of the estimate and consistency in quadratic mean

In this section, we establish the asymptotic behavior of the variance of the estimate, and also show that the estimate is consistent in quadratic mean. For this purpose, additional assumptions will be needed to be stated below. The main results will follow after a series of lemmas have been established. At this point, it would be appropriate to note that

Remark 1. Under assumptions (A.2.1) and (A.2.3), it holds

$$\int_{-\infty}^{\infty} |K(x)|^k < \infty, \qquad \text{for any integer } k \geq 1\,.$$

Lemma 1. *Let the underlying process, the sampling scheme, and h_n be as in Theorem 1. Then, under assumptions* (A.2.1)–(A.2.4), *it holds for* $x \in C(f)$,

$$h_n\, Var\,[K_n(x - X_0)] = \frac{1}{h_n}\, Var\left[K\left(\frac{x - X_0}{h_n}\right)\right] \longrightarrow f(x) \int_{-\infty}^{\infty} K^2(u)\, du\,.$$

Proof. Indeed, with $h_n = h$,

$$\begin{aligned}
&\mathrm{Var}\left[K\left(\frac{x - X_0}{h}\right)\right]\\
&= \mathcal{E}K^2\left(\frac{x - X_0}{h}\right) - \left[\mathcal{E}\left(\frac{x - X_0}{h}\right)\right]^2\\
&= \int_{-\infty}^{\infty} K^2\left(\frac{x-u}{h}\right) f(u)\,\mathrm{d}u - \left[\int_{-\infty}^{\infty} K\left(\frac{x-u}{h}\right) f(u)\,\mathrm{d}u\right]^2\\
&= h\int_{-\infty}^{\infty} K^2(v) f(x - hv)\mathrm{d}v - h^2\left[\int_{-\infty}^{\infty} K(v) f(x - hv)\,\mathrm{d}v\right]^2,
\end{aligned}$$

so that

$$\begin{aligned}
&\frac{1}{h}\mathrm{Var}\left[K\left(\frac{x - X_0}{h}\right)\right]\\
&= \int_{-\infty}^{\infty} K^2(v) f(x - hv)\,\mathrm{d}v - h\left[\int_{-\infty}^{\infty} K(v) f(x - hv)\,\mathrm{d}v\right]^2\\
&\longrightarrow f(x)\int_{-\infty}^{\infty} K^2(v)\,\mathrm{d}v,\ \text{since} \int_{-\infty}^{\infty} K(v) f(x - hv)\,\mathrm{d}v \longrightarrow f(x)\,,
\end{aligned}$$

by (5). □

Up to this point, it has been assumed that the sampling scheme is arbitrary subject to $0 = t_0 < t_1 < \cdots < t_n$. From this point on, it will be assumed that

$$t_i - t_{i-1} \geq \tau(> 0),\ \text{the } \textit{sampling rate}, \text{ for } i = 1, 2, \ldots, n.$$

It will also be assumed that:

(A.3.1) For any $0 < t_i < t_j$, the r.v.'s X_{t_i} and X_{t_j} have a joint p.d.f. $f_{X_{t_i}, X_{t_j}}$ such that

$$\left| f_{X_{t_i}, X_{t_j}}(u, v) - f_{X_{t_i}}(u) f_{X_{t_j}}(v)\right| \leq C,\ \text{for all } u, v \in \mathbb{R}.$$

Lemma 2. *Let the underlying process and h_n be as in Theorem 1, let τ be the sampling rate of the sampling scheme, and suppose that assumptions* (A.2.1), (A.2.3), *and* (A.3.1) *hold. Furthermore, suppose there is a choice of a sequence $\{c_n\}$ with $0 < c_n < n$ such that $c_n \to \infty$ and $c_n h_n \to 0$. Then*

$$\frac{h_n}{n} \sum_{i=1}^{c_n} \sum_{j=1}^{n-i} \mathrm{Cov}\left[K_n(x - X_{t_j}), K_n(x - X_{t_{j+i}})\right] \longrightarrow 0\,.$$

Proof. Indeed, with K_n as in (3) and under assumption (A.2.3),

$$\begin{aligned}
&\left| \sum_{i=1}^{c_n} \sum_{j=1}^{n-i} \mathrm{Cov}[K_n(x - X_{t_j}), K_n(x - X_{t_{j+i}})] \right| \\
&= \Bigg| \sum_{i=1}^{c_n} \sum_{j=1}^{n-i} \{\mathcal{E}\left[K_n(x - X_{t_j}) K_n(x - X_{t_{j+i}})\right] \\
&\qquad\qquad - \left[\mathcal{E} K_n(x - X_{t_j})\right]\left[\mathcal{E} K_n(x - X_{t_{j+i}})\right]\} \Bigg| \\
&= \Bigg| \sum_{i=1}^{c_n} \sum_{j=1}^{n-i} \int_{-\infty}^{\infty} \int_{-\infty}^{\infty} K_n(x-u) K_n(x-v) \Big[f_{X_{t_j} X_{t_{j+i}}}(u, v) \\
&\qquad\qquad - f_{X_{t_j}}(u) f_{X_{t_{j+i}}}(v) \Big] \, \mathrm{d}u \, \mathrm{d}v \Bigg| \\
&= \Bigg| \sum_{i=1}^{c_n} \sum_{j=1}^{n-i} \int_{-\infty}^{\infty} \int_{-\infty}^{\infty} K(y) K(z) \Big[f_{X_{t_j} X_{t_{j+i}}}(x - hy, x - hz) \\
&\qquad\qquad - f_{X_{t_j}}(x - hy) f_{X_{t_{j+i}}}(x - hz) \Big] \, \mathrm{d}y \, \mathrm{d}z \Bigg|
\end{aligned}$$

(by replacing K_n by K (see relation (3)), and by using suitable transformations)

$$\begin{aligned}
&\le C \left| \sum_{i=1}^{c_n} \sum_{j=1}^{n-i} 1 \right| \quad \text{(by assumptions (A.3.1) and (A.2.1))} \\
&= C \sum_{i=1}^{c_n} (n - i) = Cn \sum_{i=1}^{c_n} \left(1 - \frac{i}{n}\right) \le Cnc_n\,,
\end{aligned}$$

so that

$$\left| \frac{h}{n} \sum_{i=1}^{c_n} \sum_{j=1}^{n-i} \mathrm{Cov}\left[K_n(x - X_{t_j}), K_n(x - X_{t_{j+i}})\right] \right| \leq \frac{h}{n} Cnc_n = C(c_n h) \longrightarrow 0$$

by the choice of c_n. □

In the proof of the following lemma, the assumptions stated below will also be used.

(A.3.2) The derivative $\frac{\mathrm{d}}{\mathrm{d}x}K(x) = K'(x)$ exists for all $x \in \mathbb{R}$ and is bounded; i.e., $|K'(x)| \leq C$, $x \in \mathbb{R}$.

(A.3.3) The underlying strictly stationary process is PA, $\mathcal{E}X_0^2 < \infty$, and the $\mathrm{Cov}(X_0, X_t)$ is decreasing as $t \uparrow \infty$.

It is restated here that, in the sequel, all limits are taken as $n \to \infty$.

Lemma 3. *Let τ be the sampling rate of the sampling scheme, and suppose that assumptions* (A.2.1), (A.2.3), (A.3.2), *and* (A.3.3) *are satisfied. Furthermore, we suppose that the sequence $\{c_n\}$ employed in Lemma 2 also satisfies the following condition (jointly with h_n and the $Cov(X_0, X_{i\tau})$)*

$$\frac{1}{h_n^3} \sum_{i=c_n+1}^{\infty} Cov(X_0, X_{i\tau}) \longrightarrow 0. \tag{7}$$

Then, it holds,

$$\frac{1}{h_n^3} \sum_{i=c_n+1}^{\infty} \sum_{j=1}^{n-i} Cov\left[K_n(x - X_{t_j}), K_n(x - X_{t_{j+i}})\right] \longrightarrow 0,$$

where K_n is given in relation (3).

Proof. Under assumptions (A.3.1) and (A.3.3), Bulinski's lemma (Lemma 1 in Bulinski (1996); see also Theorem 5.3, page 89, in Bulinski and Shashkin (2007)) applies and yields

$$\begin{aligned}
&\left|\mathrm{Cov}\left[K_n(x - X_{t_j}), K_n(x - X_{t_{j+i}})\right]\right| \\
&= \frac{1}{h^2}\left|\mathrm{Cov}\left[K\left(\frac{x - X_{t_j}}{h}\right), K\left(\frac{x - X_{t_{j+i}}}{h}\right)\right]\right| \\
&\leq \frac{C}{h^4}\left[\mathrm{Cov}(X_{t_j}, X_{t_{j+i}})\right] \\
&= \frac{C}{h^4}\left[\mathrm{Cov}(X_0, X_{t_{j+i}-t_j})\right] \quad \text{(by stationarity)}.
\end{aligned} \tag{8}$$

But

$$t_{j+i} - t_j = (t_{j+i} - t_{j+i-1}) + (t_{j+i-1} - t_{j+i-2}) + \cdots + (t_{j+1} - t_j)$$

(i summands), and $t_{j+k} - t_{j+k-1} \geq \tau$, $k = 1, \ldots, i$, so that $t_{j+i} - t_j \geq i\tau$. Then assumption (A.3.3) implies that, for all i and j as above,

$$\mathrm{Cov}(X_0, X_{t_{j+i}-t_j}) \leq \mathrm{Cov}(X_0, X_{i\tau}) .$$

Therefore, as in the proof of Lemma 2,

$$\begin{aligned}\sum_{i=c_n+1}^{n} \sum_{j=1}^{n-i} \mathrm{Cov}(X_0, X_{t_{j+i}-t_j}) &\leq \sum_{i=c_n+1}^{n} (n-i)[\mathrm{Cov}(X_0, X_{i\tau})] \\ &= n \sum_{i=c_n+1}^{n} \left(1 - \frac{i}{n}\right) [\mathrm{Cov}(X_0, X_{i\tau})] \\ &\leq n \sum_{i=c_{n+1}}^{n} \mathrm{Cov}(X_0, X_{i\tau}), \end{aligned} \tag{9}$$

and then, by (8) and (9),

$$\begin{aligned}&\left| \frac{h}{n} \sum_{i=c_n+1}^{n} \sum_{j=1}^{n-i} \mathrm{Cov}[K_n(x - X_{t_j}), K_n(x - X_{t_{j+i}})] \right| \\ &\leq \frac{h}{n}\frac{C}{h^4} \sum_{i=c_n+1}^{n} \sum_{j=1}^{n-i} \mathrm{Cov}(X_0, X_{t_{j+i}-t_j}) \\ &\leq \frac{Cn}{nh^3} \sum_{i=c_n+1}^{n} \mathrm{Cov}(X_0, X_{i\tau}) \\ &\leq C\frac{1}{h^3} \sum_{i=c_n+1}^{n} \mathrm{Cov}(X_0, X_{i\tau}) \longrightarrow 0 \text{ (by (7))}. \end{aligned}$$

□

Now Lemmas 1–3 pave the way for the following result.

Theorem 3. *Let τ be the sampling rate of the sampling scheme, let $0 < h_n \to 0$, and suppose that assumptions (A.2.1)–(A.2.4) and (A.3.1)–(A.3.3) are satisfied. Furthermore, we assume there is a sequence $\{c_n\}$ such that $0 < c_n \to \infty$, $c_n h_n \to 0$, and relation (7) is satisfied. Then,*

$$(nh_n)\, Var[\hat{f}_n(x)] \longrightarrow f(x) \int_{-\infty}^{\infty} K^2(u)\, du, \quad x \in C(f).$$

Proof. Indeed, with K_n as defined in (3), we have,

$$
\begin{aligned}
&\mathrm{Var}[\hat{f}_n(x)] \\
&= \frac{1}{n^2}\left\{\sum_{i=1}^{n} \mathrm{Var}[K_n(x - X_{t_i})] \right. \\
&\qquad \left. + \frac{2}{n^2} \sum_{1 \le i < j \le n} \mathrm{Cov}[K_n(x - X_{t_i}), K_n(x - X_{t_j})] \right\} \\
&= \frac{1}{n}\mathrm{Var}[K_n(x - X_0)] + \frac{2}{n^2} \sum_{1 \le i < j \le n} \mathrm{Cov}[K_n(x - X_{t_i}), K_n(x - X_{t_j})]\,,
\end{aligned}
$$

so that,

$$
\begin{aligned}
&(nh)\mathrm{Var}[\hat{f}_n(x)] \\
&= h\mathrm{Var}[K_n(x - X_0)] + 2\frac{h}{n} \sum_{1 \le i < j \le n} \mathrm{Cov}[K_n(x - X_{t_i}), K_n(x - X_{t_j})]\,.
\end{aligned}
\tag{10}
$$

Since $h\mathrm{Var}[K_n(x - X_0) \to f(x) \int_{-\infty}^{\infty} K^2(u)\ du$, by Lemma 1, it suffices to show that the second term on the right-hand side in (10) tends to 0. However,

$$
\begin{aligned}
&\frac{h}{n} \sum_{1 \le i < j \le n} \mathrm{Cov}[K_n(x - X_{t_i}), K_n(x - X_{t_j})] \\
&= \frac{h}{n} \sum_{i=1}^{n-1} \sum_{j=i+1}^{n} \mathrm{Cov}[K_n(x - X_{t_i}), K_n(x - X_{t_j})] \\
&= \frac{h}{n} \sum_{i=1}^{n-1} \sum_{j=1}^{n-i} \mathrm{Cov}[K_n(x - X_{t_j}), K_n(x - X_{t_{j+i}})] \\
&\text{(by rearranging the order of the double summation)} \\
&= \frac{h}{n} \sum_{i=1}^{c_n} \sum_{j=1}^{n-i} \mathrm{Cov}[K_n(x - X_{t_j}), K_n(x - X_{t_{j+i}})] \\
&\quad + \frac{h}{n} \sum_{i=c_n+1}^{n-1} \sum_{j=1}^{n-i} \mathrm{Cov}[K_n(x - X_{t_j}), K_n(x - X_{t_{j+i}})]\,,
\end{aligned}
\tag{11}
$$

and by Lemmas 2 and 3, the expressions on the right-hand side in relation (11) tend to 0. The proof is then completed. □

As a consequence of Theorems 1 and 3, we have consistency of the estimate $\hat{f}_n(x)$ in quadratic mean, as well as in the sense of probability. That is,

Theorem 4. *Under the conditions of Theorem 3, and for $x \in C(f)$,*

$$\mathcal{E}\left[\hat{f}_n(x) - f(x)\right]^2 \longrightarrow 0, \text{ and } \hat{f}_n(x) \xrightarrow{P} f(x)\,.$$

Proof. It is immediate, since

$$\mathcal{E}\left[\hat{f}_n(x) - f(x)\right]^2 = \mathrm{Var}[\hat{f}_n(x)] + \left[\mathcal{E}\hat{f}_n(x) - f(x)\right]^2 \longrightarrow 0\,,$$

on account of Theorems 3 and 1. Also,

$$P\left[\left|\hat{f}_n(x) - f(x)\right| \geq \varepsilon\right] \leq \frac{1}{\varepsilon^2}\mathcal{E}\left[\hat{f}_n(x) - f(x)\right]^2 \longrightarrow 0$$

for every $\varepsilon > 0$. □

4. Asymptotically uncorrelated

In this section, it is shown that, for $x \neq y$, the estimates $\hat{f}_n(x)$ and $\hat{f}_n(y)$ are asymptotically uncorrelated. The proof of this result, which is stated here as Theorem 5, follows as a consequence of a series of four lemmas.

Theorem 5. *Let τ be the sampling rate of the sampling scheme, let $0 < h_n \to 0$, and suppose that assumptions* (A.2.1)–(A.2.4) *and* (A.3.1)–(A.3.3) *are satisfied. Furthermore, suppose there exists a sequence $\{c_n\}$ with $0 < c_n \to \infty$ such that $c_n h_n \to 0$ and relation* (7) *is satisfied. Then, for $x, y \in \mathbb{R}$ with $x \neq y$ and $x, y \in C(f)$, we have*

$$(nh_n)\,Cov\left[\hat{f}_n(x),\ \hat{f}_n(y)\right] \longrightarrow 0\,.$$

Lemma 4. *Let $x, y \in \mathbb{R}$ with $x \neq y$ and $x \in C(f)$, and suppose that assumptions* (A.2.1)–(A.2.4) *are satisfied. Then,*

$$\int_{-\infty}^{\infty} K(u)K\left(\frac{y-x}{h_n} + u\right) f(x - h_n u)\, du \longrightarrow 0\,.$$

Proof. Indeed, writing as usual h instead of h_n, we have,

$$0 \leq \left|K(u)K\left(\frac{y-x}{h} + u\right)\right| f(x-hu) \leq C|K(u)|f(x-hu) \qquad \text{(by (A.2.3))},$$

with $K(u)f(x - hu) \rightarrow K(u)f(x)$ (by continuity of f at x), and $\int_{-\infty}^{\infty} K(u)f(x - hu)\,\mathrm{d}u \rightarrow f(x)$. Furthermore,

$$K(u)K\left(\frac{y-x}{h}+u\right)f(x-hu) \rightarrow 0$$

(by the fact that $K(x) \rightarrow 0$ as $|x| \rightarrow \infty$, which is implied by (A.2.4)). Then using Pratt's lemma (see Pratt (1960)) (or a generalization of the DCT (see Exercise 4, Chapter 5, in Roussas (2005)), we obtain,

$$\int_{-\infty}^{\infty} K(u)K\left(\frac{y-x}{h}+u\right)f(x-hu)\,\mathrm{d}u \longrightarrow 0\,,$$

as was to be seen. □

For the formulation of the rest of the lemmas, observe that

$$\begin{aligned}
\mathrm{Cov}\left[\hat{f}_n(x),\, \hat{f}_n(y)\right] &= \frac{1}{n^2}\mathrm{Cov}\left[\sum_{i=1}^{n} K_n\left(x - X_{t_i}\right),\, \sum_{j=1}^{n} K_n\left(y - X_{t_j}\right)\right] \\
&= \frac{1}{n^2}\sum_{i=1}^{n}\sum_{j=1}^{n}\mathrm{Cov}\left[K_n\left(x - X_{t_i}\right),\, K_n\left(y - X_{t_j}\right)\right] \\
&= \frac{1}{n^2}\sum_{i=1}^{n}\mathrm{Cov}\left[K_n\left(x - X_{t_i}\right),\, K_n\left(y - X_{t_j}\right)\right] \\
&\quad + \frac{1}{n^2}\sum_{i\neq j}\mathrm{Cov}\left[K_n\left(x - X_{t_i}\right),\, K_n\left(y - X_{t_j}\right)\right] \\
&= \frac{1}{n}\mathrm{Cov}\left[K_n\left(x - X_0\right),\, K_n\left(y - X_0\right)\right] \\
&\quad + \frac{1}{n^2}\sum_{i\neq j}\mathrm{Cov}\left[K_n\left(x - X_{t_i}\right),\, K_n\left(y - X_{t_j}\right)\right],
\end{aligned} \tag{12}$$

whereas

$$
\begin{aligned}
&\sum_{i\neq j} \mathrm{Cov}\left[K_n\left(x-X_{t_i}\right),\ K_n\left(y-X_{t_j}\right)\right] \\
&= \sum_{1\leq i<j\leq n} \mathrm{Cov}\left[K_n\left(x-X_{t_i}\right),\ K_n\left(y-X_{t_j}\right)\right] \\
&\quad + \sum_{i=1}^{n-1}\sum_{j=1}^{n-i} \mathrm{Cov}\left[K_n\left(x-X_{t_{j+i}}\right),\ K_n\left(y-X_{t_i}\right)\right] \\
&= \sum_{i=1}^{n-1}\sum_{j=i+1}^{n} \mathrm{Cov}\left[K_n\left(x-X_{t_i}\right),\ K_n\left(y-X_{t_j}\right)\right] \\
&\quad + \sum_{i=1}^{n-1}\sum_{j=1}^{n-i} \mathrm{Cov}\left[K_n\left(x-X_{t_{j+i}}\right),\ K_n\left(y-X_{t_i}\right)\right] \\
&= \sum_{i=1}^{n-1}\sum_{j=1}^{n-i} \mathrm{Cov}\left[K_n\left(x-X_{t_j}\right),\ K_n\left(y-X_{t_{j+i}}\right)\right] \\
&\quad + \sum_{i=1}^{n-1}\sum_{j=1}^{n-i} \mathrm{Cov}\left[K_n\left(x-X_{t_{j+i}}\right),\ K_n\left(y-X_{t_i}\right)\right] \qquad (13)
\end{aligned}
$$

(by rearranging the order of the summation in the first double sum on the right-hand side above). Then, by (12) and (13),

$$(nh)\mathrm{Cov}\left[\hat{f}_n(x),\ \hat{f}_n(y)\right] = hI_{n_1} + \frac{h}{n}I_{n_2} + \frac{h}{n}I_{n_3}\ , \qquad (14)$$

where

$$I_{n_1} = \mathrm{Cov}\left[K_n\left(x-X_0\right),\ K_n\left(y-X_0\right)\right]\ , \qquad (15)$$

$$I_{n_2} = \sum_{i=1}^{n-1}\sum_{j=1}^{n-i} \mathrm{Cov}\left[K_n\left(x-X_{t_j}\right),\ K_n\left(y-X_{t_{j+i}}\right)\right]\ , \qquad (16)$$

$$I_{n_3} = \sum_{i=1}^{n-1}\sum_{j=1}^{n-i} \mathrm{Cov}\left[K_n\left(x-X_{t_{j+i}}\right),\ K_n\left(y-X_{t_i}\right)\right]\ . \qquad (17)$$

Now, we proceed with

Lemma 5. *Let $x, y \in \mathbb{R}$ with $x \neq y$ and $x, y \in C(f)$, and suppose that assumptions* (A.2.1)–(A.2.4) *are satisfied. Then,*

$$h_n I_{n_1} \longrightarrow 0, \qquad \text{where } I_{n_1} \text{ is given in } (15)\ .$$

Proof. Indeed, writing h instead of h_n, we have

$$\begin{aligned}
& h\mathcal{E}\left[K_n(x-X_0)\,K_n(y-X_0)\right] \\
&= \frac{1}{h}\mathcal{E}\left[K\left(\frac{x-X_0}{h}\right)K\left(\frac{y-X_0}{h}\right)\right] \\
&= \frac{1}{h}\int_{-\infty}^{\infty} K\left(\frac{x-u}{h}\right)K\left(\frac{y-u}{h}\right)f(u)\,\mathrm{d}u \\
&= \int_{-\infty}^{\infty} K(v)K\left(\frac{y-x}{h}+v\right)f(x-hv)\,\mathrm{d}v \longrightarrow 0 \quad \text{(by Lemma 4)},
\end{aligned}$$

and by relations (3) and (5),

$$\begin{aligned}
& h\left[\mathcal{E}K_n(x-X_0)\right]\left[\mathcal{E}K_n(y-X_0)\right] \\
&= \frac{1}{h}\left[\mathcal{E}K\left(\frac{x-X_0}{h}\right)\right]\left[K\left(\frac{y-X_0}{h}\right)\right] \\
&= \frac{1}{h}\left[\int_{-\infty}^{\infty} K\left(\frac{x-u}{h}\right)f(u)\,\mathrm{d}u\right]\left[\int_{-\infty}^{\infty} K\left(\frac{y-v}{h}\right)f(v)\,\mathrm{d}v\right] \\
&= h\left[\int_{-\infty}^{\infty} K(s)f(x-hs)\,\mathrm{d}s\right]\left[\int_{-\infty}^{\infty} K(t)f(y-ht)\,\mathrm{d}t\right] \\
&\longrightarrow 0\times f(x)\times f(y) = 0\,.
\end{aligned}$$

This completes the proof of the lemma. □

Lemma 6. *Let τ be the sampling rate of the sampling scheme, let $0 < h_n \to 0$, and suppose that assumptions* (A.2.1) *and* (A.3.1)–(A.3.3) *are satisfied. Furthermore, suppose there exists a sequence $\{c_n\}$ with $0 < c_n \to \infty$ such that $c_n h_n \to 0$ and relation* (7) *is satisfied. Then, for $x, y \in \mathbb{R}$,*

$$\frac{h_n}{n} I_{n_2} \longrightarrow 0\,,$$

where I_{n_2} is given in (16).

Proof. Split the summation on the right-hand side of (16), as was done in Lemmas 5 and 6, to obtain

$$\begin{aligned}
I_{n_2} &= \sum_{i=1}^{c_n}\sum_{j=1}^{n-i} \operatorname{Cov}\left[K_n\left(x-X_{t_j}\right),\, K_n\left(y-X_{t_{j+i}}\right)\right] \\
&\quad + \sum_{i=c_n+1}^{n-1}\sum_{j=1}^{n-i} \operatorname{Cov}\left[K_n\left(x-X_{t_j}\right),\, K_n\left(y-X_{t_{j+i}}\right)\right]\,, \qquad (18)
\end{aligned}$$

and observe that

$$
\begin{aligned}
&\left|\operatorname{Cov}\left[K_n\left(x-X_{t_j}\right),\ K_n\left(y-X_{t_{j+i}}\right)\right]\right| \\
&= \left|\mathcal{E}\left[K_n\left(x-X_{t_j}\right)K_n\left(y-X_{t_{j+i}}\right)\right]-\left[\mathcal{E}K_n\left(x-X_{t_j}\right)\right]\left[\mathcal{E}K_n\left(y-X_{t_{j+i}}\right)\right]\right| \\
&= \left|\int_{-\infty}^{\infty}\int_{-\infty}^{\infty}K_n(x-u)K_n(y-v)\left[f_{X_{t_j},X_{t_{j+i}}}(u,v)\right.\right. \\
&\qquad\qquad\qquad\qquad \left.\left.-f_{X_{t_j}}(u)f_{X_{t_{j+i}}}(v)\right]\,\mathrm{d}u\mathrm{d}v\right| \\
&= \left|\int_{-\infty}^{\infty}\int_{-\infty}^{\infty}K(s)K(t)\left[f_{X_{t_j},X_{t_{j+i}}}(x-hs,y-ht)\right.\right. \\
&\qquad\qquad\qquad\qquad \left.\left.-f_{X_{t_j}}(x-hs)f_{X_{t_{j+i}}}(y-ht)\right]\,\mathrm{d}s\,\mathrm{d}t\right| \\
&\quad \text{(by replacing } K_n \text{ by } K \text{ (see relation (3)) and using} \\
&\qquad\qquad\qquad\qquad \text{suitable transformations)} \\
&\le C \qquad \text{(by assumptions (A.3.1) and (A.2.1)).} \qquad (19)
\end{aligned}
$$

Therefore,

$$
\begin{aligned}
&\left|\sum_{i=1}^{c_n}\sum_{j=1}^{n-i}\operatorname{Cov}\left[K_n\left(x-X_{t_j}\right),\ K_n\left(y-X_{t_{j+i}}\right)\right]\right| \\
&\le C\sum_{i=1}^{c_n}\sum_{j=1}^{n-i}1 = C\sum_{i=1}^{c_n}(n-i)\le Cnc_n\,. \qquad (20)
\end{aligned}
$$

Next,

$$
\begin{aligned}
&\left|\operatorname{Cov}\left[K_n\left(x-X_{t_j}\right),\ K_n\left(y-X_{t_{j+i}}\right)\right]\right| \\
&= \frac{1}{h^2}\left|\operatorname{Cov}\left[K\left(\frac{x-X_{t_j}}{h}\right),\ K\left(\frac{y-X_{t_{j+i}}}{h}\right)\right]\right| \\
&\le \frac{C}{h^4}\operatorname{Cov}\left(X_{t_j},\ X_{t_{j+i}}\right) \qquad \text{(by Bulinski's lemma for PA r.v.'s)} \\
&= \frac{C}{h^4}\operatorname{Cov}\left(X_0,\ X_{t_{j+i}-t_j}\right) \qquad \text{(by stationarity)} \\
&\le \frac{C}{h^4}\operatorname{Cov}\left(X_0,\ X_{i\tau}\right) \qquad (21)
\end{aligned}
$$

(by relation (8) and the discussion immediately following it). Hence

$$\begin{aligned}
&\left|\sum_{i=c_n+1}^{n-1}\sum_{j=1}^{n-i}\mathrm{Cov}\left[K_n\left(x-X_{t_j}\right),\ K_n\left(y-X_{t_{j+i}}\right)\right]\right| \\
&\leq \frac{Cn}{h^4}\sum_{i=c_n+1}^{n-1}\mathrm{Cov}\left(X_0,\ X_{i\tau}\right) \\
&\leq \frac{Cn}{h^4}\sum_{i=c_n+1}^{\infty}\mathrm{Cov}\left(X_0,\ X_{i\tau}\right)\ . \qquad (22)
\end{aligned}$$

By means of (18), (20), and (21), we have then

$$\begin{aligned}
\frac{h}{n}I_{n_2} &\leq \frac{h}{n}\cdot Cnc_n + \frac{h}{n}\cdot\frac{Cn}{h^4}\sum_{i=c_n+1}^{\infty}\mathrm{Cov}\left(X_0,\ X_{i\tau}\right) \\
&= C(hc_n) + C\cdot\frac{1}{h^3}\sum_{i=c_n+1}^{\infty}\mathrm{Cov}\left(X_0,\ X_{i\tau}\right)\longrightarrow 0, \qquad (23)
\end{aligned}$$

by the assumed existence of $\{c_n\}$ with $0 < c_n \to \infty$ for which $c_n h \to 0$ and relation (7) is satisfied. $\square$

Lemma 7. *Under the same assumptions as those in Lemma 6, and for $x, y \in \mathbb{R}$, we have,*

$$\frac{h_n}{n}I_{n_3}\longrightarrow 0\ ,$$

where I_{n_3} is given in (17).

Proof. Split I_{n_3} in two parts as in (18) to get

$$\begin{aligned}
I_{n_3} &= \sum_{i=1}^{c_n}\sum_{j=1}^{n-i}\mathrm{Cov}\left[K_n\left(x-X_{t_{j+i}}\right),\ K_n\left(y-X_{t_i}\right)\right] \\
&\quad+\sum_{i=c_n+1}^{n-1}\sum_{j=1}^{n-i}\mathrm{Cov}\left[K_n\left(x-X_{t_{j+i}}\right),\ K_n\left(y-X_{t_i}\right)\right]\ , \qquad (24)
\end{aligned}$$

and work as in (19) to obtain

$$\begin{aligned}
&\left|\mathrm{Cov}\left[K_n\left(x-X_{t_{j+i}}\right),\ K_n\left(y-X_{t_i}\right)\right]\right| \\
&=\left|\int_{-\infty}^{\infty}\int_{-\infty}^{\infty}K(s)K(t)\left[f_{X_{t_{j+i}},X_{t_i}}(x-hs,y-ht)\right.\right. \\
&\qquad\qquad\left.\left.-f_{X_{t_{j+i}}}(x-hs)f_{X_{t_i}}(y-ht)\right]\,\mathrm{d}s\,\mathrm{d}t\right| \\
&\leq C\quad\text{(again, by assumptions (A.3.1) and (A.2.1)).} \qquad (25)
\end{aligned}$$

Therefore, as in (20),

$$\left|\sum_{i=1}^{c_n}\sum_{j=1}^{n-i}\operatorname{Cov}\left[K_n\left(x-X_{t_{j+i}}\right),\, K_n\left(y-X_{t_i}\right)\right]\right| \leq Cnc_n\,. \tag{26}$$

Next, working as in (21), we get

$$\left|\operatorname{Cov}\left[K_n\left(x-X_{t_{j+i}}\right),\, K_n\left(y-X_{t_i}\right)\right]\right| \leq \frac{C}{h^4}\operatorname{Cov}\left(X_0,\, X_{i\tau}\right)\,,$$

so that

$$\left|\sum_{i=c_n+1}^{\infty}\sum_{j=1}^{n-i}\operatorname{Cov}\left[K_n\left(x-X_{t_{j+i}}\right),\, K_n\left(y-X_{t_i}\right)\right]\right|$$

$$\leq \frac{Cn}{h^4}\sum_{i=c_n+1}^{\infty}\operatorname{Cov}\left(X_0,\, X_{i\tau}\right)\,. \tag{27}$$

Then, by means of (17), (26), and (27), we obtain

$$\frac{h}{n}I_{n_3} \leq C(hc_n) + C\cdot\frac{1}{h^3}\sum_{i=c_n+1}^{\infty}\operatorname{Cov}\left(X_0,\, X_{i\tau}\right) \longrightarrow 0$$

as in (23). □

Proof of Theorem 5. It is an immediate consequence of Lemmas 4–7, by way of relation (14). □

5. Uniform strong convergence of the estimates

The purpose of this section is to show that, under certain regularity conditions, the estimate $\hat{f}_n(x)$ converges to $f(x)$ a.s., uniformly over finite intervals which tend to cover $\mathbb{R}$ as $n \to \infty$. To be more specific, let $0 < k_n \uparrow \infty$ as $n \to \infty$, and let $J_n = [-k_n, k_n]$. Then, under suitable regularity conditions,

$$\sup\{|\hat{f}_n(x) - f(x)|;\ x \in J_n\} \xrightarrow{a.s.} 0\,. \tag{28}$$

The justification of such a statement is split in parts formulated as lemmas. To motivate the formulation of the lemmas, observe that, for $x \in \mathbb{R}$,

$$|\hat{f}_n(x) - f(x)| \leq |\hat{f}_n(x) - \mathcal{E}\hat{f}_n(x)| + |\mathcal{E}\hat{f}_n(x) - f(x)|\,. \tag{29}$$

Then we have

Lemma 8. *Under strict stationarity of the underlying process and assumptions* (A.2.2)–(A.2.3), *and* (A.2.5)–(A.2.6) *(with $r = 0$), we have,*

$$|\mathcal{E}\hat{f}_n(x) - f(x)| \leq Ch_n \quad (\textit{independently of } x \in \mathbb{R})\,. \tag{30}$$

Proof. Indeed, writing h instead of h_n, we have,

$$\begin{aligned}
|\mathcal{E}\hat{f}_n(x) - f(x)| &= \left|\frac{1}{h}\mathcal{E}K\left(\frac{x - X_0}{h}\right) - f(x)\right| \text{ (by stationarity and (A.2.3))} \\
&= \left|\frac{1}{h}\int_{-\infty}^{\infty} K\left(\frac{x-y}{h}\right) f(y)\,\mathrm{d}y - \int_{-\infty}^{\infty} K(u)f(x)\,\mathrm{d}u\right| \\
&\qquad\qquad\qquad\qquad\qquad\qquad\qquad\qquad \text{(by (A.2.2))} \\
&= \left|\int_{-\infty}^{\infty} K(u)f(x-hu)\,\mathrm{d}u - \int_{-\infty}^{\infty} K(u)f(x)\,\mathrm{d}u\right| \\
&= \left|\int_{-\infty}^{\infty} K(u)[f(x-hu) - f(x)]\,\mathrm{d}u\right| \\
&\leq Ch\int_{-\infty}^{\infty} |uK(u)|\,\mathrm{d}u \text{ (by (A.2.6) with } r = 0) \\
&\leq Ch \text{ (by (A.2.5) for } r = 0),
\end{aligned}$$

as was to be seen. □

Next, work with the first term on the right-hand side in (29). To this end, divide the interval J_n of length $2k_n$ into b_n subintervals J_{n_i}, $i = 1, \ldots, b_n$, each of length δ_n (so that $b_n = [2k_n/\delta_n]$). Then

$$\begin{aligned}
\left|\hat{f}_n(x) - \mathcal{E}\hat{f}_n(x)\right| \leq \left|\hat{f}_n(x) - \hat{f}_n(x_{n_i})\right| + \left|\hat{f}_n(x_{n_i}) - \mathcal{E}\hat{f}_n(x_{n_i})\right| \\
+ \left|\mathcal{E}\hat{f}_n(x_{n_i}) - \mathcal{E}\hat{f}_n(x)\right| ,
\end{aligned}$$

where x_{n_i} is an arbitrary point in J_{n_i}, so that

$$\begin{aligned}
\sup_{x \in J_n} \left|\hat{f}_n(x) - \mathcal{E}\hat{f}_n(x)\right| \leq& \max_{1 \leq i \leq b_n} \sup_{x \in J_{n_i}} \left|\hat{f}_n(x) - \hat{f}_n(x_{n_i})\right| \\
&+ \max_{1 \leq i \leq b_n} \left|\hat{f}_n(x_{n_i}) - \mathcal{E}\hat{f}_n(x_{n_i})\right| \\
&+ \max_{1 \leq i \leq b_n} \sup_{x \in J_{n_i}} \left|\mathcal{E}\hat{f}_n(x_{n_i}) - \mathcal{E}\hat{f}_n(x)\right| . \qquad (31)
\end{aligned}$$

However,

$$
\begin{aligned}
\left|\mathcal{E}\hat{f}_n(x_{n_i}) - \mathcal{E}\hat{f}_n(x)\right| &= \left|\int_{-\infty}^{\infty} K(u)f(x_{n_i} - hu)\,\mathrm{d}u - \int_{-\infty}^{\infty} K(u)f(x - hu)\,\mathrm{d}u\right| \\
&\qquad\qquad \text{(as was seen in the proof of Lemma 8)} \\
&= \left|\int_{-\infty}^{\infty} K(u)[f(x_{n_i} - hu) - f(x - hu)]\,\mathrm{d}u\right| \\
&\le C|x_{n_i} - x| \int_{-\infty}^{\infty} |K(u)|\,\mathrm{d}u \quad \text{((by (A.2.6) with } r = 0) \\
&\le C|x_{n_i} - x| \qquad \text{(by (A.2.1))}.
\end{aligned}
$$

Therefore,

$$
\begin{aligned}
\max_{1\le i\le b_n} \sup_{x\in J_{n_i}} \left|\mathcal{E}\hat{f}_n(x_{n_i}) - \mathcal{E}\hat{f}_n(x)\right| &\le C \max_{1\le i\le b_n} \sup_{x\in J_{n_i}} |x_{n_i} - x| \\
&\le C\delta_n\,.
\end{aligned}
$$

That is, we have

Lemma 9. *Under strict stationarity of the underlying process and assumptions* (A.2.1), (A.2.3), *and* (A.2.6) *(with $r = 0$), we have,*

$$
\max_{1\le i\le b_n} \sup_{x\in J_{n_i}} \left|\mathcal{E}\hat{f}_n(x_{n_i}) - \mathcal{E}\hat{f}_n(x)\right| \le C\delta_n\,. \tag{32}
$$

At this point, we assume that the kernel K satisfies a Lipschitz condition; i.e.,

(A.5.1) $|K(x) - K(y)| \le C|x - y|$.

Then, proceed as follows:

$$
\begin{aligned}
\left|\hat{f}_n(x) - \hat{f}_n(x_{n_i})\right| &\le \frac{1}{nh}\sum_{j=1}^{n}\left|K\left(\frac{x - X_{t_j}}{h}\right) - K\left(\frac{x_{n_i} - X_{t_j}}{h}\right)\right| \\
&\le \frac{C}{nh}\sum_{j=1}^{n}\frac{|x - x_{n_i}|}{h} = \frac{C}{nh^2}\sum_{j=1}^{n}|x - x_{n_i}|\,,
\end{aligned}
$$

so that

$$
\max_{1\le i\le b_n} \sup_{x\in J_{n_i}} \left|\hat{f}_n(x) - \hat{f}_n(x_{n_i})\right| \le \frac{C}{nh^2}\cdot n\delta_n = \frac{C\delta_n}{h^2}\,;
$$

i.e., we have the following result.

Lemma 10. *Under assumption* (A.5.1), *we have,*

$$\max_{1\le i\le b_n}\sup_{x\in J_{n_i}}\left|\hat f_n(x)-\hat f_n(x_{n_i})\right|\le\frac{C\delta_n}{h^2}\,. \tag{33}$$

Then, in reference to (31), we are left with the term $\max_{1\le i\le b_n}\left|\hat f_n(x_{n_i})-\mathcal{E}\hat f_n(x_{n_i})\right|$ alone. To this end, observe that, for $\varepsilon>0$,

$$\begin{aligned}
&P\left[\max_{1\le i\le b_n}\left|\hat f_n(x_{n_i})-\mathcal{E}\hat f_n(x_{n_i})\right|\ge\varepsilon\right]\\
&=P\left[\bigcup_{i=1}^{b_n}\left\{\left|\hat f_n(x_{n_i})-\mathcal{E}\hat f_n(x_{n_i})\right|\ge\varepsilon\right\}\right]\\
&\le\sum_{i=1}^{b_n}P\left[\left\{\left|\hat f_n(x_{n_i})-\mathcal{E}\hat f_n(x_{n_i})\right|\ge\varepsilon\right\}\right]\\
&\le\frac{1}{\varepsilon^k}\sum_{i=1}^{b_n}\mathcal{E}\left|\hat f_n(x_{n_i})-\mathcal{E}\hat f_n(x_{n_i})\right|^k\qquad(k>0)\,.
\end{aligned}\tag{34}$$

At this point, observe that, under suitable conditions to be spelled out below, Theorem 2.7, page 243, in Bulinski and Shashkin (2007) implies that

$$(nh_n)^2\mathcal{E}\left[\hat f_n(x_{n_i})-\mathcal{E}\hat f_n(x_{n_i})\right]^4\le C\,,\ \text{for all } x_{n_i}\text{'s and all } n. \tag{35}$$

Then relation (34), applied with $k=4$, yields

$$\begin{aligned}
&P\left[\max_{1\le i\le b_n}\left|\hat f_n(x_{n_i})-\mathcal{E}\hat f_n(x_{n_i})\right|\ge\varepsilon\right]\\
&\le\frac{1}{\varepsilon^4(nh_n)^2}\sum_{i=1}^{b_n}(nh_n)^2\mathcal{E}\left|\hat f_n(x_{n_i})-\mathcal{E}\hat f_n(x_{n_i})\right|^4\\
&\le\frac{cb_n}{\varepsilon^4(nh_n)^2}\,.
\end{aligned}\tag{36}$$

To fix ideas, let us take $k_n=n^\nu$ $(\nu>0)$, $b_n=\left[\left(\frac{n}{h_n}\right)^{1/2}\right]$, so that $\delta_n=\frac{2k_n}{b_n}\simeq\frac{2n^\nu h_n^{1/2}}{n^{1/2}}=\frac{2h_n^{1/2}}{n^{1/2-\nu}}$, which suggests that $(0<)\nu<\frac{1}{2}$. Then relation (36) becomes

$$P\left[\max_{1\le i\le b_n}\left|\hat f_n(x_{n_i})-\mathcal{E}\hat f_n(x_{n_i})\right|\ge\varepsilon\right]\le\frac{C}{\varepsilon^4}\cdot\frac{1}{(nh_n)^{3/2}}\,. \tag{37}$$

Now, assumptions under which (35) holds are as follows; references are made to Bulinski and Shashkin (2007).

(A.5.1) This assumption already made is assumed to hold (Theorem 2.2, page 339, and Theorem 2.8, page 344).

(A.5.2) The kernel K is a p.d.f. (page 338).

(A.5.3) $\int_{-\infty}^{\infty} |u|K(u)\, du < \infty$ (Theorem 2.4, page 341). (This is assumption (A.2.5) (with $r = 0$).)

(A.5.4) The p.d.f. f is Lipschitz; i.e., $|f(x) - f(y)| \leq C|x - y|$ (relation (2.2), page 338). (This is assumption (A.2.6) (with $r = 0$).)

(A.5.5) The joint p.d.f.'s $f_{X_0, X_{t_j}}$ exist for all $t_j > 0$, and

$$\sup_{j \geq 1} \sup_{x_0, x_{t_j} > 0} f_{X_0, X_{t_j}}(x_0, x_{t_j}) < \infty$$

(relation (2.3), page 338). (This assumption would follow from (A.3.1) and boundedness of f_{X_0}.)

(A.5.6) $0 < h_n \to 0$ is chosen, so that $nh_n^{1+\delta+3/\lambda} \geq 1$, for some $\delta > 0$, some $\lambda > 3$, and all n (Theorem 2.7, page 343, and relation on line 7 from below on page 340).

Thus, under conditions (A.5.1)–(A.5.6), relation (37) holds. If, in addition, h_n is chosen, so that $\sum_{n=1}^{\infty}(nh_n)^{-3/2} < \infty$, then the Borel–Cantelli Lemma implies that

$$\max_{1 \leq i \leq b_n} \left| \hat{f}_n(x_{n_i}) - \mathcal{E}\hat{f}_n(x_{n_i}) \right| \xrightarrow{P} 0 \,. \tag{38}$$

Furthermore, if h_n is chosen as above, and also to satisfy the requirements

$$\left(\frac{\delta_n}{2} =\right) \frac{h_n^{1/2}}{n^{(1/2)-\nu}} \longrightarrow 0 \left(0 < \nu < \frac{1}{2}\right), \qquad \left(\frac{\delta_n}{2h_n^2} =\right) \frac{1}{n^{(1/2)-\nu} h_n^{3/2}} \longrightarrow 0 \,,$$

then relations (32) and (33) imply that

$$\max_{1 \leq i \leq b_n} \sup_{x \in J_{n_i}} \left| \mathcal{E}\hat{f}_n(x_{n_i}) - \mathcal{E}\hat{f}_n(x) \right| \longrightarrow 0 \,, \tag{39}$$

$$\max_{1 \leq i \leq b_n} \sup_{x \in J_{n_i}} \left| \hat{f}_n(x) - \hat{f}_n(x_{n_i}) \right| \longrightarrow 0 \quad \text{(pointwise)}. \tag{40}$$

Since, on account of (30),

$$\sup_{x\in J_n} \left|\mathcal{E}\hat{f}_n(x) - f(x)\right| \le Ch_n \longrightarrow 0 , \tag{41}$$

then relations (29), (41); and (31) in conjunction with (38)–(40) imply relation (28). Formalizing the additional requirements on h_n as an assumption, we have,

(A.5.7) Let h_n be as in assumption (A.5.6) and also satisfy the requirement:

$$\frac{1}{n^{(1/2)-\nu}h_n^{3/2}} \longrightarrow 0 \quad \left(\text{which implies } \frac{h_n^{1/2}}{n^{(1/2)-\nu}} \longrightarrow 0\right) .$$

Then, we have the following main result in this section.

Theorem 6. *Let the underlying process be strictly stationary and PA, and suppose assumptions* (A.2.1)–(A.2.3), (A.2.5)–(A.2.6) *(both with* $r = 0$*),* (A.5.1)–(A.5.2), (A.5.5)–(A.5.7) *hold. Then, we have,*

$$\sup\left\{\left|\hat{f}_n(x) - f(x)\right| ;\ x \in [-n^\nu, n^\nu]\right\} \xrightarrow{P} 0 ,$$

with $0 < \nu < \frac{1}{2}$.

It is to be noted here that a result like this, for the i.i.d. case, was obtained in Bhattacharya (1967) by utilizing a Hoeffding-type exponential inequality.

Remark 2.

(i) At this point, it should be pointed out that the restrictions imposed upon h_n in assumptions (A.5.6) and (A.5.7) are not as stringent as they may look. Indeed, let $h_n = n^{-c}$ $(c > 0)$. Then: $nh_n^{1+\delta+3/\lambda} \ge 1$ if and only if $\lambda \ge \frac{3c}{1-c-c\delta}$, and $\frac{3c}{1-c-c\delta} < \varepsilon$ (so that $\lambda > 3$ as required) if and only if $c < \frac{1}{1+\delta}$. So, if $c < 1/(1+\delta)$, for some $\delta > 0$, then $nh_n^{1+\delta+3/\lambda} \ge 1$, for some $\delta > 0$, some $\lambda > 3$, and all n; i.e., assumption (A.5.6) is satisfied. Next, $n^{(1/2)-\nu}h_n^{3/2} = n^{(1/2)-\nu-3c/2}$, and this tends to ∞ if and only if $c < \frac{1-2\nu}{3}$ (which is doable since $\nu < 1/2$). Finally, $(nh_n)^{3/2} = n^{3(1-c)/2}$, and the inverse of it is summable if $c < 1/3$. On the other hand, $(1-2\nu)/3$ is always less than $1/3$, thus, it suffices to require only that $c < (1-2\nu)/3$. Then assumption (A.5.7) is satisfied. Combining these results, we have that assumptions (A.5.6) and (A.5.7) are both satisfied, provided $h_n = n^{-c}$ with $0 < c < \min\{1/(1+\delta), (1-2\nu)/3\}$, for some $\delta > 0$ and $0 < \nu < 1/2$.

(ii) The choice of h_n suggested in part (i) allows for a great latitude in selecting a sequence $0 < c_n \to \infty$ with $c_n h_n \to 0$ and such that relation (7) is satisfied for a broad class of covariance functions. Indeed, suppose that $\mathrm{Cov}(X_0, X_{i\tau}) = C_0 \rho^{i\tau}$ (some $C_0 > 0$ and $0 < \rho < 1$), and take $c_n = n^\alpha$ ($\alpha > 0$) and $h_n = n^{-c}$ ($c > 0$) as in part (i). Then $c_n h_n = n^{\alpha - c} \to 0$, provided $c > \alpha$. Next, we observe that

$$\sum_{i=c_n+1}^{\infty} \mathrm{Cov}(X_0, X_{i\tau}) = C_0 \sum_{i=c_n+1}^{\infty} \rho^{i\tau} = C_0 \frac{\rho^{(c_n+1)\tau}}{1-\rho^\tau},$$

so that

$$\begin{aligned} h_n^{-3} \sum_{i=c_n+1}^{\infty} \mathrm{Cov}(X_0, X_{i\tau}) &= \frac{C_0}{1-\rho^\tau} \cdot n^{3c} \rho^{(c_n+1)\tau} \\ &= \frac{C_0}{1-\rho^\tau} \cdot n^{3c} \rho^{(n^\alpha+1)\tau} \longrightarrow 0 \end{aligned}$$

as the first term on the right-hand side above tends to ∞ at an algebraic rate, and the second term tends to 0 at an exponential rate; i.e., relation (7) is satisfied.

References

Bhattacharya, P. K. (1967). Estimation of a probability density function and its derivatives, *Sankhya, Ser.A*, **29**, pp.373–382.

Bulinski, A. V. (1996). On the convergence rates in the CLT for positively and negatively dependent random fields. In: Ibragimov, I.A., Zaitsev, A. Yu (eds.), *Probability Theory and Mathematical Statistics.* Gordon and Breach Publishers, Amsterdam.

Bulinski, A. and Shashkin, A. (2007). *Limit Theorems for Associated Random Fields and Related Systems.* World Scientific Publishing Co., Singapore.

Cai, Z. and Roussas, G. G. (1992). Uniform strong estimation under α-mixing, *Statist. Probab. Lett.*, **15**, pp.47–55.

Cai, Z. and Roussas, G. G. (1998). Kaplan–Meier estimator under association, *J. Multivar. Anal.*, **67**, pp.318–348.

Dieudonné, J. (1960). *Foundations of Modern Analysis.* Academic Press.

Esary, J. D., Proschan, F. and Walkup, D. W. (1967). Association of random variables, with applications, *Ann. Math. Statist.*, **38**, pp.1466–1474.

Ioannides, D. and Roussas, G. G. (1998). Exponential inequalities for associated random variables, *Statist. Probab. Lett.*, **42**, pp.423–431.

Lenain, J.-F., Harel, M. and Puri, M. (2011). Asymptotic behavior of the kernel density estimators for nonstationary dependent random variables with binned data. In: Jiang, J., Roussas, G. G. and Samaniego, F. J. (eds.), *Nonparametric Statistical Methods and Related Topics: A Festschrift in Honor of Professor P. K. Bhattacharya on the Occasion of his 80th Birthday.* World Scientific.

Masry, E. (1983). Probability density estimation from sampled data, *IEEE Transactions on Information Theory*, **IT-29**, No.5, pp.696–709.

Oliveira, P. E. (2005). An exponential inequality for associated variables, *Statist. Probab. Lett.*, **73**, pp.189–197.

Parzen, E. (1962). On estimation of a probability density function and mode, *Ann. Math. Statist.*, **33**, pp.1065–1076.

Pratt, J. W. (1960). On interchanging limits and integrals, *Ann. Math. Statist.*, **31**, pp.74–77.

Roussas, G. G. (1991). Kernel estimates under association: Strong uniform consistency, *Statist. Probab. Lett.*, **12**, pp.393–403.

Roussas, G. G. (1993). Asymptotic normality of positively or negatively associated proceses. *Journal of Nonparametric Statistics*, **50**(1), pp.152–173.

Roussas, G. G. (1999). Positive and negative dependence with some statistical applications. In: S. Ghosh (ed.), Asymptotics, nonparametrics, and time series, *A tribute to Madan Lai Puri*, Text. Monogr., **158**, pp.757–788. Marcel Dekker, New York.

Roussas, G. G. (2000). Prediction under association. In: Limnios, et al. (eds.), *Recent Advances in Reliability Theory, Methodology, Practice and Inference*, pp.451–475. Birkhauser, Boston.

Roussas, G. G. (2005). *An Introduction to Measure-Theoretic Probability.* Elsevier/Academic Press.

Chapter 6

A Short Proof of the Feigin–Tweedie Theorem on the Existence of the Mean Functional of a Dirichlet Process

Jayaram Sethuraman

Florida State University

The Feigin–Tweedie theorem shows that the mean functional of a Dirichlet process with parameters α and β is finite if and only if $E_\beta(\log(1+|X|)) < \infty$. We give a short self-contained proof of this result using the Sethuraman constructive definition of a Dirichlet process. The problem of giving a similar short proof for the so-called two-parameter Dirichlet process is left open.

Contents

1. Introduction

Consider independent identically distributed random variables in R_1 with common distribution P. The most popular and well studied prior distribution for P is the Dirichlet process $\mathcal{D}$ introduced by Ferguson (1973). One generally wants to estimate functionals of P like the mean $\int xP(dx)$. The Bayesian estimate under square error loss will be the mean of $\int xP(dx)$ under $\mathcal{D}$. This makes sense only if $\int xP(dx)$ is finite with $\mathcal{D}$-probability 1, and if its mean and variance are finite under $\mathcal{D}$. Thus it is important to know, at a minimum, under what conditions the mean functional $\int xP(dx)$ will be finite under $\mathcal{D}$.

The Dirichlet process $\mathcal{D}$ comes with parameters α, β where $\alpha > 0$ and β is a probability measure on R_1, and so will be denoted as $\mathcal{D}(\alpha, \beta)$ hereafter.

Furthermore

$$\int P(A)\mathcal{D}(\alpha,\beta)(dP) = \beta(A).$$

Thus, the probability measure β is the mean of P under $\mathcal{D}(\alpha,\beta)$. Ferguson showed that

$$\int \left[\int xP(dx)\right] \mathcal{D}(\alpha,\beta)(dP) = \int x\beta(dx) \tag{1}$$

if

$$\int |x|\beta(dx) < \infty. \tag{2}$$

This however left open the question of finding conditions under which the mean functional $\int xP(dx)$ is finite under $\mathcal{D}(\alpha,\beta)$.

Feigin and Tweedie (1989) were the first to establish the result:

Theorem 1 (Feigin–Tweedie Theorem). *The log-moment condition*

$$\int \log(1+|x|)\beta(dx) < \infty \tag{3}$$

is a necessary and sufficient condition for

$$\mathcal{D}(\alpha,\beta)\left(P : \int xP(dx) \text{ is finite}\right) = 1. \tag{4}$$

Any proof of this result has to first give a clear definition of the Dirichlet process. Feigin and Tweedie show that the Dirichlet process can be defined as the stationary measure for a Markov chain defined by a recurrence relation and establish that an associated Markov chain is ϕ-irreducible. By further showing that this associated Markov chain is also Harris ergodic when (3) holds, they establish that this condition is sufficient for (4). They show the reverse conclusion by a refinement of the techniques of Hannum, et al. (1981).

The Dirichlet process can also be viewed as a normalized Gamma process. This can be already seen in Ferguson (1973). The Gamma process is an increasing additive process and a special case of a Lévy process with positive increments and no drift (also called subordinators). Thus more general distributions of random measures can be obtained from normalized subordinators. This has been proposed and pursued by several authors including Kingman (1975), Cifarelli and Regazzini (1990), Regazzini, et al. (2003), Lijoi and Prünster (2009). Proposition 1 of Regazzini, et al. (2003) gives a complete result on necessary and sufficient conditions for $\int xP(dx)$

to be finite under the class of random probability measures obtained in this manner. Thus, this is an alternate proof of the Feigin–Tweedie theorem.

Most of the interest in recent papers have been centered around the distribution of $\int P(dx)$ under these general classes of random probability measures. The paper of Lijoi and Prünster (2009) gives a fairly general summary of results this area. We do not deal with this very interesting and exciting area in this paper.

Instead, we will give a short self contained proof of the Feigin–Tweedie theorem. It will be based on the constructive definition of a Dirichlet process given in Sethuraman (1994). This is done in Section 2.

Another example of a random probability measure is the so-called two-parameter Dirichlet process, $\mathcal{D}(\alpha, \sigma, \beta)$. This, together with its constructive definition, is given in Pitman and Yor (1997). The necessary and sufficient conditions for $\int xP(dx)$ to be finite under $\mathcal{D}(\alpha, \sigma, \beta)$ can be derived from Proposition 1 of Regazzini, et al. (2003). This is stated as Theorem 2 in Section 3. It should be possible to give a simple proof of this result using the methods of Section 2. In Section 3 we proceed along these lines but hit a road block and leave it as an open problem.

Most papers in this area consider more general measurable spaces $\mathcal{X}$ as the basic space, and random probability measures P on $\mathcal{X}$ whose distribution is the Dirichlet process $\mathcal{D}(\alpha, \beta)$. They also consider the finiteness of $\int_{\mathcal{X}} g(x)P(dx)$ for general *real* valued measurable functions g. Since g is a real valued function, the transformation $y = g(x)$ reduces this problem to what we have described so far. This is easily seen from the identity $\int_{\mathcal{X}} g(x)P(dx) = \int_{R_1} yQ(dy)$ where $Q = Pg^{-1}$ and the fact that the distribution of Q is the Dirichlet process $\mathcal{D}(\alpha, \beta g^{-1})$. The condition (3) is in terms of $|x|$ because $\int xP(dx)$ exists only if $\int |x|P(dx) < \infty$. Also condition (3) can be replaced by $\int \log(\max(1, |x|))\beta(dx) < \infty$. One can actually take the basic space as R_1^+, the non-negative real line, or even $[1, \infty)$ in proving the Feigin–Tweedie Theorem. We adopt these simplifications in Section 2. We note, in passing, that $\int xP(dx)$ is finite under $\mathcal{D}(\alpha, \beta)$ if and only if $\int x^2 P(dx)$ is finite under $\mathcal{D}(\alpha, \beta)$.

2. Dirichlet processes

We begin by giving the definition of a Dirichlet process on the real line from Ferguson (1973). Consider a random function P defined on the Borel subsets of R_1 taking values in $[0, 1]$. For any finite measurable partition, $(B_1, \ldots, B_k)$, the let the random vector $(P(B_1), \ldots, P(B_k))$ have the finite dimensional Dirichlet distribution $\mathcal{D}(\alpha\beta(B_1),\ldots,\alpha\beta(B_k))$ where $\alpha > 0$ and β is a probability measure on R_1. The finite dimensional Dirichlet distribution $\mathcal{D}(\alpha_1,\ldots,\alpha_k)$ with $(\alpha_1,\ldots,\alpha_k) \neq (0,0,\ldots,0)$ is just the distribution of $(Z_1/Z,\ldots,Z_k/Z)$ where $Z = Z_1 + \cdots + Z_k$ and $Z_1,\ldots,Z_k$ are independent Gamma random variables with scale parameters $\alpha_1,\ldots,\alpha_k$, respectively and shape parameter 1. It can be shown that this defines a unique distribution under which P is a random probability measure and the distribution of P is called the Dirichlet process $\mathcal{D}(\alpha,\beta)$.

Ferguson established many useful properties of Dirichlet process. For instance, if $X_1,\ldots,X_n$ are i.i.d. with common probability measure P and we choose the Dirichlet process $\mathcal{D}(\alpha,\beta)$ as the prior for P, then the posterior distribution of P is $\mathcal{D}((\alpha+n),\beta_n)$ where $\beta_n = (\alpha\beta + nF_n)/(\alpha+n)$ and F_n is the empirical probability measure of $X_1,\ldots,X_n$. One could use this to obtain posterior estimates of various functionals of P like $P(B), \int xP(dx)$, the median interval of P, etc. Whereas it is clear that the functionals $P(B)$ and the median interval of P are well defined, it is not clear that $\int xP(dx)$ is finite under the prior or posterior distributions, in this instance. This raises the question about conditions under which

$$\mathcal{D}(\alpha,\beta)(P : \int xP(dx) \text{ is finite}) = 1.$$

It was known that the condition $\int |x|\beta(dx) < \infty$ was too strong for this, as can be gleaned from Doss and Selke (1982) or from Yamato (1984), where it was shown that $\int xP(dx)$ has a Cauchy distribution if β is Cauchy. It was Feigin and Tweedie (1989) who first gave the necessary and sufficient condition and proved Theorem 1.

We will use the constructive definition of Sethuraman (1994) to give a simple self contained proof of Theorem 1. Let $V_1, V_2, \ldots$ be i.i.d. random variables with Beta distribution Beta$(1,\alpha)$. We can define a discrete distribution $(p_1, p_2, \ldots)$ on $\{1,2,\ldots\}$, by setting

$$p_1 = V_1 \text{ and } p_n = \prod_{r=1}^{n-1}(1-V_r)V_n, n = 2,3,\ldots. \tag{5}$$

Conversely, the variables $V_1, V_2, \ldots$ are just the discrete random failure rates of the random discrete distribution $(p_1, p_2, \ldots)$. Let $Y_1, Y_2, \ldots$ be i.i.d. with common distribution β on R_1 and be independent of $(V_1, V_2, \ldots)$. Then δ_{Y_n} is a (measurable) random degenerate probability measure (with all its mass at Y_n) for each n and

$$P = \sum_1^\infty p_n \delta_{Y_n} \tag{6}$$

is a random probability measure on R_1. It was established in Sethuraman (1994) that the distribution of this P is the Dirichlet process $\mathcal{D}(\alpha, \beta)$. This constructive definition and has proved very useful in both theoretical and practical applications.

In recent papers, the colloquial expression "stick breaking rules" is being used for the random i.i.d. discrete failure rates $V_1, V_2, \ldots,$ when referring to the constructive definition of the Dirichlet process and its extensions.

We can give a simpler expression of $\int xP(dx)$ in terms of this constructive definition. To simplify our presentation, we will assume that P is a random probability on R_1^+, the positive half of the real line (otherwise, we will have to use the function $|x|$ instead of x). Then

$$\int xP(dx) = \sum_1^\infty p_n Y_n$$

in the sense that both sides are finite and equal to one another or both sides are equal to $+\infty$. We will now show that condition (3) is necessary and sufficient for

$$\sum_1^\infty p_n Y_n < \infty. \tag{7}$$

This will be our approach to complete the simple proof of Theorem 1.

We begin with two lemmas.

Lemma 1. *Let $Y_1, Y_2, \ldots$ be i.i.d. with common distribution β and let $0 < \rho < 1$. Then the series*

$$\sum_1^\infty \rho^n Y_n \text{ is finite with probability } 1 \tag{8}$$

if and only if

$$E(\log(\max(1, |Y_1|))) < \infty. \tag{9}$$

Proof. We can assume that β is a probability measure on R_1^+ and write Y_1 for $|Y_1|$ from here on. With no loss of generality we can also assume that $Y_1 \geq 1$ and write Y for Y_1, when convenient.

From the the Kolmogorov three-series theorem we know that the three conditions

$$\sum_n P(\rho^n Y_n > \epsilon) = \sum_n P(\rho^n Y_1 > \epsilon) < \infty,$$
$$\sum_n E(\rho^n Y_n I(\rho^n Y_n \leq \epsilon)) < \infty \text{ and}$$
$$\sum_n V(\rho^n Y_n I(\rho^n Y_n \leq \epsilon)) < \infty$$

for any $\epsilon > 0$ are necessary and sufficient for the series $\sum \rho^n Y_n$ to converge with probability 1.

We will show that the first of these conditions is equivalent to (9). We will see that the second condition holds all the time and note that the third condition also holds all the time in a similar way.

The first condition is the same as

$$\sum_n P((\log(Y_1) - \log(\epsilon)) > -n\log(\rho)) < \infty$$

and this is equivalent to condition (9) since, for any positive random variable W and $a > 0$, $E(W) < \infty$ is equivalent to $\sum_n P(W > na) < \infty$.

To show that the second condition holds all the time, we employ the usual chopping up method and check that

$$\begin{aligned}
&\sum_n E(\rho^n Y_n I(\rho^n Y_n \leq \epsilon)) \\
&= \sum_{n=1}^{\infty}\sum_{i=1}^{n} E(\rho^n Y I(\frac{\epsilon}{\rho^{i-1}} < Y \leq \frac{\epsilon}{\rho^i})) + \sum_{n=1}^{\infty} E(\rho^n Y \leq \epsilon) \\
&\leq \sum_{i=1}^{\infty}\sum_{n=i}^{\infty} E(\rho^n Y I(\frac{\epsilon}{\rho^{i-1}} < Y \leq \frac{\epsilon}{\rho^i})) + \frac{\rho\epsilon}{1-\rho} \\
&= \frac{1}{1-\rho}\sum_i E(\rho^i Y I(\frac{\epsilon}{\rho^{i-1}} < Y \leq \frac{\epsilon}{\rho^i})) + \frac{\rho\epsilon}{1-\rho} \leq \frac{2\rho\epsilon}{1-\rho}.
\end{aligned}$$

This completes the proof. □

The following is a slight generalization of the above lemma which will be required to complete the proof of Theorem 1:

Lemma 2. *Let $\{Y_1, Y_2, \ldots\}$ be i.i.d. random variables in R_1^+ with distribution β. Let $a_n \geq 0$ be constants such that*

$$\frac{1}{n}\log(a_n) \to -\log(\rho) \tag{10}$$

with $0 < \rho < 1$. *Then the series*

$$\sum_{1}^{\infty} a_n\, Y_n \textit{ is finite with probability } 1 \tag{11}$$

if and only if (9) holds.

Proof. Condition (10) implies that

$$\rho_1^n \leq a_n \leq \rho_2^n$$

for all large n for some ρ_1, ρ_2 with $0 < \rho_1 \leq \rho \leq \rho_2 < 1$. Since the Y_n's are non-negative, we also have

$$\{\sum_n \rho_1^n Y_n < \infty\} \subset \{\sum_n a_n Y_n < \infty\} \subset \{\sum_n \rho_2^n Y_n < \infty\}. \tag{12}$$

Lemma 1 shows that condition (9) implies $\text{Prob}(\sum_n \rho_1^n Y_n < \infty) = 1$ and that $\text{Prob}(\sum_n \rho_2^n Y_n < \infty) = 1$ implies condition (9). This completes the proof of this lemma. □

Proof of the Feigin–Tweedie Theorem. Recall the quantities p_n defined as $p_n = \prod_1^{n-1}(1 - V_r)V_n, n = 2, 3, \ldots$ in (5) and note that $V_1, V_2, \ldots$ are i.i.d. with Beta distribution $\text{Beta}(1, \alpha)$ with $\alpha > 0$ and that $-\infty < E(\log(1 - V_1)), E(\log(V_1)) < 0$. From the law of large numbers, $(1/n)\sum_1^{n-1} \log(1 - V_r) \to E(\log(1 - V_1))$ with probability 1. Since $\log(V_n), n = 1, 2, \ldots$ are i.i.d. with a finite mean, we also have $(1/n)\log(V_n) \to 0$ with probability 1. Thus $(1/n)\log(p_n) \to E(\log(1 - V_1))$ with probability 1. The p_n's are independent of $Y_1, Y_2, \ldots$, and thus the series $\sum_1^\infty p_n Y_n = \int x P(dx)$ is finite with probability 1 if and only if $\int \log(\max(1, |x|)\beta(dx) < \infty$, from an application of Lemma 2. □

The constructive definition of the Dirichlet random measure P in (6) can be generalized in many ways. There is no need to keep the p_n's independent of Y_n's or require the i.i.d. properties of the V_n's or Y_n's. The simple proof outlined above can be pursued for such random probability measures. The distributions of such random probability measures are generally not obtainable as that of normalized subordinators which are well understood. However, they can be used as nonparametric priors even if they cannot be described in familiar terms and the posterior distributions can be calculated only by numerical simulation. For instance, see Ishwaran and James (2001).

3. Two-parameter Dirichlet processes

The two-parameter Dirichlet process can be obtained by considering a suitable subordinator different from the Gamma process. Pitman and Yor (1997) give a constructive definition of the two-parameter Dirichlet process $\mathcal{D}(\alpha, \sigma, \beta)$ as the distribution of the random probability measure

$$P = \sum p_n \delta_{Y_n} \tag{13}$$

much like the constructive definition of the Dirichlet process. Here, the random discrete probability distribution $(p_1, p_2, \ldots)$ has random discrete failure rates $V_1, V_2, \ldots$ which are independent with the distribution of V_n equal to the Beta distribution Beta$(1-\sigma, \alpha+n\sigma)$, $n = 1, 2, \ldots$. We will also require that $0 < \sigma < 1$ and $\alpha > -\sigma$. We will not deal with another choice $\sigma < 0, \alpha = -k\sigma$, that has also been proposed, because the series in (13) becomes just a finite series. As before, the direct relationship between the V_n's and p_n's is $p_1 = V_1, p_n = \prod_1^{n-1}(1-V_r)V_n$. It is also required that $Y_1, Y_2, \ldots$ be independent of $V_1, V_2, \ldots$ and constitute an i.i.d. sequence with distribution β.

A special case of Proposition 1 of Regazzini, et al. (2003) yields the following theorem:

Theorem 2. *The moment condition*

$$\int |x|^\sigma \beta(dx) < \infty \tag{14}$$

is a necessary and sufficient condition for

$$\mathcal{D}(\alpha, \sigma, \beta)(P : \int xP(dx) \text{ is finite }) = 1. \tag{15}$$

This is analogous to the Theorem 1 of Feigin and Tweedie.

It should be possible to emulate the methods of Section 1 to exploit the constructive definition to give a simple proof of this theorem. We will list a few steps in this direction and point out that we are not able to complete a proof along these lines.

We can prove the following lemma:

Lemma 3. *Let $Y_1, Y_2, \ldots$ be i.i.d. with common distribution β and let $0 < \sigma < 1$. Let $a_n = n^{-1/\sigma}, n = 1, 2, \ldots$ Then the series*

$$\sum_1^\infty a_n Y_n \text{ is finite with probability } 1 \tag{16}$$

if and only if

$$E(\log(|Y_1|^\sigma)) < \infty. \tag{17}$$

Proof. This can be established by a proof similar to that of Lemma 1. The main difference is that the first of the Kolmogorov three-series conditions implies (17) (proving necessity), while condition (17) has to be used to prove all of the Kolmogorov three-series conditions(to prove sufficiency). □

We are not able to establish an extension similar to Lemma 2, namely that when $\log(a_n)/\log(n) \to -1/\sigma$, (16) is equivalent to (17). We are also unable to show that $\log(p_n)/\log(n) \to -1/\sigma$ with probability 1; we are able to show that $\log(p_n)+\log(n)/\sigma$ converges to a non-degenerate random variable in distribution which can only show that $\frac{1}{\log(n)}\log(p_n) \to -\frac{1}{\sigma}$ in probability. A short proof for the existence of the mean functional of a two-parameter Dirichlet process that proceeds along these lines remains an open problem.

References

Cifarelli, D. M. and Regazzini, E. (1990). Distribution functions of means of a Dirichlet process. *Ann. Statist.* **18**, 1, 439–442.

Doss, H. and Selke, T. (1982). The tails of probabilities chosen from a Dirichlet prior. *Ann. Statist.* **10**, 4, 1302–1305.

Feigin, P. D. and Tweedie, R. L. (1989). Linear functionals and Markov chains associated with Dirichlet processes, *Math. Proc. Camb. Phil. Soc.* **105**, 579–585.

Ferguson, T. S. (1973). A Bayesian analysis of some nonparametric problems. *Ann. Statist.* **1**, 2, 209–230.

Hannum, R. C., Hollander, M. and Langnerg, N. (1981). Distributional results for random functionals of a Dirichlet process. *Ann. Statist.* **9**, 4, 665–670.

Ishwaran, H. and James, L. F. (2001). Gibbs sampling methods for stick-breaking priors. *J. Amer. Statist. Assoc.* **96**, 453, 161–173.

Kingman, J. F. C. (1975). Random discrete distributions. *J. Roy. Statist Soc., Ser. B* **37**, 1–22.

Lijoi, A. and Prünster, I. (2009). Distributional properties of means of random probability measures. *Statist. Surveys* **3**, 47–95.

Pitman, J. and Yor, M. (1997). The two-parameter Poisson–Dirichlet distribution derived from a stable subordinator. *Ann. Probab.* **25**, 2, 855–900.

Regazzini, E., Lijoi, A. and Prünster, I. (2003). Distributional results for of means of normalized random probability measures with independent increments. *Ann. Statist.* **31**, 2, 560–585.

Sethuraman, J. (1994). A constructive definition of Dirichlet priors. *Statist. Sinica* **36**, 416–418.

Yamato, H. (1984). Characteristic functions of means of distributions chosen from a Dirichlet process. *Ann. Probab.* **12**, 2, 262–267.

Chapter 7

Max–Min Bernstein Polynomial Estimation of a Discontinuity in Distribution

Kai-Sheng Song

University of North Texas

We consider the problem of estimating the location of a discontinuity in a mixed type distribution and propose a nonparametric method of locating such a point of jump discontinuity. The estimator is constructed by maximizing the difference between the maximum and minimum values of the estimated cumulative distribution function (cdf) in a shrinking neighborhood. The empirically estimated cdf is based on Bernstein polynomial approximations that are free from Gibbs phenomenon. We prove that the proposed estimator is a statistically consistent estimator of the jump discontinuity. Some numerical experiments are also provided to demonstrate the finite sample performance of the proposed method.

Contents

1. Introduction

Discontinuities in distributions have arisen in a variety of contexts including earnings management, vote manipulation in elections, and corruption in sports, where highly asymmetric incentives bracket a fixed hurdle. The existence of earnings management by firms has been widely documented in the finance and accounting literature. Burgstahler and Dichev (1997) and Degeorge, et al. (1999) identify three benchmarks that indicate earnings management: firms avoid small losses, earnings decrease, and earnings that fall short of analysts' forecasts. If firms have incentives to achieve earnings above some threshhold, the distribution of reported earnings will have fewer-than-expected observations for earnings just below this threshhold and more just above. This implies that the earnings distribution would have a discontinuity around the threshhold, which have been demonstrated by many real data analysis. For example, Bollen and Pool (2009) find a significant discontinuity in the pooled distribution of monthly hedge fund returns. Detection and estimation of discontinuities in distributions have also found applications in vote manipulation in extremely close elections (Snyder (2005), Listokin (2007)) and insider manipulation of sports results (match-fixing) in sports events (Duggan and Levitt (2002), Wolfers (2006)).

There is an extensive literature on the related problems of testing and estimating change-points in distributions, regression and hazard functions, detecting and locating discontinuities in regression curves and surfaces both parametrically and nonparametrically. Bhattacharya (1994) provides a review article on the classical change-point problem. Parametric methods include the maximum likelihood estimation of a change-point studied by Hinkley (1970; 1972) and the asymptotic distribution of the MLE of a change-point derived by Bhattacharya and Brockwell (1976) in the normal means problem and by Bhattacharya (1987) in general multiparameter families. Nonparametric change-point estimators include the estimator studied by Carlstein (1988) for estimating change-points in distributions based on empirical cumulative distribution functions. Dumbgen (1991) extended Carlstein's approach to a more general framework and improved the rate of convergence in probability and derived the limiting distributions for certain models. Ferger (1994) investigated the rate of almost sure convergence of Dumbgen's estimators. More recently Hariz, Wylie and Zhang (2007) generalized these estimators to nonstationary sequences and obtained the optimal rate of convergence. Nonparametric estimators of change-points in regression and/or its derivatives were proposed and investigated by Müller (1992), Eubank and Speckman (1994), and Müller and Song (1997).

The literature on testing and estimating a discontiniuty in distribution is rather sparce, limited mainly to discontinuities in a density function rather than discontinuities in the cumulative distribution function (cdf). The problem of estimating the location of a discontinuity in a density function was first studied by Chernoff and Rubin (1956) who derived the rate of convergence of the MLE of a point of discontinuity in density and related its limiting distribution to a random walk problem. In the related finance and accounting literature, most research has actually focused on detecting and testing discontinuities in a density function and does not address the problem of discontinuities in the cdf. For example, Bollen and Pool (2009) and Lahr (2010) have proposed to construct a reference distribution by fitting the integrated kernel density function as close as possible to the empirical cdf of the data in terms of the Kolmogorov–Smirnov criterion through various choices of the bandwidth. The resulting kernel density estimate is then used as a reference distribution to locally test for discontinuities in a density function. We note that by construction, the resulting kernel density estimate is in fact a smooth function. Thus, no actual discontinuity occurs but rather a "point of most rapid change" exists. In other words, what is being tested is really kink points in the density function. Burgstahler and Dichev (1997) proposed to partition the real line into non-overlapping intervals (bins) and test whether the number of observations in bin i is significantly different from the average number of observations in the two immediately adjacent bins $i-1$ and $i+1$. Takeuchi (2004) interpreted Burgstahler and Dichev's procedure as a test for discontinuities in distribution. However, the Burgstahler–Dichev method only tests the distribution on specific intervals, not the whole distribution, and does not locate the exact point of discontinuity. We are not aware of any statistical procedure in the literature for estimating the discontinuity in a mixed-type distribution.

In this paper, we consider the problem of estimating a discontinuity in the cdf based on a random sample from a mixed-type distribution on the unit interval. One of the key motivations for constructing our estimator based on Bernstein polynomial is the absence of Gibbs phenomenon for the approximation of monotone, piece-wise smooth (deterministic) functions, having both left and right derivatives at every point, by Bernstein polynomials as demonstrated by Gzyl and Palacios (2003). This is contrary to what happens with the Fourier series or other orthogonal expansions of such a function. Because of this nice behavior of Bernstein polynomials for deterministic functions, it seems natural to expect that the method of Bernstein polynomials would also lead to good approximations for random noisy functions.

2. Max-Min Bernstein Polynomial Estimation Method

The Gibbs phenomenon can be generally described by the following behavior:

1. Let F_m be a sequence of approximants to a piece-wise smooth function F with left and right derivatives at each point. If x_0 is a discontinuity

point of F, then

$$\lim_{m\to\infty} F_m(x_0) = \frac{1}{2}(F(x_0+0) + F(x_0-0)).$$

2. We have uniform convergence on any subinterval $[x_1, x_2]$ on which the function is continuous:

$$\lim_{m\to\infty} \max_{x_1\le x\le x_2} |F_m(x) - F(x)| = 0.$$

3. On any subinterval containing a single discontinuity x_0 of F, we have Gibbs phenomenon:

$$\lim_{\epsilon\to 0+} \lim_{m\to\infty} [\max_{|x-x_0|\le\epsilon} F_m(x) - \min_{|x-x_0|\le\epsilon} F_m(x)] = C|F(x_0+0) - F(x_0-0)|,$$

where

$$C = \frac{2}{\pi}\int_0^\pi \frac{\sin x}{x}dx.$$

If the sequence of approximants is the Bernstein polynomials, Gzyl and Palacios (2003) established that the constant $C = 1$. Hence the Gibbs phenomenon does not occur in this case.

We consider the following mathematical model for a jump discontinuity in distribution:

$$F(x) = \begin{cases} (1-\delta)G_1(x) & \text{if } x < \tau_0, \\ \delta + (1-\delta)G_2(x) & \text{if } x \ge \tau_0, \end{cases} \tag{1}$$

where the $G_i(x)$ are nonnegative and nondecreasing continuous functions on $[0, 1]$ satisfying

$$\lim_{x\to 0^+} G_1(x) = 0, \lim_{x\to 1^-} G_2(x) = 1, \tag{2}$$

and

$$\delta + (1-\delta)[G_2(\tau_0) - G_1(\tau_0)] > 0, \tag{3}$$

$0 < \tau_0 < 1$ is a jump discontinuity point; and $0 < \delta < 1$ denotes the jump size when $G_1(\tau_0) = G_2(\tau_0)$. The role of assumptions (2) and (3) is to ensure that $F(x)$ is a proper cdf. Assumption (3) clearly holds for all $0 < \delta < 1$ if $G_2(\tau_0) \ge G_1(\tau_0)$, which allows $F(x)$ to model arbitrarily small jump sizes. If $G_2(\tau_0) < G_1(\tau_0)$, this constraint would impose a lower bound for the jump size δ. Assumptions (2) and (3) are automatically satisfied when both $G_1(x)$ and $G_2(x)$ are cumulative distribution functions such that $G_2(\tau_0) \geq G_1(\tau_0)$. It is worth remarking that even though only a single jump discontinuity in the distribution function $F(x)$ is considered in this paper, the proposed method can be extended to a finite number of discontinuity points without much difficulty.

The mth order Bernstein polynomial for the cdf $F(x)$ is defined as

$$B_m F(x) = \sum_{i=0}^{m} b_{i,m}(x) F\left(\frac{i}{m}\right), \ x \in [0,1],$$

where $b_{i,m}(x) := \binom{m}{i} x^i (1-x)^{m-i}$ for $i = 0, \cdots, m$ are called the mth Bernstein basis polynomials since they form a basis for the collection of all polynomials with degree less or equal to m. The observation that $b_{i,m}(x)$ is actually the binomial probability mass function is heavily used in our proofs. We note that although $F(x)$ is discontinous, $B_m F(x)$ is a polynomial in x and hence has all the derivatives. Given a random sample $\{X_1, \cdots, X_n\}$ from F, a natural estimator of $B_m F(x)$ is $B_m F_n(x)$, where F_n is the empirical cdf of the sample. In addition to having all the derivatives, $B_m F_n(x)$ is in fact nondecreasing as a function of x. To see this, let the shifting operator J be defined by $JF_{n,i} = F_{n,i+1}$, where $F_{n,i} = F_n(\frac{i}{m})$ and the forward difference operator Δ be defined by $\Delta = J - I$, were I is the identity operator. Then by the binomial theorem, $B_m F_n(x)$ can be rewritten as

$$B_m F_n(x) = (I + x\Delta)^m F_{n,0}.$$

This expression allows us to obtain a simple expression for the derivative of $B_m F_n(x)$ with respect to x as

$$\begin{aligned}\frac{dB_m F_n(x)}{dx} &= m(I + x\Delta)^{m-1} \Delta F_{n,0} \\ &= m \sum_{i=0}^{m-1} b_{i,m-1}(x)[F_n(\frac{i+1}{m}) - F_n(\frac{i}{m})] \geq 0,\end{aligned}$$

which implies that $B_m F_n(x)$ is a nondecreasing function of x. Hence

$$\begin{aligned}S_{m,n}(\tau_0) &:= \max_{|\tau-\tau_0|<h_m} B_m F_n(\tau) - \min_{|\tau-\tau_0|<h_m} B_m F_n(\tau) \\ &= B_m F_n(\tau_0 + h_m) - B_m F_n(\tau_0 - h_m).\end{aligned}$$

where h_m denotes a given sequence of positive real numbers and is chosen as a function of m. The expected value of $S_{m,n}(\tau_0)$ is then given by

$$E(S_{m,n}(\tau_0)) = B_m F(\tau_0 + h_m) - B_m F(\tau_0 - h_m).$$

One of the nice approximation properties of Bernstein polynomials is that $B_m F(x)$ does not overshoot at a jump discontinuity τ_0 as x approaches the discontinuity from the right and does not undershoot at τ_0 as x approaches τ_0 from the left. More specifically, for any τ, let $S(\tau) := F(\tau+0)-F(\tau-0)$. If $h_m \to 0$ as $m \to \infty$, then it follows from Lemma 2 that

$$E(S_{m,n}(\tau_0)) \to S(\tau_0) = |F(\tau_0+0) - F(\tau_0-0)|.$$

This fact shows that the Gibbs phenomenon does not occur for the approximation of monotone, piece-wise smooth functions with left and right derivatives at every point by Bernstein polynomials. In addition, $S(\tau_0) = max_{\tau\in[0,1]}S(\tau)$. In other words, the limit function $S(\tau)$ is maximized at the true discontinuity point $\tau = \tau_0$. This observation motivates us to consider maximizing $S_{m,n}(\tau) = B_m F_n(\tau + h_m) - B_m F_n(\tau - h_m)$ with respect to each candidate discontinuity point τ and the resulting smallest maximizer denoted by $\hat{\tau}_n := \text{argmax}_\tau\, S_{m,n}(\tau)$ is used to estimate τ_0.

3. Main Result

The main result of this paper is that under a couple of weak assumptions, the proposed estimator $\hat{\tau}_n$ is a consistent estimator of the jump discontinuity τ_0.

Theorem 1. *Assume that $h_m \to 0$ and $mh_m^2 \to \infty$ as $m \to \infty$. Then under model (1), for every small $\epsilon > 0$, as $m \to \infty$ and $n \to \infty$,*

$$P(|\hat{\tau}_n - \tau_0| > \epsilon) \to 0.$$

We present a proof of the theorem in the next section after establishing several lemmas. In the case of estimating the location of a discontinuity in a density function, Chernoff and Rubin (1956) have shown that the MLE of a point of discontinuity in density is consistent with rate $O_p(n^{-1})$. It would be interesting and useful to investigate the rate of convergence of our proposed estimator, its limiting distribution, and the optimal choice of m. These questions will be addressed in our future reseach.

4. Proofs

Lemma 1. *Assume that $h_m \to 0$ as $m \to \infty$. Then for every small $\epsilon > 0$, as $m \to \infty$, we have*

$$\sup_{|\tau-\tau_0|>\epsilon} \left|B_m F(\tau \pm h_m) - F(\tau)\right| \to 0.$$

Proof. We provide a proof only for $\sup_{|\tau-\tau_0|>\epsilon}\left|B_mF(\tau+h_m)-F(\tau)\right|$. The same argument applies to $\sup_{|\tau-\tau_0|>\epsilon}\left|B_mF(\tau-h_m)-F(\tau)\right|$. For any $x\in[0,1]$, we decompose the Bernstein polynomial for the cdf $F(x)$ as

$$B_mF(x)=\sum_{i=0}^{m}b_{i,m}(x)F(\frac{i}{m})=\sum_{i<m\tau_0}b_{i,m}(x)F(\frac{i}{m})+\sum_{i\geq m\tau_0}b_{i,m}(x)F(\frac{i}{m}).$$

Using the definition of $F(x)$ under model (1), we obtain

$$\begin{aligned}B_mF(x)&=(1-\delta)\sum_{i<m\tau_0}b_{i,m}(x)G_1(\frac{i}{m})+\sum_{i\geq m\tau_0}b_{i,m}(x)[\delta+(1-\delta)G_2(\frac{i}{m})]\\&=(1-\delta)\sum_{i=0}^{m}b_{i,m}(x)G_1(\frac{i}{m})+\delta\sum_{i\geq m\tau_0}b_{i,m}(x)\\&\quad+(1-\delta)\sum_{i\geq m\tau_0}b_{i,m}(x)[G_2(\frac{i}{m})-G_1(\frac{i}{m})]\end{aligned}$$

Since the function G_1 is continuous on $[0,1]$, it follows from Weierstrass's theorem that $\sum_{i=0}^{m}b_{i,m}(x)G_1(\frac{i}{m})$ converges as $m\to\infty$ to $G_1(x)$ uniformly in x on $[0,1]$. Hence

$$\begin{aligned}&\sup_{|\tau-\tau_0|>\epsilon}|\sum_{i=0}^{m}b_{i,m}(\tau+h_m)G_1(\frac{i}{m})-G_1(\tau)|\\=&\sup_{|\tau-\tau_0|>\epsilon}|\sum_{i=0}^{m}b_{i,m}(\tau+h_m)[G_1(\frac{i}{m})-G_1(\tau+h_m)]+G_1(\tau+h_m)-G_1(\tau)|\\&\to 0.\end{aligned}$$

Let $B(m,x)$ denote a binomial random variable with m Bernoulli trials and success probability x. For any $\tau<\tau_0-\epsilon$ with $\tau\in(0,1)$, we have

$$\begin{aligned}\sum_{i\geq m\tau_0}b_{i,m}(\tau+h_m)&=P(\frac{B(m,\tau+h_m)}{m}\geq\tau_0)\\&\leq P(\left|\frac{B(m,\tau+h_m)}{m}-(\tau+h_m)\right|\geq\tau_0-\tau-h_m).\end{aligned}$$

We note that for large m, $\tau_0-\tau-h_m>\epsilon-h_m>0$. It follows from Hoeffding's inequality for binomial random variables that

$$\begin{aligned}P\left(\left|\frac{B(m,\tau+h_m)}{m}-(\tau+h_m)\right|\geq\tau_0-\tau-h_m\right)&\leq 2e^{-2m(\tau_0-\tau-h_m)^2}\\&\leq 2e^{-2m(\epsilon-h_m)^2}.\end{aligned}$$

Hence as $m \to \infty$,

$$\sum_{i \geq m\tau_0} b_{i,m}(\tau + h_m) \to 0.$$

This result coupled with the observation that both $G_1(x)$ and $G_2(x)$ are continuous functions on $[0,1]$ implies that as $m \to \infty$,

$$(1-\delta) \sum_{i \geq m\tau_0} b_{i,m}(x)[G_2(\frac{i}{m}) - G_1(\frac{i}{m})] \to 0.$$

We conclude that

$$\sup_{\tau < \tau_0 - \epsilon} \left| B_m F(\tau + h_m) - F(\tau) \right| \to 0.$$

In the case of $\tau > \tau_0 + \epsilon$, since

$$\sum_{i \geq m\tau_0} b_{i,m}(\tau + h_m) = 1 - \sum_{i < m\tau_0} b_{i,m}(\tau + h_m),$$

and $B_m F(x)$ can be rewritten as

$$\begin{aligned} B_m F(x) = \delta + (1-\delta) \sum_{i=0}^{m} b_{i,m}(x) G_2(\frac{i}{m}) - \delta \sum_{i < m\tau_0} b_{i,m}(x) \\ + (1-\delta) \sum_{i < m\tau_0} b_{i,m}(x)[G_1(\frac{i}{m}) - G_2(\frac{i}{m})], \end{aligned}$$

It is sufficient to show that

$$\sup_{\tau > \tau_0 + \epsilon} \sum_{i < m\tau_0} b_{i,m}(\tau + h_m) \to 0.$$

But this conclusion follows readily from

$$\begin{aligned} P(\frac{B(m, \tau + h_m)}{m} < \tau_0) &= P(\frac{B(m, \tau + h_m)}{m} - (\tau + h_m) < \tau_0 - (\tau + h_m)) \\ &\leq P(\left|\frac{B(m, \tau + h_m)}{m} - (\tau + h_m)\right| > \tau - \tau_0 + h_m) \\ &\leq 2e^{-2m(\tau - \tau_0 + h_m)^2} \leq 2e^{-2m(\epsilon + h_m)^2} \to 0. \end{aligned}$$

□

Lemma 2. *Assume that $h_m \to 0$ and $mh_m^2 \to \infty$ as $m \to \infty$. Then*

$$B_m F(\tau_0 \pm h_m) \longrightarrow F(\tau_0 \pm 0).$$

Proof. By the same decomposition as given in Lemma 1, we have

$$B_mF(\tau_0 - h_m) = (1-\delta)\sum_{i=0}^{m} b_{i,m}(\tau_0 - h_m)G_1(\frac{i}{m}) + \delta \sum_{i\geq m\tau_0} b_{i,m}(\tau_0 - h_m)$$
$$+ (1-\delta)\sum_{i\geq m\tau_0} b_{i,m}(\tau_0 - h_m)[G_2(\frac{i}{m}) - G_1(\frac{i}{m})].$$

The definition of $F(x)$ and $h_m \to 0$ show that $(1-\delta)\sum_{i=0}^{m} b_{i,m}(\tau_0 - h_m)G_1(\frac{i}{m}) \to F(\tau_0 - 0)$. Since $mh_m^2 \to \infty$ as $m \to \infty$, we conclude that

$$\begin{aligned}\sum_{i\geq m\tau_0} b_{i,m}(\tau_0 - h_m) &= P(\frac{B(m, \tau_0 - h_m)}{m} \geq \tau_0)\\ &\leq P(|\frac{B(m,\tau_0 - h_m)}{m} - (\tau_0 - h_m)| \geq h_m)\\ &\leq 2e^{-2mh_m^2} \to 0.\end{aligned}$$

The same proof applies to $B_mF(\tau_0 + h_m)$. □

Lemma 3. *Let $T_{m,n}(x) := B_mF_n(x) - B_mF(x)$. Then, as $n \to \infty$,*

$$\sup_{\tau} |T_{m,n}(\tau \pm h_m)| = O_p(\frac{1}{\sqrt{n}}).$$

Proof. It follows from Dvoretsky, Kiefer, and Wolfowitz's inequality that $\sqrt{n}\sup_x |F_n(x) - F(x)| = O_p(1)$. This implies that

$$\begin{aligned}|T_{m,n}(\tau \pm h_m)| &\leq |\sum_{i=0}^{m} b_{i,m}(\tau \pm h_m)| \sup_x |F_n(x) - F(x)|\\ &= \sup_x |F_n(x) - F(x)|.\end{aligned}$$

□

Armed with these lemmas, we are ready to prove our theorem.

Proof of the Theorem. By definition $\hat{\tau}_n$ is a maximum of $S_{m,n}(\tau)$. Hence $S_{m,n}(\hat{\tau}_n) \geq S_{m,n}(\tau_0)$. This inequality can be rewritten as

$$\begin{aligned}&T_{m,n}(\hat{\tau}_n + h_m) - T_{m,n}(\hat{\tau}_n - h_m) - T_{m,n}(\tau_0 + h_m)\\ &+T_{m,n}(\tau_0 - h_m) + D_m(\hat{\tau}_n) \geq D_m(\tau_0),\end{aligned}$$

where, $D_m(\tau) := B_mF(\tau + h_m) - B_mF(\tau - h_m)$. It follows from Lemma 2 that as $m \to \infty$,

$$D_m(\tau_0) \to F(\tau_0 + 0) - F(\tau_0 - 0) = \delta + (1-\delta)[G_2(\tau_0) - G_1(\tau_1)] > 0.$$

Therefore, $D_m(\tau_0) > c\eta$ for some constant $c > 0$ when n is large enough, where $\eta := \delta + (1-\delta)[G_2(\tau_0) - G_1(\tau_1)]$. We obtain for sufficiently large n

$$\begin{aligned} &P(|\hat{\tau}_n - \tau_0| > \epsilon) = P(|\hat{\tau}_n - \tau_0| > \epsilon, S_{m,n}(\hat{\tau}_n) \geq S_{m,n}(\tau_0)) \\ \leq\ & P(|\hat{\tau}_n - \tau_0| > \epsilon, T_{m,n}(\hat{\tau}_n + h_m) - T_{m,n}(\hat{\tau}_n - h_m) - T_{m,n}(\tau_0 + h_m) \\ &+ T_{m,n}(\tau_0 - h_m) + D_m(\hat{\tau}_n) > c\eta) \\ \leq\ & P(|T_{m,n}(\hat{\tau}_n + h_m) - T_{m,n}(\hat{\tau}_n - h_m) - T_{m,n}(\tau_0 + h_m) \\ &+ T_{m,n}(\tau_0 - h_m)| > c\eta/2) + P(|\hat{\tau}_n - \tau_0| > \epsilon, |D_m(\hat{\tau}_n)| > c\eta/2) \\ :=\ & \Phi_{m,n} + \Psi_{m,n}. \end{aligned}$$

It follows from Lemma 3 that the expression in the first term $\Phi_{m,n}$

$$|T_{m,n}(\hat{\tau}_n + h_m) - T_{m,n}(\hat{\tau}_n - h_m) - T_{m,n}(\tau_0 + h_m) + T_{m,n}(\tau_0 - h_m)| = O_p(\frac{1}{\sqrt{n}}),$$

hence it is $o_p(1)$ which implies that $\Phi_{m,n} \to 0$. We note that the event $\{|\hat{\tau}_n - \tau_0| > \epsilon, |D_m(\hat{\tau}_n)| > c\eta/2)\}$ in the second term $\Psi_{m,n}$ implies the event $\{\sup_{|\tau-\tau_0|>\epsilon} |D_m(\tau)| > c\eta/2\}$. Lemma 1 and use of the triangle inequality yield

$$\begin{aligned} \sup_{|\tau-\tau_0|>\epsilon} |D_m(\tau)| \leq & \sup_{|\tau-\tau_0|>\epsilon} \left|B_m F(\tau + h_m) - F(\tau)\right| \\ & + \sup_{|\tau-\tau_0|>\epsilon} \left|B_m F(\tau - h_m) - F(\tau)\right| \to 0. \end{aligned}$$

We conclude that $\Psi_{m,n} \to 0$. These two facts complete the proof of the theorem. □

5. Numerical Experiments

To assess the performance of the proposed method, we have conducted some numerical experiments. In our simulations, the functions $G_1(x)$ and $G_2(x)$ in F are chosen to be $G_1(x) = x^2$ and $G_2(x) = \exp\{x-1\}$ on the unit interval. We generate random samples of size $n = 20$ from the cdf F with the jump size $\delta = 0.0001$ and jump discontinuity at various values of τ_0. We choose $m = n/2$ and $h_m = 1/m^4$. Figure 1(a) shows the plot of $S_{m,n}(\tau)$ versus τ based on a single random sample of size $n = 20$ from F with $\tau_0 = 0.8822$. It exhibits a sharp maximum value at $\hat{\tau}_n = 0.8653$. To allow easy comparison, Figure 1(b) displays the plot of the empirical cumulative distribution function (ecdf) for the same random sample.

Another random sample of size $n = 20$ is generated from F with $\tau_0 = 0.6544$ and its plot of $S_{m,n}(\tau)$ versus τ is presented in Figure 2(a) and its

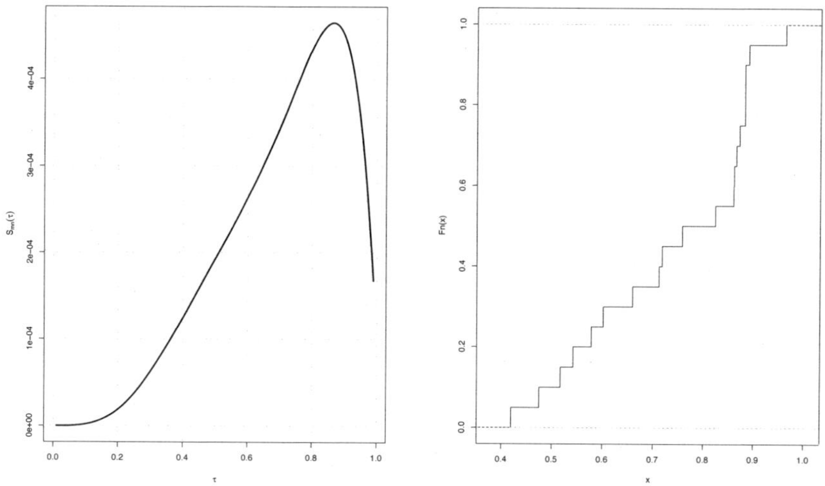

(a) (b)

Fig. 1. (a) The plot of $S_{m,n}(\tau)$ versus τ is based on a single random sample of size $n = 20$ from F with $\tau_0 = 0.8822$. (b) The plot of the empirical cumulative distribution function for the same random sample from panel (a).

ecdf is plotted in Figure 2(b). As can be seen from Figure 2(a), $S_{m,n}(\tau)$ in this case has a less sharp yet still well defined maximum value, whereas the ecdf plot from Figure 2(b) shows two jump discontinuity points with almost the same maximum jump size. The implication is that if the data are rounded, which often happens in practice, the ecdf plot could easily show multiple jump discontinuity points with the same magnitude of maximum jump size.

To evaluate the accuracy of the proposed estimator $\hat{\tau}_n$ for various values of τ_0, we carry out Monte Carlo simulations with 500 replicates and compute the point estimates and their mean squared errors based on these replicates. The results are presented in Table 1. These Monte Carlo simulations suggest that the accuracy of the proposed estimator is quite good, considering the relatively small sample size used.

Acknowledgments

The author is and will remain forever grateful to Professor P.K. Bhattacharya for bringing him to the United States and for the generous help and support during the author's graduate studies at UC Davis. Song's

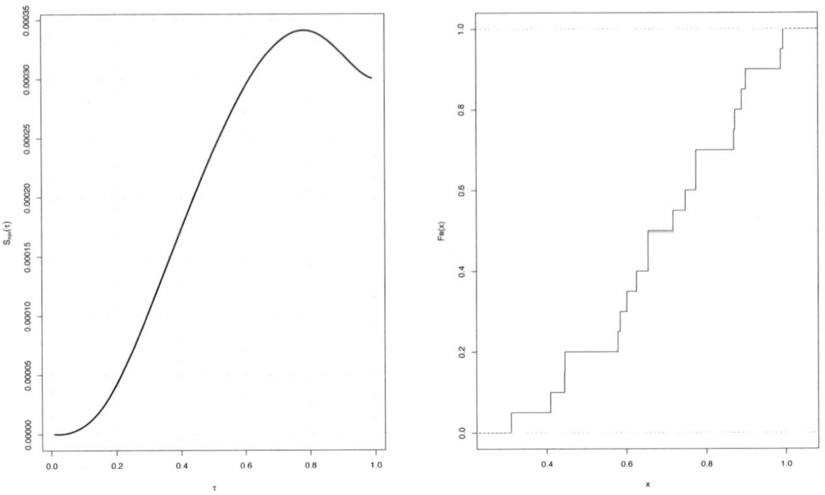

(a) (b)

Fig. 2. (a) The plot of $S_{m,n}(\tau)$ versus τ is based on a single random sample of size $n = 20$ from F with $\tau_0 = 0.6544$. (b) The plot of the empirical cumulative distribution function for the same random sample from panel (a).

Table 1. Point estimator and its root mean square error.

τ_0 (True discontinuity point)	$\hat{\tau}_n$ (Average of replicates)	RMSE Root mean square error
0.2534	0.2720	0.1324
0.4618	0.4702	0.0779
0.5582	0.5657	0.0645
0.6544	0.6590	0.0663
0.8822	0.8449	0.1008

research was supported in part by NSA Grant H98230-11-1-0140.

References

Bollen, N. P. B. and Pool, V. K. (2009). Do hedge fund managers misreport returns? Evidence from the pooled distribution. *Journal of Finance* **LXIV**, 5, 2257–2288.

Bhattacharya, P. K. (1987). Maximum likelihood estimation of a change-point in the distribution of independent random variables: General multiparameter case. *J. Multivariate Anal.* **23**, 183–208.

Bhattacharya, P. K. (1994). Some aspects of change-point analysis. In *Change-point Problems*, E. Carlstein, H.G. Muller, and D. Siegmund (eds.), IMS Lecture Notes-Monograph Series **23**, 28–56.

Bhattacharya, P. K. and Brockwell, P. J. (1976). The minimum of an additive process with applications to signal estimation and storage theory. *Z. Wahrsch. verw. Gebiete* **37**, 51–75.

Burgstahler, D. and Dichev, I. (1997). Earnings management to avoid earnings decreases and losses. *Journal of Accounting and Economics* **24**, 99–126.

Carstein, E. (1988). Nonparametric change-point estimation. *Ann. Statist.* **16**, 188–197.

Chernoff, H. and Rubin, H. (1956). The estimation of the location of a discontinuity in density. *Proc. Third Berkeley Symp. on Math. Statist. and Prob.* **1**, 19–37.

Degeorge, F., Patel, J., and Zeckhauser, R. (1999). Earnings management to exceed thresholds. *Journal of Business* **72**, 1–33.

Duggan, M. and Levitt, S. (2002). Winning isn't everything: Corruption in sumo wrestling. *American Economic Review* **92**, 1594–1605.

Dumbgen, L. (1990). The asymptotic behavior of some nonparametric change-point estimators. *Ann. Statist.* **19**, 1471–1495.

Eubank, R. L. and Speckman, P. L. (1994). Nonparametric estimation of functions with jump discontinuities. In *Change-point Problems*, E. Carlstein, H.G. Muller, and D. Siegmund (eds.), IMS Lecture Notes-Monograph Series **23**, 130–144.

Ferger, D. (1994). On the rate of almost sure convergence of Dumbgen's change-point estimators. *Stat. Probabil. Lett.* **19**, 27–31.

Gzyl, H. and Palacios, J. L. (2003). On the approximation properties of Bernstein polynomials via probabilistic tools. *Boletin de la Asociacion Matematica Venezolana* **X**, 5–13.

Hariz, S. B., Wylie, J. J., and Zhang, Q. (2007). Optimal rate of convergence for nonparametric change-point estimators for nonstationary sequences. *Ann. Statist.* **35**. 1802–1826.

Hinkley, D. V. (1970). Inference about the change-point in a sequence of random variables. *Biometrika* **57**, 1–17.

Hinkley, D. V. (1972). Time-ordered classification. *Biometrika* **59**, 509–523.

Lahr, H. (2010). Identifying discontinuities in distributions of earnings by kernel density estimation. *Working Paper, Social Science Research Network.*

Listokin, Y. (2007). Management always wins the close ones. *Working Paper, Yale Law School.*

Müller, H. G. (1992). Change-points in nonparametric regression analysis. *Ann. Statist.* **20**, 737–761.

Müller, H. G. and Song, K. S. (1997). Two-stage change-point estimators in smooth regression models. *Stat. Probabil. Lett.* **34**, 323–335.

Snyder, J. (2005). Detecting manipulation in U.S. house elections. *Working Paper, Haas School of Business, UC Berkeley.*

Takeuchi, Y. (2004). On a statistical method to detect discontinuity in the distribution function of reported earnings. *Mathematics and Computers in*

Simulation **64**, 103–111.

Wolfers, J. (2006). Point shaving: Corruption in NCAA basketball. *American Economic Review* **96**, 279–284.

Chapter 8

U-Statistics Based on Higher-Order Spacings

David D. Tung and S. Rao Jammalamadaka

ALEKS Corporation, Irvine, California, and University of California, Santa Barbara

In this paper, we investigate the asymptotic theory for U-statistics based on higher-order sample spacings. The usual asymptotic theory for U-statistics does not fully apply here because spacings are dependent variables. However, under the null hypothesis, the higher-order spacings can be represented by conditionally independent Gamma random variables. We exploit this idea to derive the relevant asymptotic theory both under the null hypothesis $H_0 : F = F_0$ and under a sequence of close alternatives.

The generalized Gini mean difference of the higher-order sample spacings is a prime example of a U-statistic of this type. It is found that the Gini mean squared difference test based on these higher-order spacings is the asymptotically locally most powerful test in this class, and has the same efficacy as the generalized Greenwood statistic based on the sum of squares of the higher-order spacings.

Contents

1. Introduction

The basic goodness-of-fit problem consists of testing whether a given data set fits a specified distribution. In particular, consider independent and identically distributed random variables $X_1, X_2, \ldots, X_{n-1}$ from a continuous distribution function F defined on the real line $\mathbb{R}$. In the statistical literature, much attention has been devoted to the nonparametric problem of simple goodness-of-fit, namely testing the null hypothesis

$$H_0 : F(x) = F_0(x),$$

where F_0 is a completely specified distribution function.

If F is assumed to be continuous as we shall do, by way of the probability integral transform, the support of F reduces to the unit interval $[0,1]$, and this also permits us to equate F_0 with the Uniform($[0,1]$) distribution. Thus, the goodness-of-fit problem reduces to one of testing uniformity, i.e. testing the null hypothesis

$$H_0 : F(x) = x \cdot I(0 \le x \le 1).$$

Let $X_{(1)}, X_{(2)}, \ldots, X_{(n-1)}$ denote the sample order statistics. Put $X_{(0)} \equiv 0$ and $X_{(n)} \equiv 1$, so that $0 = X_{(0)} \le X_{(1)} \le X_{(2)} \le \cdots \le X_{(n-1)} \le X_{(n)} = 1$. The *higher-order sample spacings* are defined by the random variables

$$D_{km}^{(m)} = X_{(km)} - X_{((k-1)m)}, \qquad \text{for } k = 1, 2, \ldots, N = \lfloor n/m \rfloor. \tag{1}$$

If F is the Uniform($[0,1]$) distribution, as under the null hypothesis, we use the notation $\{U_k\}$ for the sample observations and

$$T_{km}^{(m)} = U_{(km)} - U_{((k-1)m)}, \qquad \text{for } k = 1, 2, \ldots, N = \lfloor n/m \rfloor \tag{2}$$

for the *higher-order uniform spacings*. When $m = 1$, then $\{D_k^{(1)}\}$ or simply $\{D_k\}$ are called the *sample spacings*, and likewise $\{T_k^{(1)}\}$ or simply $\{T_k\}$ are called the *uniform spacings*. Tests based on higher-order spacings are studied here for testing the null hypothesis. Since we are mainly interested in asymptotic properties, we assume that $N = n/m$, without loss of generality.

Let $h : [0,\infty) \times [0,\infty) \to \mathbb{R}$ be a symmetric function in its arguments satisfying some regularity conditions. Consider the general test statistic

$$W_N(h) = \frac{2}{N(N-1)} \sum_{1 \le i < j \le N} h(nD_{im}^{(m)}, nD_{jm}^{(m)}) \tag{3}$$

which is a second-order U-statistic of the higher-order sample spacings and symmetric in the pairs $(D_{im}^{(m)}, D_{jm}^{(m)})$. An important example of such a statistic is the generalized Gini mean difference of the higher-order sample spacings, i.e.

$$G_N(r) = \frac{2}{N(N-1)} \sum_{1 \le i < j \le N} \left| nD_{im}^{(m)} - nD_{jm}^{(m)} \right|^r, \qquad r > 0 \tag{4}$$

which is an average of absolute pairwise differences of the higher-order sample spacings to the rth power. The special case of $G_N(1)$ is the Gini mean difference higher-order spacings test, and the special case of $G_N(2)$ will be called the Gini mean squared difference higher-order spacings test.

On the other hand, symmetric sum-functions of the higher-order spacings, i.e. general test statistics of the form

$$V_N(g) = \frac{1}{N}\sum_{k=1}^{N} g\left(nD_{km}^{(m)}\right), \tag{5}$$

where $g(\cdot)$ is a real-valued function satisfying some regularity conditions, are symmetric in $\{D_{km}^{(m)}\}$, and can also be thought of as first-order U-statistics of the higher-order sample spacings.

For the special case of $m = 1$ where $N = n$, note that the general test statistic $W_N(h)$ reduces to

$$W_n(h) = \frac{2}{n(n-1)} \sum_{1\le i<j\le n} h(nD_i, nD_j) \tag{6}$$

and the general test statistic $V_N(g)$ reduces to

$$V_n(g) = \frac{1}{n}\sum_{k=1}^{n} g(nD_k). \tag{7}$$

Likewise, the statistic $G_N(r)$ reduces to

$$G_n(r) = \frac{2}{n(n-1)} \sum_{1\le i<j\le n} |nD_i - nD_j|^r, \qquad r > 0 \tag{8}$$

the generalized Gini mean difference of the sample spacings.

The generalized Gini mean difference spacings test $G_n(1)$ was proposed in Jammalamadaka and Goria (2004) for testing goodness-of-fit. There they derive both the exact and asymptotic distribution of $G_n(1)$ under the null hypothesis, and show that it has good performance based on Monte Carlo powers. More generally, the asymptotic theory for $W_n(h)$ has been studied in Tung and Jammalamadaka (2010). They show that the asymptotically locally most powerful (ALMP) test for U-statistics based on the spacings is the Gini mean squared difference spacings test

$$G_n(2) = \frac{1}{n(n-1)}\sum_{i=1}^{n}\sum_{j=1}^{n}(nD_i - nD_j)^2 \tag{9}$$

which corresponds to the classical Greenwood statistic $\frac{1}{n}\sum_{k=1}^{n}(nD_k)^2$ based on the sum of squares of the spacings. The importance of the

Greenwood statistic is somewhat justified in view of the result established in Sethuraman and Rao (1970) that among the class of symmetric sum-functions of the spacings, i.e., for $V_n(g)$, the Greenwood statistic is the ALMP test.

There are asymptotically more efficient procedures than the Greenwood statistic if one takes into consideration test statistics based on higher-order spacings. Del Pino (1979) showed that among the class of symmetric sum-functions of the higher-order spacings, i.e., for $V_N(g)$, the generalized Greenwood statistic, i.e., $\frac{1}{N}\sum_{k=1}^{N}(nD_{km}^{(m)})^2$ based on the sum of squares of these higher-order spacings is the ALMP test, and asymptotically more efficient than the classical Greenwood statistic.

One perceived weakness of symmetric spacings tests is that they cannot discriminate alternatives converging to the null hypothesis at a rate faster than $n^{-1/4}$, and hence have poor asymptotic performance as compared to say the Kolmogorov–Smirnov test.

The asymptotic distribution for second-order U-statistics of the higher-order spacings under the null hypothesis is studied in the next section. Section 3 deals with their asymptotic behavior under a sequence of close alternatives. Section 4 contains results on the ALMP test for the class of U-statistics of the higher-order sample spacings.

2. The Asymptotic Null Distribution

In this section, we obtain the asymptotic distribution for second-order U-statistics of the higher-order uniform spacings, i.e., general test statistics of the form

$$W_N(h) = \frac{2}{N(N-1)} \sum_{1\le i<j\le N} h(nT_{im}^{(m)}, nT_{jm}^{(m)}) \tag{10}$$

under the null hypothesis. Since we are only interested in asymptotic properties, we shall assume that $N = n/m$, without loss of generality, and increases to infinity while m is fixed. Under the null hypothesis, the higher-order uniform spacings have the well-known conditional representation

$$\left(nT_{1m}^{(m)}, nT_{2m}^{(m)}, \ldots, nT_{Nm}^{(m)}\right) \simeq \left(S_1, S_2, \ldots, S_N \,\middle|\, \bar{S}_N = m\right), \tag{11}$$

where $S_1, S_2, \ldots, S_N$ are independent Gamma$(m, 1)$ random variables (cf. with Sobel and Wells (1990) and Rao and Sobel (1980)). Note that for

$m = 1$ where $N = n$, this reduces to

$$(nT_1, nT_2, \ldots, nT_n) \simeq \left(Z_1, Z_2, \ldots, Z_n \,\middle|\, \bar{Z}_n = 1 \right), \tag{12}$$

which is a conditional representation of the uniform spacings in terms of independent Exponential(1) random variables $Z_1, Z_2, \ldots, Z_n$. Here as elsewhere, we use $\simeq$ to denote the distributional equivalence of quantities on the left and right hand sides of the symbol.

There are at least two known approaches to deriving the asymptotic null distribution. One approach is by applying a conditional limit theorem for U-statistics due to Holst (1981, Theorem 6.2). A second approach is by way of the well-known Hoeffding decomposition for U-statistics, which connects the asymptotic theory for U-statistics of the higher-order uniform spacings with the asymptotic theory for symmetric sum-functions of the higher-order uniform spacings.

Let

$$U_N = \frac{2}{N(N-1)} \sum_{1 \le i < j \le N} h(S_i, S_j) \tag{13}$$

be a second-order U-statistic based on the independent Gamma$(m, 1)$ random variables $S_1, S_2, \ldots, S_N$, where the kernel $h : [0, \infty) \times [0, \infty) \to \mathbb{R}$ is a symmetric function with $\mathrm{Var}[h(S_1, S_2)] < \infty$.

For the case of independent and identically distributed random variables, the Hoeffding decomposition (cf. Lee (1990, Sec. 1.6)) asserts that a U-statistic of order k is a linear combination of uncorrelated U-statistics of order $1, 2, \ldots, k$. The case for $k = 2$ has been most studied and best understood. We state the Hoeffding decomposition for U_N in the following.

Lemma 1. *The Hoeffding decomposition of U_N has the form*

$$U_N = \theta + \frac{2}{N} \sum_{k=1}^{N} h^{(1)}(S_k) + \frac{2}{N(N-1)} \sum_{1 \le i < j \le N} h^{(2)}(S_i, S_j) \tag{14}$$

where

$$\begin{aligned} \theta &= \mathbb{E}[h(S_1, S_2)] \\ g(t) &= \mathbb{E}[h(t, S_2)] = \mathbb{E}[h(S_1, S_2) \mid S_1 = t] \\ h^{(1)}(t) &= g(t) - \theta \\ h^{(2)}(s_1, s_2) &= h(s_1, s_2) - g(s_1) - g(s_2) + \theta. \end{aligned}$$

Moreover, the normalized U-statistic

$$\sqrt{N}(U_N - \theta) = \frac{2}{\sqrt{N}} \sum_{k=1}^{N} [g(S_k) - \theta] + N^{1/2} R_N \tag{15}$$

where

$$N^{1/2} R_N = \frac{2\sqrt{N}}{N(N-1)} \sum_{1 \le i < j \le N} h^{(2)}(S_i, S_j) = o_P(1). \tag{16}$$

The following result provides some useful identities for the expectations, variances, and covariances of the U-statistic kernels in the Hoeffding decomposition.

Lemma 2. *Let S_1, S_2 and S_3 be independent Gamma$(m, 1)$ random variables, and let g be defined as in Lemma 1. Then*

$$\begin{aligned}
\mathbb{E}[h(S_1, S_2)] &= \mathbb{E}[g(S_1)] \\
Cov[h(S_1, S_2), h(S_1, S_3)] &= Var[g(S_1)] \\
Cov[h(S_1, S_2), S_1] &= Cov[g(S_1), S_1] \\
& Cov[h(S_1, S_2), (S_1 - m - 1)^2 + (S_2 - m - 1)^2] \\
&= 2 \cdot Cov[g(S_1), (S_1 - m - 1)^2].
\end{aligned}$$

The next result gives the asymptotic null distribution for symmetric sum-functions of the higher-order uniform spacings (cf. with Del Pino (1979)). We use the notation $N_1(\mu, \sigma^2)$ to denote the one-dimensional Normal distribution with mean μ and variance σ^2.

Lemma 3. *Under the null hypothesis, in the limit as $N \to \infty$,*

$$\frac{1}{\sqrt{N}} \sum_{k=1}^{N} [g(nT_{km}^{(m)}) - \mathbb{E}g(S_1)] \xrightarrow{D} N_1(0, \sigma^2(g)) \tag{17}$$

where

$$\sigma^2(g) = Var[g(S_1)] - \frac{Cov^2[g(S_1), S_1]}{Var[S_1]}. \tag{18}$$

The next result gives the asymptotic distribution for U-statistics based on the higher-order uniform spacings under the null hypothesis.

Theorem 1. *Under the null hypothesis, in the limit as $N \to \infty$,*

$$\sqrt{N} \left(\frac{2}{N(N-1)} \sum_{1 \le i < j \le N} h(nT_{im}^{(m)}, nT_{jm}^{(m)}) - \mathbb{E}[h(S_1, S_2)] \right) \xrightarrow{D} N_1(0, \sigma^2(h)), \tag{19}$$

where

$$\begin{aligned}\sigma^2(h) &= 4(\sigma_1^2 - \sigma_{12}^2)\\ \sigma_1^2 &= Cov[h(S_1, S_2), h(S_1, S_3)]\\ \sigma_{12}^2 &= \frac{Cov^2[h(S_1, S_2), S_1]}{Var[S_1]}.\end{aligned}$$

Proof. By Lemma 1, and the conditional representation of the higher-order uniform spacings (11), we have

$$\begin{aligned}&\sqrt{N}\left(\frac{2}{N(N-1)}\sum_{1\le i<j\le N} h(nT_{im}^{(m)}, nT_{jm}^{(m)}) - \mathbb{E}[h(S_1, S_2)]\right)\\ &\simeq \left(\sqrt{N}(U_N - \theta)\,\middle|\, \bar{S}_N = m\right)\\ &\simeq \left(\frac{2}{\sqrt{N}}\sum_{k=1}^{N}[g(S_k) - \mathbb{E}g(S_1)]\,\middle|\, \bar{S}_N = m\right) + o_P(1)\\ &\simeq \frac{2}{\sqrt{N}}\sum_{k=1}^{N}[g(nT_{km}^{(m)}) - \mathbb{E}g(S_1)] + o_P(1) \xrightarrow{D} N_1(0, 4\sigma^2(g)).\end{aligned}$$

The convergence in distribution to the $N_1(0, 4\sigma^2(g))$ distribution follows from Lemma 3 and Slutsky's Theorem. By Lemma 2, the asymptotic variance

$$4\sigma^2(g) = 4\left(\mathrm{Var}[g(S_1)] - \frac{\mathrm{Cov}^2[g(S_1), S_1]}{\mathrm{Var}[S_1]}\right) = 4(\sigma_1^2 - \sigma_{12}^2) = \sigma^2(h). \quad (20)$$

This completes the proof. □

3. The Asymptotic Distribution Under a Sequence of Close Alternatives

In this section, we derive the asymptotic distribution for second-order U-statistics of the higher-order sample spacings under a sequence of close alternatives. In order to study asymptotic efficiencies, one needs to obtain the asymptotic distribution of test statistics under a sequence of close alternatives (also called smooth alternatives), which converges to the null hypothesis. Thus, we specify the alternative hypothesis by a sequence of distribution functions $\{F_n(x) : n \ge 1\}$ that converges to the Uniform([0, 1]) distribution function, which corresponds to the null hypothesis, in the limit as $n \to \infty$.

For symmetric spacings tests, the appropriate sequence of close alternatives (cf. with Sethuraman and Rao (1970) and Rao and Sethuraman (1975)) is obtained by letting the distribution function

$$F_n(x) = x + \frac{L_n(x)}{n^{1/4}}, \qquad \text{for } 0 \leq x \leq 1 \tag{21}$$

where $L_n(0) = L_n(1) = 0$. We further assume that $L_n(x)$ is twice differentiable on the unit interval $[0, 1]$ and that there exists a function $L(x)$ which is twice continuously-differentiable with $L(0) = L(1) = 0$ and

$$\begin{aligned} n^{1/4} \sup_{0\leq x\leq 1} |L_n(x) - L(x)| &= o(1) \\ n^{1/4} \sup_{0\leq x\leq 1} |L_n'(x) - l(x)| &= o(1) \\ n^{1/4} \sup_{0\leq x\leq 1} |L_n''(x) - l'(x)| &= o(1), \end{aligned}$$

where $l(x)$ and $l'(x)$ are respectively the first and second derivatives of $L(x)$. Note that $L(x) = \int_0^x l(u)\,du$ and $\int_0^1 l(u)\,du = 0$ by the fundamental theorem of calculus.

The next result gives, under a sequence of close alternatives, the asymptotic distribution for symmetric sum-functions of the higher-order spacings (cf. Del Pino (1979)).

Lemma 4. *Under the close alternatives* (21), *in the limit as* $N \to \infty$,

$$\frac{1}{\sqrt{N}} \sum_{k=1}^{N} [g(nD_{km}^{(m)}) - \mathbb{E}g(S_1)] \xrightarrow{D} N_1(\mu(g), \sigma^2(g)), \tag{22}$$

where

$$\begin{aligned} \mu(g) &= \frac{m^{-1/2}}{2} \left(\int_0^1 l^2(u)\,du \right) Cov[g(S_1), (S_1 - m - 1)^2] \\ \sigma^2(g) &= Var[g(S_1)] - \frac{Cov^2[g(S_1), S_1]}{Var[S_1]}. \end{aligned}$$

Let $0 = \xi_0 < \xi_1 < \xi_2 < \cdots < \xi_n < \xi_{N+1} = 1$ form a partition of the unit interval $[0, 1]$, where

$$\xi_k = \frac{k}{N+1}, \qquad \text{for } k = 0, 1, 2, \ldots, N+1.$$

Under the close alternatives, the higher-order spacings $\{D_{km}^{(m)}\}$ are related to the uniform spacings $\{T_{km}^{(m)}\}$ by the relation

$$\begin{aligned} nD_{km}^{(m)} &= n[F_n^{-1}(U_{(km)}) - F_n^{-1}(U_{((k-1)m)})] \\ &= nT_{km}^{(m)} + \left(\frac{-l(\xi_k)}{n^{1/4}} + \frac{l^2(\xi_k) + L(\xi_k)l'(\xi_k)}{n^{1/2}}\right)(nT_{km}^{(m)}) + o_P(n^{-1/2}) \end{aligned} \tag{23}$$

where $o_P(\cdot)$ is uniform in k. This follows from the mean value theorem for differential calculus and a continuity argument found in Rao and Sethuraman (1975).

We assume that $h : [0,\infty)\times[0,\infty) \to \mathbb{R}$ is a symmetric function with first and second-order partial derivatives. We use the notation $h_x(x,y) = \frac{\partial h}{\partial x}$ and $h_y(x,y) = \frac{\partial h}{\partial y}$ to denote the first partial derivatives of h, and use $h_{xx}(x,y) = \frac{\partial^2 h}{\partial x^2}$, $h_{yy}(x,y) = \frac{\partial^2 h}{\partial y^2}$ and $h_{xy}(x,y) = \frac{\partial^2 h}{\partial x \partial y}$ to denote the second-order partial derivatives of h.

By using (23), we prove the following.

Lemma 5. *Under the close alternatives* (21), *in the limit as* $N \to \infty$,

$$\begin{aligned} &\frac{2\sqrt{N}}{N(N-1)} \sum_{1\le i<j\le N} [h(nD_{im}^{(m)}, nD_{jm}^{(m)}) - h(nT_{im}^{(m)}, nT_{jm}^{(m)})] \\ &\xrightarrow{\mathbb{P}} \frac{m^{-1/2}}{2}\left(\int_0^1 l^2(u)\,du\right)\cdot Cov[h(S_1,S_2),(S_1-m-1)^2 + (S_2-m-1)^2]. \end{aligned} \tag{24}$$

Proof. Using (23) in a two-dimensional Taylor expansion of $h(nD_{im}^{(m)}$, $nD_{jm}^{(m)})$ around $h(nT_{im}^{(m)}$, $nT_{jm}^{(m)})$ and summing over all $i<j$ gives

$$\begin{aligned} &\frac{2\sqrt{N}}{N(N-1)} \sum_{1\le i<j\le N} [h(nD_{im}^{(m)}, nD_{jm}^{(m)}) - h(nT_{im}^{(m)}, nT_{jm}^{(m)})] \\ &= -\frac{2m^{-1/4}N^{1/4}}{N(N-1)} \sum_{1\le i<j\le N} l(\xi_i)(nT_{im}^{(m)})\, h_x(nT_{im}^{(m)}, nT_{jm}^{(m)}) \\ &\quad - \frac{2m^{-1/4}N^{1/4}}{N(N-1)} \sum_{1\le i<j\le N} l(\xi_j)(nT_{jm}^{(m)})\, h_y(nT_{im}^{(m)}, nT_{jm}^{(m)}) \\ &\quad + \frac{2m^{-1/2}}{N(N-1)} \sum_{1\le i<j\le N} [l^2(\xi_i) + L(\xi_i)l'(\xi_i)](nT_{im}^{(m)})\, h_x(nT_{im}^{(m)}, nT_{jm}^{(m)}) \end{aligned}$$

$$+ \frac{2m^{-1/2}}{N(N-1)} \sum_{1 \le i < j \le N} [l^2(\xi_j) + L(\xi_j)l'(\xi_j)](nT_{jm}^{(m)})\, h_y(nT_{im}^{(m)}, nT_{jm}^{(m)})$$

$$+ \frac{2m^{-1/2}}{N(N-1)} \sum_{1 \le i < j \le N} l(\xi_i)l(\xi_j)(nT_{im}^{(m)})(nT_{jm}^{(m)})\, h_{xy}(nT_{im}^{(m)}, nT_{jm}^{(m)})$$

$$+ \frac{m^{-1/2}}{N(N-1)} \sum_{1 \le i < j \le N} l^2(\xi_i)(nT_{im}^{(m)})^2\, h_{xx}(nT_{im}^{(m)}, nT_{jm}^{(m)})$$

$$+ \frac{m^{-1/2}}{N(N-1)} \sum_{1 \le i < j \le N} l^2(\xi_j)(nT_{jm}^{(m)})^2\, h_{yy}(nT_{im}^{(m)}, nT_{jm}^{(m)}) + o_P(1).$$

The composite trapezoid rule asserts that there exists a number $c \in (0,1)$ for which

$$\int_0^1 l(u)\, du = \frac{1}{N+1} \sum_{k=1}^{N} l(\xi_k) - \frac{l''(c)}{12(N+1)^2}. \tag{25}$$

Since we have

$$\lim_{N\to\infty} N^{1/4} \left(\frac{1}{N+1} \sum_{k=1}^{N} l(\xi_k) - \int_0^1 l(u)\, du \right) = \lim_{N\to\infty} \frac{N^{1/4} \cdot l''(c)}{12(N+1)^2} = 0, \tag{26}$$

then the first two terms on the RHS of (24) converge in probability to zero.

Since $(nT_{im}^{(m)}, nT_{jm}^{(m)}) \xrightarrow{D} (S_1, S_2)$, as $n \to \infty$, observe that both

$$\frac{2m^{-1/2}}{N(N-1)} \sum_{1 \le i < j \le N} [l^2(\xi_i) + L(\xi_i)l'(\xi_i)](nT_{im}^{(m)})\, h_x(nT_{im}^{(m)}, nT_{jm}^{(m)})$$

$$\xrightarrow{\mathbb{P}} m^{-1/2} \cdot \mathbb{E}[S_1\, h_x(S_1, S_2)] \int_0^1 [l^2(x) + L(x)l'(x)]\, dx = 0$$

and

$$\frac{2m^{-1/2}}{N(N-1)} \sum_{1 \le i < j \le N} [l^2(\xi_j) + L(\xi_j)l'(\xi_j)](nT_{jm}^{(m)})\, h_y(nT_{im}^{(m)}, nT_{jm}^{(m)})$$

$$\xrightarrow{\mathbb{P}} m^{-1/2} \cdot \mathbb{E}[S_2\, h_y(S_1, S_2)] \int_0^1 [l^2(y) + L(y)l'(y)]\, dy = 0$$

because from integration by parts

$$\int_0^1 L(u)\, l'(u)\, du = - \int_0^1 l^2(u)\, du.$$

Observe also that

$$\frac{2m^{-1/2}}{N(N-1)} \sum_{1\le i<j\le N} l(\xi_i)l(\xi_j)(nT_{im}^{(m)})(nT_{jm}^{(m)})\, h_{xy}(nT_{im}^{(m)}, nT_{jm}^{(m)})$$
$$\xrightarrow{\mathbb{P}} m^{-1/2}\left(\int_0^1\int_0^1 l(x)\,l(y)\,dx\,dy\right)\cdot \mathbb{E}[S_1\cdot S_2\, h_{xy}(S_1,S_2)] = 0$$

because

$$\int_0^1\int_0^1 l(x)\,l(y)\,dx\,dy = \left(\int_0^1 l(x)\,dx\right)\left(\int_0^1 l(y)\,dy\right) = 0. \tag{27}$$

Moreover, we have

$$\frac{m^{-1/2}}{N(N-1)} \sum_{1\le i<j\le N} l^2(\xi_i)(nT_{im}^{(m)})^2\, h_{xx}(nT_{im}^{(m)}, nT_{jm}^{(m)})$$
$$\xrightarrow{\mathbb{P}} \frac{m^{-1/2}}{2}\left(\int_0^1 l^2(x)\,dx\right)\cdot \mathbb{E}[S_1^2\, h_{xx}(S_1,S_2)]$$

and

$$\frac{m^{-1/2}}{N(N-1)} \sum_{1\le i<j\le N} l^2(\xi_j)(nT_{jm}^{(m)})^2\, h_{yy}(nT_{im}^{(m)}, nT_{jm}^{(m)})$$
$$\xrightarrow{\mathbb{P}} \frac{m^{-1/2}}{2}\left(\int_0^1 l^2(y)\,dy\right)\cdot \mathbb{E}[S_2^2\, h_{yy}(S_1,S_2)]$$

with

$$\mathbb{E}[S_1^2\, h_{xx}(S_1,S_2) + S_2^2\, h_{yy}(S_1,S_2)]$$
$$= \mathrm{Cov}[h(S_1,S_2), (S_1-m-1)^2 + (S_2-m-1)^2].$$

This completes the proof. □

By combining Theorem 1 with Lemma 5, we have the following.

Theorem 2. *Under the close alternatives* (21), *in the limit as* $N\to\infty$,

$$\sqrt{N}\left(\frac{2}{N(N-1)}\sum_{1\le i<j\le N} h(nD_{im}^{(m)}, nD_{jm}^{(m)}) - \mathbb{E}[h(S_1,S_2)]\right)$$
$$\xrightarrow{D} N_1(\mu(h), \sigma^2(h)),$$

where

$$\mu(h) = \frac{m^{-1/2}}{2}\left(\int_0^1 l^2(u)\,du\right)\cdot Cov[h(S_1,S_2),(S_1-m-1)^2 + (S_2-m-1)^2]$$

$$\sigma^2(h) = 4(\sigma_1^2 - \sigma_{12}^2)$$
$$\sigma_1^2 = Cov[h(S_1,S_2),h(S_1,S_3)]$$
$$\sigma_{12}^2 = \frac{Cov^2[h(S_1,S_2),S_1]}{Var[S_1]}.$$

Proof. Put

$$\begin{aligned}
&\sqrt{N}\left(\frac{2}{N(N-1)}\sum_{1\le i<j\le N} h(nD_{im}^{(m)},nD_{jm}^{(m)}) - \mathbb{E}[h(S_1,S_2)]\right)\\
&= \sqrt{N}\left(\frac{2}{N(N-1)}\sum_{1\le i<j\le N} h(nT_{im}^{(m)},nT_{jm}^{(m)}) - \mathbb{E}[h(S_1,S_2)]\right)\\
&\quad + \frac{2\sqrt{N}}{N(N-1)}\sum_{1\le i<j\le N}[h(nD_{im}^{(m)},nD_{jm}^{(m)}) - h(nT_{im}^{(m)},nT_{jm}^{(m)})]\\
&\xrightarrow{D} N_1(\mu(h),\sigma^2(h)).
\end{aligned}$$

The convergence in distribution follows from Slutsky's Theorem, where the first term on the RHS converges to the $N_1(0,\sigma^2(h))$ distribution by Theorem 1, and the second term converges in probability to $\mu(h)$ by Lemma 5. This completes the proof. □

4. The Asymptotically Locally Most Powerful Test

Recall that $\mu(g)$ and $\sigma^2(g)$ denote the asymptotic mean and asymptotic variance corresponding to the general test statistic

$$V_N(g) = \frac{1}{N}\sum_{k=1}^{N} g(nD_{km}^{(m)})$$

under the sequence of close alternatives. Here it is assumed that $V_N(g)$ has been normalized to have asymptotic mean zero and finite variance under the null hypothesis. The *Pitman asymptotic relative efficiency* (ARE) of

$V_N(g_1)$ relative to $V_N(g_2)$ is given by

$$\mathrm{ARE}(g_1, g_2) = \left(\frac{e^2(g_1)}{e^2(g_2)}\right)^2 = \frac{\left(\frac{\mu^2(g_1)}{\sigma^2(g_1)}\right)^2}{\left(\frac{\mu^2(g_2)}{\sigma^2(g_2)}\right)^2}. \tag{28}$$

The quantity

$$e^2(g) = \frac{\mu^2(g)}{\sigma^2(g)} \tag{29}$$

is called the *efficacy* of the test $V_N(g)$. A test with maximum efficacy is the asymptotically locally most powerful (ALMP) test. In order to find the ALMP test, for symmetric sum-functions of the higher-order spacings, against the close alternatives, one needs to find a function $g(\cdot)$ which maximizes

$$e(g) = \frac{m^{-1/2}\left(\int_0^1 l^2(u)\,du\right)\mathrm{Cov}[g(S_1), (S_1 - m - 1)^2]}{2\left\{\mathrm{Var}[g(S_1)] - m^{-1}\cdot \mathrm{Cov}^2[g(S_1), S_1]\right\}^{1/2}}. \tag{30}$$

As mentioned before, the importance of the generalized Greenwood statistic is somewhat justified by the next two results, which were established by Del Pino (1979). The generalized Greenwood statistic is the ALMP test among the class of symmetric sum-functions of the higher-order spacings.

Lemma 6. *The functional $e(g)$ is maximized by taking $g(t) = t^2$, which in turn gives*

$$\max e(g) = \sqrt{\frac{m+1}{2}}\left(\int_0^1 l^2(u)\,du\right).$$

Lemma 7. *For symmetric sum-functions of the higher-order spacings, the asymptotically locally most powerful (ALMP) test of the null hypothesis against the sequence of close alternatives is to reject the null hypothesis when*

$$\sum_{k=1}^{N}\left(nD_{km}^{(m)}\right)^2 > C(\alpha), \tag{31}$$

where the critical value $C(\alpha)$ is determined by the level of significance α. The asymptotic distribution of this optimal statistic under the sequence of *close alternatives (21) is given by*

$$\frac{1}{\sqrt{N}}\sum_{k=1}^{N}\left[\left(nD_{km}^{(m)}\right)^2 - m(m+1)\right]$$

$$\xrightarrow{D} N_1\left(m^{1/2}(m+1)\left(\int_0^1 l^2(u)\,du\right), 2m(m+1)\right).$$

The asymptotic distribution under the null hypothesis is obtained by taking $l(u) = 0$ in the above.

Recall that $\mu(h)$ and $\sigma^2(h)$ denote the asymptotic mean and asymptotic variance corresponding to the general test statistic

$$W_N(h) = \frac{2}{N(N-1)}\sum_{1\le i<j\le N} h(nD_{im}^{(m)}, nD_{jm}^{(m)}) \tag{32}$$

under the sequence of close alternatives. It is assumed that $W_N(h)$ has been normalized to have asymptotic mean zero and finite variance under the null hypothesis. The Pitman ARE of $W_N(h_1)$ relative to $W_N(h_2)$ is given by

$$\mathrm{ARE}(h_1, h_2) = \left(\frac{e^2(h_1)}{e^2(h_2)}\right)^2 = \frac{\left(\frac{\mu^2(h_1)}{\sigma^2(h_1)}\right)^2}{\left(\frac{\mu^2(h_2)}{\sigma^2(h_2)}\right)^2} \tag{33}$$

and the quantity

$$e^2(h) = \frac{\mu^2(h)}{\sigma^2(h)} \tag{34}$$

is the efficacy of the test $W_N(h)$.

In order to find the ALMP test, for U-statistics of the higher-order spacings, against the sequence of close alternatives, we need to find a function h which maximizes the functional

$$e(h) = \frac{m^{-1/2}\left(\int_0^1 l^2(u)\,du\right)\cdot \mathrm{Cov}[h(S_1,S_2),(S_1-m-1)^2+(S_2-m-1)^2]}{4\left\{\mathrm{Cov}[h(S_1,S_2),h(S_1,S_3)] - m^{-1}\cdot\mathrm{Cov}^2[h(S_1,S_2),S_1]\right\}^{1/2}}. \tag{35}$$

Lemma 8. *The functional $e(h)$ is maximized by taking the symmetric function $h(s_1, s_2) = (s_1 - s_2)^2$, which in turn gives*

$$\max e(h) = \sqrt{\frac{m+1}{2}}\left(\int_0^1 l^2(u)\,du\right) \tag{36}$$

which corresponds to that of the generalized Greenwood statistic.

Proof. It is enough to find a function h which maximizes the numerator in (35). By the Cauchy–Bunyakovsky–Schwarz Inequality, we have

$$\begin{aligned}
&e(h)\\
&= \frac{m^{-1/2}\left(\int_0^1 l^2(u)\,du\right)\cdot \mathrm{Cov}[h(S_1,S_2),(S_1-m-1)^2+(S_2-m-1)^2]}{4\left\{\mathrm{Cov}[h(S_1,S_2),h(S_1,S_3)]-m^{-1}\cdot \mathrm{Cov}^2[h(S_1,S_2),S_1]\right\}^{1/2}}\\
&\le m^{-1/2}\left(\int_0^1 l^2(u)\,du\right)\\
&\qquad\times\frac{\sqrt{\mathrm{Var}[h(S_1,S_2)]}\sqrt{\mathrm{Var}[(S_1-m-1)^2+(S_2-m-1)^2]}}{4\left\{\mathrm{Cov}[h(S_1,S_2),h(S_1,S_3)]-m^{-1}\cdot \mathrm{Cov}^2[h(S_1,S_2),S_1]\right\}^{1/2}}.
\end{aligned} \tag{37}$$

The inequalities become equalities if and only if $h(s_1,s_2)=a[(s_1-m-1)^2+(s_2-m-1)^2]+b$, for some real numbers $a\neq 0$ and b. In this particular case, the functional $e(h)$ attains the upper bound in (37), i.e.

$$\begin{aligned}
\max e(h) &= \frac{m^{-1/2}\left(\int_0^1 l^2(u)\,du\right)a\cdot \mathrm{Var}[(S_1-m-1)^2+(S_2-m-1)^2]}{4\sqrt{a^2\cdot \mathrm{Var}(S_1-m-1)^2-m^{-1}\cdot a^2\cdot \mathrm{Cov}^2[(S_1-m-1)^2,S_1]}}\\
&= \frac{m^{-1/2}\cdot\left(\int_0^1 l^2(u)\,du\right)a\cdot 4\cdot \mathbb{E}[S_1^2]}{4\sqrt{2m(m+1)a^2}}\\
&= \frac{m^{-1/2}m(m+1)\cdot\left(\int_0^1 l^2(u)\,du\right)}{\sqrt{2m(m+1)}}\\
&= \sqrt{\frac{m+1}{2}}\left(\int_0^1 l^2(u)\,du\right).
\end{aligned}$$

On the other hand, since $\mathrm{Cov}[S_1S_2,(S_1-m-1)^2+(S_2-m-1)^2]=0$, the maximum of the functional $e(h)$ is attained by taking $h(s_1,s_2)=(s_1-s_2)^2$, and directly from (35) we have

$$
\begin{aligned}
&\max e(h) \\
&= \frac{m^{-1/2}\left(\int_0^1 l^2(u)\,du\right)\cdot \mathrm{Cov}[(S_1-S_2)^2,(S_1-m-1)^2+(S_2-m-1)^2]}{4\left\{\mathrm{Cov}[(S_1-S_2)^2,(S_1-S_3)^2]-m^{-1}\cdot\mathrm{Cov}^2[(S_1-S_2)^2,S_1]\right\}^{1/2}} \\
&= \frac{m^{-1/2}\left(\int_0^1 l^2(u)\,du\right)\cdot 2\cdot \mathbb{E}[S_1^2+S_2^2]}{4\sqrt{2m(3+m)-4m}} \\
&= \frac{4m^{1/2}(m+1)\left(\int_0^1 l^2(u)\,du\right)}{4\sqrt{2m(m+1)}} \\
&= \sqrt{\frac{m+1}{2}}\left(\int_0^1 l^2(u)\,du\right).
\end{aligned}
$$

This completes the proof. □

By combining Theorem 2, and Lemma 8, we have the following.

Theorem 3. *For U-statistics of the higher-order spacings, the asymptotically locally most powerful (ALMP) test of the null hypothesis against the sequence of close alternatives is to reject the null hypothesis when*

$$\sum_{1\le i<j\le N}\left(nD_{im}^{(m)}-nD_{jm}^{(m)}\right)^2 > C(\alpha) \tag{38}$$

where the critical value $C(\alpha)$ is determined by the level of significance α. The asymptotic distribution of this optimal statistic under the sequence of close alternatives (21) is given by

$$
\begin{aligned}
&\sqrt{N}\left(\frac{2}{N(N-1)}\sum_{1\le i<j\le N}\left(nD_{im}^{(m)}-nD_{jm}^{(m)}\right)^2-2m\right) \\
&\xrightarrow{D} N_1\left(2m^{1/2}(m+1)\left(\int_0^1 l^2(u)\,du\right),8m(m+1)\right).
\end{aligned}
$$

The asymptotic distribution under the null hypothesis is obtained by taking $l(u)=0$ in the above.

From Theorem 3, under a sequence of close alternatives, the Gini mean squared difference higher-order spacings test

$$G_N(2)=\frac{2}{N(N-1)}\sum_{1\le i<j\le N}\left(nD_{im}^{(m)}-nD_{jm}^{(m)}\right)^2 \tag{39}$$

is the ALMP test for U-statistics of the higher-order spacings. This also means that $G_N(2)$ is the best test among the generalized Gini mean difference higher-order spacings tests $G_N(r), r > 0$. Moreover, by Lemma 8, the Gini mean squared difference higher-order spacings test has efficacy

$$e^2(h) = \frac{m+1}{2}\left(\int_0^1 l^2(u)\,du\right)^2$$

which is the same as that of the generalized Greenwood statistic.

We compare the efficacies of the Gini mean squared difference higher-order spacings test, and the Kullback-Leibler divergence higher-order spacings test

$$\frac{1}{\sqrt{N}}\sum_{k=1}^{N}\left[\left(nD_{km}^{(m)}\right)\log\left(nD_{km}^{(m)}\right) - m\cdot\psi(m+1)\right] \tag{40}$$

which under the alternatives, has the asymptotic Normal distribution, with asymptotic mean

$$\mu(g) = \frac{m^{1/2}}{2}\left(\int_0^1 l^2(u)\,du\right) \tag{41}$$

and asymptotic variance

$$\begin{aligned}\sigma^2(g) = -m(m+1)[m\cdot\psi^2(m+1) - 2m\cdot\psi(m+1)\cdot\psi(m+2)\\ +m\cdot\psi^2(m+2) - \psi'(m+2)].\end{aligned}$$

Here, the function $\psi(m) = \frac{\Gamma'(m)}{\Gamma(m)} = \mathbb{E}[\log(S_1)]$ is the digamma function, and its derivative

$$\psi'(m) = \sum_{n=0}^{\infty}(m+n)^{-2} = \mathrm{Var}[\log(S_1)]$$

is the trigamma function. Some notable constants are

$$\psi(1) = -\gamma = -0.57721\ldots = \mathbb{E}[\log(Z_1)]$$

which is the negative of the Euler–Mascheroni constant, and $\psi'(1) = \pi^2/6$.

It will be convenient to define the modified efficacy of a test as

$$e_M^2(\cdot) = \frac{e^2(\cdot)}{\left(\int_0^1 l^2(u)\,du\right)^2}.$$

The table lists, for various values of m, the modified efficacies of the higher-order Gini mean squared difference test, the Kullback–Leibler divergence based on higher-order spacings, and the Pitman's ARE of the Gini mean squared difference test relative to the Kullback–Leibler divergence. It is seen that the Gini mean squared difference higher-order spacings test has the most efficacy and is more Pitman efficient. As expected from earlier literature, the efficacies of both tests increase with m for any finite m, but the latter test catches up with the former with increasing m.

Table 1. Modified Efficacies for the Gini Mean Squared Difference Higher-Order Spacings Test and the Kullback–Leibler Divergence Higher-order Spacings Test with Pitman ARE.

m	Gini Mean Squared Difference	Kullback–Leibler	Pitman ARE
1	1.000	0.862461	1.344377
2	1.500	1.3528	1.229463
3	2.000	1.84786	1.171445
4	2.500	2.34489	1.136672
5	3.000	2.84292	1.113559
10	5.500	5.33849	1.061423
20	10.500	10.336	1.031986
30	15.500	15.3351	1.021622

5. Conclusion

We derived the general asymptotic theory for U-statistics based on higher-order spacings and found the Asymptotically Locally Most Powerful test in this class under a sequence of close alternatives. The results established here extend those obtained in Tung and Jammalamadaka (2010) to the case of higher-order spacings.

References

Del Pino, G. E. (1979). On the asymptotic distribution of k-Spacings with applications to goodness-of-fit tests, *The Annals of Statistics* **7**, 5, pp. 1058–1065.

Holst, L. (1981). Some conditional limit theorems in exponential families, *The Annals of Probability* **9**, 5, pp. 818–830.

Jammalamadaka, S. R. and Goria, M. N. (2004). A test of goodness of fit based on Gini's index of spacings, *Statistics and Probability Letters* **68**, pp. 177–187.

Lee, A. J. (1990) *U-Statistics* (Marcel Dekker, Statistics: Textbooks and Monographs).

Rao, J. S. and Sethuraman, J. (1975). Weak convergence of empirical distribution functions of random variables subject to perturbations and scale factors, *The Annals of Statistics* **3**, 2, pp. 299–213.

Rao, J. S. and Sobel, M. (1980) Incomplete Dirichlet integrals with applications to ordered uniform spacings, *Journal of Multivariate Analysis* **10**, 4, pp. 603–610.

Sethuraman, J. and Rao, J. S. (1970). Pitman Efficiencies of Tests Based in Spacings, in M. L. Puri (ed.) *Nonparametric Techniques in Statistical Inference*, Cambridge University Press, pp. 405–415.

Sobel, M. and Wells, M. (1990). Dirichlet integrals and moments of gamma distribution order statistics, *Statistics and Probability Letters* **9**, pp. 431–437.

Tung, D. D. and Jammalamadaka, S. R. (2010). *U*-statistics based on spacings. Submitted for publication.

Chapter 9

Nonparametric Models for Non-Gaussian Longitudinal Data

Nan Zhang, Hans-Georg Müller and Jane-Ling Wang

University of California, Davis

We discuss nonparametric methods for functional principal components analysis in the case of non-Gaussian longitudinal data. In this framework, a sample of realizations of a hidden process gives rise to noisy non-Gaussian repeated measurements, such as Poisson counts or Binomial data. The repeated measurements obtained for each subject are assumed to be determined by a subject-specific smooth random trajectory plus measurement errors. A link function relates the subject-specific trajectories to an underlying latent Gaussian process, which is represented by an eigenfunction expansion with functional principal component scores as random coefficients. The eigenfunction basis is estimated from the data, and the functional principal component score estimates are obtained by maximizing a quasi-likelihood. We develop a functional version of the Akaike information criterion for model selection, corresponding to the choice of the number of principal components for the eigenfunction expansion. The proposed framework is compared to approaches based on Character Process Models and Cubic B-spline Models in a simulation study.

Contents

1. Introduction

Scientists in the life sciences often collect data that consist of repeated measurements taken at various time points from the same subject. Such longitudinal data are essential to the understanding of trends and variation for time-dynamic biological processes. A distinctive feature of such data is that they are time correlated. Observations obtained from a single individual at different occasions typically exhibit positive correlation, and this correlation must be accounted for in the analysis. If longitudinal observations are measured on a suitably dense grid, they can be interpreted as a sample of random curves or functional data.

The functional data analysis of samples of repeatedly observed vectors of noisy non-Gaussian observations, whose distributions may not belong to the exponential family, such as Poisson counts or binary observations, presents challenging issues. To facilitate the modeling of such data, we assume that the measurements for each subject are determined by a subject-specific smooth random trajectory (in time) plus measurement errors, and that this random trajectory is the realization of a latent (Gaussian) stochastic process. This assumption has also been made in related work by Hall, et al. (2008), while most previous work in this area had focused on models which assumed that the underlying processes are directly observed. While Hall, et al. (2008) focused on the case where the times at which measurements are obtained are sparse and irregular, we consider the case where the measurement times are regular and densely spaced.

A widely used parametric approach for non-Gaussian longitudinal data is the Generalized Linear Mixed Model (GLMM) (Stiratelli (1984)). The GLMM extends the framework of linear mixed models (LMMs) and of generalized linear models (GLMs), by allowing for non-Gaussian observations, nonlinear link functions and the inclusion of random effects and of correlated errors. Implementing GLMMs involves complex specification issues, such as the choice of the link function and specification of the covariance or correlation structure. In addition, one has to handle thorny computational issues, such as integration over the distributions of the random effects. A variety of computationally attractive approaches have been proposed to improve inference and estimation procedures for the fixed effects in GLMM. These include Gibbs sampling (Zeger and Karim (1991)), penalized quasi-likelihood and marginal quasi-likelihood (Breslow and Clayton

(1993)), pseudo-likelihood based on approximate marginal models (Wolfinger and O'Connell (1993)), and Monte Carlo Expectation-Maximization (MCEM), Monte Carlo Newton-Raphson (MCNR) and simulated maximum likelihood (SML) algorithms (McCulloch (1997), Booth and Hobert (1999)). However, these nonlinear methods often prove to be computationally impractical when the number of random effects is large (Diggle, et al. (2002)).

A different approach that assumes a parametric connection between the observed non-Gaussian data and a latent Gaussian process has been developed in work by Pletcher and Geyer (1999), and Pletcher and Jaffrézic (2002), known as the Character Process Model. This model leads to a maximum likelihood method that can be implemented via Markov Chain Monte Carlo estimation. Since a parametric model will only find features in the data which are already incorporated *a priori* in the model, such parametric approaches might be less adequate if there is little prior knowledge about the nature of the time courses. Nonparametric functional data analysis is becoming a standard approach when one has dense designs of Gaussian repeated measurements for the trajectory of each subject. A key method is Functional Principal Component Analysis (FPCA), in which one attempts to find the dominant "modes of variation" around an overall smooth trend function. Alternative series expansions with B-spline bases with random coefficients also have become quite popular (Rice and Wu (2000)).

The main objective of this paper is to study an extension of FPCA to the situation of repeatedly observed non-Gaussian data. The nonparametric approach of functional data analysis has several desirable properties compared to parametric approaches: Its flexibility in covariance modeling, as the covariance surface is a nonparametric component of the model; and computational simplicity, as no high-dimensional integration step is needed even for a large number of covariates. The paper is organized as follows: In Section 2, we introduce the proposed functional principal component analysis model via Quasi-likelihood maximization (FPCA-Q) method. In Section 3, two alternative models, the Character Process Model (CPM) and a cubic B-spline model (BS), are introduced and the numerical challenges that one faces when fitting these various random effects models are discussed. Simulation results that illustrate the usefulness of the proposed methodology and provide comparisons of various approaches are presented in Section 4.

2. Functional Principal Component Analysis via Quasi-likelihood Maximization (FPCA-Q)

2.1. *Model and notation*

Assume data are available for n subjects with m repeated observations for each subject, which are made at times t_{ij}, $i = 1, \ldots, n$ and $j = 1, \ldots, m$, where $t_{ij} \in \mathcal{T}$ and $\mathcal{T}$ is an interval, while i is the subject and j the measurement index. We assume the measurement times are evenly spaced over $\mathcal{T}$. For simplicity, we assume that the number of measurements per subject is the same across subjects, but this is not essential and could be relaxed, as long as some regularity conditions are satisfied. The random trajectory for the i-th subject $X_i(t)$ is assumed to be an independent realization of a latent Gaussian process $Z_i(t)$ as follows,

$$X_i(t) = g^{-1}(Z_i(t)), \tag{1}$$

where $g(\cdot)$ is a link function. The latent Gaussian process $Z_i(t)$ us characterized by unknown mean function $E[Z_i(t)] = \mu(t)$ and covariance function $\mathrm{Cov}(Z_i(s), Z_i(t)) = G(s,t)$, both of which are assumed to be smooth functions, for all $i = 1, \ldots, n$, and $t, s \in \mathcal{T}$.

We assume that the generalized responses $Y_{ij} = Y_i(t_{ij})$ for the i-th subject are generated by a random function $X_i(t_{ij})$, measured at time points t_{ij}, such that

$$\begin{aligned} \mathrm{E}(Y_i(t)|Z_i) &= X_i(t) = g^{-1}(Z_i(t)), \\ \mathrm{Var}(Y_i(t)|Z_i) &= V(g^{-1}(Z_i(t)) = \widetilde{V}(Z_i(t)). \end{aligned} \tag{2}$$

Here $\widetilde{V} = V \circ g^{-1}$, where V is a variance function, assumed to be strictly positive, smooth and defined on the range of the link function (in analogy to McCullagh (1983), McCullagh and Nelder (1989)). Considering potential variation in the measurements, assume $\epsilon_{ij} = Y_{ij} - \mathrm{E}[Y_i(t_{ij})|Z_i)]$ to be independent with mean zero, leading to $\mathrm{Var}(\epsilon_{ij}|Z_i) = \widetilde{V}(Z_i(t_{ij}))$ and with $\tilde{\sigma}^2(t_{ij}) = \mathrm{E}\{\widetilde{V}(Z_i(t_{ij}))\}$ to $\mathrm{Var}(\epsilon_{ij}) = \mathrm{E}\{\widetilde{V}(Z_i(t_{ij}))\} = \tilde{\sigma}^2(t_{ij})$.

Because of the unexplained random variation in response probabilities and correlation between responses, it is often necessary for realistic modeling to allow for overdispersion. Then the conditional variance of Y_i given Z_i takes the form $\mathrm{Var}(Y_i(t)|Z_i) = \tau^2\widetilde{V}(Z_i(t))$, where τ^2 is the overdispersion parameter. For example, if Y_{ij} is a Binomial outcome,

$$\begin{aligned}\operatorname{Var}(Y_i(t)|Z_i) &= \tau^2 g^{-1}(Z_i(t))(1-g^{-1}(Z_i(t))),\\ \operatorname{Var}(Y_i(t)) &= \operatorname{Var}\{E(Y_i(t)|Z_i)\} + E\{\operatorname{Var}(Y_i(t)|Z_i)\}\\ &= \operatorname{Var}(g^{-1}(Z_i(t))) + \tau^2 E\{g^{-1}(Z_i(t))(1-g^{-1}(Z_i(t)))\}.\end{aligned} \tag{3}$$

Similarly, for observations which are Poisson counts,

$$\begin{aligned}\operatorname{Var}(Y_i(t)|Z_i) &= \tau^2 g^{-1}(Z_i(t)),\\ \operatorname{Var}[Y_i(t)] &= \operatorname{Var}(g^{-1}(Z_i(t))) + \tau^2\ \mathrm{E}(g^{-1}(Z_i(t))).\end{aligned} \tag{4}$$

We also note that in model (2),

$$\begin{aligned}\operatorname{Cov}(Y_i(s), Y_i(t)) =&\ \operatorname{Cov}(E[Y_i(s)|Z_i], E[Y_i(t)|Z_i]) +\ \mathrm{E}[\operatorname{Cov}(Y_i(s), Y_i(t)|Z_i)]\\ =&\ \operatorname{Cov}(g^{-1}(Z_i(s)), g^{-1}(Z_i(t))) + \tilde{\sigma}^2(t)\delta_{st},\end{aligned} \tag{5}$$

where δ_{st} is 1 if $s = t$ and 0 otherwise.

2.2. *FPCA-Q Model*

Recall that in model (2), $E[Y_{ij}|Z_i)] = X_i(t_{ij}) = g^{-1}(Z_i(t_{ij})),\ \ t_{ij} \in \mathcal{T}$, and $\epsilon_{ij} = Y_{ij} - E[Y_{ij}|Z_i)]$ are assumed to be independent, with $E(\epsilon_{ij}) = 0$ and $\operatorname{Var}(\epsilon_{ij}) = \tilde{\sigma}^2(t_{ij})$, $i = 1, \ldots, n$, $j = 1, \ldots, m$. Using the Karhunen–Loève representation (Ash and Gardner (1975)), the latent Gaussian process $\{Z(t)\}$ can be expanded in the eigenfunction basis ϕ_k, $k \geq 1$ of the autocovariance operator, with associated ordered eigenvalues $\lambda_1 \geq \lambda_2 \geq \ldots$, such that

$$\begin{aligned}Z(t) &= \mu(t) + \sum_{k=1}^{\infty} \xi_k \phi_k(t),\\ \operatorname{Cov}(Z(s), Z(t)) = \Sigma_Z(s,t) &= \sum_{k=1}^{\infty} \lambda_k \phi_k(s)\phi_k(t),\end{aligned} \tag{6}$$

where the ξ_k are the functional principal component scores of processes Z.

The observation Y_{ij} may then be expressed as

$$E[Y_{ij}|Z_i)] = g^{-1}\{Z_i(t_{ij})\} = g^{-1}\{\mu_Z(t_{ij}) + \sum_{k=1}^{\infty} \xi_{ik}\phi_k(t_{ij})\}, \tag{7}$$

where the scores ξ_{ik} are independent Gaussian random variables with zero mean and variances $E[\xi_{ik}^2] = \lambda_k$. We approximate model (7) by a series of models where the number of included eigencomponents is truncated at

$p = p(n)$ and the dimension $p(n)$ increases asymptotically as $n \to \infty$. Thus, the p-truncated model becomes

$$E[Y_{ij}|Z_i)] = g^{-1}\{Z_i(t_{ij})\} = g^{-1}\{\mu_Z(t_{ij}) + \sum_{k=1}^{p} \xi_{ik}\phi_k(t_{ij})\}. \tag{8}$$

2.3. *Estimation of Model Components*

2.3.1. *Estimation of mean and covariance function*

We use local linear smoothers (Fan and Gijbels (1996)) to obtain estimates $\hat{\mu}_X(t)$ of the mean function and $\widehat{\Sigma}_X(s,t)$ of the covariance surface of the random trajectories X, by pooling the available data. The smoothing parameters are chosen by leave-one-curve-out cross-validation; for details see Yao, et al. (2003). Assuming that g has two bounded derivatives, a second-order Taylor series of Z about $\mu_X(t)$ yields

$$\begin{aligned} Z(t) = g(X(t)) &= g(\mu_X(t)) + g^{(1)}(\mu_X(t))(X(t) - \mu_X(t)) \\ &\quad + \frac{1}{2} g^{(2)}(\tilde{\mu}_X(t))(X(t) - \mu_X(t))^2, \end{aligned}$$

where $\tilde{\mu}_X(t)$ is a random variable between $\mu_X(t)$ and $X(t)$. One can show that the above equation leads to

$$\mu_Z(t) = E\{g(X(t))\} = g(\mu_X(t)) + O(\delta^2), \tag{9}$$

and

$$\begin{aligned} \Sigma_Z(s,t) = &\ \mathrm{Cov}\{g(X(s)), g(X(t))\} \\ = &\ g^{(1)}(\mu_X(s))g^{(1)}(\mu_X(t))\Sigma_X(s,t) + O(\delta^2), \end{aligned} \tag{10}$$

where $\delta > 0$ is a relatively small constant. These formulas suggest estimators

$$\hat{\mu}_Z(t) = g(\hat{\mu}_X(t)), \tag{11}$$

of the mean function $\mu(t)$ and

$$\widehat{\Sigma}_Z(s,t) = g^{(1)}(\hat{\mu}_X(s))g^{(1)}(\hat{\mu}_X(t))\widehat{\Sigma}_X(s,t) \tag{12}$$

of the covariance function $\Sigma_Z(s,t)$.

2.3.2. *Estimation of overdispersion and error variance*

Note that in model (8), $\text{cov}(Y_{ij}, Y_{il}) = \text{cov}(X(t_{ij}), X(t_{il})) + \tilde{\sigma}^2(t_{ij})\delta_{jl}$, where δ_{jl} is 1 if $j = l$ and 0 otherwise. Let $C_i(t_{ij}, t_{il}) = (Y_{ij} - \hat{\mu}_X(t_{ij}))(Y_{il} - \hat{\mu}_X(t_{il}))$ be the raw covariances, where $\hat{\mu}_X(t)$ is the estimated mean function obtained from the previous step. It is easy to see that $E[C_i(t_{ij}, t_{il})] \approx \text{cov}(X(t_{ij}), X(t_{il})) + \tilde{\sigma}^2(t_{ij})\delta_{jl}$. Therefore, we remove the diagonal of the raw covariances, and only the elements $C_i(t_{ij}, t_{il})$, $j \neq l$, are included as predictors in the smoothing step to obtain the estimator $\widehat{\Sigma}_X(s, t)$ of the covariance function of $X_i(t)$ (Staniswalis and Lee (1998), Yao, et al. (2005)). Denote the diagonal of the resulting surface estimate by $\widetilde{\Sigma}_X(t)$.

Prior to the surface smoothing step to obtain the estimate of the covariance surface, the diagonal elements of the raw covariances, that is $C_i(t_{ij}, t_{ij})$, $i = 1, \ldots, n, j = 1, \ldots, m$, are removed. An estimate $\widehat{W}(t)$ of $\{\Sigma_X(t, t) + \tilde{\sigma}^2(t)\}$ is then obtained in a second local weighted linear smoothing step, applied to the scatter plot $(t_{ij}, C_i(t_{ij}, t_{ij}))$ with pooled raw variances $C_i(t_{ij}, t_{ij})$, $i = 1, \ldots, n$, $j = 1, \ldots, m$. To mitigate against boundary effects, we cut off the two ends of the interval to get a more stable estimate of σ^2. The resulting estimate of $\sigma^2 = \int_{\mathcal{T}} \tilde{\sigma}^2(t)dt$, suitably truncated to avoid the possibility of negative estimates, is

$$\hat{\sigma}^2 = \frac{2}{T} \int_{T/4}^{3T/4} \{\widehat{W}(t) - \widetilde{\Sigma}_X(t)\}dt, \tag{13}$$

if $\hat{\sigma}^2 > 0$ and $\hat{\sigma}^2 = 0$ otherwise.

The overdispersion parameter τ^2 is of special interest, where

$$\begin{aligned} \text{Var}(Y(t)) &= \text{Var}\{E(Y_i(t)|Z_i)\} + E\{\text{Var}(Y_i(t)|Z_i)\} \\ &= \text{Var}(g^{-1}(Z_i(t))) + \tau^2 E\{V(g^{-1}(Z_i(t)))\} \\ &= \text{Var}(X_i(t)) + \tau^2 E\{V(X_i(t))\} \end{aligned} \tag{14}$$

and estimates $\widehat{W}(t)$ for $\text{Var}(Y(t))$ and $\widetilde{\Sigma}_X(t)$ for $\text{Var}(X_i(t))$ have already been constructed, as described above. Consider the class of variance functions where $V(X_i(t)) = \alpha_0 X_i(t)) + \alpha_1 X_i^2(t)$ for suitable constants α_0, α_1. Then there exists a function $\tilde{h}$ such that $U(t) = E\{V(X_i(t))\} = \tilde{h}(\mu_X(t), \widetilde{\Sigma}_X(t))$. One can then plug in estimators to obtain the estimate $\widehat{U}(t) = \tilde{h}(\hat{\mu}_X(t), \widetilde{\Sigma}_X(t))$. For example, if the observations Y_{ij} are binomial, the variance function of $X_i(t)$ is $V(X_i(t)) = X_i(t)(1 - X_i(t))$ and $\text{E}\{V(X_i(t))\} = EX_i(t) - [\ \text{Var}(X_i(t)) + (EX_i(t))^2]$, leading to the estimator $\widehat{U}(t) = \hat{\mu}_X(t) - (\widetilde{\Sigma}_X(t) + \hat{\mu}_X^2(t))$.

These considerations motivate the following estimator for τ^2, adjusted in analogy to (13),

$$\hat{\tau}^2 = \frac{2}{T}\int_{T/4}^{3T/4} \frac{\widehat{W}(t) - \widetilde{\Sigma}_X(t)}{\widehat{U}(t)}\,dt. \tag{15}$$

2.3.3. *Estimation of eigenfunctions and functional principal components*

The eigenvalues λ_k and eigenfunctions ϕ_k in model (8) are estimated by discretizing the smoothed covariance, as previously described in Rice and Silverman (1991) and Capra and Müller (1997), obtaining $\hat{\lambda}_k$ and $\hat{\phi}_k$ as the estimates of the eigenvalues and eigenfunctions.

Let $\xi_{ik} = \xi_{ik}(\omega_1)$, where ω_1 is an element of the probability space $(\Omega_1, \mathcal{F}_1, \mathbb{P}_1)$ so that $X_i(\omega_1, t) = g^{-1}(\mu_Z(t_{ij}) + \sum_{k=1}^{p} \xi_{ik}(\omega_1)\phi_k(t_{ij}))$ is a realization of the random trajectory $X_i(t)$. Analogously, $\epsilon_{ij} = \epsilon_{ij}(\omega_2)$, where ω_2 is an element of a probability space $(\Omega_2, F_2, \mathbb{P}_2)$. Since the ξ_{ik} are independent of the ϵ_{ij}, $\Omega = \Omega_1 \times \Omega_2$ and for fixed $\omega_1 \in \Omega_1$, $\mathrm{Cov}(Y_{ij}, Y_{ij'}) = 0$ for $j \neq j'$. Hence, for fixed $\omega_1 \in \Omega_1$, once the number of included components p has been chosen, one is in the situation of a traditional estimating equation approach and may estimate the coefficient vector $\xi_i = (\xi_{i1}, \ldots, \xi_{ip})^T$ by solving a score or estimating equation, conditional on each subject,

$$U(\xi_i) = 0. \tag{16}$$

Setting the linear predictor $\hat{\eta}_{ij} = \hat{\mu}_Z(t_{ij}) + \sum_{k=1}^{p} \xi_{ik}\hat{\phi}_k(t_{ij})$, $\hat{\mu}_{ij} = g^{-1}(\hat{\eta}_{ij})$, the vector of eigenfunctions $\hat{\phi}(t) = (\hat{\phi}_1(t), \ldots, \hat{\phi}_p(t))^T$, $i = 1, \ldots, n$ and $j = 1, \ldots, m$, we define the vector valued score function, conditional on the i-th subject, as

$$U(\xi_i) = \sum_{j=1}^{m} \frac{(Y_{ij} - \hat{\mu}_{ij})(g^{-1})^{(1)}(\hat{\eta}_{ij})}{V(\hat{\mu}_{ij})}\hat{\phi}(t_j). \tag{17}$$

With matrices

$$\begin{aligned}\mathbf{D} &= \mathbf{D}_{m,p} = ((g^{-1})^{(1)}(\hat{\eta}_{ij})\hat{\phi}_k(t_j))_{1\le j\le m, 1\le k\le p}, \mathbf{V} \\ &= \mathbf{V}_{m,m} = \mathrm{diag}(V(\hat{\mu}_{i1}), \ldots, V(\hat{\mu}_{im}))_{1\le j\le m}\end{aligned}$$

and vectors $\mathbf{Y}_i = (Y_{i1}, \ldots, Y_{im})^T$ and $\hat{\mu}_i = (\hat{\mu}_{i1}, \ldots, \hat{\mu}_{im})^T$, one may rewrite the estimating equation $U(\xi_i) = 0$ (16) as

$$\mathbf{D}^T\mathbf{V}^{-1}(\mathbf{Y}_i - \hat{\mu}_i) = 0, \tag{18}$$

with

$$E(U(\xi_i)) = 0, \qquad \mathbf{Cov}(U(\xi_i)) = \mathbf{D}^T\mathbf{V}^{-1}\mathbf{D}/\hat{\tau}^2,$$

where $\hat{\tau}^2$ is the estimate of the overdispersion parameter as described above. This equation can be solved iteratively by the method of iterated weighted least squares. Under the basic assumptions, as $\frac{1}{m}E(\mathbf{D}^T\mathbf{D}) = \Gamma_p$ is a fixed positive definite matrix for each p, the existence of a unique solution for each fixed p is guaranteed asymptotically. We denote the solution of the estimating equation $U = 0$ (17) for a given subject by $\hat{\xi}_i = (\hat{\xi}_{i1}, \ldots, \hat{\xi}_{ip})^T$.

2.3.4. *Selection of the number of eigenfunctions*

We describe here an AIC type criterion, based on a pseudo-Gaussian likelihood conditional on estimated random coefficients, for choosing the number of eigenfunctions p. Let $\widehat{Z}_i(t_j) = \hat{\mu}(t_j) + \sum_{k=1}^p \hat{\xi}_{ik}\hat{\phi}_k(t_j)$, $\widehat{Y}_i(t_j) = g(\hat{\mu}(t_j) + \sum_{k=1}^p \hat{\xi}_{ik}\hat{\phi}_k(t_j))$ and $\mathbf{Y}_i = (Y_{i1}, \ldots, Y_{im})^T$.

Consider the first-order Taylor expansion

$$g(Z_i(t_j)) \approx g(\widehat{Z}_i(t_j)) + g^{(1)}(\widehat{Z}_i(t_j))(Z_i(t_j) - \widehat{Z}_i(t_j)). \tag{19}$$

Rearranging terms yields the expression

$$Z_i(t_j) \approx [g(Z_i(t_j)) - g(\widehat{Z}_i(t_j))]/g^{(1)}(\widehat{Z}_i(t_j)) + \widehat{Z}_i(t_j).$$

The right-hand side is the expected value, conditional on ξ_i, of $[Y_{ij} - \widehat{Y}_i(t_j)]/g^{(1)}(\widehat{Z}_i(t_j)) + \widehat{Z}_i(t_j)$. Hence, we can obtain pseudo-Gaussian observations $\widetilde{Z}_i(t)$ by

$$\widetilde{Z}_i(t_j) \doteq [Y_{ij} - \widehat{Y}_i(t_j)]/g^{(1)}(\widehat{Z}_i(t_j)) + \widehat{Z}_i(t_j). \tag{20}$$

Accordingly, an estimated pseudo-Gaussian log-likelihood, summing the contributions from all subjects and conditional on the estimated FPC scores $\hat{\xi}_i$, may be defined as,

$$\begin{aligned}\widehat{L} = \sum_{i=1}^n \{ &-\frac{1}{2}(\widetilde{\mathbf{Z}}_\mathbf{i} - \hat{\mu}_Z - \sum_{k=1}^p \hat{\xi}_{ik}\hat{\phi}_k)^T \widehat{\mathbf{\Sigma}}_\mathbf{Z}^{-1}(\widetilde{\mathbf{Z}}_\mathbf{i} - \hat{\mu}_Z - \sum_{k=1}^p \hat{\xi}_{ik}\hat{\phi}_k) \\ &-\frac{m}{2}\log(2\pi) - \frac{1}{2}\log(|\widehat{\mathbf{\Sigma}}_\mathbf{Z}|)\},\end{aligned} \tag{21}$$

where $\widetilde{\mathbf{Z}}_\mathbf{i} = (\widetilde{Z}_i(t_{i1}), \ldots, \widetilde{Z}_i(t_{im}))^T$ and $\widehat{\mathbf{\Sigma}}_\mathbf{Z}$ is the estimated covariance surface of $Z_i(t)$ as shown in (12). This suggests to define a version of conditional AIC, given by $\text{AIC}(p) = -2\widehat{L} + 2p$, choosing the minimizing argument p as the number of included components.

3. Other Modeling Approaches

3.1. *The Character Process Model (CPM)*

In quantitative genetics, the analysis of functional data (also called function-valued characters) has received increasing attention. A function-valued character being considered can naturally be described as a mathematical function of some continuous index. Kirkpatrick and collaborators (Kirkpatrick and Heckman (1989), Kirkpatrick, et al. (1990)) proposed a framework for analyzing growth trajectories as a function of temperature. Meyer and Hill (Meyer and Hill (1997), Meyer; Meyer (1998; 2000)) further developed a random regression approach to model agricultural breeding systems. As with traditional quantitative genetics, it is assumed that the observed sample of curves (phenotypic trajectories) changes with age or environmental conditions and is influenced by an unobservable latent Gaussian process, such as genetic variation in age-specific mortality rates (Shaw, et al. (1999)).

For example, one might have a simple case where many genes along with environmental factors affect the observations but where these predictors are not observed. Pletcher and Geyer (1999) suggest a decomposition of the underlying unobserved Gaussian process $Z(t)$, including parametric forms for its covariance function $G(s,t)$, as follows,

$$Z_i(t) = \mu(t) + d_i(t) + e_i(t) + \varepsilon_i(t), \qquad t \in \mathcal{T}, \tag{22}$$

where $d_i(t)$ and $e_i(t)$ are Gaussian random functions, which are independent of one another and have an expected value of zero at each time t (Kirkpatrick and Heckman (1989), Pletcher and Geyer (1999)). These functions represent the i-th subject's genetic and environmental deviations at time t. Here $\mu(t)$ is a fixed effect which corresponds to the common mean function of $Z_i(t)$, i.e., $E[Z_i(t)] = \mu(t)$. Furthermore, $\varepsilon_i(t)$ is assumed to be the independent measurement error with mean 0 and variance σ_e^2, i.e., $\mathrm{E}[\varepsilon_i(t)] = 0$ and $\mathrm{Cov}(\varepsilon_i(s), \varepsilon_i(t)) = \sigma_e^2\delta_{st}$ where δ_{st} is 1 if $s = t$ and 0 otherwise. Combining with (2), this model becomes

$$\mathrm{E}[Y_{ij}|Z_i)] = g^{-1}(\mu(t_{ij}) + d_i(t_{ij}) + e_i(t_{ij}) + \varepsilon_i(t_{ij})), \qquad t_{ij} \in \mathcal{T}. \tag{23}$$

Then for any $t, s \in \mathcal{T}$, the covariance function of $Z_i(t)$ can be expressed by

$$G(s,t) = \mathrm{Cov}(d_i(s), d_i(t)) + \mathrm{Cov}(e_i(s), e_i(t)) + \sigma_e^2\delta_{st}, \tag{24}$$

where δ_{st} is 1 if $s = t$ and 0 otherwise.

Taking the environmental covariance function $\text{Cov}(e_i(t), e_i(s)) = H(s,t)$ as an example, character process models parameterize the covariance functions as

$$H(s,t) = v(s)v(t)\rho(|s-t|), \tag{25}$$

where $v^2(t)$ describes how the environmental variance changes with time t and $\rho(|s-t|)$ describes the environmental correlation between two times. There are no restrictions on the form of $v^2(\cdot)$, and this function is often modelled using simple polynomials (linear, quadratic, etc). either on the natural or the log scale. If the correlation between two times is a function only of the time distance ($|s-t|$) between them (correlational stationarity), then several choices for $\rho(\cdot)$ are available (for details, see Pletcher and Geyer (1999)).

3.2. *Cubic B-spline Models (BS)*

Under the cubic B-spline model, each underlying hidden Gaussian process $Z_i(t)$ is decomposed into a mean function of time $\mu(t)$ and a random deviation from the mean. These components are modeled as linear combinations of B-spline functions with random coefficients. We model the mean function as

$$E[Z_i(t)] = \mu(t) = \sum_{k=1}^{p} \beta_k \bar{B}_k(t), \qquad t \in \mathcal{T}, \tag{26}$$

where β_k are unknown fixed coefficients, $\{\bar{B}_k, k \geq 1\}$ is a cubic B-spline basis on $\mathcal{T}$ with a fixed knot sequence, usually chosen equidistantly, and p is the corresponding number of included spline basis functions for this representation.

The deviation of the i-th subject curve from the mean can be similarly modelled as the spline function $\sum_{k=1}^{q} \gamma_{ik} B_k(t)$. Here $\{B_k, k \geq 1\}$ is another B-spline basis and q the number of included basis functions, where γ_{ik} are normally distributed random coefficients with mean 0 and covariance matrix Γ (in analogy to Rice and Wu (2000)). Here one needs to choose p and q data-adaptively. Then, for each subject,

$$E[Y_{ij}|Z_i)] = g^{-1}(\sum_{k=1}^{p} \beta_k \bar{B}_k(t_{ij}) + \sum_{k=1}^{q} \gamma_{ik} B_k(t_{ij})), \qquad t_{ij} \in \mathcal{T}, \tag{27}$$

and the covariance function of $Z_i(t)$ can be expressed as

$$G(s,t) = \sum_{k=1}^{q}\sum_{l=1}^{q} \Gamma_{kl} B_k(s) B_l(t). \tag{28}$$

Taking the vector observations made for the i-th subject as the response, one may express (27) in the form of a classical GLMM,

$$\mathrm{E}[\mathbf{Y_i}|\mathbf{Z_i}] = g^{-1}(\mathbf{X_i}\beta + \mathbf{R_i}\gamma_\mathbf{i}), \tag{29}$$

where

$$\mathbf{X}_i = \begin{pmatrix} \bar{B}_1(t_{i1}) & \dots & \bar{B}_p(t_{i1}) \\ \dots & \dots & \dots \\ \bar{B}_1(t_{im}) & \dots & \bar{B}_p(t_{im}) \end{pmatrix}, \qquad \mathbf{R}_i = \begin{pmatrix} B_1(t_{i1}) & \dots & B_q(t_{i1}) \\ \dots & \dots & \dots \\ B_1(t_{im}) & \dots & B_q(t_{im}) \end{pmatrix},$$

$\beta = (\beta_1, \dots, \beta_p)^T$, $\gamma_i = (\gamma_1, \dots, \gamma_q)^T$ and $\mathbf{Z_i} = (Z_i(t_{i1}), \dots, Z_i(t_{im}))^T$. Therefore, we have a GLMM with β as fixed effects and γ as random effects, and one can apply the methodology that has been developed for GLMM (Rice and Wu (2000), Shi, et al. (1996)).

3.3. *Estimation for Character Process and spline models*

All of the parameters involved in the Character Process Model (23) and the cubic B-spline model (29 can be estimated by procedures that are suitable for the general GLMM by maximizing a log-likelihood through a hybrid algorithm that is composed of a Markov chain Monte Carlo EM algorithm (MCEM) and a simulated maximum likelihood algorithm (SML) (see McCulloch (1997), Shaw, et al. (1999)).

3.3.1. *Markov chain Monte Carlo EM algorithm (MCEM)*

MCEM originated from Chan and Ledholter (1995) as an estimation tool for time series models. McCulloch (1997) constructed an MCEM algorithm for maximum likelihood estimation in GLMM by incorporating a Metropolis–Hastings step. Without loss of generality, denote by β the fixed and by γ the random effects, and assume that the distribution of γ only depends on parameters $\mathbf{\Gamma}$, i.e. $f_\gamma(\gamma|\mathbf{\Gamma})$. Let $\mathbf{Y} = (\mathbf{Y}_1, \dots, \mathbf{Y}_n)^T$ denote the concatenated vector of all non-Gaussian observations. The likelihood is given by

$$L(\beta, \mathbf{\Gamma}|\mathbf{Y}) = \int \prod_{i=1}^{n} f_{Y_i|\gamma}(\mathbf{Y}_i|\beta,\gamma) f_\gamma(\gamma|\mathbf{\Gamma}) d\gamma. \tag{30}$$

It cannot usually be evaluated in closed form as the dimension of the integral is equal to the number of random effects. By equating the random effects γ to missing observations, the log-likelihood for the data $\mathbf{W} = (\mathbf{Y}, \gamma)$ is as follows,

$$\ln L_W = \sum_{i=1}^{n} \ln f_{Y_i|\gamma}(\mathbf{Y}_i|\beta, \gamma) + \ln f_\gamma(\gamma|\mathbf{\Gamma}). \tag{31}$$

It can be advantageous to equate γ to missing data: Given γ, the $\mathbf{Y}_i$'s are independent, and one can use $f_{Y_i|\gamma}$ to find the maximum likelihood estimators (MLE) of β; the MLE of $\mathbf{\Gamma}$ will only depend on f_γ.

Since the conditional distribution of $\gamma|\mathbf{Y}$ involves f_Y, the distribution that we are trying to avoid calculating directly, neither of the two expectations in the EM algorithm can be evaluated in analytical form. However, the Metropolis–Hastings algorithm (Metropolis, et al. (1953), Hastings (1970)), can produce random draws from the conditional distribution $\gamma|\mathbf{Y}$ without specifying f_Y. Choosing f_γ as the candidate distribution, a new value γ^* can be generated from $\gamma|\mathbf{Y}$ with acceptance probability

$$A_k(\gamma^*, \gamma) = \min\left\{1, \frac{\prod_{i=1}^{n} f_{Y_i|\gamma}(\mathbf{Y}_i|\gamma^*)}{\prod_{i=1}^{n} f_{Y_i|\gamma}(\mathbf{Y}_i|\gamma)}\right\}, \tag{32}$$

where non-acceptance means to retain γ. This calculation involves only the specification of the conditional distribution of $f_{Y_i|\gamma}$.

Combining the EM and Metropolis–Hastings algorithms leads to an MCEM algorithm with the following steps:

(1) Choose initial values, $\beta^{(0)}, \mathbf{\Gamma}^{(0)}$ and set $l = 0$;

(2) Generate k random draws, $\gamma^{(1)}, \ldots, \gamma^{(k)}$, from $f_{\gamma|Y_i}(\gamma|\mathbf{Y}_i, \beta^{(l)}, \Gamma^{(l)})$ by the Metropolis–Hastings algorithm.

   (a) Choose $\beta^{(l+1)}$ to maximize the expectation $\frac{1}{k}\sum_{i=1}^{k} \ln f_{Y|\gamma}(\mathbf{Y}|\beta, \gamma^{(i)})$;

   (b) Choose $\mathbf{\Gamma}^{(l+1)}$ to maximize $\frac{1}{k}\sum_{i=1}^{k} \ln f_\gamma(\gamma^{(i)}|\mathbf{\Gamma})$;

   (c) $l = l + 1$;

(3) If $\|\frac{\beta^{(l)} - \beta^{(l-1)}}{\beta^{(l)}}\|^2 \leq 10^{-3}$ and $\|\frac{\mathbf{\Gamma}^{(l)} - \mathbf{\Gamma}^{(l-1)}}{\mathbf{\Gamma}^{(l)}}\|^2 \leq 10^{-3}$, then declare $\beta^{(l)}$ and $\mathbf{\Gamma}^{(l)}$ to be the MLE's (convergence criterion); otherwise, go back to step 2.

3.3.2. *Simulated Maximum Likelihood Algorithm (SML)*

Geyer and Thompson (1992) have suggested implementing a simulation strategy to estimate the likelihood in (30) directly. According to (30), we have

$$L(\beta, \mathbf{\Gamma}|\mathbf{Y}) = \int f_{Y|\gamma}(\mathbf{Y}|\beta, \gamma) f_\gamma(\gamma|\mathbf{\Gamma}) d\gamma \\ \cong \frac{1}{k} \sum_{i=1}^{k} \frac{f_{Y|\gamma}(\mathbf{Y}|\beta, \gamma^{(i)}) f_\gamma(\gamma^{(i)}|\mathbf{\Gamma})}{h_\gamma(\gamma^{(i)})}, \tag{33}$$

where $\gamma^{(1)}, \ldots, \gamma^{(k)}$ are simulated values selected from the importance sampling distribution $h_\gamma(\gamma)$. Browsing a large number of histograms and scatterplots of the simulated values from $f_{\gamma|Y}$ provides empirical justification for approximating the optimal importance sampling distribution $h_\gamma(\gamma)$ by a Gaussian distribution, then using the Metropolis algorithm to estimate its mean and variance. Maximization of (33) was implemented by quasi-Newton maximization with non-linear constraints in the Numerical Algorithms Group Library, algorithm E04UCF (also available under the name NPSOL from Stanford University).

MCEM algorithms may converge to a local, rather than a global, maxima (Chan and Ledholter (1995)), as the likelihood surfaces need not be unimodal (Searle, et al. (1992)). SML is slower than MCEM, but attains estimates with higher accuracy (McCulloch (1997)). To refine the estimates, and to achieve an estimator closer to the true maximum of the likelihood, we found that an initial MCEM step, followed by SML, gave good results. This hybrid algorithm is summarized as follows,

(1) Use MCEM to get rough MLE's β^* and $\mathbf{\Gamma}^*$;
(2) Produce a series of random draws from $h_\gamma(\gamma)$ with respect to β^* and $\mathbf{\Gamma}^*$;
(3) Maximize (33) to obtain $\widehat{\beta}$ and $\widehat{\mathbf{\Gamma}}$.

For the implementation of cubic B-splines, we choose all the knots as equidistant and use the same cubic B-spline basis $\{\bar{B}_k, 1 \leq k \leq p\}$ to model both the mean and covariance functions of the underlying Gaussian process. Estimators for the number of included components in model (29) are obtained by minimizing the one-curve-leave-out cross-validation score, minimizing squared prediction error

$$\mathrm{CV}(p) = \sum_{i=1}^{n} \sum_{j=1}^{m} \{Y_{ij} - \widehat{X}_i^{(-i)}(t_{ij})\}^2, \tag{34}$$

where $\widehat{X}_i^{(-i)}$ is the predicted curve for the i-th subject after removing the i-th sample curve from fitting model (29), i.e., $\widehat{X}_i^{(-i)}(t) =$

$g^{-1}(\sum_{k=1}^{p}\hat{\beta}_k^{(-i)}B_k(t_{ij})+\sum_{k=1}^{q}\hat{\gamma}_{ik}^{(-i)}B_k(t_{ij}))$, where $\hat{\beta}_k^{(-i)}, \hat{\gamma}_{ik}^{(-i)}$ are estimated as described above, after removing the data for the i-th subject.

4. Simulation Studies

To compare the different approaches under a sampling design scenario with densely sampled data, we consider two simulation settings, binary longitudinal data generated with a logit link and Poisson longitudinal data generated with a log link. For each simulation setting, we generated 200 i.i.d. regular samples consisting of $n = 400$ subjects each.

4.1. *Design of the simulations*

For the simulation scenario with binary designs, we assume that the data for each subject are generated by an underlying latent Gaussian process $Z_i(t) = \mu(t) + \sum_{k=1}^{11}\xi_{ik}\phi_k(t)$ with mean function $\mu(t) = -0.0064(t-25)^2+2$ and covariance function derived from eleven eigenfunctions corresponding to a Fourier basis on the time interval $[0, 40]$, given as follows,

$$\begin{aligned}
\phi_1(t) &= \sqrt{1/40}, \qquad \phi_2(t) = \sqrt{2/40}\sin((t-40/2)2\pi/40),\\
\phi_3(t) &= \sqrt{2/40}\cos((t-40/2)2\pi/40),\\
\phi_4(t) &= \sqrt{2/40}\sin((t-40/2)4\pi/40), \ldots,\\
\phi_{11}(t) &= \sqrt{2/40}\cos((t-40/2)10\pi/40), \qquad 0 \le t \le 40.
\end{aligned}$$

We choose eigenvalues as $\{\lambda_k\}_{k=1}^{11} = (4\cdot 0.4^j), j = 0, 1, \ldots, 10$ and $\{\lambda_k\}_{k>11} = 0$. The functional principal component scores ξ_{ik} are generated as independent $\mathcal{N}(0, \lambda_k)$ r.v.s. For the i-th subject, the measurement times are equally spaced and the number of observations is chosen as $m = 41$. We define $p(Z_i(t)) = g^{-1}(Z_i(t))$, choosing a logit link, i.e., $g(x) = \log(x/(1-x))$. Then the binomial longitudinal observations for the i-th subject are generated as $Y_i(t) \sim \text{Binomial}(p(Z_i(t)), 1)$, with probability $p(Z_i(t)) = g^{-1}(Z_i(t)) = \exp(Z_i(t))/(1+\exp(Z_i(t)))$ of observing a response $Y_{ij} = 1$ at measurement time $t = t_{ij}$.

For the case of Poisson longitudinal measurements, the simulated latent Gaussian processes have mean function $\mu(t) = -0.004(t-25)^2 + 4$ and the same eigenfunctions, eigenvalues and functional principal components as well as measurement designs as those for the binary response case. We then choose Poisson parameters $\lambda(Z_i(t)) = g^{-1}(Z_i(t))$, choosing the log link function $g(x) = \log(x)$ and generating Poisson longitudinal observations for the i-th subject $Y_i(t) \sim \text{Poisson}(\lambda(Z_i(t)))$ with means $\lambda(Z_i(t)) = \exp(Z_i(t))$.

For simulating the Character Process Model (CPM) (23), we only consider the environmental effect and parameterize its covariance as

$$\text{Cov}(e_i(t), e_i(s)) = H(s,t) = \sqrt{a_0 + a_1 t}\sqrt{a_0 + a_1 s}\exp(-a_C(s-t)^2),$$

modeling the environmental variance $v^2(t)$ as a linear function $a_0 + a_1 t$ and the environmental correlation as a standard normal function $\rho(s,t) = e^{-a_C(s-t)^2}$. When fitting CPM and Cubic B-spline models (BS) (29), the MCEM routine was run with 5000 replications in each of the 500 iterations. The initial estimates for SML were obtained by averaging the MCEM estimates over the last 200 iterations. The SML estimates were based on 1000 replications, sampled from the Markov Chain every 200 steps.

The number p of included components in the FPCA-Q model (8) for each simulation run was chosen by AIC, based on (21). Whenever the matrix corresponding to the discretized covariance surface $\widehat{\Sigma}_Z$ was found to be close to singular, we used a Moore-Penrose pseudo-inverse for matrix inversion.

4.2. *Simulation results and discussion*

The results of the simulations are reported in Table 1, and are evaluated using three criteria: First, the leave-one-curve-out average mean squared prediction error for the observed non-Gaussian data,

$$\text{MSPE1} = \frac{1}{nm}\sum_{i=1}^{n} \|Y_i - \widehat{X}_i^{(-i)}\|^2, \tag{35}$$

where $\|\cdot\|$ is defined by approximating the usual L^2 distance through $\|Y_i - \widehat{X}_i^{(-i)}\|^2 = \sum_{j=1}^{m}(Y_{ij} - \widehat{X}_i^{(-i)}(t_{ij}))^2(t_{ij} - t_{i,j-1})$, where we chose $m = 41$. Recall that $\widehat{X}_i^{(-i)}$ is the predicted curve for the i-th subject after removing the subject's sample curve from the sample used to fit the model. As a second criterion we report average mean squared prediction error for the latent Gaussian process Z_i,

$$\text{MSPE2} = \frac{1}{nm}\sum_{i=1}^{n} \|Z_i - \widehat{Z}_i^{(-i)}\|^2. \tag{36}$$

The third evaluation metric is the Pearson Criterion,

$$\text{PE} = \frac{1}{nm}\sum_{i=1}^{n}\sum_{j=1}^{m} \frac{\{Y_{ij} - \widehat{X}_i^{(-i)}(t_{ij})\}^2}{V(\widehat{X}_i^{(-i)}(t_{ij}))}, \tag{37}$$

where $V(\cdot)$ is the variance function of $Y_i(t)$ defined in (2). This criterion weighs down differences where variances are large.

Table 1. Comparisons of the one-curve-leave-out average mean squared predicted error for the observed non-Gaussian data Y_i (MSPE1) (35), the average mean squared predicted error for the latent Gaussian process Z_i (MSPE2) (36) and Pearson criterion (PE) (37) for the Character Process Model (CPM) (23), cubic B-spline model (BS) (29) and FPCA via Quasi-likelihood (FPCA-Q) (8).

Model	Binomial			Poisson		
	MSPE1	MSPE2	PE	MSPE1	MSPE2	PE
CPM	0.332	0.679	2.974	194.200	0.281	3.87
BS with 3 knots	0.161	0.414	0.962	142.880	0.070	1.118
BS with 6 knots	0.161	0.431	0.957	142.280	0.070	3.180
FPCA-Q	0.118	0.215	0.588	27.512	0.011	0.762

In both Binomial and Poisson scenarios, we find that FPCA-Q outperforms the other methods in predicting the observed non-Gaussian data and in estimating the underlying latent Gaussian process as well. FPCA-Q results in dramatic decreases in MSPE1 of about 65% and 27% for the Binomial case, when comparing its performance with that of CPM and BS with three knots, respectively. The corresponding improvements for the Poisson case are about 86% and 81%.

Even more remarkably, MSPE2 is reduced by 68% and 48% in the Binomial case and by 96% and 84% in the Poisson case, when using FPCA-Q as compared to CPM and BS with three knots, respectively. The results from BS with six knots are nearly the same as those for BS with three knots since the structure of the simulated latent Gaussian process is relatively simple and three knots suffice to approximate it. With regard to all three measures in Table 1, the gains for FPCA-Q are substantial throughout when comparing its performance with CPM, or B-splines. The gains of FPCA-Q for Poisson data are particularly large.

We conclude that there are sizable differences in performance between various functional approaches for modeling longitudinal data with generalized responses that are densely sampled. Among the methods considered here, FPCA-Q, a proposed quasi-likelihood based version of functional principal component analysis that is specially adjusted to the case of generalized responses outperforms previously established methods. It will be interesting to broaden these comparisons in future work and to investigate the theoretical properties of these methods.

Acknowledgments

This work was supported in part by NSF grants DMS08-06199 and DMS09-06813. The paper was finalized during a visit of H.G.M. and J.L.W. at the Statistical and Applied Mathematical Sciences Institute (SAMSI), Research Triangle Park, North Carolina. The support of this visit through SAMSI is gratefully acknowledged.

References

Ash, R. B. and Gardner, M. F. (1975). *Topics in Stochastic Processes.* Academic Press, New York.

Booth, J. G. and Hobert and J. H. (1999). Maximizing generalized linear mixed model likelihoods with an automated Monte Carlo EM algorithm. *Journal of the Royal Statistical Society B,* **62**, 265–285.

Breslow, N. E. and Clayton, D. G. (1993). Approximate inference in generalized linear mixed models. *Journal of the American Statistical Association*, **88**, 9–25.

Capra, W. B. and Müller, H.-G. (1997). An Accelerated-Time Model for Response Curves. *Journal of the American Statistical Association*, **92**, 72–83.

Chan, K. S. and Ledholter, J. (1995). Monte Carlo EM Estimation for Time Series Models Involving Counts. *Journal of the American Statistical Association*, **90**, 242–252.

Diggle, P. J., Heagerty, P. J., Liang, K.-Y. and Zeger, S. L. (2002). *Analysis of Longitudinal Data*, Oxford University Press.

Fan, J. and Gijbels, I. (1996). *Local Polynomial Modelling and Its Application.* London: Chapman and Hall.

Geyer, C. J. and Thompson, E. A. (1992). Constrained Monte Carlo Likelihood for Dependent Data (with discussion). *Journal of the Royal Statistical Society* **B 54**, 657–699.

Hall, P., Müller, H.-G. and Yao, F. (2008). Functional Data Analysis for Generalized Longitudinal Observations via Latent Gaussian Processes. *Journal of the Royal Statistical Society* **B 70**, 703–723.

Hastings, W.K. (1970). Monte Carlo sampling methods Using Markov chains and their applications. *Biometrika*, **57**, 97–109.

Kirkpatrick, M. and Heckman, N. (1989). A quantitative genetic model for growth, shape, reaction norms, and other infinite-dimensional characters. *Journal of Mathematical Biology*, **27**, 429–450.

Kirkpatrick, M., Lofsvold, D. and Bulmer, M. (1990). Analysis of inheritance, selection and evolution of growth trajectories. *Genetics*, **124**, 979–993.

Laird, N. M., and Ware, J. H. (1982). Random-effects models for longitudinal data. *Biometrics*, **38**, 963–974.

McCullagh, P. (1983). Quasi-likelihood functions. *Ann. Statist.*, **11**, 59–67.

McCullagh, P. and Nelder, J. A. (1989). *Generalized Linear Models* (2nd ed.). London: Chapman and Hall.

McCulloch, C. E. (1997). Maximum likelihood algorithms for generalized linear mixed models. *Journal of the American Statistical Association*, **92**, 162–170.

Metropolis, N., Rosenbluth, A.W., Rosenbluth, M.N., Teller, A.H., and Teller, E. (1953). Equations of state calculations by fast computing machines. *Journal of Chemical Physics*, **21**, 1087–1092.

Meyer, K. (1998). Estimating covariance functions for longitudinal data using a random regression model. *Genetics, Selection, and Evolution*, **30**, 221–240.

Meyer, K. (2000). Random regressions to model phenotypic variation in monthly weights of Australian beef cows. *Livestock Production Science*, **65**, 19–38.

Meyer, K. and Hill, W. G. (1997). Estimation of genetic and phenotypic covariance functions for longitudinal or 'repeated' records by restricted maximum likelihood. *Livestock Production Science*, **47**, 185–200.

Pletcher, S. D. and Geyer, C. J. (1999). The genetic analysis of age-dependent traits: Modelling a character process. *Genetics*, **153**, 825–833.

Pletcher, S. D. and Jaffrézic, F. (2002). Generalized character process models: estimating the genetic basis of traits that cannot be observed and that change with age or environmental conditions. *Biometrics*, **58**, 157–162.

Rice, J. and Silverman, B. (1991). Estimating the Mean and Covariance Structure Nonparametrically When the Data are Curves. *Journal of the Royal Statistical Society*, **53**, 233–243.

Rice, J. and Wu, C. (2000). Nonparametric Mixed Effects Models for Unequally Sampled Noisy Curves. *Biometrics*, **57**, 253–259.

Searle, S. R., Casella, G. and McCulloch, C. E. (1992). *Variance Components*. New York: Wiley.

Shaw, F., Promislow, D.E.L., Tatar, M., Hughes, K. and Geyer, C. J. (1999). Towards reconciling inferences concerning genetic variation in senescence. *Genetics*, **152**, 553–566.

Shi, M., Weiss, R. E. and Taylor, J. M. G. (1996). An Analysis of Paediatric CD4 counts for Acquired Immune Deficiency Syndrome using Flexible Random Curves. *Applied Statistics*, **45**, 151–163.

Staniswalis, J. G., and Lee, J. J. (1998). Nonparametric Regression Analysis of Longitudinal Data. *Journal of the American Statistical Association*, **93**, 1403–1418.

Stiratelli, R., Laird, N. and War, J. H. (1984). Random Effects Models for Serial Observations with Binary Responses. *Biometrics*, **40**, 961–971.

Wolfinger, R. and O'Connell, M. (1993). Generalized linear mixed models: A pseudo-likelihood approach. *The Journal of Statistical Computation and Simulation*, **48**, 233–243.

Yao, F., Müller, H.-G., Clifford, A. J., Dueker, S. R., Follett, J., Lin, Y., Buchholz, B. A. and Vogel, J. S. (2003). Shrinkage estimation for functional principal component scores, with application to the population kinetics of plasma folate. *Biometrics*, **59**, 676–685.

Yao, F., Müller, H.-G. and Wang, J.-L. (2005). Functional data analysis for sparse longitudinal data. *Journal of the American Statistical Association* **100**, 577–590.

Zeger, S. L. and Karim, M. R. (1991). Generalized linear models with random effects: A Gibbs sampling approach. *Journal of the American Statistical Association*, **82**, 231–238.

PART 3

Papers on Aspects of Linear or Generalized Linear Models

Chapter 10

Better Residuals

Rudolf Beran

University of California, Davis

Suppose that p unknown constants are observed with random errors. How should the random errors incurred be predicted from the data? A plausible answer—use the residuals from a classically efficient large-model estimator of the constants—is wrong, both experimentally and theoretically. When p is not small, residuals from an adaptive regularized estimator of the constants can predict the errors far better and never worse.

Contents

1. Introduction

The case for using large statistical models was stated with characteristic insight by Hájek (1973): "Robust methods are based on models involving large families of distributions and thus having greater chance to include the true situation. Any gain in robustness, however, has to be paid for by losses in efficiency, even if in some cases large gains may be bargained for small losses." During the 1960's and 70's, robustness formulations sought to achieve relatively high efficiency at or near a conjectured small model while preserving acceptable performance in a much larger model.

Consider a mirror formulation: achieving high efficiency in a large model while achieving superior performance in smaller models conjectured to be adequate. The feasibility and potential of this alternative formulation are suggested strongly by Stein's results on shrinkage estimation of a mean vector (Stein (1956)). The James–Stein estimator performs well over the entire parameter space of mean vectors (the large model in this instance) while achieving strikingly superior performance when the mean vector is restricted to small balls about the origin (the small models). That the James–Stein estimator addresses a simple estimation scenario in no way lessens its import.

This paper explores the question of constructing trustworthy residuals when estimating a vector of unknown means. Residuals are obtained by comparing the data vector with a fit to the means. Typically there are many plausible estimates of the means, according to the generality of the model fitted. Each estimated mean vector generates a set of residuals. Which of these residuals are most trustworthy in predicting the observation errors under the largest model considered? As we will see, it is usually not the residuals generated by classically efficient fitting of the largest model.

Figure 1 is a scatterplot of the Canadian Earnings Data considered by Ullah (1985) and by Chu and Marron (1991). The plot displays log(incomes) for $n = 205$ individuals versus their $p = 45$ ages. A general model for the data is

$$y = Cm + e. \tag{1}$$

Here y is the $n \times 1$ vector of observed log(incomes), m is the $p \times 1$ vector of mean log(incomes), C is the $n \times p$ data-incidence matrix that expresses the replication pattern in observing the means, and e is the $n \times 1$ error vector. The error vector is taken to satisfy the *strong Gauss–Markov condition*: the components of e are independent, identically distributed random variables whose unknown distribution has mean 0, variance σ^2, and finite fourth moment.

Competing submodels to model (1) impose restrictions on the unknown mean vector m. Linear submodels postulate that

$$m = V\gamma, \tag{2}$$

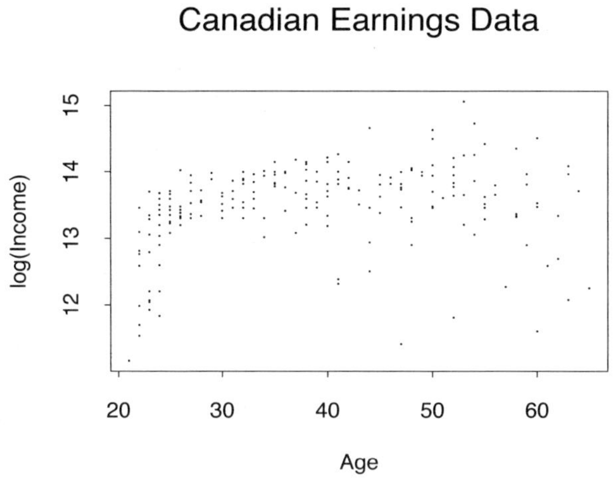

Fig. 1. Scatterplot of log(incomes) for 205 individuals of ages 21 to 65

where V is a specified $p \times r$ matrix of rank $r \leq p$ and γ is an unknown $r \times 1$ vector. For example, by taking successive orthogonal polynomials as the columns of V, we may construct competing least squares polynomial fits of degrees 0 to $p-1$ to $\mathrm{E}(y) = Cm$. The least squares polynomial fit of degree $p-1$ coincides with the least squares fit to model (1).

For any estimator $\hat{m}$ of m, the corresponding residual vector is $\hat{e} = y - C\hat{m}$. Figure 2 displays, for the Canadian Earnings Data, the least squares polynomial fits of degrees 2 to 9 and of degree 44, The last submodel coincides with general model (1). The least squares fit to this largest submodel yields the least squares variance estimate $\hat{\sigma}_{ls}^2 = .295$. Figure 3 exhibits the corresponding residuals. The dashed lines in the plots are at $\pm 2\hat{\sigma}_{ls}$. Note that:

- The fitted values at each age are interpolated linearly, to make them visible against the data. The age covariate is discrete here.
- Consider the usual F-test for the null hypothesis H: degree 2 submodel is true versus K: not so but the general model (degree 44) is true. The p-value for the test statistic is .12.
- Nevertheless, the degree 2 and degree 3 least squares fits appear to the eye to underfit the data while the general model fit (degree 44) may overfit.

- The fits of degrees 4 to 9 differ modestly to the eye.
- The residual plots from competing fits have a parallel visual grouping. There is little apparent difference among the residuals from the fits of degrees 4 to 9.

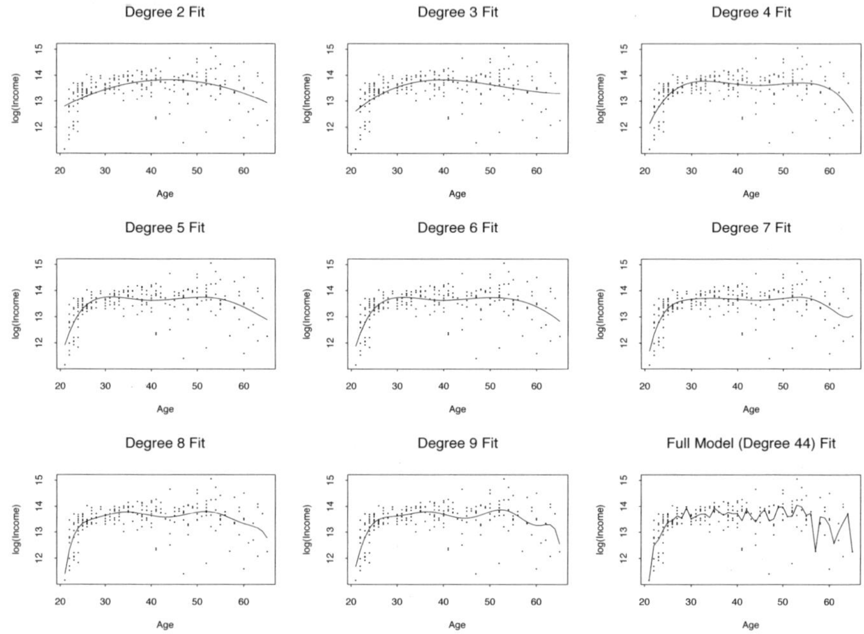

Fig. 2. Interpolated polynomial submodel fits to the Canadian Earnings Data

Assuming that the model (1) is sufficiently general, which of the least squares polynomial fits yields residuals that best predict the error vector e? The answer is not apparent from the plots. Nor is it necessarily the case that the residuals from the highest degree polynomial fit are the answer. This paper develops a risk-based methodology for identifying best predictors of the errors e.

2. An Experiment

Artificial Data is constructed to roughly mimic the Canadian Earnings Data. The $p = 100$ components of the mean vector m are given by the

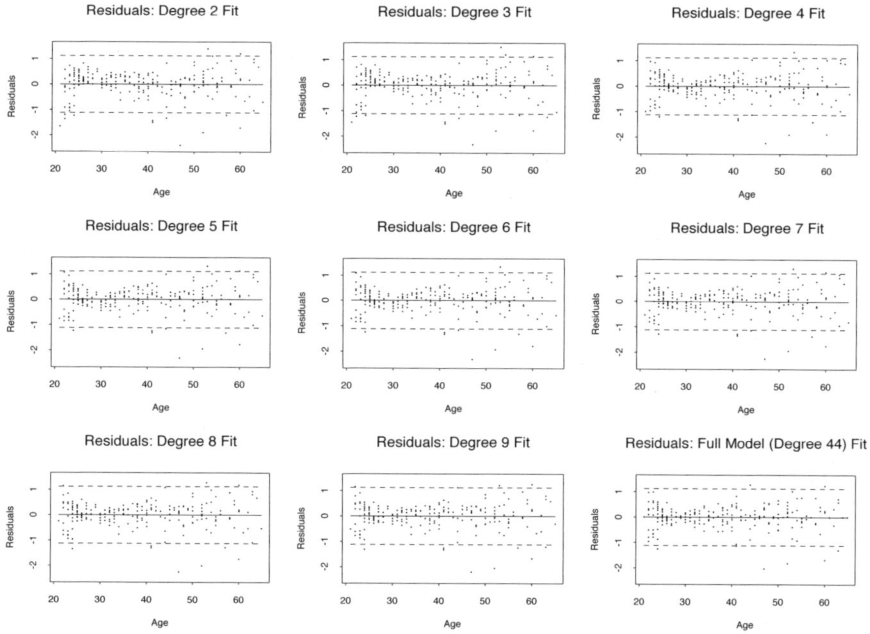

Fig. 3. Residuals from polynomial submodel fits to the Canadian Earnings Data

formula

$$m_i = 15 - 100[(x_i - 1/4)(x_i - 3/4)]^2 \quad \text{with} \quad x_i = i/(p+1), \quad 1 \leq i \leq p. \tag{3}$$

Two observations are made on each m_i by adding pseudorandom $N(0, .25)$ errors. Thus, y has $n = 200$ components. Figure 4 plots the components of y against the corresponding values of x_i.

Figure 5 displays, for the Artificial Data, the least squares polynomial fits of degrees 2 to 9 and of degree 99. The last submodel coincides with general model (1). The least squares fit to this largest submodel yields the least squares variance estimate $\hat{\sigma}_{ls}^2 = .265$, which compares well with the actual $\sigma^2 = .25$ used to construct the Artificial Data. Figure 6 exhibits the corresponding residuals. The dashed lines in the plots are again at $\pm 2\hat{\sigma}_{ls}$. Much as for the Canadian Earnings Data:

- The fitted values in Figure 5 are interpolated linearly, to make them visible against the Artificial Data.
- The degree 2 and degree 3 least squares fits appear to the eye to underfit

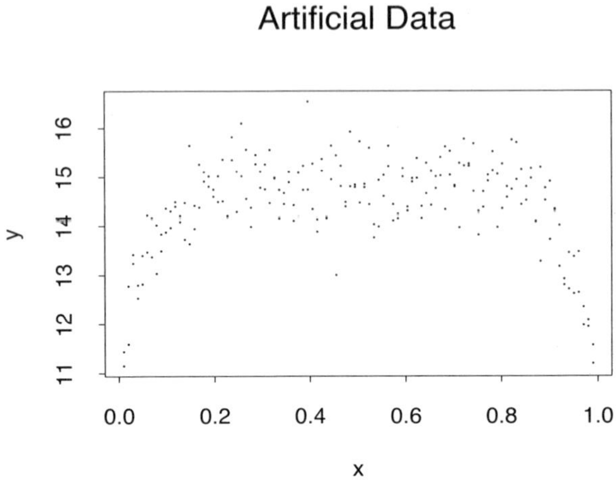

Fig. 4. Scatterplot of the Artificial Data

the data while the general model fit, of degree 99, may overfit.

- The fits of degrees 4 to 9 appear similar to the eye.
- The residual plots from competing fits have a parallel visual grouping. There is little apparent difference among the residuals from the fits of degrees 4 to 9.

For the Artificial Data, we have the advantage of knowing both the mean vector m and the error vector e. Figure 7 plots the residuals from the various least squares fits against the actual errors.

A one-number summary of how well the residuals $\hat{e}$ predict the errors e in model (1) is the *normalized quadratic loss*

$$L(\hat{e}, e) = p^{-1}|\hat{e} - e|^2. \qquad (4)$$

For the Artificial Data, the observed losses for residuals from the polynomial fits of degrees 2 to 10 and from the general model, of degree 99, are:

Degree	2	3	4	5	6
Loss	.419	.423	**.006**	.009	.018

Degree	7	8	9	10	99
Loss	.019	.020	.025	.028	.312

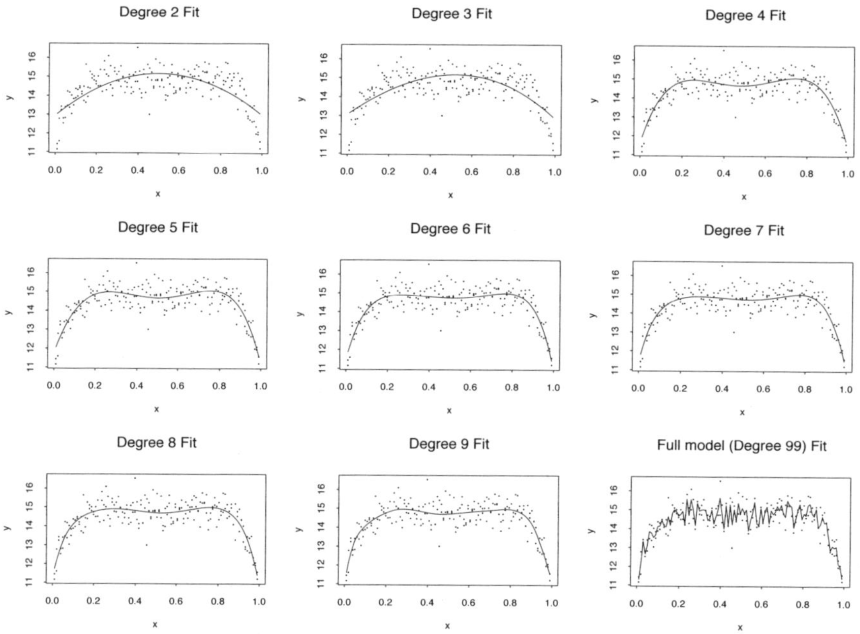

Fig. 5. Interpolated polynomial submodel fits to the Artificial Data

It is clear now that the degree of the fit matters when using the residuals to predict the errors. Specifically:

- In Figure 7, the residuals from the degree 4 fit best predict the errors e. They are visibly a better predictor than the residuals from fits of lower or much higher degree.
- The fit with smallest loss is also of degree 4. When assessed through their losses, the quality of the residuals as a predictor or e declines rapidly as the degree of the fit drops below 4 but only gradually as the degree of the fit climbs above 4.
- The residuals from the least squares fit to model (1), that is, from the polynomial fit of degree 99, are significantly inferior as a predictor of the errors. This is so even though model (1) governs the manner in which the Artificial Data was generated.

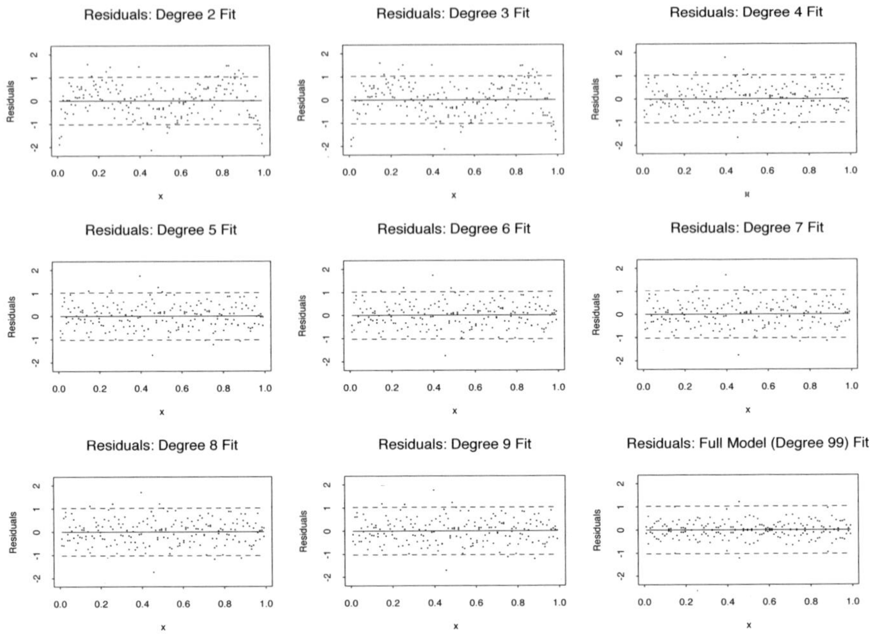

Fig. 6. Residuals from polynomial submodel fits to the Artificial Data

3. Theoretical Framework

For theoretical analysis, we formalize the preceding discussion. As in (1), the data vector y is governed by the model

$$y = \eta + e \qquad \text{with} \qquad \eta = Cm. \tag{5}$$

The components of e are independent, identically distributed with unknown distribution P that has mean 0, variance $\sigma^2 = \sigma^2(P) < \infty$, and finite 4th moment. Consequently, $\eta = \mathrm{E}(y)$. The vectors y and e are $n \times 1$; m is $p \times 1$ with $p \leq n$; and the $n \times p$ data-incidence matrix C is of full rank, with pseudoinverse $C^+ = (C'C)^{-1}C'$.

Define the *loss* $L(\hat{e}, e)$ of any predictor $\hat{e}$ of e by (4). The corresponding *risk* is

$$R(\hat{e}, P) = \mathrm{E}[L(\hat{e}, e)] = p^{-1}\mathrm{E}|\hat{e} - e|^2. \tag{6}$$

The expectation E is *always* computed under model (5).

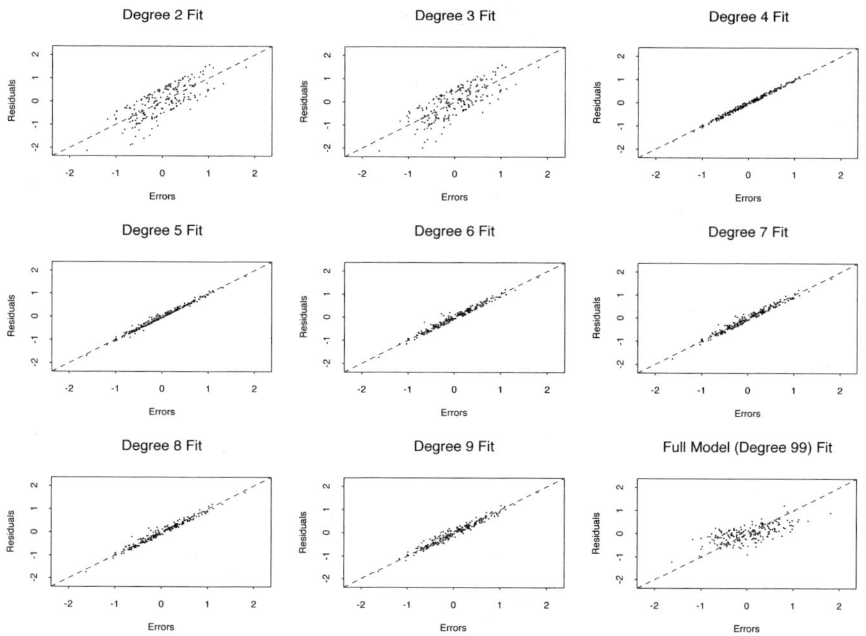

Fig. 7. Residuals from polynomial submodel fits to the Artificial Data versus the errors

A predictor $\hat{e}$ is *unbiased* for e if $\mathrm{E}(\hat{e}) = \mathrm{E}(e) = 0$ for every m and P in model (5). The *least squares predictor*

$$\hat{e}_{ls} = y - CC^{+}y = [I_n - CC^{+}]y \tag{7}$$

is unbiased and is a linear function of y. Two classical results concerning $\hat{e}_{ls}$ are:

- Suppose that P satisfies the Gauss–Markov conditions. Among all unbiased predictors of e that are linear in y, $\hat{e}_{ls}$ minimizes risk for all m and P (from the Gauss–Markov theorem).
- Suppose that $P = N(0, \sigma^2)$. Among all unbiased predictors of e, $\hat{e}_{ls}$ minimizes risk for all m and σ^2 (from the Lehmann–Scheffé theorem).

The experiment in Section 2 suggests, correctly, that the least squares predictor $\hat{e}_{ls}$ is not optimal within the class of *all* predictors. Limiting attention to unbiased predictors is far from harmless in its consequences. The better predictors found in the experiment are biased under model (5).

3.1. *Candidate Linear Predictors of Error*

For any $p \times p$ matrix A, consider the *linear estimator* Ay for η in model (5). The associated *candidate linear predictor* $\hat{e}(A) = y - Ay$ for e has loss $L(\hat{e}(A), e) = p^{-1}|\hat{e}(A) - e|^2 = p^{-1}|Ay - \eta|^2$. The risk

$$R(\hat{e}(A), P) = \mathrm{E}[L(\hat{e}(A), e)] = p^{-1}[\mathrm{tr}(A'A) + \eta'(I_n - A)'(I_n - A)\eta] \qquad (8)$$

is minimized when A equals

$$\tilde{A} = I_n - (I_n + \sigma^{-2}\eta\eta')^{-1} = (\sigma^2 + |\eta|^2)^{-1}\eta\eta'. \qquad (9)$$

Let $R = C'C$ (the data replication matrix), $U = CR^{-1/2}$, $\xi = R^{1/2}m$. Then

$$\tilde{A} = U\tilde{S}U', \qquad \tilde{S} = (\sigma^2 + |\xi|^2)^{-1}\xi\xi', \qquad U'U = I_p. \qquad (10)$$

The oracle predictor is thus $\hat{e}(\tilde{A}) = [I_n - U\tilde{S}U']y$. The matrix $\tilde{S}$ is $p \times p$ symmetric with all eigenvalues in $[0, 1]$.

Successful estimation of $\tilde{S}$ from the data proves to be impossible—the class of all such matrices is too large. Let $\{S(t) \colon t \in \mathcal{T}\}$ be a *restricted* family of symmetric $p \times p$ matrices whose eigenvalues all lie in $[0, 1]$. Consider only the *candidate symmetric linear predictors* of e

$$\hat{e}(t) = [I_n - US(t)U']y, \qquad t \in \mathcal{T}. \qquad (11)$$

When the family $\{S(t) \colon t \in \mathcal{T}\}$ is suitably restricted, it proves possible to construct an adaptive estimator whose loss and risk converge, as p tends to infinity, to that of the restricted oracle estimator that minimizes risk over the candidate estimator class (11).

3.2. *Some Classes of Candidate Symmetric Linear Predictors*

The following examples of interesting candidate symmetric linear predictors motivate the asymptotic theory described in section 4.

Example 1 (Least squares predictor). *The less than optimal least squares predictor $\hat{e}_{ls}$, defined in (7), is of the form (11). The family $\{S(t) \colon t \in \mathcal{T}\}$ consists in this case of the single matrix I_p.*

Example 2 (Multiply penalized least squares). *Suppose the matrices $\{Q_i \colon 1 \leq i \leq s\}$ are positive semidefinite, with s fixed independently* *of p. The penalty weights are $t = (t_1, t_2, \ldots, t_s) \in [0, c]^s$, with c fixed and large. Let*

$$\hat{m}(t) = \underset{m}{\operatorname{argmin}}[|y - Cm|^2 + \sum_{i=1}^{s} t_i m' Q_i m] = (C'C + \sum_{i=1}^{s} t_i Q_i)^{-1} C'y. \quad (12)$$

The candidate penalized least squares predictor (PLS) of e is

$$\hat{e}(t) = y - C\hat{m}(t) = [I_n - US(t)U']y, \quad (13)$$

with

$$S(t) = [I_p + \sum_{i=1}^{s} t_i V_i]^{-1}, \qquad V_i = R^{-1/2} Q_i R^{-1/2}. \quad (14)$$

Example 3 (Running weighted means of span 3). *Consider a balanced one-way layout with ordinal factor, in which r scalar observations are made at each of p equally spaced factor levels. Thus $C = I_p \otimes u_r$, with u_r being the $r \times 1$ vector $(1, 1, \ldots 1)$, and $U = r^{-1/2} C$. For $\mathcal{T} = \{(t_1, t_2) \in [0, 1]^2 \colon t_1 + 2t_2 = 1, t_1 \geq .5\}$, define the $p \times p$ symmetric matrix*

$$S(t) = \begin{pmatrix} t_1 + t_2 & t_2 & 0 & 0 & \ldots & 0 & 0 & 0 \\ t_2 & t_1 & t_2 & 0 & \ldots & 0 & 0 & 0 \\ 0 & t_2 & t_1 & t_2 & \ldots & 0 & 0 & 0 \\ \vdots & \vdots & \vdots & \vdots & \ddots & \vdots & \vdots & \vdots \\ 0 & 0 & 0 & 0 & \ldots & t_2 & t_1 & t_2 \\ 0 & 0 & 0 & 0 & \ldots & 0 & t_2 & t_1 + t_2 \end{pmatrix} \quad (15)$$

The eigenvalues of $S(t)$ lie in $[0, 1]$ when $t \in \mathcal{T}$. The candidate running weighted mean *(RWM) predictor is $\hat{e}(t) = [I_n - US(t)U']y = y - C\hat{m}(t)$, where $\hat{m}(t)$ is the running weighted mean estimator of m that uses the weights t.*

Example 4 (Projective submodel fits to a 2-way layout). *The model is $y_{ijk} = m_{ij} + e_{ijk}$, $1 \leq i \leq p_1$, $1 \leq j \leq p_2$, $1 \leq k \leq n_{ij}$. Here i, j label the levels of factors 1 and 2, k labels the replications, and $p = p_1 p_2$. In vector form, $y = Cm + e$ with*

$$\begin{aligned} m &= \{\{m_{ij} \colon 1 \leq i \leq p_1\}, 1 \leq j \leq p_2\} \quad (16) \\ y &= \{\{\{y_{ijk} \colon 1 \leq k \leq n_{ij}\}, 1 \leq i \leq p_1\}, 1 \leq j \leq p_2\}. \end{aligned}$$

Let Q be a $p \times p$ symmetric idempotent matrix with spectral representation $Q = VV'$. Submodel Q asserts $m = Q\beta$ for some $\beta \in R^p$. The least squares *estimator of m in Submodel Q is*

$$\hat{m}(Q) = (CQ)^+ y = V(V'RV)^{-1}V'C'y, \quad R = C'C. \tag{17}$$

Consequently, the candidate Submodel Q predictor is $\hat{e}(Q) = y - C\hat{m}(Q) = [I_n - US(Q)U']y$, with $S(Q) = R^{1/2}V(V'RV)^{-1}V'R^{1/2}$. Q ranges over a specified finite set of projections.

Example 5 (Monotone penalized least squares). *In the notation of Example 2, set the penalty matrix $Q_i = P_i$, where $\{P_i\colon 1 \le i \le p\}$ are symmetric, idempotent, mutually orthogonal $p \times p$ matrices such that $\sum_{i=1}^p P_i = I_p$. Let $\mathcal{T} = \{t \in [0,1]^p\colon t_1 \le t_2 \le \ldots \le t_p\}$. Note that the fixed s in Example 2 is replaced here by p, which will be allowed to increase in the asymptotic theory. Define*

$$S(t) = \sum_{i=1}^{p} t_i P_i, \qquad t \in \mathcal{T}. \tag{18}$$

The candidate monotone shrinkage predictor of e is defined by (13). The formula simplifies algebraically for balanced designs. In that case, $\hat{e}(t)$ is the residual vector obtained from a monotone shrinkage estimator of m. See Beran and Dümbgen (1998) for the latter.

Remark 1. Further examples of candidate symmetric linear estimators of m were considered by Buja, et al. (1995) and Kneip (1994). Related candidate estimators for mean functions on continuous domains were treated by Wahba (1990), Lin (2000), and Heckman and Ramsay (2000).

4. Adaptive Symmetric Linear Predictors

As in the previous section, let $\eta = Cm$, $R = C'C$, $U = CR^{-1/2}$, $\xi = R^{1/2}m = U'\eta$. Then $\eta = U\xi$. Let $z = U'y$. Under the strong Markov assumption on the distribution of e, $\mathrm{E}(z) = \xi$ and $\mathrm{Cov}(z) = \sigma^2 I_p$.

The loss of the candidate symmetric linear predictor (11) is

$$L(\hat{e}(t), e) = p^{-1}|\hat{e}(t) - e|^2 = p^{-1}|S(t)z - \xi|^2. \tag{19}$$

The corresponding risk is

$$r(t) = R(\hat{e}(t), P) = \mathrm{E}[L(\hat{e}(t), e)] = p^{-1}[\sigma^2 \operatorname{tr}\{T(t)\} + \operatorname{tr}\{\bar{T}(t)\xi\xi'\}], \tag{20}$$

where $T(t) = [S(t)]^2$ and $\bar{T}(t) = [I_p - S(t)]^2$. This risk is unknown because it depends on the unknown mean vector m as well as σ^2. Ways of estimating it successfully from the data are the topic of this section.

4.1. *Estimating risk*

Let $\hat{\sigma}^2$ be an L_1-consistent estimator of σ^2 as p tends to infinite. Possible constructions of $\hat{\sigma}^2$ include: the least squares estimator when there is adequate replication; first- or higher-order difference variance estimators; pooling estimators; combinations of these methods. For specific constructions and results, see Beran (2000; 2002). Most of these variance estimators exhibit upward bias at finite p.

Define the *estimated risk*

$$\hat{r}(t) = p^{-1}[\hat{\sigma}^2 \operatorname{tr}\{T(t)\} + \operatorname{tr}\{\bar{T}(t)(zz' - \hat{\sigma}^2 I_p)\}]. \tag{21}$$

This formula is derived through the arguments for Mallows C_p. If $\mathrm{E}(\hat{\sigma}^2) = \sigma^2$, then $\mathrm{E}[\hat{r}(t)] = r(t)$ for every $t \in \mathcal{T}$. It is plausible to use the estimated risk function $\hat{r}(\cdot)$ as a surrogate for the actual unknown risk function $r(\cdot)$. When does this approach work?

4.2. *Asymptotic Convergence of Loss, Risk, and Estimated Risk*

Let $|\cdot|$ denote Euclidean norm. The spectral norm $|A|_{spec}$ of any matrix A is given by

$$|A|_{spec} = \sup_{x \neq 0} \frac{|Ax|}{|x|}. \tag{22}$$

The following assumptions support the theorems in this paper:

A1 Either $\mathcal{T} = [0,1]^k$, with k not depending on p, or the cardinality of $\mathcal{T}$ is finite, not depending on p. The family of symmetric matrices $\{S(t) \colon t \in \mathcal{T}\}$ is such that $\sup_p \sup_{t \in \mathcal{T}} |S(t)|_{spec} < \infty$. When $\mathcal{T} = [0,1]^k$, $S(t)$ is continuous on $\mathcal{T}$ and is differentiable on the interior of $\mathcal{T}$ with partial derivatives $\{\nabla_i S(t) = \partial S(t)/\partial t_i \colon 1 \leq i \leq k\}$ such that $\sup_{i,p} \sup_{t \in \mathcal{T}} |\nabla_i S(t)|_{spec} < \infty$.

A2 The strong Gauss–Markov model (5) holds.

A3 Under the strong Gauss–Markov model, the variance estimator $\hat{\sigma}^2$ is L_1-consistent:

$$\lim_{p \to \infty} \sup_{p^{-1}|\eta|^2 \leq a} \mathrm{E}|\hat{\sigma}^2 - \sigma^2| = 0 \tag{23}$$

for every finite $a > 0$ and $\sigma^2 > 0$.

Theorem 1. *Suppose Assumptions A1 to A3 hold. Let $W(t)$ denote either the loss $L(\hat{e}(t), e)$ or the estimated risk $\hat{r}(t)$. Then, for every finite $a > 0$ and $\sigma^2 > 0$,*

$$\lim_{p\to\infty} \sup_{p^{-1}|\eta|^2 \le a} \mathrm{E}[\sup_{t\in\mathcal{T}} |W(t) - r(t)|] = 0. \tag{24}$$

Thus, the loss, risk, and estimated risk of candidate predictor $\hat{e}(t)$ converge together asymptotically as p tends to infinity. Uniformity of the convergence over all $t \in \mathcal{T}$ makes $\hat{r}(\cdot)$ a trustworthy surrogate for the unknown loss or risk function. The proof of Theorem 1 relies on empirical process arguments and follows from Theorem 1 in Beran (2007).

4.3. *Successful Adaptation*

Let

$$\hat{t} = \operatorname*{argmin}_{t\in\mathcal{T}} \hat{r}(t), \qquad \tilde{t} = \operatorname*{argmin}_{t\in\mathcal{T}} r(t). \tag{25}$$

Within the candidate class $\{\hat{e}(t) \colon t \in \mathcal{T}\}$ of predictors, $\hat{e}(\hat{t})$ is empirically best while $\hat{e}(\tilde{t})$ is the oracle. Theorem 1 readily implies:

Theorem 2. *Suppose Assumptions A1 to A3 hold. Let V denote either the loss or risk of $\hat{e}(\hat{t})$. For every finite $a > 0$ and $\sigma^2 > 0$,*

$$\lim_{p\to\infty} \sup_{p^{-1}|\eta|^2 \le a} |V - r(\tilde{t})| = 0 \tag{26}$$

and

$$\lim_{p\to\infty} \sup_{p^{-1}|\eta|^2 \le a} |V - \hat{r}(\hat{t})| = 0. \tag{27}$$

Thus, the loss and risk of the empirically best predictor $\hat{e}(\hat{t})$ converge, respectively, to those of the oracle predictor $\hat{e}(\tilde{t})$. The plug-in risk estimator $\hat{r}(\hat{t})$ converges to the actual loss or risk of $\hat{e}(\hat{t})$. These convergences are uniform over balls in η which bound the signal-to-noise ratio.

Remark 2. Assumptions A1 and A2 are satisfied for Examples 1 through 4. Treating Example 5 is analogous but requires deeper empirical process theory, of the type used in Beran and Dümbgen (1998).

5. Case Studies

5.1. *Losses and Estimated Losses in the Experiment*

For the Artificial Data, $p = 100$, $n = 200$, the error variance is $\sigma^2 = .25$, and the least squares variance estimator is $\hat{\sigma}^2_{ls} = .265$. We assess the asymptotic results in section 4 by comparing observed losses with estimated risks/losses for the predictors based on the polynomial fits of degrees 2 to 10 and degree 99:

Degree	2	3	4	5	6
Loss	.419	.423	**.006**	.009	.018
Estimated Risk/Loss	.457	.458	**.068**	.070	**.066**

Degree	7	8	9	10	99
Loss	.019	.020	.025	.028	.312
Estimated Risk/Loss	.071	.075	.075	.077	.265

Estimated losses are higher than actual losses in this realization of the experiment. The two smallest estimated losses occur for the degree 6 and the degree 4 fits, are close in value, and approximate the truth in Figure 7. That the actual loss is minimized by the residuals from the degree 4 fit and increases only gradually for residuals from higher degree fits is recovered reasonably through the data analysis.

5.2. *Canadian Earnings Data*

For the Canadian Earnings Data, $p = 45$, $n = 205$, and the least squares variance estimator is $\hat{\sigma}^2_{ls} = .295$. The estimated losses/risks for the predictors based on the polynomial fits of degrees 2 to 10 and degree 44 are:

Degree	2	3	4	5	6
Estimated Risk	.106	.084	**-.023**	**-.029**	-.017

Degree	7	8	9	10	44
Estimated Risk	-.013	-.015	-.016	-.006	.295

The negative loss/risk estimates are not a concern. The risk estimator $\hat{r}(\cdot)$ is not constrained to be non-negative. Because adding a large positive constant to the loss function is harmless, it is clear that the most negative estimated loss/risk in the table is the best. The degree 5 error predictor in Figure 3 is best, followed by the degree 4 predictor. The associated fits to the data in Figure 2 appear to conform visually with this preference ordering.

5.3. *Plasma Citrate Data*

Data-set 41 in Andrews and Herzberg (1985) reports the following clinical experiment. For each of 10 patients, the concentration of citrate in plasma was measured hourly (in μmil per liter) at 14 times (8 to 21 hours). Meals were given at 8, 12 and 17 hours. The concentration measurements for each patient are linearly interpolated, to guide the eye, in the plot on the left side of Figure 8. There is considerable variability among the individual patient curves. It is not clear whether or how meal-times influence plasma citrate concentration.

The data-set is analyzed as a balanced two-way layout with one observation per mean. Factor 1 (patient number) is nominal with $p_1 = 10$ levels. Factor 2 (hour) is ordinal with $p_2 = 14$ equally spaced values. Here the general model (5) reduces to $y = m + e$, with

$$m = \{\{m_{ij} \colon 1 \le i \le p_1\}, 1 \le j \le p_2\} \qquad (28)$$
$$y = \{\{y_{ij} \colon 1 \le i \le p_1\}, 1 \le j \le p_2\}.$$

The dimension of m is $p = p_1 p_2 = 140$.

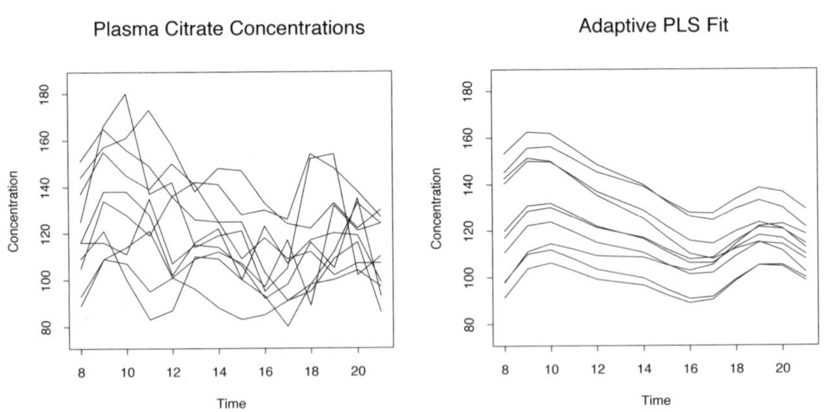

Fig. 8. The Interpolated Plasma Citrate Data and Adaptive PLS Fit

Using a strategy described in Beran (2005), candidate PLS predictors for the two-way layout are constructed as follows. For $k = 1, 2$, let $u_k = (1, 1, \ldots, 1)'$ be a $p_k \times 1$ vector. Let A_k be an *annihilator* matrix such that $A_k u_k = 0$. Let $J_k = u_k u_k'$ and $B_k = A_k A_k'$. Define the penalty matrices

$$Q_1 = J_2 \otimes B_1, \quad Q_2 = B_2 \otimes J_1, \quad Q_3 = B_2 \otimes B_1. \tag{29}$$

For $t = (t_1, t_2, t_3) \in R_+^3$, define $\hat{m}(t)$ through the PLS criterion

$$\hat{m}(t) = \underset{m}{\operatorname{argmin}}[|y - m|^2 + \sum_{i=1}^{3} t_i m' Q_i m] = (I_p + \sum_{i=1}^{3} t_i Q_i)^{-1} y_i. \tag{30}$$

Matrices Q_1 and Q_2 penalize, respectively. the fits to the main effects of Factors 1 and 2. Matrix Q_3 penalizes the fit to the interaction of Factors 1 and 2. The candidate PLS predictor is then $\hat{e}(t) = y - \hat{m}(t)$.

Choice of the annihilators A_1 and A_2 is influenced by the following considerations:

- Because Factor 1 (patient number) is nominal, we set $A_1 = I_{p_1} - u_1 u_1'$, as discussed in Beran (2005). This choice relates classical ANOVA fits to a larger class of PLS fits.
- Because Factor 2 (hour) is ordinal and the plasma citrate means are expected (but not assumed) to vary slowly as a function of hour, we set A_2 to be the r-th difference matrix. This choice encourages fits that are locally of polynomial degree $r - 1$ in hour.

There being no replication, third differencing of the measurements for each patient is used to estimate the error variance σ^2. The tacit assumption is that the running third differences of each patient's plasma citrate means are small. Thus,

$$\hat{\sigma}^2 = p_1^{-1} \sum_{i=1}^{p_1} \hat{\sigma}_i^2, \qquad \hat{\sigma}_i^2 = [20(p_2 - 3)]^{-1} \sum_{j=4}^{p_2} (y_{i,j} - 3y_{i,j-1} + 3y_{i,j-2} - y_{i,j-3})^2. \tag{31}$$

For the Plasma Citrate Data, $\hat{\sigma}^2 = 151.4$.

Estimated risk is used to determine the best order r for defining the annihilator A_2. The estimated losses of the candidate error predictors for orders $r = 1$ to 4 are:

Order r of A_2	1	2	3	4
Estimated Risk	11.9	**8.5**	22.1	23.4

Thus, the choice $r = 2$ in defining annihilator A_2 minimizes estimated predictor risk among $1 \leq r \leq 4$. The plot on the right side of Figure 8 displays, with linear interpolation, the fitted means. For each patient, the estimated plasma citrate means rise after breakfast and dinner but drop after lunch. Evidently factors other than meal times influence a patient's plasma citrate concentrations.

Figure 9 displays the corresponding empirically best predictor of the observation errors. It is noteworthy that this error predictor has sound theoretical support even though the two-way layout of data lacks any replication.

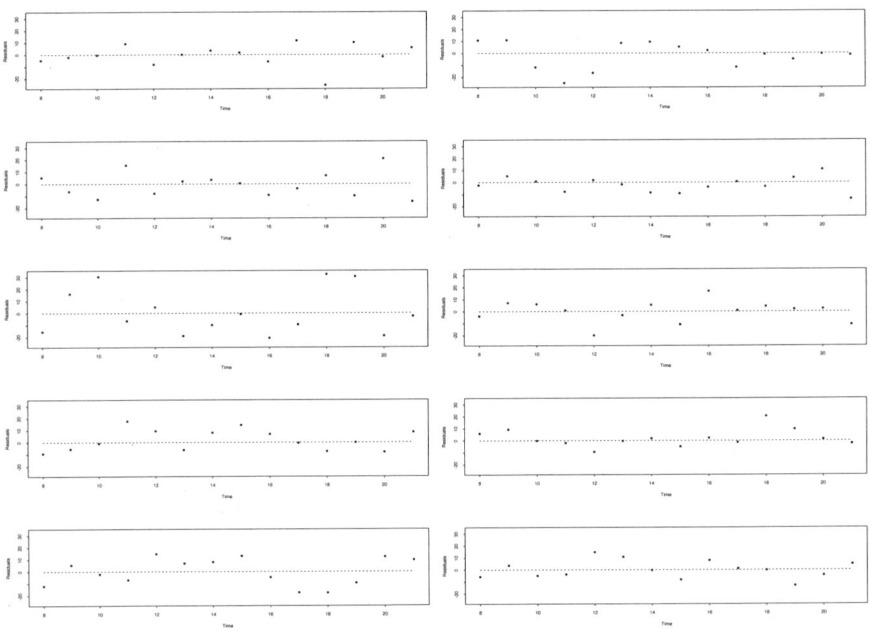

Fig. 9. Residuals from the Adaptive PLS fit to the Plasma Citrate Data

Remark 3. Tukey (1980) commented: "In practice, methodologies have no assumptions and deliver no certainties." Such is the situation in the second and third case studies just considered. The role of statistical theory is to test methodologies extensively on virtual data expressed in probability models or otherwise. Such is the function of the theorems presented earlier and the experiment with artificial data.

References

Andrews, D. F. and Herzberg, A. M. (1985) *Data: A Collection of Problems from Many Fields for the Student and Research Worker.* Springer, New York.

Beran, R. (2000). REACT scatterplot smoothers: superefficiency through basis economy. *Journal of the American Statistical Association.* **63**, 155–171.

Beran, R. (2002). Improving penalized least squares through adaptive selection of penalty and shrinkage. *Annals of the Institute of Statististical Mathematics.* **54**, 900–917.

Beran, R. (2005). ASP fits to multi-way layouts. *Annals of the Institute of Statistical Mathematics.* **57**, 201-220.

Beran, R. (2007) Adaptation over parametric families of symmetric linear estimators. *Journal of Statistical Planning and Inference.* (Special issue on Nonparametric Statistics and Related Topics). **137**, 684–696.

Beran, R. and Dümbgen, L. (1998). Modulation of estimators and confidence sets. *Annals of Statistics.* **26**, 1826–1856.

Buja, A., Hastie, T., and Tibshirani, R. (1989). Linear smoothers and additive models (with discussion). *Annals of Statistics.* **17**, 453–555.

Chu, C.-K. and Marron, J. S. (1991). Choosing a kernel regression estimator. *Statistical Science.* **6**, 404–436.

Hájek, J. (1973). Opening address. In *Proceedings of the Prague Symposium on Asymptotic Statistics* **1**, (J. Hájek, ed.) 7–11. Charles University, Prague.

Heckman, N. E. and Ramsay, J. O. (2000). Penalized regression with model-based penalties. *Canadian Journal of Statistics.* **28**, 241–258.

Kneip, A. (1994). Ordered linear smoothers. *Annals of Statistics.* **22**, 835–866.

Lin, Yi (2000). Tensor Product Space ANOVA Fits. *Annals of Statistics* **28**, 734–755.

Stein, C. (1956). Inadmissibility of the usual estimator for the mean of a multivariate normal distribution. In *Proceedings of the Third Berkeley Symposium on Mathematical Statistics and Probabability* **I** (J. Neyman, ed.) 197–206. University of California Press, Berkeley.

Tukey, J. (1980). Methodological comments focused on opportunities. In *Multivariate Techniques in Human Communication Research* (P. R. Monge and J. N. Cappella, eds.) 490–528. Academic Press, New York.

Ullah, A. (1985). Specification analysis of econometric models. *Journal of Quantitative Economics.* **2**, 187–209.

Wahba, G. (1990) *Spline Models for Observational Data.* SIAM, Philadelphia.

Chapter 11

The Use of Peters–Belson Regression in Legal Cases

Efstathia Bura, Joseph L. Gastwirth and Hiroyuki Hikawa

The George Washington University and Ernst & Young LLP, Washington, DC*

In equal employment cases it is important to compare the salary, hiring or promotion status of minority employees or applicants to that of similarly qualified majority members. Standard regression methods include a binary variable indicating minority status along with major job-related variables to assess whether belonging to a minority has a negative effect. Implicitly, this method assumes that any discriminatory policy has the same effect on every minority applicant or employee. Peters–Belson regression is a form of statistical matching, akin in spirit to Bhattacharya's band-width matching. In the context of studying possible sex discrimination in pay, one fits a regression to the salaries of employees of the males and then predicts the salary of females from the male equation. This prediction estimates the salary a female employee would receive assuming her job-related characteristics were rewarded in a similar way as males, i.e. it provides a "statistical match" for the actual salary of each female. The difference, if any, between the actual and predicted salary of each female employee estimates her under-compensation. Both parametric and non-parametric approaches to Peters–Belson regression are described and their use illustrated on salary data based on the *EEOC v Shelby County* equal pay case. Implications for the analysis of the data considered by courts to decide whether a class action is appropriate, as in *Dukes v. Walmart*, are discussed.

Contents

*The views expressed are solely of the authors and do not necessarily reflect the views of Ernst & Young LLP.

1. Introduction

The Peters–Belson (PB) regression method was first introduced by Peters (1941) and Belson (1956) for conducting treatment-control comparisons that accounted for relevant covariates by creating statistical matches for the treatment group observations. Blinder (1973) and Oaxaca (1973) used this idea to decompose the difference between the means of the two groups into components. The PB regression method fits a regression model to the control group and estimates the response a member of the protected group would have had they been a control from the regression fitted to the control group. The difference between the value of each treatment group member's response and its prediction obtained from the control group equation estimates the effect of the treatment.

In employment discrimination cases (e.g., equal pay cases), where members of the protected group (e.g., racial minority, females) are compared to those of the unprotected group (e.g., racial majority, males), the PB regression method as well as ordinary regression with an indicator representing group status have been accepted by courts (Gray (1993)). The PB method offers some advantages compared to standard regression with a dummy or indicator variable. First, the method is intuitive and comparatively easy to understand for a general audience (e.g., judges, juries, etc.). For example, in the context of sex discrimination cases, PB regression estimates the salary equation for the majority group incorporating related covariates and then takes the difference between the minority's actual salary and the estimated salary that the minority employee would have received if s/he were paid according to the equation for majority employees. Moreover, the estimated pay differential obtained from the PB approach is individualized for each member of the protected group. In contrast, ordinary least squares linear regression with an indicator variable estimates a common overall effect of being minority, after adjusting for the relevant covariates. This approach assumes that any differential is the same across the entire range of covariate values. Another advantage of the PB method is that the females whose salaries are higher than predicted and were not discriminated against are readily identifiable.

This paper reviews the Peters–Belson regression approach in the context of legal cases and also describes a recent extension of the method, similar in spirit to Bhattacharya (1989), by the authors. In Section 2 we review parametric PB regression, based on parametric ordinary linear regression. Section 3 introduces a recent nonparametric version that increases the applicability of the PB approach. In Section 4 we apply all the methods we discuss to data from a sex discrimination case, and Section 5 shows how the PB approach is superior to the regression approaches of both parties in the *Dukes v. Walmart* sex discrimination case. We present concluding remarks in Section 6.

2. Parametric Peters–Belson Regression

2.1. *Review of the method*

Suppose the salaries, denoted by Y, for minority and majority employees are given by

$$\text{Minority:}\quad Y_{1i} = \mathbf{X}_{1i}^T\boldsymbol{\beta}_1 + \epsilon_{1i} \qquad \text{for } i = 1, \ldots, n_1 \tag{1}$$

$$\text{Majority:}\quad Y_{2j} = \mathbf{X}_{2j}^T\boldsymbol{\beta}_2 + \epsilon_{2j} \qquad \text{for } j = 1, \ldots, n_2 \tag{2}$$

$\mathbf{X}$ denotes the covariate vector and $\boldsymbol{\beta}$ the corresponding coefficient vector. The errors, ϵ, in each equation are assumed to be normally distributed with mean zero and variance σ_1^2 and σ_2^2, respectively. When both groups are paid according to the same system, the regression coefficients in the two models are equal, that is $\boldsymbol{\beta}_1 = \boldsymbol{\beta}_2$.

In parametric PB, a linear regression model is fitted to the data for the majority employees. Then each minority member's salary is predicted by $\mathbf{x}_{1i}^T\hat{\boldsymbol{\beta}}_2$, where $\mathbf{x}_{1i}$ is the covariate vector for the ith minority member and $\hat{\boldsymbol{\beta}}_2$ is the least squares estimate of $\boldsymbol{\beta}_2$. The difference, $D_i = y_{1i} - \mathbf{x}_{1i}^T\hat{\boldsymbol{\beta}}_2$, between the actual and predicted salaries is the estimate of the pay differential of the i-th minority employee relative to a similarly qualified majority employee. Thus, D_i is the parametric PB estimate of $\delta_i = \mathbf{x}_{1i}^T\boldsymbol{\beta}_1 - \mathbf{x}_{1i}^T\boldsymbol{\beta}_2$, the difference in pay between the i-th minority employee and a similar majority member. When the model is correct, D_i is unbiased for δ_i and the corresponding unbiased estimator for the average disparity over all minority employees

$$\delta = \frac{1}{n_1}\sum_{i=1}^{n_1}[\mathbf{x}_{1i}^T\boldsymbol{\beta}_1 - \mathbf{x}_{1i}^T\boldsymbol{\beta}_2] = \bar{\mathbf{x}}_1^T(\boldsymbol{\beta}_1 - \boldsymbol{\beta}_2) \tag{3}$$

equals

$$\bar{D} = \frac{\sum_{i=1}^{n_1} D_i}{n_1} = \frac{\sum_{i=1}^{n_1}[y_{1i} - \mathbf{x}_{1i}^T\hat{\boldsymbol{\beta}}_2]}{n_1} \tag{4}$$

where $\bar{\mathbf{x}}_1$ is the mean vector of the minority covariate values. The variance of $\bar{D}$ is

$$\mathrm{Var}(\bar{D}) = \frac{\sigma_1^2}{n_1} + \sigma_2^2\bar{\mathbf{x}}_1^T(\mathbf{x}_2^T\mathbf{x}_2)^{-1}\bar{\mathbf{x}}_1 \tag{5}$$

where $\mathbf{x}_2 = (\mathbf{x}_{21}, \mathbf{x}_{22}, \ldots, \mathbf{x}_{2n_2})^T$ is the usual design matrix of the majority group.

When $\sigma_1^2 = \sigma_2^2 = \sigma^2$, the test statistic

$$t = \frac{\bar{D}}{\sqrt{\hat{\sigma}^2(1/n_1 + \bar{\mathbf{x}}_1^T(\mathbf{x}_2^T\mathbf{x}_2)^{-1}\bar{\mathbf{x}}_1)}} \tag{6}$$

can be used to test $\delta = 0$. Gastwirth (1989) suggested that majority observations be used to estimate the common variance σ^2 because under the hypothesis of no discrimination both majority and minority should be paid under the same system and hence the variances are the same as well. If we use $\hat{\sigma}^2 = \hat{\sigma}_2^2$, under the null hypothesis of $\delta = 0$, the test statistic in (6) is t-distributed with $n_2 - p_2$ degrees of freedom, where p_2 is the number of parameters (coefficients) in the majority model.

When the error variances are different, one can approximate the distribution of the test statistic using Welch's approximation (Welch (1949), Scheffe (1970), Nayak and Gastwirth (1997)). Under this approximation, the test statistic distribution under the null is approximated by a t distribution with degrees of freedom:

$$\mathrm{df} = \frac{(\frac{1}{n_1}\sigma_1^2 + \bar{\mathbf{x}}_1^T(\mathbf{x}_2^T\mathbf{x}_2)^{-1}\bar{\mathbf{x}}_1\sigma_2^2)^2}{\sigma_1^4/(n_1^2(n_1 - p_1)) + \sigma_2^4/(n_2 - p_2)\bar{\mathbf{x}}_1^T(\mathbf{x}_2^T\mathbf{x}_2)^{-1}\bar{\mathbf{x}}_1\bar{\mathbf{x}}_1^T(\mathbf{x}_2^T\mathbf{x}_2)^{-1}\bar{\mathbf{x}}_1} \tag{7}$$

Since σ_1^2 and σ_2^2 are unknown, the degrees of freedom are estimated by substituting $\hat{\sigma}_1^2$ and $\hat{\sigma}_2^2$ in (7).

Gastwirth and Greenhouse (1995) applied the PB method to salary data as well as to logistic regression for binary responses in order to analyze the data arising from a case involving promotion decisions (*Capaci v. Katz and Besthoff*[a]). Furthermore, Nayak and Gastwirth (1997) extended the method to generalized linear models. Hikawa, et al. (2010a) introduced nonparametric PB in regressions with a binary response as an alternative to logistic regression.

3. Local Linear Peters–Belson Regression

There are two problems an analyst of pay discrimination data often encounters (Hikawa, et al. (2010b)). The first is the difficulty in estimating the salary equation when it does not appear to follow any usual parametric forms (e.g., linear, quadratic). The second problem pertains to determining who the relevant male/majority employees are to be compared against female/minority employees of interest. Including too many irrelevant majority employees in the comparisons (e.g., male employees who are too senior compared to the target female employees) may introduce serious bias in the estimated disparity (Greiner (2008)).

To address these problems, Hikawa, et al. (2010b) introduced local linear regression in PB. Local linear regression fits a linear regression in the *neighborhood* of the covariate values of each minority member. The method is well suited for equal pay cases since the estimation/prediction of the salary of a minority employee is based on majority employees whose qualifications are closest to those of the minority employee and thus should receive the greatest weight. Furthermore, the similarity of this method to matched-pairs is expected to make the results more understandable to judges and juries. Local linear regression is similar in spirit to bandwidth matching introduced by Bhattacharya (1989). However, the weight given to each majority observation decreases with the distance of the covariate values from the target minority member.

3.1. *Description of Local Linear PB*

Suppose we have d covariates and the data consist of n_1 minority observations, $(\mathbf{x}_{11}, y_{11}), \ldots, (\mathbf{x}_{1n_1}, y_{1n_1})$, and n_2 majority observations, $(\mathbf{x}_{21}, y_{21}), \ldots, (\mathbf{x}_{2n_2}, y_{2n_2})$, where $\mathbf{x}$ is a vector of d fixed covariate values. The response in the two groups can always be modeled as follows

$$\text{Minority:} \quad Y_{1i} = m_1(\mathbf{x}_{1i}) + \epsilon_{1i}, \qquad \epsilon_{1i} \sim N(0, \sigma_1^2) \tag{8}$$

$$\text{Majority:} \quad Y_{2j} = m_2(\mathbf{x}_{2j}) + \epsilon_{2j}, \qquad \epsilon_{2j} \sim N(0, \sigma_2^2) \tag{9}$$

[a] *Capaci v. Katz and Besthoff* 711 F. 2d 647, 5th Cir., 1983

with $E(\epsilon_1) = E(\epsilon_2) = 0$. The only assumption we make on the unknown functions modeling the mean response in the two groups, $m_1(\mathbf{x})$ and $m_2(\mathbf{x})$, is that they are twice differentiable.

As in the parametric PB definition of disparity, the pay disparity for the i-th minority member is

$$\delta_i = m_1(\mathbf{x}_{1i}) - m_2(\mathbf{x}_{1i}), \tag{10}$$

and the average disparity of all minority members is

$$\delta = \frac{\sum_{i=1}^{n_1}(m_1(\mathbf{x}_{1i}) - m_2(\mathbf{x}_{1i}))}{n_1}. \tag{11}$$

Let

$$\mathbf{Z}_i = \begin{pmatrix} 1 & (x_{211} - x_{11i}) & \cdots & (x_{2d1} - x_{1di}) \\ 1 & (x_{212} - x_{11i}) & \cdots & (x_{2d2} - x_{1di}) \\ \vdots & \vdots & & \vdots \\ 1 & (x_{21n_2} - x_{11i}) & \cdots & (x_{2dn_2} - x_{1di}) \end{pmatrix} \tag{12}$$

and $\mathbf{W}_i = diag\big(W(||\mathbf{x}_{2j} - \mathbf{x}_{1i}||/h)\big)$, where W is a kernel weight function and $||\cdot||$ is a norm. Denoting the elements of the first row of $(\mathbf{Z}_i^T\mathbf{W}_i\mathbf{Z}_i)^{-1}\mathbf{Z}_i^T\mathbf{W}_i$ by $S_{i1}, S_{i2}, \ldots, S_{in_2}$, the fitted value for the design point $\mathbf{x}_{1i}$ is given by

$$\hat{m}_2(\mathbf{x}_{1i}) = \sum_{j=1}^{n_2} S_{ij}y_{2j} \tag{13}$$

The estimated disparity of the i-th minority member is

$$D_i = y_{1i} - \hat{m}_2(\mathbf{x}_{1i}) \tag{14}$$

The estimators of the overall average disparity and its variance are

$$\bar{D}_{LOC} = \frac{\sum_{i=1}^{n_1} D_i}{n_1} = \frac{\sum_{i=1}^{n_1}(y_{1i} - \sum_{j=1}^{n_2} S_{ij}y_{2j})}{n_1} \tag{15}$$

$$\mathrm{Var}(\bar{D}_{LOC}) = \mathrm{Var}(\bar{D}) = \frac{1}{n_1^2}\Big(n_1\sigma_1^2 + \sum_{i=1}^{n_1}\sum_{j=1}^{n_2} S_{ij}^2\sigma_2^2 + \sum_{i\neq l}^{n_1}\sum^{n_1}\sum_{j=1}^{n_2} S_{ij}S_{lj}\sigma_2^2\Big) \tag{16}$$

A nearest neighbor bandwidth that fixes the fraction of data that contribute to the estimation (Cleveland and Devlin (1988), Loader (1999)) is used in fitting the local linear regression model. The Epanechnikov kernel, defined by

$$W(u) = \frac{3}{4}(1-u^2)I_{(|u|\leq 1)}, \tag{17}$$

where $u = ||\mathbf{x}_{2j} - \mathbf{x}_{1i}||/h$ and h is the bandwidth, is used as the weight function. It is the optimal kernel in the sense that roughly both the asymptotic mean squared error and the mean integrated squared error are minimized (see Thm. 3.4 in Fan and Gijbels (1996), p.75). Loader (1999), p.20, suggests scaling the vector norm. We use

$$||\mathbf{x}_{2j} - \mathbf{x}_{1i}||^2 = \sum_{k=1}^{d} \left(\frac{x_{2kj} - x_{1ki}}{s_k} \right)^2,$$

where $s_k = \sqrt{(\sum_{j=1}^{n_2}(x_{2kj} - x_{1ki} - \sum_{j=1}^{n_2}(x_{2kj} - x_{1ki})/n_2)^2)/(n_2 - 1)}$. Since x_{1ki} is fixed, s_k is simply the sample standard deviation of x_{2k} about x_{1ki} for the majority members for the k-th covariate.

Since σ_1^2 and σ_2^2 are usually unknown, the estimated variance of $\bar{D}_{LOC}$ in (16) can be obtained by using estimates $\hat{\sigma}_1^2$ and $\hat{\sigma}_2^2$ which can be obtained from the residuals of the separate local linear regression models within each group. Details of the estimation approach are given in Hikawa, et al. (2010b). The estimated variance of $\bar{D}_{LOC}$ is given by

$$\widehat{\mathrm{Var}}(\bar{D}_{LOC}) = \frac{1}{n_1^2}\left(n_1 \hat{\sigma}_1^2 + \sum_{i=1}^{n_1}\sum_{j=1}^{n_2} S_{ij}^2 \hat{\sigma}_2^2 + \sum_{i \neq l}^{n_1}\sum_{j=1}^{n_2} S_{ij} S_{lj} \hat{\sigma}_2^2 \right) \tag{18}$$

Cleveland and Devlin (1988) showed that, under the assumption of normal errors and negligible bias of $\hat{m}_j(\mathbf{x})$[b], the distribution of $(\eta_j^2 \hat{\sigma}_j^2)/(\kappa_j \sigma_j^2)$ can be approximated by a χ^2 distribution with degrees of freedom η_j^2/κ_j, where

$$\eta_j = \mathrm{tr}(\mathbf{I} - \mathbf{S}_j)(\mathbf{I} - \mathbf{S}_j)^T \tag{19}$$

$$\kappa_j = \mathrm{tr}[(\mathbf{I} - \mathbf{S}_j)(\mathbf{I} - \mathbf{S}_j)^T]^2 \tag{20}$$

for $j = 1, 2$. Let $\mathbf{S}_j$ be the $n_j \times n_j$ matrix whose $(i,k)^{th}$ element is S_{ik} obtained from fitting a separate smooth curve for the minority and majority groups (i.e., separately estimating $m_1(x_i)$ for $i = 1, \ldots n_1$ and $m_2(x_i)$ for $i = 1, \ldots n_2$). The test statistic for testing for lack of disparity ($H_0 : \delta = 0$) is defined to be

$$t = \frac{\bar{D}_{LOC}}{\sqrt{\hat{\mathrm{Var}}(\bar{D}_{LOC})}} \tag{21}$$

When the two variances are equal (i.e., $\sigma_1^2 = \sigma_2^2 = \sigma^2$), we estimate the common variance by the majority variance estimate $\hat{\sigma}^2 = \hat{\sigma}_2^2$, and the

[b]Cleveland and Devlin (1988) state that $\hat{m}(\mathbf{x})$ in (13) is unbiased only when the underlying mean function $m(\cdot)$ is linear. In general, the bias goes to zero asymptotically at the rate of h^2 as $h \to 0$ and $nh \to \infty$ (Fan and Gijbels (1996), p.62).

estimated variance of $\bar{D}_{LOC}$ in (18) becomes

$$\widehat{\mathrm{Var}}(\bar{D}_{LOC}) = \frac{\hat{\sigma}^2}{n_1^2}\Big(n_1 + \sum_{i=1}^{n_1}\sum_{j=1}^{n_2} S_{ij}^2 + \sum_{i\neq l}^{n_1}\sum_{j=1}^{n_2} S_{ij}S_{lj}\Big)$$

Therefore, under the assumption of equal variances, the test statistic for group disparity is t distributed with η^2/κ degrees of freedom, where $\eta = \eta_2$ and $\kappa = \kappa_2$ as defined in (19) and (20). The distribution of the test statistic when the variances are unequal is given in the Appendix .

4. Application: Reanalysis of Data from *EEOC v. Shelby County Government*

In this section, we illustrate the methods on a data set from the pay discrimination case, *EEOC v. Shelby County Government.* In 1983, a female clerical employee of Shelby County Criminal Court brought a charge of salary discrimination. The data, presented in Table 1, consist of salaries per pay period as the response and seniority measured in months since hired as the covariate with 16 male and 15 female clerical employees as of February 1, 1984. Fitting a simple linear regression to each gender separately yields:

$$\hat{Y}_f = 885.1 + 4.3X_f \tag{22}$$

$$\hat{Y}_m = 913.9 + 6.0X_m \tag{23}$$

where the subscripts m and f stand for male and female. The estimated variances of the error terms are $\hat{\sigma}_m^2 = 7494.5$ and $\hat{\sigma}_f^2 = 8757.1$, and the sample means and variances of seniority are: $\bar{X}_m = 81$, $S_m^2 = 160.3.3$, $\bar{X}_f = 69.3$, and $S_f^2 = 2195.5$. Table 2 summarizes the ANOVA tables for both the male and female models.

Since the sample size of 16 males is too small to fit a local linear regression model, we augmented the data set following the method used by Bhattacharya (1989) and Bhattacharya and Gastwirth (1989), where they analyze data from *Berger v. Iron Workers Local 201*[c]. Fitting a Gamma distribution to the seniority data yielded $Gamma(4, 20)$ for males and $Gamma(2, 33)$ for females. Then, we generated additional salary data for 84 males and 85 females according to the fitted models in (22). The error variances were set equal to $\hat{\sigma}_m^2 = 7494.5$ and $\hat{\sigma}_f^2 = 8757.1$ in order to match the estimated variances of the error terms from the fitted regression models.

[c] *Berger v. Iron Workers Local 201*, 42 FEP Cases 1161 (D.D.C. 1985), 843 F.2d 1395 (D.C. Cir. 1988)

Table 1. Shelby County pay discrimination case data.

Male		Female	
Salary (Y)	Seniority (X)	Salary (Y)	Seniority (X)
1666	125	1474	128
1666	124	1474	96
1666	124	1403	121
1666	120	1403	117
1548	117	1403	114
1548	116	1336	89
1548	113	1336	97
1548	95	1157	105
1548	73	1000	64
1306	69	1000	51
1157	51	1000	13
1157	47	1000	17
1157	46	929	13
1157	18	929	8
1000	41	929	6
1000	17		

Table 2. ANOVA tables for male and female models from the Shelby County data.

Male				
Sources	DF	SS	MS	F
Seniority	1	852412	852412	113.74*
Error	14	104924	7495	
Total	15	957336		
$R^2 = .89$				
Female				
Sources	DF	SS	MS	F
Seniority	1	575666	575666	65.74*
Error	13	113842	8757	
Total	14	689508		
$R^2 = .83$				

* P-value was smaller than .0001.

The data were simulated based on unequal variances for males and females. In consequence, we compute the variance of $\bar{D}$ based on the assumption of unequal variances and approximate the degrees of freedom of the test statistics using the Welch's approximation approach for Local Linear PB by (26) and parametric PB by (7). Table 3 summarizes the results from applying the four methods: (1) Local Linear PB, (2) parametric PB, (3) ordinary regression with and without interaction between seniority and gender indicator variables, and (4) propensity score (Rosenbaum and Rubin (1983), Zanutto (2006)). This last method creates subgroups in which males and females have similar covariate values. This is accomplished by creating a logistic regression predicting gender from the covariates. The quintiles of the females' estimated probabilities or propensity scores are used to form the comparable subgroups. Within each subgroup, the average salaries of males and females are compared and the subgroup differences are aggregated to form the estimate of the overall disparity.

Table 4 presents the predicted salaries of female employees that were in the original data before augmentation.

Table 3. Analysis of the augmented Shelby County Pay Discrimination data.

	Local PB	Parametric PB	Ordinary Reg. w\o Int.	Ordinary Reg. w\ Int.	Propensity Score
$\bar{D}$	-128.5	-126.6	-140.3	-126.6	-123.6
$\sqrt{\widehat{\text{Var}}(\bar{D})}$	15.4	13.8	14.8	13.7	23.1
$\bar{D}/\sqrt{\widehat{\text{Var}}(\bar{D})}$	-8.3	-9.2	-9.5	-9.2	-5.4
P-value	$< .001$	$< .001$	$< .001$	$< .001$	$< .001$

Table 4. Actual and predicted salary of female employees.

Actual	Local PB	Parametric PB	Ordinary Reg. w\o Int.	Ordinary Reg. w\ Int.
1474	1681	1686	1629	1686
1474	1496	1493	1475	1493
1403	1639	1643	1595	1643
1403	1616	1619	1576	1619
1403	1598	1601	1561	1601
1336	1460	1450	1441	1450
1336	1501	1499	1479	1499
1157	1544	1547	1518	1547
1000	1297	1300	1320	1300
1000	1205	1221	1258	1221
1000	1026	992	1075	992
1000	1046	1016	1094	1016
929	1026	992	1075	992
929	1003	962	1051	962
929	993	950	1041	950

The negative values of $\bar{D}$ indicate that female employees were underpaid on average compared to their similarly qualified male counterparts. Local Linear PB (-128.5), parametric PB (-126.6), and ordinary first order regression with interaction (-126.6) give similar estimated average pay differential. From Table 4 we can see that OLS without interaction yields a higher average estimate than the others as it predicts higher salaries for the less senior female employees and underpredicts the salaries more senior females should receive. The propensity score method yielded a slightly smaller estimate (−123.6) and its standard error (23.1) is much larger than the other methods.

Since the estimated amount of pay differential from all the methods is quite large, the *p*-values of all test statistics are very small. Hence, all methods would reject the null hypothesis of no pay differential and confirm the court's conclusion that the female employees were discriminated in their pay with an average differential about $126-128 per pay period.

5. The Suitability of Peters–Belson Regression for Studying the Class Certification Issue

In order for the claims of many individual plaintiffs, who are represented by a relatively small number of named plaintiffs, to be combined into a single class action, four criteria need to be met (see Fed. R. Civ. P. 23). Statistical studies are often submitted to satisfy the criterion (23(a)) that there are questions of law or fact *common* to the members of the class. In *Caridad v. Metro-North Commuter Railroad* plaintiffs used a multiple regression controlling for job related factors, to demonstrate that African American employees received fewer promotions but were disciplined more often than similar whites and that this occurred in many locations of the defendant's operation. Perhaps the best known controversy over the evidence plaintiffs need to submit to support a class action is a lower court's certification of a class of female employees in *Dukes et al. v. Walmart*, which will be considered by the U.S. Supreme Court in 2011.

A statistical issue in Walmart, noted in the 6-5 split decision in the 9th Circuit upholding the District court's certification of the class, concerned the appropriate degree of aggregation of the employee data. The plaintiffs analyzed the data on all employees in each of the 41 regions. Each region contains about 80 to 85 stores and the number of employees in a store ranges from 200 to 500. The defendant argued that the data should be analyzed store by store and before the store-wide data are combined into a single sample the plaintiffs' expert should have applied a Chow (1960) or Rao (1952)[d] test of the equality of the regression equations before pooling the data. The defendants' expert actually analyzed the data at the sub-store level, i.e. similar departments within each store were combined. This

[d]While Chow was the first economist to consider testing whether two regressions were identical, the problem had also been considered in the statistical literature.

led to about three regression analyses being conducted per store. While the Walmart case is used as a familiar back-drop, the salary data are not available for independent reanalysis. Thus, only general statistical implications are discussed here.

As noted in Gastwirth, et al. (2003) in the context of pooling 2×2 tables, requiring that the odds ratios in all the tables be equal is too strong a requirement. The Cochran–Mantel–Haenszel test is consistent for detecting an overall shortfall across the strata provided there is no interaction, i.e. individual odds ratios are consistent with all being less (greater) than one. Thus, that summary test is applicable in many situations when the odds ratios in each table are not equal. The Gail–Simon test (Gail, M. H. and Simon (1985)) for interaction can be used in place of the test of equality of the odds ratios. Similarly, the null hypothesis tested in the Chow–Rao test is whether the coefficients for every predictor, e.g. seniority, gender, are identical in all the strata (stores in Walmart). Again, this is too strong a requirement. For example, if the data for each store fit a model relating salary to gender and seniority, the coefficient reflecting the added salary increment of each additional year of employment, could vary among the stores but the coefficient for being female could be negative in every store.

Because the PB approach provides an estimate of the underpayment for each female employee compared to a statistically matched male, it is appropriate for examining the issue of whether female employees are generally underpaid relative to similarly qualified males. Whether one fits the regression to males in each store or to stores in an entire region or to stores in a sub-region, e.g. those located in the same or an economically similar labor market, one has an estimate D_i of the difference between each female employee's salary and her "statistical match." Under the hypothesis of equal treatment approximately half of the D_i will be positive and half negative and the average of them will be near zero. A pattern of negative D_i suggests that the issue of underpayment of female employees, relative to comparable males, is a common situation. Of course, in Walmart and similar cases, the regression equation should include information on the major predictors that were considered in the salary setting process, e.g. seniority, job type, special education or experience, etc.

In the Walmart case both parties could present their estimates of the D_i for their preferred models, i.e. at the level of aggregation they believe is appropriate. When looking at highly disaggregated data, such as examining each store by itself, the power of a test of significance of the average difference is likely to be low. Thus, a finding of non-significance in each of many small strata or sub-groups of a large data set (stores in Walmart) may have little meaning unless a power analysis shows that one could detect a legally meaningful difference (see Bura and Gastwirth (2009) for a discussion of the power of regression analyses in a securities law case and Gastwirth and Pan (2011) give an example where a sample of over 900 venire members was insufficient to reliably detect a legally meaningful under-representation of minority in a jury pool). By examining the pattern of negative D_i obtained from the sub-store analyses advocated by Walmart or 41 regional regressions based on more aggregated data, advocated by the plaintiffs, one can assess whether there is a general pattern of underpayment of women. Thus, it is possible that when the estimated salary obtained from a PB regression analysis falls short for each female employee it could support the plaintiffs' position. On the other hand, if a pattern of predominantly negative differences is not seen or only occurs in a small subset of stores, this would be inconsistent with a nation-wide practice of underpaying females.

The issue of homogeneity of the regression models in the individual stratum (the plaintiffs' regions or the defendant's sub-store units) becomes less important when the PB method is used because one is working with the individualized estimates, D_i of underpayment of each female employee obtained from a regression based on comparable males. If the differences observed from the plaintiffs' analysis of the much larger regional data sets are "explained" or substantially decreased, when the defendant incorporates local factors, the latter would explain the pattern of differences observed in the more highly aggregated analysis.

So far in the Walmart proceedings, it appears that neither side has submitted a PB regression. Both submitted regressions in which the gender effect is reflected by an "indicator" variable, along with other predictors. As noted previously, this approach assumes that the expected underpayment of a female employee is the same, regardless of the values of the other predictors and often it is less powerful than PB when the data does not follow this assumption. A pattern of discrimination in pay may exist even if the coefficients of the regression models are not identical in each store or stratum. Thus, the Chow test should not be required before pooling the data. For example, if some stores started men and women at the same salary but gave men a higher percentage raise or other stores gave similar pay raises but started women at a lower salary the coefficients of the fitted equations would not be equal. Yet, there is a common pattern of underpayment of females. To illustrate these issues we conducted a small simulation study in a simple situation. While the concepts discussed are general and applicable to class certification cases, they need to be interpreted in the context of a particular case.

5.1. *Statistical Properties of the two Regression Approaches when the Data are Stratified*

The issue of how to analyze data when a firm has many locations has arisen in a number of major discrimination cases. In particular, the brief to the appeals court filed by Walmart essentially claims that the Chow–Rao test must be used before the data are pooled. Here we present a simple example showing that this is too stringent. Combining the estimated differentials obtained by the Peters–Belson approach in the strata is more appropriate than ordinary least squares applied to the pooled data. Nonetheless, it will be seen that one can reach a correct inference when analyzing the pooled data with either OLS or PB.

Consider a company that has two stores with employees of both sexes. Women are underpaid relative to comparable men in both stores but the system leading to the disparity differ in the two stores. In the first, men and women start at the same salary but men receive better raises over time while in the second store men start at a slightly higher salary and also receive higher raises over time. Formally the salary equations are

$$y_i = \begin{cases} 20000 + 200x_i + \epsilon_{1i} & \text{when } i \text{ is female} \\ 20000 + 250x_i + \epsilon_{2i} & \text{when } i \text{ is male} \end{cases} \tag{24}$$

for store 1 and

$$y_i = \begin{cases} 20000 + 200x_i + \epsilon_{1i} & \text{when } i \text{ is female} \\ 20500 + 250x_i + \epsilon_{2i} & \text{when } i \text{ is male} \end{cases} \tag{25}$$

for store 2. We fitted PB regression and ordinary least squares with an indicator variable for gender. The values of the seniority predictor variable for females and males were generated from the Gamma distribution with scale parameters 3 and 2 for females and males, respectively, and shape 2 for both sexes. In both stores, the number of males is 40 and the number of females is 30. The error variances were the same. One thousand replicates were used in the simulation.

First, we analyzed the data for each store using PB and OLS with an indicator for Gender regression. In store 1 the PB method detected a significant disparity at the .05 level 82.15% of the time while OLS found a statistically significant disparity 75.5% of the time. In store two both methods had 100% power.

Combining both stores into a single sample, we followed the approach of the plaintiffs' expert in Walmart who used OLS with indicators for store and gender. The Gujarati implementation (Gujarati (1970a), Gujarati (1970b)) of the Chow–Rao test for the equality of the coefficients in two regressions was carried out by fitting a model that included gender by store and gender by seniority interactions. Then the partial F test for both coefficients is used to assess whether they are needed. Although the average p-value of the test was 0.0164, indicating that most of the time the Chow–Rao test would find that the coefficients are statistically significantly different at the 0.05 level, both the PB and OLS applied to the pooled sample detected an overall negative effect on female employees every time.

In Table 5 the estimated disparities obtained by the different methods are reported. Because each store employed the same number of males and females, the PB estimate of the overall disparity is the average of the individual store disparities. Notice that using OLS with a gender indicator, yielded a smaller disparity than PB in each store. When the pooled data were analyzed, the estimates obtained from both regression approaches were close to the average of those obtained from the analysis of the stores individually. While this may not be a general phenomenon, the example shows that fitting PB or OLS regression to data pooling over strata or stores where the regressions are not identical but reflect a similar type of disparity in the stores, still detected the general pattern that females were disadvantaged. Analogous to checking for a cross-over effect when the results from several two by two contingency tables are combined (Gail, M. H. and Simon (1985), Gastwirth, et al. (2003)) one should check that the regressions in a substantial majority of stores follow similar pattern of disparity against one group. In stores where the sample sizes are small, especially when there is an imbalance between the number of employees of the two sexes, the power of any regression type test to find a statistically significant disparity in a single store is likely to be small.

6. Concluding Remarks

Even though the PB approach is related to estimating the "counterfactual" in modern treatments of causality (Pearl (2000), p. 308; Heckman (2005), Rubin (2005)) it is not widely used outside the area of measuring minority disparities (Sinclair and Pan (2009), Graubard (2009)). Depending on

Table 5. Estimated Disparities.

	Ave. $\bar{D}$	Ave. $\hat{\beta}_2$	St. Dev.	Ave. $\bar{D} - \delta$	Ave. $\hat{\beta}_2 - \delta$
Store 1 - PB	-303.561		99.914	-1.198	
Store 1 - OLS		-271.098	98.34		31.265
Store 2 - PB	-803.570		98.962	-1.207	
Store 2 - OLS		-771.184	97.059		31.179
Pooled - PB	-553.566		71.856	-1.202	
Pooled - OLS		-521.141	70.613		31.223
OLS with Store Effect		-521.134	70.612		31.222
OLS with Store by x and Store by Gender Effect		-271.098	98.34		281.265

one's prior knowledge concerning the relationship between the response of interest and the predictor variables one can estimate the regression from a parametric model or a non-parametric one.

Bhattacharya's bandwidth matching approach computes two estimates of a differential based on the Mann–Whitney–Wilcoxon measure. The first matches each female to a set of male "near-neighbors" and the second matches each male to a set of "nearby" females. This approach may well have greater statistical power in some situations than the PB method. In the context of the Equal Pay Act, however, the law is asymmetric in the sense that the salaries of the underpaid group need to be increased to those of the majority group, but the employer cannot decrease the salaries of the employees of the majority group.

Regression and matching procedures lose statistical power and efficiency when the distribution of predictors varies noticeably in the two groups. The article by Hall, et al. (2009) provides a useful approach to this issue, which deserves further study.

Using the PB estimates of the pay differential in each stratum and combining them to obtain a summary test of equality avoids the problem of aggregation that occurs in many class action cases. While we have seen that the Chow–Rao test is too stringent, further research is needed to develop a test for "no interaction" between the strata and the response of interest, i.e., to ensure that the pattern of signs and values of the stratum specific estimated D_i is consistent with a common general policy.

7. Appendix: The Distribution of (21) when the Variances are Unequal

As in parametric PB, Welch's approximation can be used to find the approximate distribution of the test statistic. The expression for the variance of $\bar{D}_{LOC}$ in (16) can be simplified as follows:

$$\begin{aligned}
\mathrm{Var}(\bar{D}_{LOC}) &= \frac{1}{n_1^2}\Big(n_1\sigma_1^2 + \sum_{i=1}^{n_1}\sum_{j=1}^{n_2} S_{ij}^2\sigma_2^2 + \sum_{i\neq l}^{n_1}\sum_{j=1}^{n_2} S_{ij}S_{lj}\sigma_2^2\Big) \\
&= \sigma_1^2\frac{1}{n_1} + \sigma_2^2\frac{1}{n_1^2}\Big(\sum_{i=1}^{n_1}\sum_{j=1}^{n_2} S_{ij}^2 + \sum_{i\neq l}^{n_1}\sum_{j=1}^{n_2} S_{ij}S_{lj}\Big) \\
&= \sigma_1^2 C_1 + \sigma_2^2 C_2
\end{aligned}$$

When $\delta = 0$ is true, the test statistic in (21) is approximately t distributed with degrees of freedom

$$df_{LOC} = \frac{(\sigma_1^2 C_1 + \sigma_2^2 C_2)^2}{\sigma_1^4\kappa_1/\eta_1^2 C_1^2 + \sigma_2^4\kappa_2/\eta_2^2 C_2^2} \tag{26}$$

In most practical situations σ_1^2 and σ_2^2 are unknown and the degrees of freedom are approximated by plugging $\hat{\sigma}_1^2$ and $\hat{\sigma}_2^2$ in (26).

References

Bhattacharya, P. K. (1989). Estimation of treatment main effect and treatment-covariate interaction in observational studies using bandwidth-matching. Technical Report 188, Division of Statistics, University of California, Davis.

Bhattacharya, P. K. and Gastwirth, J. L. (1999). Estimation of the Odds-Ratio in an Observational Study Using Bandwidth-Matching. *Journal of Nonparametric Statistics*, **11**, 1–12.

Belson, W. A. (1956). A technique for studying the effects of a television broadcast. *Applied Statistics*, **5**, 195–202.

Blinder, A. S. (1973). Wage discrimination: reduced form and structural estimates. *Journal of Human Resources*, **8**, 436–455.

Bura, E. and Gastwirth, J. L. (2009). How accurate are the power calculations relied upon the SEC? *Law, Probability and Risk*, **8**, 277–288.

Chow, G. C. (1960). Tests of equality between sets of coefficients in two linear regressions. *Econometrica*, **28**, 591–603.

Cleveland, W. S. and Devlin, S. J. (1988). Locally-weighted regression: an approach to regression analysis by local fitting. *J. Amer. Statist. Assoc.*, **83**, 597–610.

Fan, J. and Gijbels, I. (1996). *Local Polynomial Modelling and Its Applications.* London: Chapman and Hall.

Gail, M. H. and Simon, R. (1985). Testing for qualitative interactions between treatment effects and patient subjects. *Biometrics*, **41**, 361–372.

Gastwirth, J. L. (1989). A clarification of some statistical issues in Watson vs. Fort Worth Bank and Trust. *Jurimetrics Journal*, **29**, 267–284.

Gastwirth, J. L. and Greenhouse, S. W. (1995). Biostatistical Concepts and Methods in the Legal Setting. *Statistics in Medicine*, **14**, 1641–1653.

Gastwirth, J. L., Miao, W. and Zheng, G. (2003). Statistical issues arising in disparate impact cases and the use of the expectancy curve in assessing the validity of pre-employment tests. *International Statistical Review*, **71**, 565–580.

Gastwirth, J. L. and Pan, Q. Statistical Measures and Methods for Assessing the Representativeness of Juries: A Reanalysis of the data in Berghuis v. Smith. To appear in *Law, Probability and Risk.*

Graubard, B. I. (2009). Comment on "Using the Peters–Belson method in equal employment opportunity personnel evaluations" by Sinclair and Pan. *Law, Probability and Risk*, **8**, 119–122.

Gray, M. W. (1993). Can statistics tell us what we do not want to hear? The case of complex salary structures. *Statistical Science*, **8**, 144-179.

Greiner, D. J. (2008). Causal inference in civil rights litigation. *Harvard Law Review*, **122**, 533-598.

Gujarati, D. (1970a). Use of dummy variables in testing for equality between sets of coefficients in two linear regressions: A note. *The American Statistician*, **24**, 50–52.

Gujarati, D. (1970b). Use of dummy variables for equality between sets of coefficients in linear regressions: A generalization. *The American Statistician*, **24**, 18–22.

Hall, P., Leng, X. Y. and Müller H. G. (2009). Weighted-bootstrap alignment of explanatory variables. *Journal of Statistical Planning and Inference*, **138**, 1817–1827.

Heckman, J. J. (2005). The scientific model of causality. *Sociological Methodology*, **35**, 1–97.

Hikawa, H., Bura, E. and Gastwirth, J. L. (2010a). Local Linear Logistic Peters–Belson Regression and its application in employment discrimination cases. *Statistics and its Interface*, **3**, 125-144.

Hikawa, H., Bura, E. and Gastwirth, J. L. (2010b). Robust Peters–Belson Type Estimators of Measures of Disparity and their Applications in Employment Discrimination Cases. *Technical Report*, Department of Statistics, George Washington University.

Loader, C. (1999). *Local Regression and Likelihood.* New York: Springer-Verlag.

Nayak, T. K. and Gastwirth, J. L. (1997). The Peters–Belson approach to measures of economic and legal discrimination. In: Johnson N. L. and Balakrishnan N. (eds.) *Advances in the Theory and Practice of Statistics: a Volume in Honor of Samuel Kotz.* Hoboken, NJ: Wiley.

Oaxaca, R. (1973). Male-Female differentials in urban labor markets. *International Economic Review*, **14**, 693–709.

Pearl, J. (2000). *Causality: Models, Reasoning and Inference.* Cambridge: Cambridge University Press.

Peters, C. C. (1941). A method of matching groups for experiments with no loss of populations. *Journal of Educational Research*, **34**, 606–612.

Rao, C. R. (1952). *Advanced Statistical Methods in Biometric Research.* New York: John Wiley & Sons.

Rosenbaum, P. and Rubin, D. B. (1983). Central role of the propensity score in observational studies for causal effects. *Biometrika*, **70**, 41–55.

Rubin, D. B. (2005). Causal inference using potential outcomes: design, modeling, decisions. *Journal of the American Statistical Association*, **100**, 322–331.

Scheffe, H. (1970). Practical solutions of the Behrens-Fisher problem. *J. Amer. Statist. Assoc.*, **65**, 1501–1508.

Sinclair, M. D. and Pan, Q. (2009). Using the Peters–Belson Method in EEO Personnel Evaluations. *Law, Probability & Risk*, **8**, 2, 95–117 with Comments from Dr. Barry Graubard: 119–122.

Welch, B. L. (1949). Further note on Mrs. Aspin's tables and on certain approximations to the tabled function. *Biometrika*, **36**, 293–296.

Zanutto, E. L. (2006). A comparison of propensity score and linear regression: analysis of complex survey data. *Journal of Data Science*, **4**, 67–91.

Chapter 12

On a Hybrid Approach to Parametric and Nonparametric Regression

Prabir Burman and Probal Chaudhuri

University of California, Davis, and Indian Statistical Institute, Calcutta

A linear combination of a parametric and a nonparametric estimate of an unknown regression function is considered as a hybrid estimate. The technique can be viewed as a functional version of the famous James–Stein approach used in parameter estimation. The optimal linear combination is estimated from the data, and numerical implementation of the methodology in real data is discussed. Performance of the optimal hybrid estimate in terms of mean squared error is investigated and its rate of convergence is derived. This rate is a function of the distance between the true regression function and the parametric model under consideration. In particular, when the true regression function is sufficiently close to the fitted parametric model, the hybrid estimate converges at the $n^{-1/2}$ rate. On the other hand, when the actual regression function differs significantly from the chosen parametric model, the rate of convergence of the hybrid estimate is the same as that of its nonparametric component.

Contents

1. Introduction

Consider a real valued response Y, a real or vector valued regressor x and a regression model connecting them through the equation $Y = \theta(x) + \epsilon$. Here the function θ is the systematic component of the regression relationship and ϵ is the random component with zero conditional mean given any value of the regressor. The regressor x can be random or deterministic depending on the situation under consideration. In a parametric set up, one assumes that $\theta(x) = f(\beta, x)$, where $f(\beta, x)$ is a real valued function with a known form, and β is an unknown k-dimensional Euclidean parameter, which is to be estimated from a given set of observations (Y_i, x_i), $i = 1, \ldots, n$. Note that here the dimension k of the parameter β is a fixed positive integer that does not depend on the sample size n. On the other hand, in a nonparametric set up, no finite dimensional parametric structure is imposed on the regression function, and it is only assumed to be a smooth function. In this case, θ is estimated by fitting a smooth function (a "smoother") to the data by some standard nonparametric procedure (e.g., kernel or spline smoothing). Silverman (1985), Eubank (1988), Müller (1988), Buja, et al. (1989), Wahba (1990), Rosenblatt (1990), Hardle (1990), Wand and Jones (1995), Fan and Gijbels (1996), etc., have extensively discussed several popular nonparametric regression techniques.

Clearly, a finite dimensional parametric model for θ has the advantage of being simple and easily interpretable. Perhaps linear parametric models are by far the most popular ones in regression analysis for several good practical reasons. Besides, many parametric regression models have special and well-established scientific meaning. The well-known nonlinear parametric regression models used in chemical reaction kinetics [see, e.g., Bates and Watts (1988) for several examples of such models] and various parametric versions of the 'Engel curve' [Engel (1895)] used by economists and econometricians for modeling the expenditure patterns of people in different income groups are two examples of such popular regression models. However, often in practice, data sets (particularly large ones) bear many important features that can hardly be captured by a low dimensional parametric model. As a matter of fact, the rejection of simple parametric models by standard goodness of fit tests applied to large data sets is not at all uncommon, and this is caused primarily by the bias in the selected parametric models. A nonparametric smoother is free from the constraints associated with a parametric structure, and therefore tries to follow the data as much as possible by picking up features that might be missed by a parametric estimate. In the process, however, nonparametric estimates may overfit the data and give rise to smooth curves and surfaces with characteristics that are artifacts of the noise term ϵ in the regression model and are not related to the regression function θ. This causes the difficulty frequently encountered by statisticians in finding good scientific interpretation for some of the features of the data reflected in a nonparametric estimate.

Nonparametric regression estimates do not suffer from the biases that might be associated with a parametric structure due to model misspecification. As a result, nonparametric estimates are usually consistent in a very general set up [see e.g., Stone (1977)]. However, a nonparametric estimate will typically converge to the true regression function at a rate much slower than $n^{1/2}$-consistent parametric estimates [see, e.g., Stone (1980; 1982), Hall (1989)]. This is largely due to the statistical variability in the data affecting a nonparametric estimate, and the degree of this variability becomes quite significant when the dimension d (say) of the regressor is large. This phenomenon is popularly described as data sparseness in high dimension or the curse of dimensionality in the literature on nonparametric regression. Besides, a parametric model may have its origin in the prior belief of an expert, and therefore it may not be wise to ignore it completely.

In this paper, we aim at developing and exploring a procedure to combine a parametric and a nonparametric estimate of the same regression function in a prudent way [see also Wooldridge (1992), Ullah and Vinod (1993), Fan and Ullah (1998) and Glad (1998)]. Our principal objective here is to construct and investigate certain hybrid estimates that will possess adequate strength to cope simultaneously with the bias intrinsically associated with a finite dimensional parametric model and the inherent statistical instability of purely nonparametric smoothers (Sections 2 and 3). The hybridization technique considered here can be viewed as a functional version of the famous James–Stein approach [see Stein (1956), James and Stein (1961)] in the sense that we are trying to borrow strength from a low dimensional model in order to improve the performance of a nonparametric smoother. In Section 4, we highlight and present a new mathematical formulation to establish how our hybrid estimate derives strength from the chosen parametric model even if it is incorrect but the true regression function θ is not too far away from that model. Wolpert (1992), who considered "neural networks", and Breiman (1995), who considered "stacked regression", remarked that when there are more than one competing estimates of an unknown regression function, one can gain substantially by combining the estimates instead of choosing a single best one. They provided some heuristics and numerical results to justify their claim. Section 5 contains some general remarks and discussion. Some technical results are given in Section 6 and proofs appear in the Appendix.

2. Hybridization of the parametric and the nonparametric estimation procedures

Olkin and Spiegelman (1987) proposed and thoroughly investigated a semiparametric approach for estimating density functions. They combined a parametric and a nonparametric estimate by forming a mixture of the two, and the value of the mixing proportion was determined by means of a maximum likelihood type technique. They succeeded in establishing that such a compromise estimate converges to the true unknown density at $n^{-1/2}$ rate if the chosen parametric model is valid, and otherwise it has a nonparametric rate of convergence, which is slower than $n^{-1/2}$. Methods for combining parametric models with nonparametric ones have been explored in the literature in the context of goodness-of-fit tests [see, e.g., Eubank (1988), Aerts, et al. (2000), Claeskens and Hjort (2004)]. Readers are referred to Hjort and Glad (1995) for another closely related semiparametric approach in density estimation. The procedure, which we will describe here, can be viewed as an attempt to carry this semiparametric approach in a different direction. Specifically, we intend to modify the basic idea in Olkin and Spiegelman (1987) in an appropriate way so that it becomes fit for application in a regression set up [see also Wooldridge (1992), Ullah and Vinod (1993), Fan and Ullah (1998) and Glad (1998)].

Our basic mathematical model is $Y_i = \theta(x_i) + \epsilon_i, i = 1, \ldots, n$, where the ϵ_i's are i.i.d. with $E(\epsilon_i) = 0$ and $Var(\epsilon_i) = \sigma^2$. We will assume that the x_i's are fixed, taking values in a compact set in R^d. However, we would like to point out that all the results in this paper can be generalized to the case where the x_i's are random, but that can be done at the cost of increasing mathematical technicalities. Let $(Y_i, x_i), i = 1, \ldots, n$ be a set of observations, and $\hat{\beta}$ be an estimate of β computed under the assumption that the parametric model $Y_i = f(\beta, x_i) + \epsilon_i$ holds. Also, let $\hat{g}$ be the nonparametric estimate (e.g., a kernel or a spline smoother) of the regression function θ when no parametric structure is specified for it. After obtaining $\hat{\beta}$ and $\hat{g}$ separately, we will construct an estimate of the unknown regression function θ by forming a linear combination $\lambda f(\hat{\beta}, \cdot) + (1 - \lambda)\hat{g}$, where λ is a constant. A key issue that arises at this point is how to select an appropriate value of λ. In practice, it is expected that the parametric model will either come from some scientific consideration or from past experience of the scientist interested in the data. If the variance σ^2 is estimated on the basis of the parametric fit, then how good the estimate is depends on how close the parametric model is to the unknown regression.

We will now introduce some convenient notations. For any two functions h_1 and h_2, we define the norm and the inner product as

$$\|h_1\|^2 = n^{-1}\sum_{i=1}^{n} h_1^2(x_i) \ , \quad \langle h_1, h_2\rangle = n^{-1}\sum_{i=1}^{n} h_1(x_i)h_2(x_i) \ .$$

Ideally, we would like to choose λ in such a way that the distance between the unknown regression function θ and the hybrid estimator $\lambda f(\hat{\beta},\cdot)+(1-\lambda)\hat{g}$ is minimized. It can be shown by straight-forward algebra that the quantity $\|\theta - \lambda f(\hat{\beta},\cdot) - (1-\lambda)\hat{g}\|$ attains its minimum at

$$\lambda^* = \frac{\langle f(\hat{\beta},\cdot) - \hat{g}, \theta - \hat{g}\rangle}{\|f(\hat{\beta},\cdot) - \hat{g}\|^2} = \frac{\langle f(\hat{\beta},\cdot) - \hat{g}, \theta\rangle \ - \ \langle f(\hat{\beta},\cdot) - \hat{g}, \hat{g}\rangle}{\|f(\hat{\beta},\cdot) - \hat{g}\|^2} \ .$$

Note that the denominator and the second term in the numerator are known (i.e. they can be computed from the data). The first term in the numerator is not known and has to be estimated with a bit of care. The straightforward estimate of this first term based on the idea of ordinary least squares will be $n^{-1}\sum_{i=1}^{n}\{f(\hat{\beta},x_i) - \hat{g}(x_i)\}Y_i$. However, this leads to a rather poor performance of the hybrid estimate especially when the true regression function θ happens to be in the vicinity of the parametric model $f(\beta,\cdot)$. The primary reason for this is that such a hybridization tends to put more weight on the nonparametric estimate $\hat{g}$ than what is appropriate. We will use a certain type of cross-validated estimate for this first term in the numerator to prevent the hybridization process from putting an excessive amount of weight on $\hat{g}$ [for a discussion of leave-one-out cross-validation procedures, see Stone (1974), Clark (1975), Hardle and Marron (1985), etc.]. As we have already noted in the preceding section, the nonparametric estimate $\hat{g}$ tends to overfit the data in many situations, and cross-validation is needed as a safeguard against such overfitting in the hybridization process. Let $f(\hat{\beta}^{(i)},\cdot)$ and $\hat{g}^{(i)}$ be the parametric and the nonparametric estimates based on all the observations except the ith case. Then an estimate of $\langle f(\hat{\beta},\cdot) - \hat{g}, \theta\rangle$ is $n^{-1}\sum_{i=1}^{n}\{f(\hat{\beta}^{(i)},x_i) - \hat{g}^{(i)}(x_i)\}Y_i$. Combining all these together we get the following estimate of λ^*

$$\lambda_0^* = \frac{n^{-1}\sum_{i=1}^{n}\{f(\hat{\beta}^{(i)},x_i) - \hat{g}^{(i)}(x_i)\}Y_i \ - \ \langle f(\hat{\beta},\cdot) - \hat{g}, \hat{g}\rangle}{\|f(\hat{\beta},\cdot) - \hat{g}\|^2} \ .$$

Note that

$$\lambda_0^* - \lambda^* = \frac{n^{-1}\sum_{i=1}^{n}\{f(\hat{\beta}^{(i)},x_i) - \hat{g}^{(i)}(x_i)\}Y_i - \langle f(\hat{\beta},\cdot) - \hat{g}, \theta\rangle}{\|f(\hat{\beta},\cdot) - \hat{g}\|^2} \ .$$

The term in the numerator of $\lambda_0^* - \lambda^*$ can be rewritten as

$$n^{-1}\sum_{i=1}^{n}\{f(\hat{\beta}^{(i)}, x_i) - \hat{g}^{(i)}(x_i)\}\epsilon_i$$
$$+\, n^{-1}\sum_{i=1}^{n}[\{f(\hat{\beta}^{(i)}, x_i) - \hat{g}^{(i)}(x_i)\} - \{f(\hat{\beta}, x_i) - \hat{g}(x_i)\}]\theta(x_i)\,.$$

It can be shown that the first term in the last expression is asymptotically negligible. However, the second term is not necessarily very small if the parametric model is correct. To take care of this, we need to subtract the term

$$n^{-1}\sum_{i=1}^{n}[\,\{f(\hat{\beta}^{(i)}, x_i) - \hat{g}^{(i)}(x_i)\} \;-\; \{f(\hat{\beta}, x_i) - \hat{g}(x_i)\}\,]f(\hat{\beta}, x_i)$$

from the numerator of $\lambda_0^* - \lambda^*$. So a reasonable estimate of λ^* is given by

$$\hat{\lambda}^* \;=\; \frac{n^{-1}\sum_{i=1}^{n}\{f(\hat{\beta}^{(i)}, x_i) - \hat{g}^{(i)}(x_i)\}\{Y_i - f(\hat{\beta}, x_i)\} \;+\; \|f(\hat{\beta}, \cdot) - \hat{g}\|^2}{\|f(\hat{\beta}, \cdot) - \hat{g}\|^2}\,.$$

Since $\hat{\lambda}^*$ has a nice closed form expression, it is quite easy to compute and this makes implementation of the methodology very convenient. Note that computation of $\hat{\beta}$ and $\hat{\beta}^{(i)}$ (using ordinary least squares or non-linear least squares depending on whether $f(\beta, \cdot)$ is linear or non-linear in β) will involve inversion of matrices of the form

$$\sum_{j=1}^{n}\{\nabla_\beta f(\beta, x_j)\}\{\nabla_\beta f(\beta, x_j)\}^T \;\textit{and}\; \sum_{1\le j\le n; j\ne i}\{\nabla_\beta f(\beta, x_j)\}\{\nabla_\beta f(\beta, x_j)\}^T$$

respectively, where $\nabla_\beta f(\beta, \cdot)$ denote the gradient vector obtained by differentiating $f(\beta, \cdot)$ w.r.t. the parameter vector β, and it is a column vector with the same dimension as that of β. For this purpose, one can conveniently make use of the result :

$$\left[\sum_{1\le j\le n; j\ne i}\{\nabla_\beta f(\beta, x_j)\}\{\nabla_\beta f(\beta, x_j)\}^T\right]^{-1}$$
$$= \left[\sum_{j=1}^{n}\{\nabla_\beta f(\beta, x_j)\}\{\nabla_\beta f(\beta, x_j)\}^T \;-\; \{\nabla_\beta f(\beta, x_i)\}\{\nabla_\beta f(\beta, x_i)\}^T\right]^{-1}$$
$$= \left[A \;-\; U_i U_i^T\right]^{-1}\;(say) \;=\; A^{-1} \;+\; \left(1 - U_i^T A^{-1} U_i\right)^{-1} A^{-1} U_i U_i^T A^{-1}\,,$$

where $U_j = \nabla_\beta f(\beta, x_j)$ and $A = \sum_{j=1}^n U_j U_j^T$. Further, to reduce the amount of computation, one can use the quick method of one-step iteration which is quite popular and effective in non-linear least squares and maximum likelihood estimation. Also, note at this point that many nonparametric regression estimates are expressible in the form of a weighted sum of the Y_i's, i.e. $\hat{g}(x) = \sum_{j=1}^n w(x, x_j)Y_j$, and the leave-one-out nonparametric estimate is expressible in the form

$$\hat{g}^{(i)}(x_i) = \frac{\sum_{1 \le j \le n; j \ne i} w(x_i, x_j)Y_j}{\sum_{1 \le j \le n; j \ne i} w(x_i, x_j)}$$

(cf. Condition A6 in Section 3 and the discussion around it). All these enable us to obtain $\hat{\lambda}^*$ without much computational difficulty.

Table 1. Values of $\hat{\lambda}^*$ for the PCB data corresponding to different choices of the parametric model and the bandwidth of the nonparametric estimate.

Values of γ	Bandwidth 0.10	0.13	0.15	0.20
3/2	0.35	0.16	0.10	0.02
1	0.62	0.43	0.35	0.28
1/3	1.00	0.98	0.97	0.95

We will now consider a simple data analytic example to illustrate the methodology. Bache, et al. (1972) reported data on the concentrations of polychlorinated biphenyl (PCB) residues in lake trout caught from Cayuga Lake, New York. The ages (x) of the fish were accurately known because they are annually stocked as yearlings and marked specifically for their year class. After being caught, each whole fish was mechanically chopped, ground and thoroughly mixed, and 5-gram samples were taken. After a careful treatment of the samples, the PCB residues in parts per million (Y) were estimated using column chromatography. There are 28 observations, and Bates and Watts (1988) have analyzed the data using some parametric regression models. For demonstrating the hybridization of parametric and non-parametric regression estimates, we will consider three parametric models $f(\beta, x) = \alpha_0 + \alpha_1 x^\gamma$ with $\beta = (\alpha_0, \alpha_1)$ and $\gamma = 3/2$, 1 and 1/3, and nonparametric Nadarya–Watson kernel estimates based on the quartic kernel $K(t) = (15/16)(1-t^2)^2 1_{\{|t| \le 1\}}$ with four different bandwidths, namely $b = 0.10$, 0.13, 0.15 and 0.2, where $b = 0.13$ happens to be the optimal leave-one-out cross-validation bandwidth for this data set. We rescaled all the x values to make them lie in the unit interval $[0, 1]$ before computing the regression estimates. Different values of $\hat{\lambda}^*$ computed from the data are reported in the following table. It is worth noting the fact that the value of $\hat{\lambda}^*$ increases greatly and becomes very close to 1.0 as the value of γ decreases from 3/2 to 1/3. This is consistent with the analysis and results reported in Bates and Watts (1988), who strongly recommended the parametric model $f(\beta, x) = \alpha_0 + \alpha_1 x^{1/3}$ based on an extensive graphical and numerical analysis.

3. Properties of the hybrid estimate

Before we present the main results on our optimal hybrid estimate defined as $\hat{\theta} = \hat{\lambda}^* f(\hat{\beta}, \cdot) + (1 - \hat{\lambda}^*)\hat{g}$, we need to introduce some concepts and state a few assumptions. Let δ_n be the distance between the unknown regression function and the parametric family of regression functions under consideration. In other words

$$\delta_n = \inf\{ \|\theta - f(\beta, \cdot)\| : \beta \in \Gamma \} .$$

Here $\Gamma \subseteq R^k$ is the parameter space. If we assume that the infimum is attained at a unique β_n^*, we can write $\delta_n = \|\theta - f(\beta_n^*, \cdot)\|$. Since we are dealing with a nonrandom design $\{x_i\}$ in this work, the distance between any two models depends on this design (see Section 2). For this reason, the distance δ_n between the parametric model and the true regression has a subscript n. Clearly, if the parametric model happens to be correct and $\theta = f(\beta, \cdot)$ for some $\beta \in \Gamma$, we have $\delta_n = 0$ and $\beta_n^* = \beta$. For notational simplicity, from now on, we will drop the subscript n and write β^* in place of β_n^*. If the parametric model is not correct then $\delta_n > 0$. However, the parametric model, though not correct, may not be too be far away from the the true regression. In such a case, the asymptotic mean square error of the hybrid estimate depends on δ_n. For this reason, we have introduced new asymptotic results whereby δ_n is allowed to decrease to zero as the sample size n grows to infinity. The results in Section 4 shed more lights on this issue. We now write down an assumption on the behavior of the parametric estimate.

A1. There exists a function W_1 of two variables which is defined and bounded on $\mathcal{X} \times \mathcal{X}$, where $\mathcal{X}$ is the regressor space, and the following holds

$$\|f(\hat{\beta}, \cdot) - f(\beta^*, \cdot) - n^{-1} \sum_{i=1}^{n} W_1(\cdot, x_i)\epsilon_i\| = O_P(n^{-1}) .$$

Observe that A1 is trivially satisfied when $f(\beta, \cdot)$ is linear in β (i.e. when we are fitting a linear parametric model to the data). As a matter of fact, if $f(\beta, \cdot) = \beta_1 h_1(\cdot) + \cdots + \beta_k h_k(\cdot) = \beta^T H(\cdot)$, where $h_1(\cdot), \ldots, h_k(\cdot)$ are real valued functions defined on $\mathcal{X}$, $\beta = (\beta_1, \ldots, \beta_k)^T$ and $H(\cdot) = (h_1(\cdot), \ldots, h_k(\cdot))^T$, the quantity

$$f(\hat{\beta}, \cdot) - f(\beta^*, \cdot) - n^{-1} \sum_{i=1}^{n} W_1(\cdot, x_i)\epsilon_i$$

will be exactly equal to zero provided that we choose

$$W_1(\cdot, x_i) \;=\; H(\cdot)^T \left[\sum_{j=1}^{n} H(x_j)H(x_j)^T \right]^{-1} H(x_i)\,.$$

Consequently, when each of the functions $h_r(\cdot)$'s for $1 \leq r \leq k$ is bounded on $\mathcal{X}$, and the smallest eigenvalue of the $k \times k$ matrix $\sum_{j=1}^{n} H(x_j)H(x_j)^T$ remains positive and bounded away from zero as n tends to infinity, A1 will hold. In the case of nonlinear models, we can take

$$W_1(\cdot, x_i) \;=\; \{\nabla_{\beta^*} f(\beta^*, \cdot)\}^T A^{-1} \{\nabla_{\beta^*} f(\beta^*, x_i)\},$$

where

$$A = \sum_{j=1}^{n} \left[\{\nabla_{\beta^*} f(\beta^*, x_j)\}\{\nabla_{\beta^*} f(\beta^*, x_j)\}^T \right. \\ \left. + \{g(x_j) - f(\beta^*, x_j)\}\mathcal{H}_{\beta^*} f(\beta^*, x_j) \right]\,,$$

and $\mathcal{H}_\beta f(\beta, \cdot)$ denotes the $k \times k$ Hessian matrix of $f(\beta, \cdot)$ based on the second order partial derivatives of $f(\beta, \cdot)$ w.r.t. β. So, in this case A1 will hold with appropriate regularity conditions on $f(\beta, \cdot)$ and the true regression function $\theta(x)$. Those conditions can be easily worked out following the standard analysis involved in the derivation of the asymptotic properties (e.g. $n^{1/2}$-consistency and asymptotic normality) of nonlinear least squares estimates [see, e.g., Jenrich (1969), Wu (1981), Gallant (1987), Seber and Wild (1989)].

Next, as in Burman and Chen (1989), let us write down the nonparametric estimate $\hat{g}$ as $\hat{g} = n^{-1} \sum_{i=1}^{n} W_2(\cdot, x_i)Y_i$, which is a locally weighted average of the Y-values such that the weight function satisfies $n^{-1} \sum_{i=1}^{n} W_2(\cdot, x_i) = 1$. This framework covers all the well known nonparametric methods including kernel smoothing, nearest neighbor method and the techniques based on splines. Associated with the weight function W_2, there is a smoothing parameter τ_n, and the description of it varies from method to method. For example, τ_n is the reciprocal of the d-th power of the bandwidth in kernel regression, and in the case of spline regression with univariate regressor, τ_n is the number of knots. For nearest neighbor regression, τ_n equals n divided by the number of nearest neighbors. We will make the usual assumption that τ_n tends to infinity and $\tau_n n^{-1}$ tends to zero as the sample size n grows to infinity. Throughout this paper, we will write

$$r_n^2 = E\|\hat{g} - \theta\|^2,$$

and assume that r_n tends to zero as n goes to infinity. Here are a few assumptions concerning the nonparametric estimate $\hat{g}$.

A2 : The ratio $\frac{\|\hat{g}-\theta\|^2 - E\|\hat{g}-\theta\|^2}{E\|\hat{g}-\theta\|^2}$ tends to zero in probability as n tends to infinity.

A3 : There exist two positive constants c_1 and c_2 such that for all i we have $c_1\tau_n \le W_2(x_i, x_i) \le c_2\tau_n$.

A4 : There exists a constant $c > 0$ such that $n^{-2}\sum_{i=1}^n \sum_{j=1}^n W_2^2(x_i, x_j) \le c\tau_n$.

A5 : $n^{-1/2} r_n^{-1}$ tends to zero as n tends to infinity.
Note that assumption A2 guarantees that $\|\hat{g} - \theta\| = O_P(r_n)$. Assumption A5 says that the nonparametric estimate $\hat{g}$ converges at a rate slower than $n^{-1/2}$.

A6 : The leave-one-out nonparametric estimate has the form

$$\hat{g}^{(i)}(x_i) = \frac{\sum_{j:j\neq i} W_2(x_i, x_j) Y_j}{\sum_{j:j\neq i} W_2(x_i, x_j)} .$$

Note that the mean square error $r_n^2 = E\|\hat{g} - \theta\|^2$ of the nonparametric estimate $\hat{g}$ has the usual variance-bias decomposition. The variance part is equal to $Var(\varepsilon) n^{-3} \sum_{1\le i,j\le n} W_2(x_i, x_j)^2$. Assumption A4 guarantees that the variance part is no larger than $cVar(\varepsilon)\tau_n/n$. Assumption A6 is automatically satisfied for kernel and spline smoothers. However, for nearest neighbor regression, it is only approximately true. As a matter of fact, what we really need is that the leave-one-out nonparametric estimate is approximately of the form described in A6. But, we will not worry about such mathematical refinements here.

With the preceding assumptions in hand, we can now state the following Theorem, which describes the asymptotic behavior of the ideal linear combination of the parametric and the nonparametric estimates.

Theorem 1. *Assume that conditions A1, A2 and A5 hold.*

a) $\|\lambda^* f(\hat{\beta}, \cdot) + (1 - \lambda^*)\hat{g} - \theta\| = O_P(r_n)$ *if* δ_n *stays away from zero as* n *tends to infinity (i.e. if the parametric model is incorrect).*

b) $\|\lambda^* f(\hat{\beta}, \cdot) + (1 - \lambda^*)\hat{g} - \theta\| = O_P(n^{-1/2})$ *if* $\delta_n = 0$ *(i.e. if the parametric model is correct).*

The question that naturally arises after Theorem 1 is: how does the optimal hybrid estimate $\hat{\theta}$ behave? Interestingly, its behavior is not at all different from that of the ideal linear combination, at least in the asymptotic sense, even though λ^* is replaced by its estimate $\hat{\lambda}^*$ in $\hat{\theta}$.

Theorem 2. *Assume that conditions A1 through A6 hold.*

a) $\|\hat{\lambda}^* f(\hat{\beta}, \cdot) + (1 - \hat{\lambda}^*)\hat{g} - \theta\| = O_P(r_n)$ *if* δ_n *stays away from zero as* n *tends to infinity.*

b) $\|\hat{\lambda}^* f(\hat{\beta}, \cdot) + (1 - \hat{\lambda}^*)\hat{g} - \theta\| = O_P(n^{-1/2})$ *if* $\delta_n = 0$.

Note that the preceding Theorem ensures that the optimal hybrid estimate converges at the $n^{-1/2}$ rate when the parametric model is correct, and when the parametric model is false, its rate of convergence is same as that of the nonparametric estimate. Olkin and Spiegelman (1987) assumed that $n^{-1/2}\tau_n$ tends to zero as n tends to infinity. However, we need to assume that $n^{-1}\tau_n$ tends to zero. This is a considerably weaker condition particularly when the dimension d of the regressor x is large.

We conclude this Section with a small numerical study. Let us once again consider the data on PCB residues in lake trout discussed at the end of Section 2 and the parametric models considered there. In order to compare the performance of the hybrid estimate $\hat{\theta} = \hat{\lambda}^* f(\hat{\beta}, \cdot) + (1 - \hat{\lambda}^*)\hat{g}$ with the performance of each of the parametric least squares estimates and the nonparametric kernel estimates that are used in hybridization, we define the following efficiency measures e_1 and e_2 based on leave-one-out estimates of the prediction mean squared errors of those estimates

$$e_1 = \frac{\sum_{i=1}^n \{Y_i - f(\hat{\beta}^{(i)}, x_i)\}^2}{\sum_{i=1}^n \{Y_i - \hat{\theta}^{(i)}(x_i)\}^2}, \quad e_2 = \frac{\sum_{i=1}^n \{Y_i - \hat{g}^{(i)}(x_i)\}^2}{\sum_{i=1}^n \{Y_i - \hat{\theta}^{(i)}(x_i)\}^2}.$$

The values of e_1 and e_2 computed from the PCB data for different choices of the parametric models and nonparametric kernel estimates with different bandwidth values are reported in the following table. These efficiencies turn out to be larger than 100% in most of the cases, and they are consistent with the findings of Bates and Watts (1988) whose elaborate parametric analysis recommends in favor of the parametric model $f(\beta, x) = \alpha_0 + \alpha_1 x^{1/3}$.

Table 2. Efficiencies of the hybrid estimate for the PCB data as compared with different parametric and nonparametric estimates.

Values of γ	Bandwidth 0.10	0.13	0.15	0.20
3/2	$e_1 = 130.44\%$	$e_1 = 137.65\%$	$e_1 = 139.40\%$	$e_1 = 140.31\%$
	$e_2 = 105.97\%$	$e_2 = 101.55\%$	$e_2 = 100.76\%$	$e_2 = 100.14\%$
1	$e_1 = 107.03\%$	$e_1 = 109.74\%$	$e_1 = 110.46\%$	$e_1 = 110.75\%$
	$e_2 = 112.82\%$	$e_2 = 105.05\%$	$e_2 = 103.60\%$	$e_2 = 102.55\%$
1/3	$e_1 = 100.00\%$	$e_1 = 99.63\%$	$e_1 = 99.47\%$	$e_1 = 99.33\%$
	$e_2 = 140.03\%$	$e_2 = 126.69\%$	$e_2 = 123.94\%$	$e_2 = 122.19\%$

4. Effects of mis-specification in the parametric model and related asymptotics

Recall that δ_n is the distance between the unknown regression function and the parametric class of models. In the last section, we have studied the properties of the hybrid estimate under two distinct conditions on δ_n:

i) δ_n stays away from zero, i.e., when the parametric model is incorrect.
ii) $\delta_n = 0$, i.e., when the parametric model is correct.

However, the dichotomy, that the parametric model is correct versus that the parametric model is incorrect, is not necessarily an appropriate way to study this problem since the parametric model is not likely to be exactly correct in any practical situation. Intuitively, we understand that even if the parametric model is not correct, it may still be worthwhile to borrow weight from it as long as the true regression function is not far away from the parametric class of models in relation to the sample size. In other words, as long as δ_n is not too far away from zero, one can profitably use hybrid estimate for moderate sample sizes. The next two results formalize this intuition in terms of the rates of convergence of the ideal linear combination and the optimal hybrid estimate, which are shown to depend directly on δ_n.

Recall that the rate of convergence associated with the nonparametric estimate is r_n. We will now summarize the main ideas contained in the Theorems that are presented below. Because of assumption A5, $n^{1/2}r_n$ tends to infinity as n tends to infinity. Let us assume that δ_n tends to zero as n tends to infinity. If the ratio $\delta_n r_n^{-1}$ stays above 1, the hybrid estimate has the nonparametric rate of convergence. On the other hand, if $\delta_n r_n^{-1}$ remains smaller than $n^{-1/2}r_n^{-1}$, the hybrid estimate converges at the parametric rate. If, however, the ratio $\delta_n r_n^{-1}$ stays between 1 and $n^{-1/2} r_n^{-1}$, the rate of convergence is δ_n itself.

Theorem 3. *Assume that conditions A1 through A5 hold, and δ_n tends to zero as n tends to infinity. Then*

$$\|\lambda^* f(\hat{\beta}, \cdot) + (1-\lambda^*)\hat{g} - \theta\| = \begin{cases} O_P(r_n) & \text{if } \delta_n r_n^{-1} \text{ stays above 1} \\ O_P(\delta_n) & \text{if } \delta_n r_n^{-1} \text{ stays between } n^{-1/2} r_n^{-1} \text{ and 1} \\ O_P(n^{-1/2}) & \text{if } \delta_n r_n^{-1} \text{ stays below } n^{-1/2} r_n^{-1}\,. \end{cases}$$

Theorem 4. *As in Theorem 3, assume that conditions A1 through A6 hold, and δ_n tends to zero as n tends to infinity. Then*

$$\|\hat{\lambda}^* f(\hat{\beta}, \cdot) + (1-\hat{\lambda}^*)\hat{g} - \theta\| = \begin{cases} O_P(r_n) & \text{if } \delta_n r_n^{-1} \text{ stays above 1} \\ O_P(\delta_n) & \text{if } \delta_n r_n^{-1} \text{ stays between } n^{-1/2} r_n^{-1} \text{ and 1} \\ O_P(n^{-1/2}) & \text{if } \delta_n r_n^{-1} \text{ stays below } n^{-1/2} r_n^{-1}\,. \end{cases}$$

We close this section with a simulation study to compare the performance of the hybrid estimate $\hat{\theta} = \hat{\lambda}^* f(\hat{\beta}, \cdot) + (1-\hat{\lambda}^*)\hat{g}$ with the performance of each of the parametric and the nonparametric estimates that are used in hybridization. We generated data $(Y_1, x_1), (Y_2, x_2), \ldots, (Y_n, x_n)$ satisfying the model $Y_i = 1 + 2x + \delta\{\sin(12\pi x_i)\} + e_i$, where we had $x_i = i/(n+1)$ and the e_i's were i.i.d standard normal variables. Here the term $\delta\{\sin(12\pi x_i)\}$ represent the deviation from the linear model and the amount of model mis-specification is determined by different values of the parameter δ. We define the following efficiency measures e_1^* and e_2^* based on leave-one-out estimates of the prediction mean squared errors of those estimates

$$e_1^* = \frac{E\left[\sum_{i=1}^n \{Y_i - f(\hat{\beta}^{(i)}, x_i)\}^2\right]}{E\left[\sum_{i=1}^n \{Y_i - \hat{\theta}^{(i)}(x_i)\}^2\right]}, \quad e_2^* = \frac{E\left[\sum_{i=1}^n \{Y_i - \hat{g}^{(i)}(x_i)\}^2\right]}{E\left[\sum_{i=1}^n \{Y_i - \hat{\theta}^{(i)}(x_i)\}^2\right]}\,.$$

The expectations that appear in the definitions of the efficiency measures above were estimated based on 5000 Monte Carlo replications in each case. Here also, the parametric estimate is the ordinary least squares estimate, and the nonparametric estimate is the Nadarya–Watson estimate based on the quartic kernel $(1-t^2)^2 1_{\{|t|\le 1\}}$. The values of e_1^* and e_2^* computed from the simulations for different choices of the value of the parameter δ (i.e. for different degrees of model mis-specification) and for different choices of the bandwidth used in the nonparametric kernel estimate are reported in the following two tables one of which corresponds to $n = 50$ and the other one to $n = 100$. Efficiency figures are larger than 100% in all cases, and this is indicative of clearly superior performance of the hybrid estimate compared to both of the purely parametric and the purely nonparametric regression estimates.

Table 3. Estimated efficiencies of the hybrid estimate compared with parametric and nonparametric estimates for sample size $n = 50$.

Bandwidth	The Values of δ 0.00	0.50	1.00	1.50
0.075	$e_1^* = 102.94\%$ $e_2^* = 122.09\%$	$e_1^* = 104.68\%$ $e_2^* = 113.32\%$	$e_1^* = 115.58\%$ $e_2^* = 103.02\%$	$e_1^* = 140.08\%$ $e_2^* = 100.82\%$
0.085	$e_1^* = 103.11\%$ $e_2^* = 119.57\%$	$e_1^* = 104.24\%$ $e_2^* = 111.84\%$	$e_1^* = 113.00\%$ $e_2^* = 102.50\%$	$e_1^* = 132.62\%$ $e_2^* = 101.21\%$
0.100	$e_1^* = 103.32\%$ $e_2^* = 116.23\%$	$e_1^* = 103.67\%$ $e_2^* = 110.40\%$	$e_1^* = 108.05\%$ $e_2^* = 102.46\%$	$e_1^* = 119.17\%$ $e_2^* = 100.99\%$

Table 4. Estimated efficiencies of the hybrid estimate compared with parametric and nonparametric estimates for sample size $n = 100$.

Bandwidth	The Values of δ 0.00	0.50	1.00	1.50
0.075	$e_1^* = 102.10\%$ $e_2^* = 110.84\%$	$e_1^* = 104.56\%$ $e_2^* = 103.70\%$	$e_1^* = 124.56\%$ $e_2^* = 100.65\%$	$e_1^* = 163.09\%$ $e_2^* = 105.13\%$
0.085	$e_1^* = 102.02\%$ $e_2^* = 109.40\%$	$e_1^* = 103.98\%$ $e_2^* = 103.13\%$	$e_1^* = 121.09\%$ $e_2^* = 100.98\%$	$e_1^* = 154.67\%$ $e_2^* = 106.98\%$
0.100	$e_1^* = 102.09\%$ $e_2^* = 108.03\%$	$e_1^* = 103.01\%$ $e_2^* = 102.78\%$	$e_1^* = 115.07\%$ $e_2^* = 101.10\%$	$e_1^* = 138.41\%$ $e_2^* = 107.23\%$

5. Remarks and discussion

Remark 1. We have already mentioned (see Sections 1 and 2) that the nonparametric estimate $\hat{g}$ will tend to follow the data more closely (i.e. it may overfit the data) than the parametric estimate $f(\hat{\beta}, \cdot)$. As a result, if hybridization is done without the cross-validation and the subsequent adjustment, $\hat{g}$ is very likely to receive more weight than what it deserves.

This will produce an estimate biased towards $\hat{g}$, and it will inherit the statistical instability of $\hat{g}$ to such an extent that its convergence rate will be substantially slower than that for the hybrid estimate $\hat{\theta}$ when θ happens to be close to $f(\beta, \cdot)$ (i.e. when the magnitude of δ_n is of a small order, see Section 4). The leave-one-out cross-validation and the subsequent correction used in estimating the hybridization parameter λ^* are safeguards against such possible bias towards the nonparametric estimate $\hat{g}$. In different situations, other authors [see, e.g., Akaike (1969; 1970), Mallows (1973), Wahba (1990)] have used certain other techniques to cope with biases of a somewhat similar nature. A major advantage in our approach is that we get a simple and closed form expression for $\hat{\lambda}^*$, which can be conveniently computed from the data, and no numerical optimization will be required. Further, the simple expression for $\hat{\lambda}^*$ enables us to work out the asymptotic properties of $\hat{\theta}$, and it eventually leads to the results on convergence rates presented in Sections 3 and 4. Fan and Ullah (1998) have established asymptotic normality of certain hybrid estimates of regression functions.

Remark 2. Hardle and Mammen (1993) have explored some tests for parametric null models of very general nature against nonparametric alternatives. It is clear from their detail analytic and numerical investigations that one has to face several practical difficulties when it comes to actual implementation of those tests. These difficulties arise primarily from complex distributions of the proposed test statistics and slow convergence to the asymptotic distributions. The intractability of finite sample distributions of test statistics and the inadequacy of asymptotic approximations caused them to try some computationally intensive resampling techniques including some non-standard variations of the usual bootstrap method. In general, any preliminary test for checking the validity of a given parametric null model will require an appropriate test statistic as well as a level of significance. The choices for both of them are bound to be very subjective if not completely arbitrary [see also Section 3 in Leonard (1977) for a related discussion]. Our hybridization technique, which is based on the fundamental idea lying at the root of the James–Stein approach, is completely free from all of these subjective and controversial elements, and it can be implemented without any difficulty in very general situations as we have amply demonstrated in Section 2. While a preliminary test will lead to either the acceptance or the rejection of a given parametric model, we have explicitly indicated in Section 4 that the hybrid estimate $\hat{\theta}$ can potentially borrow strength from a low dimensional model $f(\beta, \cdot)$ even if such a model is incorrect as long as the true regression function θ is not too far away from $f(\beta, \cdot)$. In a sense, our hybridization technique is an effective way [Wolpert's (1992) "black art"] of utilizing what Wolpert (1992) and Breiman (1995) describes as "more information" in the "level one data" arising from several competing estimates of a common regression function. [see also LeBlanc and Tibshirani (1996)]. Note that the nonparametric component of $\hat{\theta}$ can be viewed as a correction to adjust for the deviation from the parametric model observed in the data, while the magnitude of $\hat{\lambda}^*$ (i.e. how close it is to zero) can be used to get an idea about the degree of such departure from the parametric model. Since the hybrid estimate is capable of deriving strength from both of the parametric and the nonparametric estimate, and its construction is adaptive without involving much subjectivity in it, we hope that it will become an attractive addition to the existing collection of exploratory regression tools.

Remark 3. We should point out that it is possible to use an adaptive data based smoothing parameter while constructing $\hat{g}$ in practice though we have not done that here in order to avoid an excessive amount of mathematical details lengthening the proofs. The smoothing parameter τ_n can be estimated from the data using any standard procedure, and that will lead to only some standard and routine modifications in the proofs without adding any further insights into the main problem. However, even if τ_n is chosen deterministically, in view of the results presented in Sections 3 and 4, the mean squared error of the hybrid estimate $\hat{\theta}$ will converge to zero at a rapid rate when the parametric model $f(\beta, \cdot)$ happens to be at least approximately valid so that δ_n is of a small order of magnitude. In a sense, the hybridization parameter λ^* plays the role of some kind of a smoothing parameter here, and we can view hybridization as an interesting way of smoothing a less smooth (and possibly overfitting) nonparametric estimate by penalizing it with a much smoother (and possibly underfitting) parametric estimate. Recall that the computation of $\hat{\lambda}^*$ is quite straight forward and does not involve any numerical optimization, while any standard technique for adaptive estimation of the smoothing parameter τ_n will almost inevitably require some form of numerical optimization. Hence, hybridization technique can be used as a convenient and effective alternative to other data based smoothing techniques commonly used in nonparametric regression whenever a reasonable (but not necessarily exact) parametric model for the regression function is available, and particularly when the dimension of the regressor is large affecting any nonparametric smoother through the curse of dimensionality.

Remark 4. We would like to conclude this Section by pointing out that in order to avoid the "curse of dimensionality", a semiparametric approach such as additive regression or projection pursuit regression is sometimes used to model a regression function. If a semiparametric model is quite close to the unknown regression, then the semiparametric fit should be superior to the method proposed here. On the other hand, if a parametric model is a good guess for the regression function, then the results given in this paper (Theorems 4.1 and 4.2) suggest that the hybrid approach described here may be the preferred method. However, more investigations are needed in order to settle these issues.

6. Some technical results

Here we will write down a few technical results which will help us in proving the Theorems stated in Sections 3 and 4. In order to simplify notation, we will write $\hat{f}$ for $f(\hat{\beta}, \cdot)$.

Lemma 1. *Let us assume that conditions A1, A2 and A5 hold.*

a) If δ_n stays away from zero, $\|\hat{f} - \hat{g}\|$ remains bounded and bounded away from zero in probability.
b) If $\delta_n = 0$, $\|\hat{f} - \hat{g}\|$ stays between $c_1 r_n$ and $c_2 r_n$ with probability approaching one for some constants $0 < c_1 < c_2$.

Lemma 2. *Assume that conditions A1, A2 and A5 hold.*

a) If δ_n stays away from zero, we have $\lambda^* = O_P(r_n)$.
b) If $\delta_n = 0$, we have $\lambda^* = O_P(n^{-1/2} r_n^{-1})$.

Lemma 3. *Assume that conditions A1 through A6 hold.*

a) If δ_n stays away from zero, we have $\hat{\lambda}^* - \lambda^* = O_P(r_n)$.
b) If $\delta_n = 0$, we have $\hat{\lambda}^* - \lambda^* = O_P(n^{-1/2} r_n^{-1})$.

The following Lemma is needed to prove the results in Section 4.

Lemma 4. *Assume that δ_n tends to zero as n tends to infinity.*

a) Under assumptions A1, A2 and A5, we have

$$\|\hat{f}-\hat{g}\| = \begin{cases} O_P(\delta_n) \text{ if } \delta_n r_n^{-1} \text{ stays above 1} \\ O_P(r_n) \text{ if } \delta_n r_n^{-1} \text{ stays below 1 .} \end{cases}$$

b) Under assumptions A1, A2 and A5, we have

$$\lambda^* = \begin{cases} O_P(r_n \delta_n^{-1}) & \text{if } \delta_n r_n^{-1} \text{ stays above 1} \\ 1 + O_P(\delta_n r_n^{-1}) & \text{if } \delta_n r_n^{-1} \text{ stays between } n^{-1/2} r_n^{-1} \text{ and 1} \\ 1 + O_P(n^{-1/2} r_n^{-1}) & \text{if } \delta_n r_n^{-1} \text{ stays below } n^{-1/2} r_n^{-1} . \end{cases}$$

c) Under assumptions A1 through A6, we have

$$\hat{\lambda}^* - \lambda^* = \begin{cases} O_P(n^{-1/2}\delta_n^{-1}) + O_P(r_n^2 \delta_n^{-1}) & \text{if } \delta_n r_n^{-1} \text{ stays above 1} \\ O_P(n^{-1/2} r_n^{-1}) + O_P(\delta_n) & \text{if } \delta_n r_n^{-1} \text{ stays below 1 .} \end{cases}$$

Appendix A. The proofs

Throughout this section, generic constants will be denoted by c. For notational simplicity, we will write $f(\beta^*, \cdot)$ as f and $f(\hat{\beta}^{(i)}, \cdot)$ as $\hat{f}^{\,(i)}$.

Proof of Theorem 1. Note that in view of assumption A1

$$\begin{aligned} \|\lambda^* \hat{f} + (1-\lambda^*)\hat{g} - \theta\| &\leq |\lambda^*| \|\hat{f} - \theta\| + |1-\lambda^*| \|\hat{g} - \theta\| \\ &\leq |\lambda^*| \{\|\hat{f} - f\| + \|f - \theta\|\} + |1-\lambda^*| \|\hat{g} - \theta\| \\ &= |\lambda^*| \{O_P(n^{-1/2}) + \delta_n\} + |1-\lambda^*| O_P(r_n). \quad \text{(A.1)} \end{aligned}$$

Both parts of the Theorem now follow from Lemma 2. □

Proof of Theorem 2. Observe that

$$\| \hat{\lambda}^* \hat{f} + (1-\hat{\lambda}^*)\hat{g} - \theta \|^2 - \| \lambda^* \hat{f} + (1-\lambda^*)\hat{g} - \theta \|^2 = (\hat{\lambda}^* - \lambda^*)^2 \| \hat{f} - \hat{g} \|^2 . \quad \text{(A.2)}$$

Both parts of the Theorem now follow by applications of Lemmas 1 and 3. □

Proof of Theorem 3. The Theorem follows from inequality (A.1) by an application of Lemma 4. □

Proof of Theorem 4. The Theorem follows from equation (A.2) by an application of Lemma 4. □

Proof of Lemma 1. Note that

$$\begin{aligned}
&\| \hat{f} - \hat{g} \|^2 = \| (\hat{f} - f) - (\hat{g} - \theta) + (f - \theta) \|^2 \\
&= \| \hat{f} - f \|^2 + \| \hat{g} - \theta \|^2 + \| f - \theta \|^2 - 2\langle \hat{f} - f, \hat{g} - \theta \rangle \\
&+ 2\langle \hat{f} - f, f - \theta \rangle - 2\langle \hat{g} - \theta, f - \theta \rangle \\
&= I_1 + I_2 + I_3 - 2I_4 + 2I_5 - 2I_6 \text{ (say)}. \qquad \text{(A.3)}
\end{aligned}$$

It is quite clear from A1 that $I_1 = O_P(1)$ irrespective of the behavior of δ_n, and $I_1 = O_P(n^{-1})$ if $\delta_n = 0$. Note that $I_2 = O_P(r_n^2)$. By Cauchy-Schwartz inequality

$$\begin{aligned}
|I_4| &\leq \|\hat{f} - f\| \|\hat{g} - \theta\| = O_P(n^{-1/2} r_n), \\
|I_5| &\leq \|\hat{f} - f\| \|f - \theta\| = O_P(n^{-1/2} \delta_n), \\
|I_6| &\leq \|\hat{g} - \theta\| \|f - \theta\| = O_P(r_n \delta_n) . \qquad \text{(A.4)}
\end{aligned}$$

From the inequalities (A.4) and assumption A5, it is clear that I_2 dominates I_1 and I_4, and I_6 dominates I_5. Consequently,

$$\|\hat{f} - \hat{g}\|^2 = O_P(r_n^2) + \|f - \theta\|^2 + O_P(r_n)\|f - \theta\| . \qquad \text{(A.5)}$$

It is quite easy to see that both parts of the Lemma now follow from equation (A.5). □

Proof of Lemma 2. Let us recall that $\lambda^* = \langle \hat{g} - \theta, \hat{g} - \hat{f} \rangle \|\hat{f} - \hat{g}\|^{-2}$.

a) Note that

$$\begin{aligned}
\left|\langle \hat{g} - \theta, \hat{g} - \hat{f} \rangle\right| &\leq |\langle \hat{g} - \theta, \theta - f \rangle| + \|\hat{g} - \theta\|^2 + \left|\langle \hat{g} - \theta, \hat{f} - f \rangle\right| \\
&\leq \|\hat{g} - \theta\| \|\theta - f\| + \|\hat{g} - \theta\|^2 + \|\hat{g} - \theta\| \|\hat{f} - f\| \\
&= O_P(r_n)\delta_n + O_P(r_n^2) + O_P(r_n n^{-1/2}) \\
&= O_P(r_n)\delta_n + O_P(r_n^2) . \qquad \text{(A.6)}
\end{aligned}$$

If δ_n stays away from zero, it is clear from (A.6) that $\langle \hat{g} - \theta, \hat{g} - \hat{f} \rangle = O_P(r_n)$. Then by an application of part (a) of Lemma 1, we conclude that $\lambda^* = O_P(r_n)$.

b) To prove this part, we begin by observing that

$$\begin{aligned}
|1 - \lambda^*| &= \left|\langle \hat{f} - \theta, \hat{g} - \hat{f} \rangle\right| \|\hat{f} - \hat{g}\|^{-2} \\
&\leq \|\hat{f} - \theta\| \|\hat{g} - \hat{f}\| \|\hat{f} - \hat{g}\|^{-2} \\
&\leq \{\|\hat{f} - f\| + \|f - \theta\|\} \|\hat{f} - \hat{g}\|^{-1} \\
&= \{O_P(n^{-1/2}) + \delta_n\} \|\hat{f} - \hat{g}\|^{-1}. \qquad \text{(A.7)}
\end{aligned}$$

If $\delta_n = 0$, using part (b) of Lemma 6.1, we get $1 - \lambda^* = O_P(r_n^{-1} n^{-1/2})$.

□

Proof of Lemma 3. First note that

$$\lambda^* - \hat{\lambda}^*$$
$$= \left[n^{-1} \sum_{i=1}^{n} \{Y_i - \hat{f}(x_i)\}\{\hat{g}^{(i)}(x_i) - \hat{f}^{\,(i)}(x_i)\} \,+\, \langle \hat{f} - \hat{g}, \theta - \hat{f} \rangle \right] \|\hat{f} - \hat{g}\|^{-2}. \tag{A.8}$$

We will show below that

$$\lambda^* - \hat{\lambda}^* \;=\; \|\hat{f} - \hat{g}\|^{-2} \{O_P(r_n n^{-1/2}) \,+\, O_P(n^{-1/2}\delta_n) \,+\, O_P(r_n^2 \delta_n)\}. \tag{A.9}$$

It is fairly easy to see that parts (a) and (b) of this lemma follow from (A.9) by using Lemma 1.

Let us now show that the expression in (A.9) is valid. The expression within [:] on the right hand side of (A.8) can be rewritten as

$$n^{-1} \sum_{i=1}^{n} \epsilon_i \{\hat{g}^{(i)}(x_i) - \hat{f}^{\,(i)}(x_i)\} \,+\, n^{-1} \sum_{i=1}^{n} \{\theta(x_i) - \hat{f}(x_i)\}\{\hat{g}^{(i)}(x_i) - \hat{g}(x_i)\}$$
$$-\; n^{-1} \sum_{i=1}^{n} \{\theta(x_i) - \hat{f}(x_i)\}\{\hat{f}^{\,(i)}(x_i) - \hat{f}(x_i)\}$$
$$= \; J_1 \,+\, J_2 \,-\, J_3 \; (say). \tag{A.10}$$

Note that the validity of (A.9) is guaranteed if we are able to show that

$$J_1 = \; O_P(n^{-1/2} r_n) \,+\, O_P(n^{-1/2}\delta_n), \tag{A.11}$$
$$J_2 = O_P(\delta_n r_n^2) \,+\, O_P(n^{-1/2} r_n^2), \text{ and} \tag{A.12}$$
$$J_3 = O_P(\delta_n n^{-1}) \,+\, O_P(n^{-3/2}). \tag{A.13}$$

The rest of the proof is devoted to proving (A.11), (A.12) and (A.13).

Denote $W_2(x_i, x_i)$ by h_i. Then by assumption A6,

$$\hat{g}^{(i)}(x_i) \;=\; \left\{\hat{g}(x_i) - n^{-1} h_i Y_i\right\} (1 - n^{-1} h_i)^{-1}, \tag{A.14}$$

Write $\bar{g}(x_i) = E[\hat{g}(x_i)]$ and $\bar{g}^{(i)}(x_i) = E[\hat{g}^{(i)}(x_i)]$. Then from (A.14) it follows that

$$\hat{g}^{(i)}(x_i) - \hat{g}(x_i) \;=\; n^{-1} h_i \left\{Y_i - \hat{g}(x_i)\right\} (1 - n^{-1} h_i)^{-1}, \text{ and}$$
$$\bar{g}^{(i)}(x_i) - \bar{g}(x_i) \;=\; n^{-1} h_i \left\{\theta(x_i) - \bar{g}(x_i)\right\} (1 - n^{-1} h_i)^{-1}. \tag{A.15}$$

Now,

$$\begin{aligned}J_1 &= n^{-1}\sum_{i=1}^{n}\epsilon_i\{\hat{g}^{(i)}(x_i)-\bar{g}^{(i)}(x_i)\} + n^{-1}\sum_{i=1}^{n}\epsilon_i\{\hat{f}^{\,(i)}(x_i)-f(x_i)\}\\ &\quad + n^{-1}\sum_{i=1}^{n}\epsilon_i\{\bar{g}^{(i)}(x_i)-\bar{g}(x_i)\} + n^{-1}\sum_{i=1}^{n}\epsilon_i\,\{\bar{g}(x_i)-\theta(x_i)\}\\ &\qquad + n^{-1}\sum_{i=1}^{n}\epsilon_i\,\{\theta(x_i)-f(x_i)\}\\ &= J_{11} + J_{12} + J_{13} + J_{14} + J_{15}\ (say).\end{aligned}$$

We will show that (A.11) is true by proving that each of the five terms J_{11} through J_{15} is either $O_P(n^{-1/2}r_n)$ or $O_P(n^{-1/2}\delta_n)$.

Note that

$$\hat{g}^{(i)}(x_i)-\bar{g}^{(i)}(x_i) = n^{-1}(1-n^{-1}h_i)^{-1}\sum_{j:j\neq i}W_2(x_i,x_j)\epsilon_j.$$

From this it is easy to see that

$$J_{11} = n^{-2}\sum\sum_{i,j::\ i\neq j}W_2(x_i,x_j)\epsilon_i\epsilon_j(1-n^{-1}h_i)^{-1}.$$

After observing that $E(J_{11})=0$, an application of Whittle's inequality [see Whittle(1960)] together with assumption A4 will yield

$$E(J_{11}^2) \le cn^{-4}\sum\sum_{i,j::\ i\neq j}W_2{}^2(x_i,x_j)(1-n^{-1}h_i)^{-2} \le c\tau_n n^{-2} = O(n^{-1}r_n^2)\,.$$

This proves that $J_{11} = O_P(n^{-1/2}r_n)$. It is quite easy to see using A1 that $J_{12} = O_P(n^{-1})$.

Using (A.15) we get

$$\begin{aligned}E(J_{13}^2) &\le cn^{-2}\sum_{i=1}^{n}\{\bar{g}^{(i)}(x_i)-\bar{g}(x_i)\}^2 \le cn^{-4}\tau_n^2\sum_{i=1}^{n}\{\theta(x_i)-\bar{g}(x_i)\}^2\\ &\le cn^{-3}\tau_n^2 r_n^2 \le cn^{-1}r_n^6.\end{aligned}$$

Hence $J_{13} = O_P(n^{-1/2}r_n^3) = o(n^{-1/2}r_n)$.

The fact that $J_{14} = O_P(n^{-1/2}r_n)$ is a consequence of the following inequality

$$E(J_{14}^2) \le cn^{-2}\sum_{i=1}^{n}\{\theta(x_i)-\bar{g}(x_i)\}^2 \le cn^{-1}r_n^2.$$

Similarly, the fact that $J_{15} = O_P(n^{-1/2}\delta_n)$ follows from the inequality

$$E(J_{15}^2) \le cn^{-2}\sum_{i=1}^{n}\{\theta(x_i) - f(x_i)\}^2 = cn^{-1}\delta_n^2.$$

Next, let us handle the term J_2. Using the Cauchy–Schwartz inequality and (A.14), we get

$$\begin{aligned} |J_2|^2 &\le \|\theta - \hat{f}\|^2 \left[n^{-1}\sum_{i=1}^{n}\{\hat{g}^{(i)}(x_i) - \hat{g}(x_i)\}^2 \right] \\ &\le c\left\{\|\theta - f\|^2 + \|f - \hat{f}\|^2\right\}\left[\tau_n^2 n^{-3}\sum_{i=1}^{n}\{Y_i{}^2 + \hat{g}^2(x_i)\}\right]. \end{aligned}$$

It is easy to see that $n^{-1}\sum_{i=1}^{n} Y_i{}^2 = O_P(1)$ and $n^{-1}\sum_{i=1}^{n}\hat{g}^2(x_i) = O_P(1)$. Consequently, $|J_2|^2 = \left[\delta_n^2 + O_P(n^{-1})\right] O_P(r_n^4)$ and hence we conclude that $J_2 = O_P(\delta_n r_n^2) + O_P(n^{-1/2}r_n^2)$.

The argument used in handling the term J_2 can be used to prove that $J_3 = O_P(\delta_n n^{-1}) + O_P(n^{-3/2})$. This completes the proof of this lemma. □

Proof of Lemma 4. Part (a) of this Lemma follows from (A.5). Part (b) is a consequence of (A.6) and (A.7). Finally, part (c) follows from (A.9). □

Acknowledgments

Supported in part by NSF grant DMS 09-07622 and NSA grant H98230-04-1-0109, and by a grant from The Indian Statistical Institute. The authors wish to thank a referee for helpful comments and suggestions which have led to an improvement in the presentation of this paper.

References

Aerts, M., Claeskens, G. and Hart, J. D. (2000). Testing lack of fit in multiple regression. *Biometrika* **87**, 405–424.

Akaike, H. (1969). Fitting autoregressive models for prediction. *Ann. Instit. Statist. Math* **21**, 243–247.

Akaike, H. (1970). Statistical predictor identification. *Ann. Instit. Statist. Math* **22**, 203–217.

Bache, C. A., Serum, J. W., Youngs, W. D. and Lisk, D. J. (1972). Polychlorinated biphenyl residues: accumulation in Cayuga Lake trout wih age. *Science* **117**, 1192–1193.

Bates, D. M. and Watts, D. G. (1988). *Nonlinear Regression Analysis and Its Applications.* Wiley: New York.

Breiman, L. (1995). Stacked regressions. *Machine Learning* **24**, 49–64.

Buja, A., Hastie, T. and Tibshirani, R. (1989). Linear smoothers and additive models (with discussion). *Ann. Statist.* **17**, 453–555.

Burman, P. and Chen, K. W. (1989). Nonparametric estimation of a regression function. *Ann. Statist.* **17**, 1567–1596.

Claeskens, G. and Hjort, N. L. (2004) Goodness of fit via non-parametric likelihood ratios. *Scand. J. Statist.* **31**, 487–513.

Clark, R. M. (1975). A calibration curve for radiocarbon data. *Antiquity* **49**, 251–266.

Engel, E. (1895). Die produktions – und consumptionsverhaltnisse des konigsreichs sachsen. *Abdruk in Bulletin de l'Institut International de Statistique* **9**, 1–54.

Eubank, R. L. (1988). *Spline Smoothing and Nonparametric Regression.* New York : Mercel Dekker.

Eubank, R. L. and Hart, J. D. (1992). Testing goodness-of-fit in regression via order selection criteria. *Ann. Statist.* **20**, 1412–1425.

Fan, J. and Gijbels, I. (1996). *Local Polynomial Modeling and Its Applications.* Chapman and Hall : London.

Fan, Y. and Ullah, A. (1998). Asymptotic normality of a combined regression estimator. Unpublished manuscript.

Gallant, A. R. (1987). *Nonlinear Statistical Models.* Wiley: New York.

Glad, I. K. (1998). Nonparametric correction of parametric regression estimates. *Scand. J. Statist.* **25**, 649–668.

Hall, P. (1989). On convergence rates in nonparametric problems. *Inter. Statist. Rev.* **57**, 45–58.

Hardle, W. (1990). *Applied Nonparametric Regression.* Cambridge University Press.

Hardle, W. and Mammen, E. (1993). Comparing nonprametric versus parametric regression fits. *Ann. Statist.* **21**, 1926–1947.

Hardle, W. and Marron, J. S. (1985). Optimal bandwidth selection in nonparametric regression function estimation. *Ann. Statist* **13**, 1465–1482.

Hjort, N. L. and Glad, I. K. (1995). Nonparametric density estimation with a parametric start. *Ann. Statist.* **23**, 882–904.

James, W. and Stein, C. (1961). Estimation with quadratic loss. *Proceedings of the Fourth Berkeley Symposium on Mathematical Statistics and Probability* **1**, 361–379. University of California, Berkeley.

Jenrich, R. J. (1969). Asymptotic properties of nonlinear least squares estimators. *Ann. Math. Statist.* **40**, 633–643.

LeBlanc, M. and Tibshirani, R. (1996). Combining estimates in regression and classification. *J. Amer. Statist. Assoc.* **91**, 1641–1650.

Leonard, T. (1977). A Bayesian approach to some multinomial estimation and pretesting problems. *J. Amer. Statist. Assoc* **72**, 869–874.

Mallows, C. L. (1973). Some comments on C_P. *Technometrics* **15**, 661–675.

Müller, H. G. (1988). *Nonparametric Regression Analysis of Longitudinal Data.*

Lecture notes in Statistics 16. New York: Springer-Verlag.

Olkin, I. and Spiegelman, C. (1987). A semiparametric approach to density estimation. *J. Amer. Statist. Assoc* **82**, 858–865.

Rosenblatt, M. (1990). *Stochastic Curve Estimation.* NSF-CBMS Regional Conference Series in Probability and Statistics. Volume 3.

Seber, G. A. F. and Wild, C. J. (1989). *Nonlinear Regression.* New York : John Wiley.

Silverman, B. W. (1985). Some aspects of the spline smoothing approach to nonparametric regression curve fitting (with discussion). *J. Roy. Statist. Soc., Ser B.* **47**, 1–52.

Stein, C. (1956). The inadmissibility of the usual estimator for the mean of a multivariate normal distribution. *Proceedings of the Third Berkeley Symposium on Mathematical Statistics and Probability* **1**, 197–206. University of California, Berkeley.

Stone, C. (1977). Consistent nonparametric regression(with discussion). *Ann. Statist.* **5**, 549–645.

Stone, C. (1980). Optimal rates of convergence of nonparametric estimators. *Ann. Statist.* **8**, 1348–1360.

Stone, C. (1982). Optimal global rates of convergence for nonparametric regression. *Ann. Statist.* **10**, 1040–1053.

Stone, M. (1974). Cross-validatory choice and assessment of statistical predictions. *J. Roy. Statist. Soc., Ser B* **36**, 111–133.

Ullah, A. and Vinod, H. D. (1993). General nonparametric regression estimation and testing in econometrics. In Handbook of Statistics (Eds. Maddala, G. S., Rao, C. R. and Vinod, H. D.). Elsevier Science Publisher: New York.

Wand, M. P. and Jones, M. C. (1995). *Kernel Smoothing.* Chapman and Hall : London.

Wahba, G. (1990). *Spline Models for Observational Data.* CBMS-NSF regional conference series in applied mathematics. Philadelphia: SIAM.

Whittle, P. (1960). Bounds for the moments of linear and quadratic forms in independent variables. *Theory Probab. Appl.* **3**, 302–305.

Wolpert, D. (1992). Stacked generalization. *Neural Networks* **5**, 241–259.

Wooldridge, J. M. (1992). A test for functional form against nonparametric alternatives. *Econometric Theory* **8**, 452–475.

Wu, C. F. J. (1981). Asymptotic theory of nonlinear least squares estimation. *Ann. Statist.* **9**, 501–513.

Chapter 13

Nonparametric Regression Models with Integrated Covariates

Zongwu Cai

University of North Carolina at Charlotte,
Wang Yanan Institute for Studies in Economics,
MOE Key Laboratory of Econometrics,
Fujian Key Laboratory of Statistical Sciences, Xiamen University

This paper studies nonparametric estimation of a regression function with integrated covariates. Both the local linear and local constant methods are developed to estimate the unknown regression function. Consistency and the asymptotic distribution of the resulting estimators are obtained, showing different convergence rates for the stationary and integrated covariates. The new findings are as follows. Asymptotic bias is the same as that for the case when covariates are stationary but the convergence rate differs, and the asymptotic distribution is a mixed normal, associated with the local time of a standard Brownian motion. Asymptotic results at boundaries are also investigated. Finally, an empirical application of real data is presented to examine if there exists a nonlinear relationship between the interest rate differential between U.S. and Canada and the exchange rate between the two nations and to demonstrate the potential of the proposed modeling procedure.

Contents

1. Introduction

Nonparametric estimation techniques have become cornerstone research topics in statistics for the last three decades. They offer numerous advantages relative to parametric techniques and have more flexibility and robustness to functional form misspecification, and have been embraced by applied researchers in many fields; see the books by Fan and Gijbels (1996) and Fan and Yao (2003). Asymptotic theory underlying various nonparametric estimators and test statistics for many commonly used models have been well established for independent and identically distributed (iid) data and some weak and strong dependent time series. The only nonparametric asymptotic analysis when covariates are integrated or unit root (denoted by I(1)) time series that I am aware of includes the papers, to name just a few, by Phillips and Park (1998), Park and Hahn (1999), Chang and Martinez-Chombo (2003), Chang and Park (2003), Juhl (2005), Cai, et al. (2009), Xiao (2009), Wang and Phillips (2009a; 2009b), and Phillips (2009). Particularly, Phillips and Park (1998), Juhl (2005), Phillips (2009), and Wang and Phillips (2009a; 2009b) considered the case when the true data generating process is a linear unit root process, while Park and Hahn (1999), Chang and Martinez-Chombo (2003), Chang and Park (2003), Cai, et al. (2009), Xiao (2009), Cai and Wang (2010) and Sun, et al. (2010) studied the models linearized in the nonstationary variables.

In this paper, for the observed data $\{(Y_t, Z_t)\}$ for $t = 1, \ldots, n$, I study a nonparametric regression function with an integrated covariate as follows,

$$Y_t = \beta(Z_t) + \varepsilon_t, \tag{1}$$

where $E(\varepsilon_t|Z_t) = 0$, $\{\varepsilon_t\}$ is stationary and $\beta(\cdot)$ is an unknown regression function. Here Z_t is an integrated process satisfying

$$Z_t = \rho\, Z_{t-1} + u_t, \tag{2}$$

where $\rho = 1$ and $\{u_t\}$ is a stationary sequence. Clearly, Z_t is persistent and nonstationary. Indeed, model (1) is not new in literature but its asymptotics developed in the present paper is novel when Z_t is persistent and nonstationary. For example, if Z_t is stationary, model (1) has been studied extensively in the literature; see Fan and Gijbels (1996) and Fan and Yao (2003) for details, while it was investigated by Karlsen and Tjøstheim (2001) for Z_t being a null recurrent time series and Karlsen, et al. (2007) for the ϕ-irreducible Markov chain time series and by Bandi (2002) and Cai, et al. (2010) for nearly integrated time series (ρ in (2) is assumed to be $\rho = 1 + c/n$ with $c < 0$). A functional coefficient type model with I(1) covariates and nonlinear cointegration is investigated by Cai, et al. (2009), Xiao (2009), Cai and Wang (2010), and Sun, et al. (2010), respectively. Finally, note that Wang and Phillips (2009a; 2009b) considered the case which allows $E(\varepsilon_t|Z_t) \neq 0$ and Cai, et al. (2010) extended the situation to allow $\{u_t\}$ in (2) to be a long memory process and ρ to be nearly one. For simplicity of notation, I consider only the one-dimensional case since extension to multivariate Z_t involves fundamentally no new ideas but some quite complicated notation.

Model (1) might have a great potential in many applications. For example, in macroeconomics, a particular parametric form of (1) can be used for forecasting inflation rate based on some persistent and nonstationary covariates such as velocity of monetary supply; see Bachmeier, et al. (2006), which showed that the velocity is an I(1) process. Also, using a semiparametric regression model with integrated covariates, Sun, et al. (2010) considered the purchasing power parity hypothesis using Canadian and U.S. price and exchange rate data. Indeed, they showed that the difference between the two countries' 10-year Treasury bond rates is an I(1) process. Finally, it can be employed for the predictability of stock returns using various lagged financial variables, such as the dividend yield, term and default premia, the dividend-price ratio, the earning-price ratio, the book-to-market ratio, and interest rates; see Elliott and Stock (1994), Cavanagh, et al. (1995), Bandi (2002), Torous, et al. (2004), Campbell and Yogo (2006), Polk, et al. (2006), Rossi (2007), Cai and Wang (2010; 2011), and Cai, et al. (2010), among others. In fact, Campbell and Yogo (2006) showed that the 95% confidence intervals for ρ in (2) are $[0.957, 1.007]$ and $[0.939, 1.000]$ for the log dividend-price ratio and the log earnings-price ratio, respectively; see Panel A in Table 4 of Campbell and Yogo (2006). As advocated by Campbell and Yogo (2006), Bachmeier, et al. (2006), and Cai, et al. (2009), the predictive power of using integrated or nearly integrated (highly persistent) covariates in a regression model can be improved significantly due to less noise.

The main purpose of the present paper is to estimate the nonparametric regression $\beta(\cdot)$ by using the local linear (polynomial) and local constant (Nadaraya–Watson) fitting schemes. The main contribution of the present paper to the literature is to derive the asymptotic theory for both estimators. For simplicity, the main results can be summarized as follows. First, the optimal rate of convergence is $n^{1/5}$, slower than the usual $n^{2/5}$ rate for stationary case. Consequently, the order of the asymptotic mean-squared error (AMSE) is $n^{-2/5}$ rather than the standard rate $n^{-4/5}$. The intuitive explanation of this phenomenon is that an I(1) time series takes longer to revisit levels in its range. Secondly, the asymptotic bias term, similar to the stationary case, is independent of the stationary density of the regressor and is due to the linear approximation, which is typical for a local linear fitting scheme; see, for example, Fan and Gijbels (1996) for details. Thirdly, the limiting distribution is a mixed-normal (conditional normal) with the asymptotic variance depending inversely on the local time of a Brownian motion in which the unit root series can be embedded. Furthermore, the integrated covariate requires the larger bandwidths. Indeed, the optimal (in the AMSE sense) bandwidth is $O_p(n^{-1/10})$ implying a larger optimal bandwidth than in conventional kernel regressions with stationary regressors where the optimal bandwidth is known to be $O(n^{-1/5})$. Clearly, the use of conventional bandwidth has the theoretical potential of under-smoothing in the presence of I(1) covariates. Finally, it is very interesting that both local linear and local constant estimators share exactly same asymptotic properties at both interior and boundary points.

The remainder of the paper is organized as follows. Section 2 is devoted to presenting nonparametric kernel estimators of $\beta(\cdot)$ using both local linear and Nadaraya–Watson (local constant) estimation methods. Their asymptotic behaviors for both interior and boundary points are obtained and our results are compared with conventional findings. In Section 3, I illustrate the finite sample performance of the estimators with an application to real data for showing that there exists a nonlinear relationship between the interest rate differential between U.S. and Canada and the exchange rate between two nations. Concluding remarks are presented in Section 4. Finally, proofs of the main results of the paper are relegated to the Appendix.

2. Statistical Properties

2.1. *Local Linear Estimation*

$\beta(\cdot)$ is estimated using local linear fitting from the observations $\{(Y_t, Z_t)\}_{t=1}^n$. The motivation for using local linear fitting is its high statistical efficiency in an asymptotic minimax sense, design adaptation and automatic correction for edge effects, as discussed in Fan and Gijbels (1996). Although a general local polynomial technique is applicable as well, it is well known that the local linear fitting will suffice for many applications;

see Fan and Gijbels (1996) for a very comprehensive discussion. The theory developed for the local linear estimator continues to hold for the local polynomial estimator with only slight modification. Another virtue of using local polynomials is that both the unknown functions as well as their derivatives can be estimated simultaneously. For simplicity, my focus is only on local linear estimation and leave the generalization for additional research.

It is assumed throughout the paper that $\beta(\cdot)$ is twice continuously differentiable, so that at any given z, a local approximation is used as $\beta(Z_t) \simeq \beta(z) + \beta'(z)\,(Z_t - z)$, when Z_t is a neighborhood of z, where $\simeq$ denotes the first order Taylor approximation and $\beta'(z)$ is the first derivative of $\beta(z)$. Hence, (1) is approximated by

$$Y_t \simeq \theta_0 + (Z_t - z)\,\theta_1 + \varepsilon_t,$$

and it becomes a local linear model. Therefore, the locally weighted sum of squares is

$$\sum_{t=1}^{n} \left[Y_t - \theta_0 - (Z_t - z)\,\theta_1\right]^2 K_h(Z_t - z), \tag{3}$$

where $K_h(z) = K(z/h)/h$, $K(\cdot)$ is the kernel function, and $h = h_n > 0$ is the bandwidth satisfying $h \to 0$ and $n\,h \to \infty$ as $n \to \infty$, which controls the amount of smoothing used in the estimation. By minimizing (3) with respect to θ_0 and θ_1, the local linear estimate of $\beta(z)$ is obtained and is denoted by $\hat{\beta}(z)$, and the local linear estimator of the derivative of $\beta(z)$ is denoted by $\hat{\beta}'(z)$. It is easy to show that the minimizer of (3) is given by

$$\begin{pmatrix} \hat{\beta}(z) \\ \hat{\beta}'(z) \end{pmatrix} = \left[\sum_{t=1}^{n} \begin{pmatrix} 1 \\ Z_t - z \end{pmatrix}^{\otimes 2} K_h(Z_t - z)\right]^{-1} \sum_{t=1}^{n} \begin{pmatrix} 1 \\ Z_t - z \end{pmatrix} Y_t\, K_h(Z_t - z), \tag{4}$$

where $A^{\otimes 2} = A\,A^T$ ($A^{\otimes 1} = A$) for a vector or matrix A.

2.2. *Notation and Assumptions*

Since Z_t is an I(1) process, it can be re-expressed as $Z_t = Z_0 + \sum_{s=1}^{t} u_s$, where $\{u_s\}$ is a stationary process with mean zero and variance σ_u^2. In what follows, it is assumed that the process $\{u_t\}$ is a stationary linear process as $u_s = \sum_{j=0}^{\infty} c_j\, \omega_{s-j}$, where ω_j is a white noise with mean zero and $\sigma_\omega^2 = \text{Var}(\omega_j) < \infty$, and $\{c_j\}$ satisfies for some $0 < \tau \le 1$,

$$\sum_{j=0}^{\infty} |c_j|^\tau < \infty, \qquad \text{and} \qquad \sum_{j=0}^{\infty} c_j = 1. \tag{5}$$

Then, $\sigma_u^2 = \text{Var}(u_s) = \sigma_\omega^2 \sum_{j=0}^{\infty} c_j^2$ and $\text{Cov}(u_s, u_{s+t}) = \sigma_\omega^2 \sum_{j=0}^{\infty} c_j\, c_{j+t}$ for any s and t. Note that the assumption on $\{u_t\}$ being a linear process is due to an application of some results from Jeganathan (2004). Of course, it can be relaxed at the cost of involving lengthier mathematical proofs. Clearly, one has

$$Z_t/\sqrt{n} = Z_0/\sqrt{n} + \frac{1}{\sqrt{n}} \sum_{s=1}^{[nr]} u_s$$

for $r = t/n$. An application of Donsker's theorem (see, for example, Theorem 14.1 in Billingsley (1999) for iid $\{u_t\}$ with the existence of the second moment of u_t) leads to

$$Z_t/\sqrt{n} \implies W_u(r), \tag{6}$$

where "$\Longrightarrow$" represents weak convergence, $W_u(\cdot) = \sigma_0\, W(\cdot)$ with $W(\cdot)$ being a standard Brownian motion on $[0,\ 1]$ and $\sigma_0^2 = \lim_{n\to\infty} \text{Var}(n^{-1/2} \sum_{t=1}^{n} u_t)$, which is assumed to exist and be finite. In particular, it follows from Merlevéde, et al. (2006) that (6) holds if $\{u_t\}$ is a stationary strong mixing sequence and satisfies, for some $\delta_0 > 0$,

$$E|u_t|^{2+\delta_0} < \infty, \quad \text{and} \quad \sum_{k=1}^{\infty} k^{(2+\delta_0)/\delta_0}\, \alpha(k) < \infty, \tag{7}$$

where $\alpha(\cdot)$ is the mixing coefficient; see, e.g., Hall and Heyde (1980) for the definition.

Define $\eta_{t,z} = (Z_t - z)/\sqrt{t}$ for any z and let $f_{t,z}(\cdot)$ denote the density of $\eta_{t,z}$. Also, let $f_{t,s,z}(\cdot,\cdot)$ represent the joint density function of $(\eta_{t,z}, \eta_{s,z})$. Furthermore, let $\mathcal{F}_t$ be the smallest sigma field generated by $\{(Y_s,\, Z_s)\}_{s=-\infty}^{t}$. The following assumptions are made.

Assumptions:

(C1) $E(\varepsilon_t|Z_t, \mathcal{F}_{t-1}) = 0$, $E(\varepsilon_t^2|Z_t, \mathcal{F}_{t-1}) = \sigma_\epsilon^2$, $E(\varepsilon_t^4|Z_t, \mathcal{F}_{t-1}) < C$ a.s., and $\{u_t\}$ is a stationary and mixing process satisfying constraints as imposed by (5) and (7), where σ_ϵ^2, C and σ_0^2 are finite positive constants.

(C2) Both $f_{t,z}(\cdot)$ and $f_{t,s,z}(\cdot,\cdot)$ have bounded continuous derivative functions (for all t, s and fixed z).

(C3) $K(\cdot)$ is a kernel function with a finite support, say $[-1,\ 1]$ and it is symmetric.

(C4) $n\,h \to \infty$ and $n\,h^{10} = O(1)$.

Next, I discuss the above conditions. Condition C1 requires that $\{\varepsilon_t\}$ is a martingale difference process with conditional homogenous variance and a finite fourth moment. The martingale difference assumption can be relaxed to a mixing process, and the assumption on the conditional homogenous error can be relaxed to the case that $E(\varepsilon_t^2|Z_t)$ is nonconstant, with a lengthier proof. C2 is a very mild assumption. Indeed, it is satisfied if $\{u_t\}$ is commonly assumed to be iid normal. Finally, Assumptions C3 and C4 are commonly imposed in the kernel estimation literature and Assumption C4 is satisfied for the optimal bandwidth $h = O(n^{-1/10})$ (see later).

Finally, the local time $L(t,x)$ for a standard Brownian motion is defined as

$$L(t,x) = \lim_{\Delta\to 0} \frac{1}{2\Delta} \int_0^t I_{\{|W(s)-x|\le\Delta\}} ds, \quad 0 \le t \le 1, \quad \text{and} \quad x \in \mathcal{R}, \tag{8}$$

where I_A is the indicator function of an event A and $W(\cdot)$ is a standard Brownian motion; see Karatzas and Shreve (1991), p.202, Phillips and Park (1998), and Park and Phillips (1999) for details. Finally, define

$$\mu_j(K) = \int u^j\, K(u)du \quad \text{and} \quad \nu_j(K) = \int u^j\, K^2(u)du.$$

Note that $L(t,z)$ can be consistently estimated by $S_{n,0}(z)$; see Lemma 1 in the appendix.

2.3. *Asymptotic Results*

Now, the main result is stated below and the proof is relegated to the appendix.

Theorem 1.. *Under Assumptions (C1)–(C4), one has*

$$\sqrt{n^{1/2}\,h}\,\left[\hat\beta(z) - \beta(z) - h^2 B(z)\right] \xrightarrow{\ \mathrm{d}\ } MN(\sigma_\beta^2),$$

where $B(z) = \mu_2(K)\beta''(z)/2$ and $MN(\sigma_\beta^2)$ is a mixed normal distribution with mean zero and variance $\sigma_\beta^2 = \sigma_\varepsilon^2\,\sigma_0\,\nu_0(K)\,/L(1,0)$ with σ_ε^2 defined in Assumption C1.

Remark 1. First, ξ_t is called a mixed normal with mean μ_t and covariance Σ_t if the conditional distribution of ξ_t given μ_t and Σ_t is $N(\mu_t, \Sigma_t)$; see Phillips and Park (1998) for details. Note that the asymptotic properties for $\hat{\beta}'(z)$ can be obtained in the same fashion as those in Theorem 1 and is omitted. By comparing the results in Theorem 1 and conventional findings in Fan and Gijbels (1996) and Fan and Yao (2003) for the stationary covariates, our new results can be summarized as follows. Clearly, $h^2\mu_2(K)\beta''(z)/2$ serves as the asymptotic bias, which is the same as that for stationary case when one uses a local linear estimation method; see Fan and Yao (2003). However, the convergence rate is the order of $n^{1/4}h^{1/2}$ much lower with a factor $n^{1/4}$ as compared to that for stationary covariates. Also, the stochastic asymptotic variance is independent of the grid point z. Indeed, one can show that the results in Theorem 1 hold true as long as any $z = z_n$ satisfies $z_n/\sqrt{n} \to 0$ and $n^{1/4}\,h^{5/2}\,\beta''(z_n) = O(1)$; see Theorem 2 below. Furthermore, from the asymptotic bias and variance presented in Theorem 1, the stochastic AMSE is given by

$$\text{AMSE} = \text{Var} + \text{bias}^2 = \sigma_\beta^2\, n^{-1/2}h^{-1} + \frac{h^4}{4}\,\mu_2^2(K)\,[\beta''(z)]^2\,.$$

The minimization of the AMSE with respect to h yields the optimal bandwidth

$$h_{opt} = \left(\frac{\sigma_\beta}{\mu_2(K)|\beta''(z)|}\right)^{2/5} n^{-1/10} = O_p(n^{-1/10}),$$

which is stochastic and much larger than the conventional optimal bandwidth $h_{opt,s} = O(n^{-1/5})$ for the stationary case; see Fan and Yao (2003). Therefore, if $h_{opt,s}$ were to be used in estimating $\beta(\cdot)$ in (1), the nonparametric estimator given in (6) would result in undersmoothing. Hence, it would be a very interesting future research topic to determine how to select the data-driven (optimal) bandwidth theoretically and empirically.

Now, I embark on investigating the asymptotic behaviors at boundaries. When Z_t is $I(1)$, it follows from (6) that when $z = a\sqrt{n}$ $(a \neq 0)$ and $r = t/n$,

$$P(Z_t \geq z) = P(Z_t \geq a\sqrt{n}) \;\to\; P(W_u(r) \geq a/\sigma_u) = 1 - \Phi(a/\sqrt{r}\sigma_0) > 0,$$

where $\Phi(\cdot)$ is the distribution of the standard normal random variable. This means that there is a great chance that $|Z_t|$ can take large values. In other words, an I(1) time series takes longer to revisit levels in its range. Now the question is how the asymptotic behaviors of the estimator look like when z is large like $z = a\sqrt{n}$ for any fixed a. To this end, the following asymptotic results are obtained at the boundary $z = a\sqrt{n}$ for any fixed a. However, the detailed proofs are not provided since they follow closely the same arguments as those used in the proof of Theorem 1.

Theorem 2.. *If Assumptions (C1) – (C4) hold and* $n^{1/4} h^{5/2} \beta''(a\sqrt{n}) = O(1)$ *for any* a*, then, one has*

$$\sqrt{n^{1/2} h}\left[\hat{\beta}(a\sqrt{n}) - \beta(a\sqrt{n}) - h^2 B(a\sqrt{n})\right] \xrightarrow{\text{d}} MN(\sigma_{\text{a}}^2),$$

where $MN(\sigma_{\text{a}}^2)$ *is a mixed normal distribution with mean zero and variance* $\sigma_{\text{a}}^2 = \sigma_\varepsilon^2\, \sigma_0\, \nu_0(K)/L(1,\, a/\sigma_0)$.

Remark 2. Comparing Theorem 2 with Theorem 1, one can observe that the magnitude of the asymptotic variance of $\hat{\beta}(\cdot)$ at the boundary points ($z = O(n^{1/2})$) differs from that for the interior points ($z = o(n^{1/2})$). This finding is different from its stationary counterpart; see Fan and Gijbels (1996) for the stationary case.

2.4. *Nadaraya–Watson Estimation*

I now turn to discussing the asymptotic properties for the local constant estimator of $\beta(\cdot)$. It is well documented that the Nadaraya–Watson estimator is given by

$$\widetilde{\beta}(z) = \sum_{t=1}^{n} Y_t\, K_h(Z_t - z)/\sum_{t=1}^{n} K_h(Z_t - z). \tag{9}$$

For $\widetilde{\beta}(z)$, the following theorem can be established.

Theorem 3.. *Under the assumptions of Theorem 1, both* $\widetilde{\beta}(z)$ *and* $\hat{\beta}(z)$ *share the exact same asymptotic properties. That is,*

$$\sqrt{n^{1/2} h}\left[\widetilde{\beta}(z) - \beta(z) - h^2 B(z)\right] \xrightarrow{\text{d}} MN(\sigma_\beta^2),$$

where $B(z) = \mu_2(K)\beta''(z)/2$ *and* $MN(\sigma_\beta^2)$ *is a mixed normal distribution with mean zero and variance* $\sigma_\beta^2 = \sigma_\varepsilon^2\, \sigma_0\, \nu_0(K)\,/L(1,0)$. *Further, Theorem 2 holds for* $\widetilde{\beta}(z)$.

Remark 3. It is clear that $h^2\mu_2(K)\beta''(z)/2$ serves as the asymptotic bias, which is the same as in the case when one uses a local linear estimation method (see Theorem 1). However, for the stationary Z_t case with a local constant estimation method, there is an additional leading bias term which has the form of $h^2\mu_2(K)f_z'(z)\beta'(z)/2f_z(z)$ where $f_z(\cdot)$ is the stationary density of Z_t when Z_t is stationary; see Fan and Gijbels (1996). Theorem 3 shows that for non-stationary Z_t, the local constant estimator has the same leading bias as that of a local linear method. This is an interesting new finding that is not shared by a local constant estimator if Z_t is stationary. It can be shown that with nonstationary Z_t, the bias term associated with $f_{t,z}'(z)\beta'(z)$ has an order of $h^2n^{-1/2}\ln(n)$, which is smaller than h^2; see Lemma 4 in Appendix. Therefore, the leading bias contains only one term associated with $\beta''(z)$ with the order h^2. Interestingly, as in the case of standard local polynomial methods, the Nadaraya–Watson estimator is *design-adaptive* too in the sense of Fan and Gijbels (1996). Clearly, this property should be interpreted as follows. The clustered designs are not expected to occur in the presence of integrated (highly persistent) processes. Therefore, the theoretical relevance of the design-adaptation property and the theoretical appeal of local polynomial methods over the standard Nadaraya–Watson kernel estimates seem to vanish.

3. An Illustrative Empirical Application

Sun, et al. (2010) investigated the purchasing power parity (PPP) hypothesis using Canadian and U.S. price index and exchange rate data. The PPP theory says that the following setup holds: $s_t = \beta_1 + \beta_1\, p_t + \beta_2\, p_t^* + u_t$, where s_t, p_t, and p_t^* are the logarithm of the nominal exchange rate expressed as Canadian dollars per unit of U.S. dollar, the Canadian and U.S. aggregate price levels, respectively. The aggregate price index is measured by the producer price index (PPI) base-weighted to the year 2000. Sun, et al. (2010) used monthly data for the period from January 1974 to December 2009 so that there are 432 observations.

Sun, et al. (2010) argued that based on the sticky-price theory of exchange rate determination, exchange rate movements also respond to monetary shocks. Due to sticky prices, the goods markets adjust to the monetary shocks slower than asset markets. Hence, in addition to the aggregate price levels, some other economic variables, such as interest rate differentials between two nations, also affect exchange rate formation and adjust more quickly to monetary shocks than the aggregate price indexes do. Therefore, to verify this economic theory, I will examine whether the exchange rate depends on the interest rate differential between U.S. and Canada. Specifically, I denote by $Z_t = T_{US,t} - T_{CN,t}$ the difference between the two countries' 10-year Treasury bond rates. The time series plot of Z_t is given in Figure 1(a) and its autocorrelation function (ACF) plot is displayed in Figure 1(b). Applying the augmented Dickey–Fuller (ADF) test statistic to

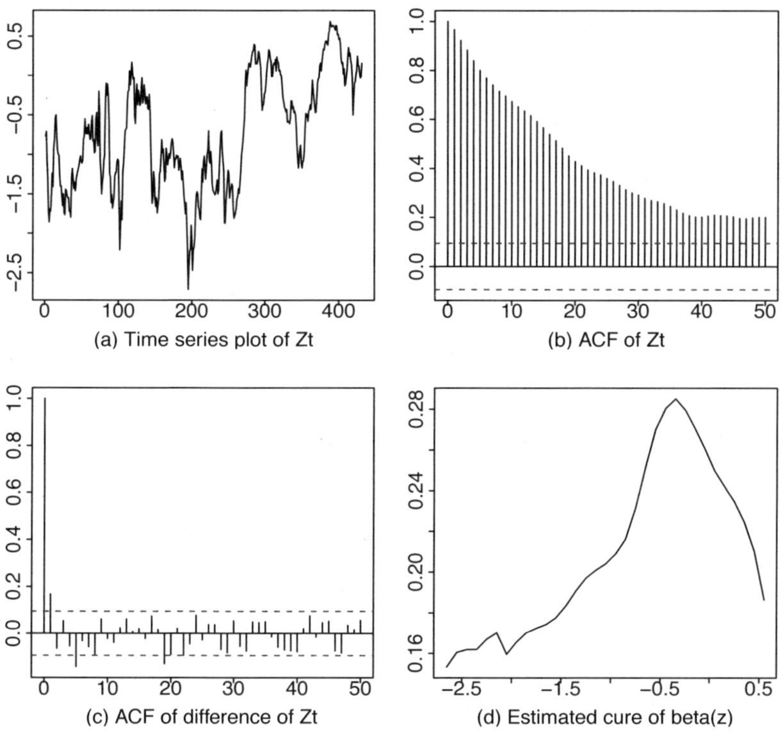

Fig. 1. (a) Time series plot of Z_t; (b) ACF plot of Z_t; (c) ACF plot of $Z_t - Z_{t-1}$; (d) Estimated curve of $\beta(z)$.

the interest rate differential, I cannot reject the null hypothesis $H_0 : \rho = 1$, so that Z_t follows a unit root process at the 5% significance level. Therefore, I treat Z_t as an integrated series. Indeed, one might infer visually from Figures 1(a) and 1(b) that Z_t is a unit root process. Also, one can see from Figure 1(c) that $u_t = Z_t - Z_{t-1}$ is autocorrelated. Indeed, the Ljung–Box test rejects the null hypothesis of no autocorrelation of u_t.

Thus, for simplicity, I consider the following nonparametric model by ignoring the price indices (p_t and p_t^*) for two countries

$$s_t = \beta(Z_t) + \varepsilon_t.$$

The Epanechinikov kernel $K(u) = 0.75(1-u^2)_+$ is used, and the smoothing parameter h is selected by the least squares cross-validation method so that $h = 0.425$. Figure 1(d) depicts the nonparametric estimate of $\beta(z)$. From Figure 1(d), it is very interesting to learn that $\hat{\beta}(z)$ is nonlinear and reaches its max when $z = -0.35$. Further, $\hat{\beta}(z)$ is increasing if $z \leq -0.35$ and then it is decreasing when $z > -0.35$. Also, it is asymmetric and the left side has a longer tail. The reaction of exchange rate to the the 10-year Treasury bond rate differential between the two nations is different based on the differential value. This means that when the U.S. 10-year Treasury bond rate becomes much lower or higher than that for the Canadian bond rate, the exchange rate between two nations becomes lower. In other words, the Canadian dollar is appreciated. Therefore, my analysis confirms that exchange rate between U.S. and Canada depends on the interest rate differential between two nations.

4. Discussion

In this paper, I have studied a nonparametric regression model for integrated time series data. I suggested using the local polynomial and local constant fitting schemes to estimate the nonparametric function and derived the asymptotic properties of the proposed estimators. Our theoretical results show that the asymptotic bias is the same as that for stationary covariates. However, the convergence rate for the nonstationary covariates is slower than that for the stationary covariates by a factor of $n^{-1/4}$. Further, the asymptotic distribution is not normal any more but just a mixed normal associated with the local time of a standard Brownian motion. Moreover, I showed that the asymptotic properties for both the local linear and local constant estimators are exactly same.

I would like to mention some interesting future research topics related to this paper. First, it would be very useful and important to discuss how to select the data-driven (optimal) bandwidth theoretically and empirically. Secondly, the model should allow the errors $\{\varepsilon_t\}$ to be a serially correlated time series, say α-mixing, to be heteroscedastic, and to be correlated with covariates as in Wang and Phillips (2009a; 2009b). Thirdly, the model should include both stationary and nonstationary covariates. Finally, it is warranted to consider some extensions to other types of models like additive models, index models and varying coefficient models, and other types of nonstationarity such as nearly integrated processes; see, e.g., Bandi (2002), Torous, et al. (2004), Campbell and Yogo (2006), Polk, et al. (2006), Rossi (2007), Cai and Wang (2010; 2011), and Cai, et al. (2010), which have a potential application in applied fields like economics and finance.

Appendix A. Proofs

Before I prove the main results of this paper, I first establish three lemmas that will be used frequently in the proofs below. Throughout this section, C denotes a generic positive constant and it may take different values in different contexts.

To prove Theorem 1, define $G_j(u) = u^j\, K(u)$ for any $j \geq 0$. Then, it is easy to verify that $G_j(\cdot)$ is continuous and has a compact support. Also, both $G_j(\cdot)$ and $G_j^2(\cdot)$ are integrable. Also, define $S_n(z)$ as follows

$$S_n(z) = n^{-1/2} \sum_{t=1}^{n} K_h(Z_t - z) \begin{pmatrix} 1 \\ Z_{t,z,h} \end{pmatrix}^{\otimes 2} = \begin{pmatrix} S_{n,0}(z) & S_{n,1}(z) \\ S_{n,1}(z) & S_{n,2}(z) \end{pmatrix},$$

where $Z_{t,z,h} = (Z_t - z)/h$ and $0 \leq j \leq 2$,

$$S_{n,j}(z) = \frac{1}{\sqrt{n}} \sum_{t=1}^{n} K_{j,h}(Z_t - z)$$

with $K_{j,h}(u) = G_j(u/h)/h$. Then, re-express $S_{n,j}(z)$ as

$$S_{n,j}(z) = \frac{\beta_n}{n} \sum_{t=1}^{n} G_j(\beta_n(\gamma_n^{-1}\, Z_t + x_n)),$$

where $\beta_n = \sqrt{n}/h$, $\gamma_n = \sqrt{n}$, and $x_n = -z/\sqrt{n}$. Clearly, $x_n \to 0$ for any fixed z and $x_n = -a$ if $z = a\sqrt{n}$. Finally, let $\phi_\delta(x) = \exp\left(-x^2/2\,\delta^2\right)/\sqrt{2\,\pi\,\delta^2}$ for any $\delta > 0$ and $o_{L_2}(1)$ denote the convergence in L_2. Before proving the theorem, I first present some preliminary results. In what follows, it is assumed that Z_t satisfies (5).

Lemma 1. *Under assumption that the density of $\eta_{t,z}$ is bounded for all t,*

$$(i) \quad S_{n,j}(z) \xrightarrow{\text{P}} \begin{cases} \mu_j(K)\, L(1,0)/\sigma_0, & \text{if } z \text{ is fixed}, \\ \mu_j(K)\, L(1, a/\sigma_0)/\sigma_0, & \text{if } z = a\sqrt{n}, \end{cases}$$

and for any $p > 0$ and z,

$$(ii)\ E\left[S_{n,j}(z)\right] = O(1), \quad \text{and} \quad (iii)\ E\left[|K_{j,h}(Z_t - z)|^p\right] = O(t^{-1/2}\, h^{1-p}).$$

Note that the above results still hold if fixed z is changed to be any z_n satisfying $z_n/\sqrt{n} \to 0$.

Proof. To establish the first assertion, I use some results from Jeganathan (2004). Indeed, by Proposition 6 and Lemma 7 of Jeganathan (2004), for each $\delta > 0$,

$$S_{n,j}(z) = \frac{\mu_j(K)}{n} \sum_{t=1}^{n} \phi_\delta(\gamma_n^{-1} Z_t + x_n) + o_{L_2}(1).$$

Since $\phi_\delta(z)$ satisfies the Lipschitz condition and $x_n \to 0$,

$$S_{n,j}(z) = \frac{\mu_j(K)}{n} \sum_{t=1}^{n} \phi_\delta(\gamma_n^{-1} Z_t) + o_{L_2}(1) = \frac{\mu_j(K)}{n} \sum_{t=1}^{n} \phi_\delta(W_u(t/n)) + o_{L_2}(1)$$

in view of (6) and (7). By Lemma 9 of Jeganathan (2004), one has

$$S_{n,j}(z) = \mu_j(K) \int_0^1 \phi_\delta(W_u(s))ds + o_{L_2}(1).$$

An application of Proposition 11 of Jeganathan (2004) gives

$$S_{n,j}(z) = \mu_j(K)\, L(1,0)/\sigma_0 + o_{L_2}(1)$$

as $\delta \downarrow 0$. By the same token, it is easy to show the case of $x_n = -a$ ($z = a\sqrt{n}$). For assertion (ii), one has

$$\begin{aligned} E\left[S_{n,j}(z)\right] &= n^{-1/2} \sum_{t=1}^{n} E\left[K_{j,h}(Z_t - z)\right] \\ &= n^{-1/2} h^{-1} \sum_{t=1}^{n} \int G_j(t^{1/2}u/h)\, f_{t,z}(u)du \\ &= n^{-1/2} \sum_{t=1}^{n} t^{-1/2} \int G_j(v)\, f_{t,z}(ht^{-1/2}v)dv \\ &\le C\, n^{-1/2} \sum_{t=1}^{n} t^{-1/2} = O(1). \end{aligned}$$

Finally, recall that $K_{j,h}(u) = h^{-1}G_j(u/h)$ and $G_j(u) = u^j K(u)$. It can be shown easily by the boundedness of $f_{t,z}(\cdot)$ that

$$E\left[|K_{j,h}(Z_t - z)|^p\right] = h^{-p} \int |K_{j,h}(t^{1/2}u/h)|^p f_t(u)du$$

$$= t^{-1/2}h^{1-p} \int |G_j(v)|^p f_{t,z}(t^{-1/2}hv)dv \le Ct^{-1/2}h^{1-p}.$$

This proves the lemma. □

By Lemma 1, one has

$$S_n(z) = \begin{pmatrix} S_{n,0}(z) \; S_{n,1}(z) \\ S_{n,1}(z) \; S_{n,2}(z) \end{pmatrix} = \begin{pmatrix} 1 & 0 \\ 0 \; h^2 & \mu_2(K) \end{pmatrix} L(1,0)/\sigma_0 \{1 + o_p(1)\},$$

which, by replacing Y_t in (4) by $Y_t = \beta(Z_t) + \varepsilon_t$, implies that

$$\hat{\beta}(z) - \beta(z) \equiv [L(1,0)/\sigma_0]^{-1} \{B_n + C_n\} \{1 + o_p(1)\}, \qquad \text{(A.1)}$$

where $B_n = n^{-1/2} \sum_{t=1}^{n} [\beta(Z_t) - \beta(z) - \beta'(z)(Z_t - z)] K_h(Z_t - z)$ and $C_n = n^{-1/2} \sum_{t=1}^{n} \varepsilon_t \; K_h(Z_t - z)$. I analyze B_n and C_n in Lemma B.2 and Lemma B.3 below.

Lemma 2. *Under Assumptions given in Theorem 1, then,*

$$B_n = h^2 B(z)[S_{n,2}(z)] + o_p(h^2) = h^2 B(z) L(1,0)/\sigma_0 + o_p(h^2).$$

Proof. Note that the proof is as in that for Lemma 1. Similar to Lemma 1, one can show that

$$\begin{aligned} B_n &= n^{-1/2} \sum_{t=1}^{n} [\beta(Z_t) - \beta(z) - \beta'(z)(Z_t - z)] \, K_h(Z_t - z) \\ &= \frac{h^2}{2} \beta''(z) \, S_{n,2}(z)\{1 + o_p(1)\} \\ &= \frac{h^2}{2} \; L(1,0) \, \beta''(z) \, \mu_2(K)/\sigma_0\{1 + o_p(1)\}. \end{aligned}$$

This completes the proof of Lemma 2. □

Lemma 3. *Under Assumptions given in Theorem 1, then,*

$$n^{3/4} h^{1/2} C_n \;\xrightarrow{\text{d}}\; MN(\sigma_1^2),$$

where $MN(\sigma_1^2)$ is a mixed normal with mean zero and covariance matrix $\sigma_1^2 = \sigma_\varepsilon^2 \nu_0(K) \; L(1,0)/\sigma_0$.

Proof. Clearly, $E[C_n] = 0$ since $E(\varepsilon_t | Z_t) = 0$. Also, by the assumptions that $\{\varepsilon_t\}$ is a martingale difference and $E(\varepsilon_t^2 | Z_t) = \sigma_\varepsilon^2$, one can conclude that the conditional variance of $n^{1/4} h^{1/2} C_n$, given $\{Z_t\}$, is

$$D_n = \frac{\sigma_\varepsilon^2 \, h}{\sqrt{n}} \sum_{t=1}^{n} K_h^2(Z_t - z).$$

Similar to the proof of Lemma 1, one can show that

$$D_n = \sigma_\varepsilon^2 \, \nu_0(K) \, L(1,0)/\sigma_0 + o_p(1).$$

Finally, by the central limit theorem for a martingale difference (see, e.g., Hall and Heyde (1980), p.58), one obtains the conditional limiting distribution of C_n given $\{Z_t\}$,

$$n^{1/4}h^{1/2}C_n \stackrel{\mathrm{d}}{\longrightarrow} MN(\sigma_1^2).$$

This proves Lemma 3. □

Proof of Theorem 1. It is easy to check from Lemmas 1 and 2 that

$$C_n = h^2 B(z)\, L(1,0)/\sigma_0 + o_p(h^2).$$

Therefore, by (A.1) and Lemma 3, one has

$$\begin{aligned} &n^{1/4}h^{1/2}\left[\hat{\beta}(z) - \beta(z) - h^2 B(z) + o_p(h^2)\right] \\ &= \sigma_0\, [L(1,0)]^{-1}\; n^{1/4}h^{1/2}\, C_n\{1 + o_p(1)\} \stackrel{\mathrm{d}}{\longrightarrow} MN(\sigma_\beta^2), \end{aligned}$$

which concludes the proof of the theorem. □

Proof of Theorem 3. It is easy to see from Lemma 1 that

$$\widetilde{\beta}(z) - \beta(z) \equiv \{E_n + C_n\}\,/S_{n,0}(z) = [L(1,0)/\sigma_0\,]^{-1}\,\{E_n + C_n\}\,\{1 + o_p(1)\},$$

where $E_n = n^{-1/2}\sum_{t=1}^n [\beta(Z_t) - \beta(z)]\, K_h(Z_t - z)$. Similar to Lemma 2, one has

$$\begin{aligned} E_n &= \left[h\,\beta'(z)\, S_{n,1}(z) + \frac{h^2}{2}\,\beta''(z)\, S_{n,2}(z)\right]\{1 + o_p(1)\} \\ &= \frac{h^2}{2}\, L(1,0)\,\beta''(z)\,\mu_2(K)/\sigma_0\{1 + o_p(1)\} \end{aligned}$$

by Lemma 4 below. By Lemma 3, similar to the proof of Theorem 1, Theorem 3 is proved. □

Lemma 4. *Under Assumptions given in Theorem 1, then,*

$$E[E_n] = O(h^2 n^{-1/2}\ln(n)) + O(h^2).$$

Proof. I first compute the following intermediate quantity. A simple calculation leads to

$$\begin{aligned}
&E[(\beta(Z_t) - \beta(z))K_h(Z_t - z)] \\
&= t^{-1/2} \int [\beta(z + h\,v) - \beta(z)]\, K(v)\, f_{t,z}(t^{-1/2}hv)dv. \\
&\approx t^{-1/2} \int [\beta'(z)hv + h^2\beta''(z)v^2/2][f_{t,z}(0) + f'_{t,z}(0)t^{-1/2}hv]K(v)dv \\
&= h^2t^{-1}\beta'(z)f'_{t,z}(0)\mu_2(K) + \frac{1}{2}\, h^2t^{-1/2}\beta''(z)f_{t,z}(0)\mu_2(K),
\end{aligned}$$

which implies that the order of the second term dominates the order of the first term. Therefore,

$$\begin{aligned}
E[E_n] &\approx h^2\beta'(z)\mu_2(K)n^{-1/2}\sum_{t=1}^{n} t^{-1}f'_{t,z}(0) \\
&\qquad + \frac{1}{2}\, h^2\mu_2(K)\beta''(z)n^{-1/2}\sum_{t=1}^{n} t^{-1/2}f_{t,z}(0) \\
&= h^2\beta'(z)\mu_2(K)^{-1/2}n^{-1/2}O(\ln(n)) + \frac{1}{2}\, h^2\mu_2(K)\beta''(z)n^{-1/2}O(n^{1/2}) \\
&= O(h^2n^{-1/2}\ln(n)) + O(h^2).
\end{aligned}$$

This concludes the proof of the lemma. □

Acknowledgments

I thank the referee and Professor George Roussas for their very helpful comments and suggestions. This paper was supported, in part, by the National Science Foundation of China grants #70871003 and #70971113, and funds provided by the "1000 Talents Project" from Chinese National Government, Fujian Province and Xiamen City Governments, the "Cheung Kong Scholars Project" from Chinese Ministry of Education, the "Minjiang Scholars Project" from Fujian Province, China, China and Xiamen University.

References

Bachmeier, L., Leelahanon, S. and Li, Q. (2006). Money growth and inflation in the United States. *Macroeconomic Dynamics*, **11**, 113–127.

Bandi, F. M. (2002). On persistence and nonparametric estimation (with an application to stock return predictability). *Working paper*, Graduate School of Business, University of Chicago.

Billingsley, P. (1999). *Convergence of Probability Measures*, 2nd Edition. Wiley: New York.

Cai, Z., Jing, B.-Y., Kong, X.-B. and Liu, Z. (2010). Nonparametric regression with nearly integrated regressors under long run dependence. Submitted for publication.

Cai, Z., Li, Q. and Park, J. Y. (2009). Functional-coefficient models for nonstationary time series data. *Journal of Econometrics*, **148**, 101–113.

Cai, Z. and Wang, Y. (2010). Instability of predictability of asset returns. *Working paper*, Department of Mathematics and Statistics, University of North Carolina at Charlotte.

Cai, Z. and Wang, Y. (2011). Testing predictive regression models with nonstationary regressors. Submitted for publication.

Campbell, J.Y. and Yogo, M. (2006). Efficient tests of stock return predictability. *Journal of Financial Economics*, **81**, 27–60.

Cavanagh, C. L., Elliott, G. and Stock, J. H. (1995). Inference in models with nearly integrated regressors. *Econometric Theory*, **11**, 1131–1147.

Chang, Y. and Martinez-Chombo, E. (2003). Electricity demand analysis using cointegration and error-correction models with time varying parameters: The Mexican case. *Working paper*, Department of Economics, Texas A&M University.

Chang, Y. and Park, J. (2003). Index models with integrated time series. *Journal of Econometrics*, **114**, 73–106.

Elliott, G. and Stock, J. H. (1994). Inference in time series regression when the order of integration of a regressor is unknown. *Econometric Theory*, **10**, 672–700.

Fan, J. and Gijbels, I. (1996). *Local Polynomial Modeling and Its Applications.* Chapman and Hall, London.

Fan, J. and Yao, Q. (2003). *Nonlinear Time Series: Nonparametric and Parametric Methods.* Springer: New York.

Hall, P. and Heyde, C. C. (1980). *Martingale Limit Theory and its Applications.* Academic Press: New York.

Jeganathan, P. (2004). Convergence of functionals of sums of random variables to local times of fractional stable motions. *The Annals of Probability*, **32**, 1771–1795.

Juhl, T. (2005). Functional coefficient models under unit root behavior. *Econometrics Journal*, **8**, 197–213.

Karatzas, I. and Shreve, S. E. (1991). *Brownian Motion and Stochastic Calculus*, Second Edition. Springer–Verlag: New York.

Karlsen, H. A., Myklebust, T. and Tjøstheim, D. (2007). Nonparametric estimation in a nonlinear cointegration type model. *Annals of Statistics*, **35**, 252–299.

Karlsen, H. and Tjøstheim, D. (2001). Nonparametric estimation in null recurrent time series. *Annals of Statistics*, **29**, 372–416.

Merlevéde, F., Peligrad, M. and Utev, S. (2006). Recent advances in invariance principles for stationary sequences. *Probability Surveys*, **3**, 1–36.

Park, J. Y. and Hahn, S. B. (1999). Cointegrating regressions with time varying coefficients. *Econometric Theory* **15**, 664–703.

Park, J. Y. and Phillips, P. C. B. (1999). Asymptotics for nonlinear transformations of integrated time series. *Econometric Theory*, **15**, 269–298.

Phillips, P. C. B. (2009). Local limit theory and spurious nonparametric regression. *Econometric Theory*, **25**, 1466–1497.

Phillips, P. C. B. and Park, J. (1998). Nonstationary density and kernel autoregression. *Cowles Foundation discuss paper No. 1181.*

Polk, C., Thompson, S. and Vuolteenaho, T. (2006). Cross-sectional forecasts of the equity premium. *Journal of Financial Economics*, **81**, 101–141.

Rossi, B. (2007). Expectation hypothesis tests and predictive regressions at long horizons. *Econometrics Journal*, **10**, 1–26.

Sun, Y., Cai, Z. and Li, Q. (2010). Semiparametric functional coefficient models with integrated covariates. Revised for *Econometric Theory.*

Torous, W., Valkanov, R. and Yan, S. (2004). On predicting stock returns with nearly integrated explanatory variables. *The Journal of Business*, **77**, 937–966.

Wang, Q. and Phillips, P. C. B. (2009a). Asymptotic theory for local time density estimation and nonparametric cointegrating regression. *Econometric Theory*, **25**, 710–738.

Wang, Q. and Phillips, P. C. B. (2009b). Structural nonparametric cointegrating regression. *Econometrica*, **77**, 1901–1948.

Xiao, Z. (2009). Functional-coefficient cointegration models. *Journal of Econometrics*, **152**, 81–92.

Chapter 14

A Dynamic Test for Misspecification of a Linear Model

Michael P. McAssey and Fushing Hsieh

University of California, Davis

Various linear models may be specified when attempting to estimate an apparent linear trend in real data sets, based on assumptions about the possibly unknown mechanism for generating the data. However, in many cases the best-fit line obtained from implementation of the specified model does not agree at all with the apparent trend in a scatterplot of the data, based on the evident geometric structure. We use both real and simulated data to illustrate the consequences of misspecifying a particular linear model for situations in which that model is inappropriate, and propose a dynamic test which may be used to determine when its specification is appropriate.

Contents

1. Introduction: Modeling the linear trend in real bivariate data

The World Health Organization (WHO) established the Multinational MONItoring of trends and determinants of CArdiovascular disease (MONICA) during the 1980s to study the association between known risk factors, like smoking and obesity, and trends in cardiovascular disease. The linear association between data on the average annual change in the observed risk score (X) and the average annual change in event rate (Y), both given as percentages, should be modeled using a measurement error structure, with the sampling error in the statistics taken as the heteroscedastic measurement error (Kulathinal, et al. (2002)). However, whether or not we account for the measurement error, we obtain about the same estimate of the slope of this linear trend when the specified model includes an additive equation error component, as in the simple linear model. In this case, the resulting best-fit line does not correspond with the perceived trend based on the geometric structure of the data cloud. Figure 1 displays the data scatterplots for males ($N = 38$) and for females ($N = 36$) separately, along with the best-fit lines derived under the ordinary least squares (OLS) method. The OLS method is based on the simple linear model (SLM), in which the ith response Y_i is associated with the covariate X_i by the relation $Y_i = \alpha + \beta X_i + \varepsilon_i$, for $i = 1, \ldots, n$. The additive equation error component ε_i is a (usually Gaussian) random variable with mean zero and variance σ^2, with $\varepsilon_1, \ldots, \varepsilon_n$ independent. If one examines either data scatterplot and

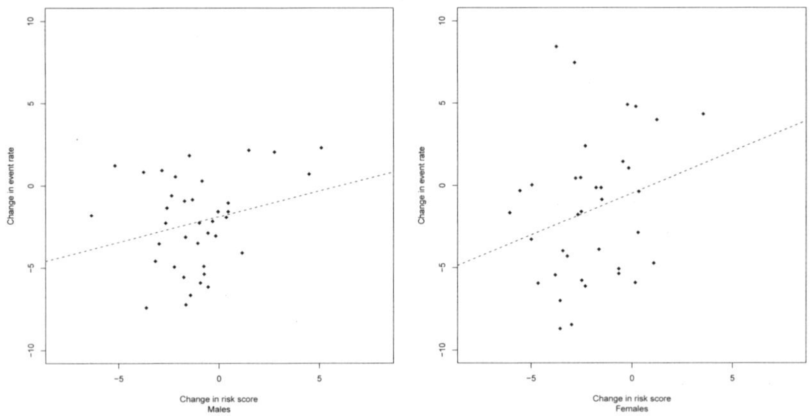

Fig. 1. Scatterplot of change in event rate versus change in risk score, from WHO MONICA project, and best-fit lines based on OLS regression, for males and females.

uses raw intuition—based on the apparent geometry of the data scatter—to project the location of the line which most suitably captures the linear association between the two variables, rather than using mathematical rigor, one would likely conclude that the slope of the plotted best-fit line based on the SLM is significantly smaller than the projected slope. A similar best-fit line is obtained when the measurement error is incorporated into the model using approaches whose structures are akin to that of the SLM. This apparent underestimation of the slope occurs because specification of the SLM for these data fails to capture the underlying mechanism which generated the data.

Since the data-generation mechanism which produced the geometric structure of the data cloud is often unknown, an investigator who wishes to derive reliable estimates for a linear trend in the data requires simple but effective tools to determine the most appropriate model for that trend among those available. In this paper, we propose such a tool and demonstrate its implementation for identifying when a proposed linear model is inappropriate. While the examples presented here involve misspecification of the SLM, this dynamic test is intended as a tool for identifying misspecification of any linear model.

2. Illustration: Slope underestimation with the SLM

To further illustrate the necessity for such a tool, we generate a random sample of 100 points scattered uniformly between -2 and 2 about the line $Y = 0$ in a thick cloud, so that the trendline has a slope of zero. In this case, the data-generation mechanism is inconsistent with the SLM. We rotate the trendline and each point in the data cloud about the origin through an angle greater than $45°$, say $51.3°$, so that the resulting data cloud should properly be perceived to follow a linear trend whose slope is $\tan(51.3°) = 1.25$, based on the underlying data-generation mechanism we have specified. However, if OLS is applied to estimate the slope of the trend, the OLS slope estimate is only 0.98 — a significant underestimate. As we increase the angle of rotation, the underestimation of the slope by OLS becomes more pronounced. Figure 2 displays the data cloud along with its actual trendline and the OLS best-fit line for six different progressively steeper linear trends. Clearly, if a researcher attempted to estimate the linear association between the two variables in any of the displayed scenarios using OLS, the result would not guide the researcher to the truth about that association. While subsequent diagnostic tools might pick up the lack-of-fit in the most extreme cases, they may well miss it in more subtle cases. The data-generation mechanism which produces the geometric structure of a data cloud must not be ignored, and any model which does not properly capture the linear trend in light of that structure must be rejected.

In Figure 3, the slope of the data-generating trendline as it makes a complete rotation through 2π radians is shown, along with the corresponding estimated slope based on OLS regression on the generated data. The larger the magnitude of the slope of the actual trend, the greater the underestimation when OLS is applied. This illustrates a common outcome

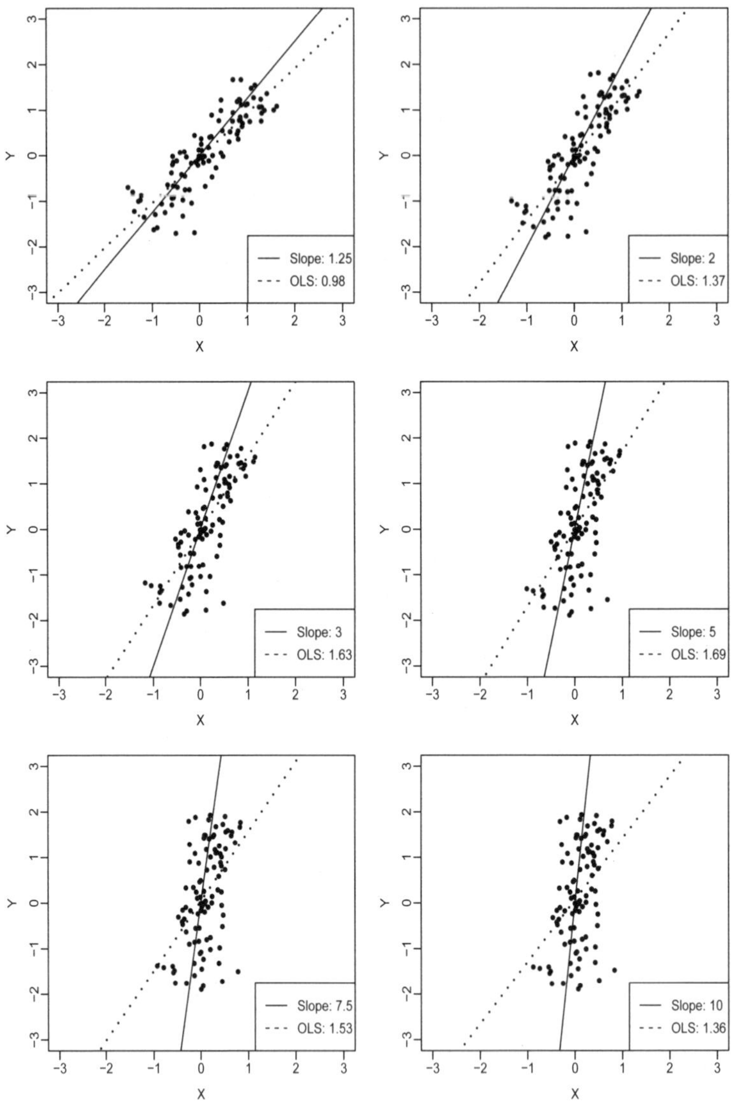

Fig. 2. Rotating data cloud, with the underlying linear trend and the estimated trend under OLS regression.

of misspecifying the SLM: the slope estimate is pulled toward zero as the trend becomes steeper. The behavior of the estimated slope under OLS as the actual trendline rotates depends on the extent to which the data-generation mechanism differes from the SLM. In a scenario in which the

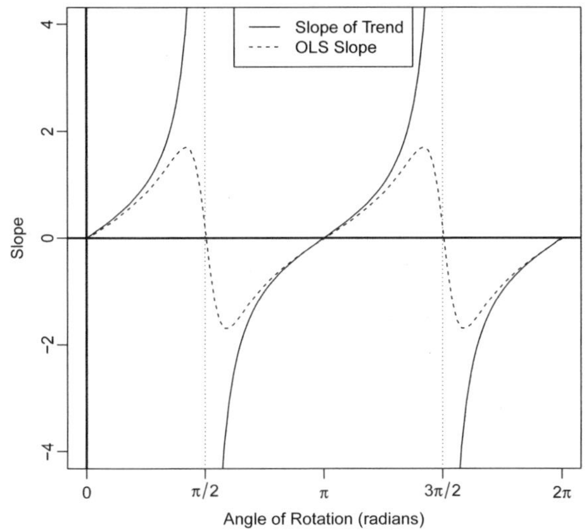

Fig. 3. Progression of the slope of the trendline for the rotating centered data cloud, and the corresponding estimated slope using OLS regression for data which are not generated from the SLM, as the plane rotates about the origin.

data are indeed generated from a SLM (with a reasonably small error variance) and the OLS slope estimate is computed as the plane rotates, the plot of the slope estimate follows that of the trendline slope very closely even when it becomes quite steep, as displayed in Figure 4. This is how the plot of slope estimates for rotating centered data should appear when the model is specified correctly—like the plot of the tangent function. Of course, as the error variance increases this ideal result will deteriorate even when the model is properly specified.

3. A test for misspecification of a linear model for real data

This illustration suggests a dynamic hypothesis test for misspecification of any linear model. Suppose we are presented with bivariate data, and we hypothesize that the association between the variables might be adequately explained by a particular linear model M_0 among linear models in a class

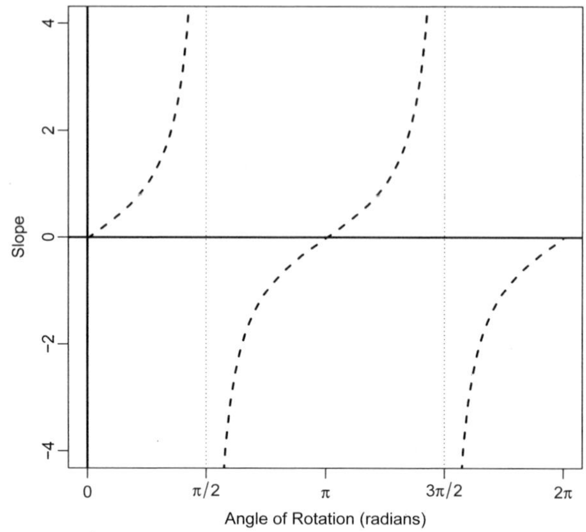

Fig. 4. Progression of the value of the slope estimate using OLS regression as the plane rotates about the origin when the centered data are generated from the SLM with a small error variance.

$\mathcal{M}$.

- Center the data about zero for both variables, select one variable as the covariate, apply an estimation procedure consistent with M_0, and obtain a slope estimate $\widehat{\beta}$ and an estimate $\widehat{\sigma}^2$ for the error variance.
- Rotate the plane completely about the origin in small increments and compute a new slope estimate at each step. Plot the progression of the slope estimates against the angle of rotation, as in Figures 3 and 4.
- Meanwhile, generate a test data set from M, using the same centered values of the covariate, and the initial estimates $\widehat{\beta}$ for the slope and $\widehat{\sigma}^2$ for the error variance.
- Apply the same estimation procedure to compute slope estimates for the test data as the plane is rotated about the origin, and add the progression of these slope estimates to the previous plot.

If the original data were generated in a manner consistent with M, the two plots should be similar in amplitude, phase and shape.

Of course, variability is expected in the generated test data, especially when $\widehat{\sigma}^2$ is large. Thus we generate N test data sets in the same way, and repeat the rotating slope estimation procedure on each set. At each increment of the rotation we will have a distribution of N slope estimates. We may select an upper and lower quantile for this distribution at each angle and thereby obtain an empirical pointwise confidence band for the path of the slope estimate if the original data may indeed be regarded as having been generated from the M. If the path of slope estimates for the original data does not lie entirely within this band, we may reject the null hypothesis: that M is a suitable model for the generation mechanism of the original data, and conclude at the corresponding significance level the alternative hypothesis: that M is not a suitable model for the generation of the data. Note that, in the event the null hypothesis is rejected, this hypothesis test does not recommend an alternative model for consideration.

For instance, consider again the WHO MONICA data for males. When we implement the procedure described above, hypothesizing that M is the SLM, we obtain the results shown in the left panel of Figure 5. The progression of OLS slope estimates as the centered real data rotate about the origin is represented by the solid line. We generate $N = 1000$ data sets under the SLM, using the OLS slope estimate for the unrotated real data, and a robust estimate of the error variance based on the middle 50% of the ordered residuals. The pair of dashed lines represents the middle 50% of OLS slope estimates for the rotating generated data, and the pair of dotted lines represents the middle 90%. Clearly, the path for the real data does not fall within either band, since its amplitude and phase are quite different, so that we may conclude at the 0.1 significance level that the SLM would be a misspecification for these data. When we perform the same experiment with the centered WHO MONICA data for females, we find that the progression of OLS slope estimates also lies outside the confidence bands, as shown in the right panel of Figure 5. Hence the null hypothesis would also be rejected for the female data. We may therefore further suspect the suitability to these data of any measurement error model which involves an equation error component, as such models produce slope estimates very close to those obtained using OLS (as in Kulathinal, et al. (2002), and Patriota, et al. (2009)).

Now consider an alternative data-generating mechanism for these data. Suppose we use orthogonal least squares to estimate the slopes for each data set, based on the linear model which regards the observations as perpendicular deviations from the trendline. In this model, $Y_i = \beta X_i + \nu_i$ for $i = 1, \ldots, n$, with independent errors $\nu_i = \varepsilon_i \sqrt{\beta^2 + 1}$, where $\varepsilon_i \sim \mathcal{N}(0, \sigma^2)$. The progression of these slope estimates as the centered real data for males

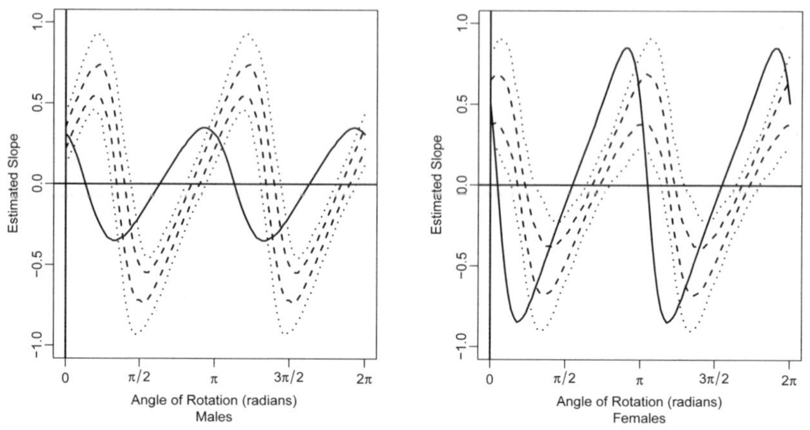

Fig. 5. Progression of the value of the slope estimate using OLS regression on the centered WHO MONICA data for males (left) and for females (right) as the plane rotates about the origin, along with 50% (dashed) and 90% (dotted) pointwise confidence bands for a specified SLM based on 1000 generated data sets.

and for females rotate about the origin is represented by the solid line in each plot of Figure 6. We also generate $N = 1000$ data sets under this model, using the orthogonal least squares slope estimate for the unrotated real data, and an estimate of the error variance based on the entire collection of residuals. The pairs of dashed lines represents the middle 50% and 90% of slope estimates for the rotating generated data. But since there is almost no variability, both of these confidence bands are very narrow—almost superimposed. In this scenario, the path for the real data comes very close to falling within the bands, especially for the females, despite the narrowness of the bands. Moreover, the paths of the estimates have the tangent-function appearance that we expect when the model is specified well. Hence the slope estimate is slightly biased in each case, so that a bias correction of appropriate size should be added to the slope estimate to bring it within the confidence bands. This would produce a model that will pass the dynamic test proposed here.

Once we make the necessary bias corrections to the slope estimates, the amended model passes our hypothesis test. Hence when the measurement error for the observations on each variable is considered for these data, it would be wise to implement a model whose structure is more akin to

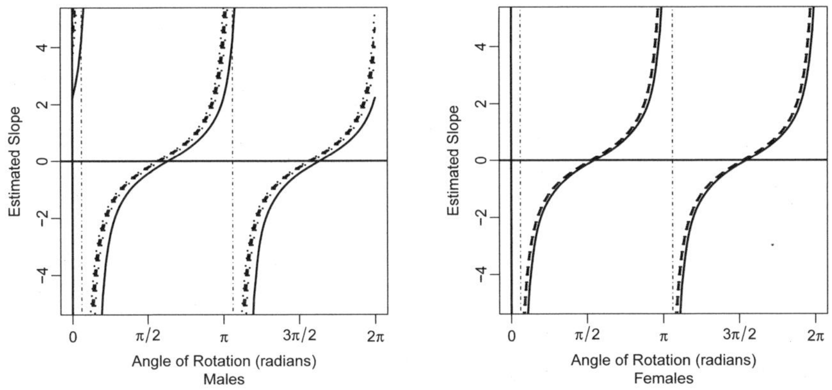

Fig. 6. Progression of the value of the slope estimate using orthogonal least-squares regression on the centered WHO MONICA data for males (left) and for females (right) as the plane rotates about the origin, along with 50% (dashed) and 90% (dotted) pointwise confidence bands for a specified perpendicular-deviation model based on 1000 generated data sets.

this latter model, and less like the SLM, as presented in McAssey and Hsieh (2010). In Figure 7, the trendlines for males and females based on the slope estimates derived using orthogonal least squares (OM) are added to the scatterplots given in Figure 1. It is intuitively apparent that the trajectories of these two trendlines correspond much more closely to the geometric structure of the respective data clouds.

4. Discussion

In many data analyses one has limited knowledge about the underlying data generation mechanism, so that the specification of a linear model may become an arbitrary choice. The above examples demonstrate that the cost of misspecification can be high. Hence researchers must be wary of specifying a particular linear model in the analysis of the linear association between two variables unless there is convincing evidence that the data were generated in a manner consistent with that model. We present a simple test that may be used to determine whether the specified model is a plausible representation of the underlying data generating mechanism for the observed data. A careful researcher may implement this simple test

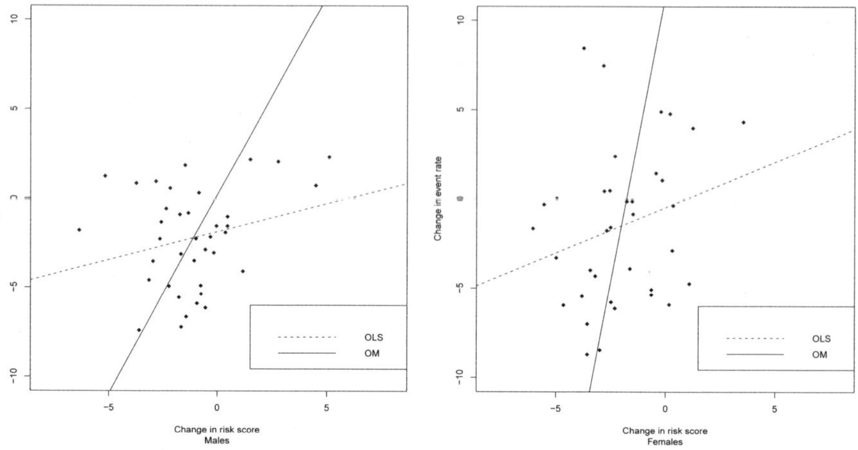

Fig. 7. Scatterplot of change in event rate versus change in risk score, from WHO MONICA project, and best-fit lines based on both orthogonal least-squares (OM) and ordinary least-squares (OLS) regression, for males and females.

as a tool to assist in discriminating among proposed linear models and corresponding estimation procedures for analysis of real data.

Acknowledgments

This research is partially supported by the National Science Foundation under Grant No: HSD 0826844, and by the National Institutes of Health under Grant No: 1R01AG025218-01A2.

The authors thank Kari Kuulasmaa for making the WHO MONICA data available, and Alexandre Patriota for sharing the data.

References

Kulathinal, S. B., Kuulasmaa, K. and Gasbarra, D. (2002). Estimation of an errors-in-variables regression model when the variances of the measurement errors vary between the observations. *Statistics in Medicine* **21**, 1089–1101.

McAssey, M. and Hsieh, F. (2010). Slope estimation in structural line-segment heteroscedastic measurement error models. *Statistics in Medicine* **29**, 2631–2642.

Patriota, A. G., Bolfarine, H. and de Castro, M. (2009). A heteroscedastic structural errors-in-variables model with equation error. *Statistical Methodology* **6**, 408–423.

Chapter 15

The Principal Component Decomposition of the Basic Martingale

Winfried Stute

University of Giessen, Germany

Contents

1. Introduction

Let $X_1, \ldots, X_n$ be a sample of independent observations on the real line from the same but unknown distribution function (d.f.) F. To test a hypothesis about F, one often encounters the empirical d.f. F_n of the sample, i.e.,

$$F_n(x) = n^{-1} \sum_{i=1}^{n} 1_{\{X_i \leq x\}}, \qquad x \in \mathbb{R}.$$

Here 1_A denotes the indicator function of a set A. $F_n(x)$ is an unbiased estimator of $F(x)$ and uniformly consistent as $n \to \infty$. Classical goodness-of-fit tests are based on the deviation between F_n and a hypothetical F_0. Two of the most famous tests are the Kolmogorov–Smirnov (KS) and the Cramér–von Mises (CvM) test:

$$\mathrm{KS} = n^{1/2} \sup_x |F_n(x) - F_0(x)|$$

incorporates the supremum distance between F_n and F_0, while

$$\mathrm{CvM} = n \int [F_n(x) - F_0(x)]^2 \, F_0(\mathrm{d}x)$$

takes into account their L^2-distance.

If F_0 is continuous and the null hypothesis

$$H_0 : F = F_0$$

holds true, we may transform each X_i into $U_i = F_0(X_i)$ to come up with a uniform random variable. Denote with $\bar{F}_n$ the empirical d.f. of $U_1, \ldots, U_n$. Then

$$\mathrm{KS} = n^{1/2} \sup_{0 \le t \le 1} |\bar{F}_n(t) - t|$$

and

$$\mathrm{CvM} = n \int_0^1 [\bar{F}_n(t) - t]^2 \mathrm{d}t.$$

Finally, introducing the empirical processes

$$\alpha_n(x) = n^{1/2}[F_n(x) - F(x)], \quad x \in \mathbb{R},$$

and

$$\bar{\alpha}_n(t) = n^{1/2}[\bar{F}_n(t) - t], \quad 0 \le t \le 1,$$

the convergence of $\bar{\alpha}_n$ to a Brownian bridge B^0 together with the continuous mapping theorem yield

$$\mathrm{KS} \to \sup_{0 \le t \le 1} |B^0(t)| \text{ and } \mathrm{CvM} \to \int_0^1 [B^0(t)]^2 \mathrm{d}t$$

in distribution.

For a deeper analysis of the tests, it turns out to be useful to expand $\bar{\alpha}_n$ into its principal components. This means we are looking for an appropriate orthonormal complete system $\{f_j\}$ of functions in $L^2[0, 1]$ such that

$$\bar{\alpha}_n = \sum_{j=1}^{\infty} \langle \bar{\alpha}_n, f_j \rangle f_j, \tag{1}$$

where

$$\langle \bar{\alpha}_n, f_j \rangle = \int_0^1 \bar{\alpha}_n(t) f_j(t) \mathrm{d}t.$$

If the f_j are taken to be the eigenfunctions of the integral operator

$$(Tg)(s) = \int_0^1 g(t) K(s, t) \mathrm{d}t$$

associated with the covariance kernel

$$K(s,t) = \min(s,t) - st$$

of $\bar{\alpha}_n$, then the $\langle \bar{\alpha}_n, f_j \rangle$ are centered and uncorrelated for $j = 1, 2, \ldots$ In this case, $\langle \bar{\alpha}_n, f_j \rangle$ is called the j-th principal component of $\bar{\alpha}_n$, and (1) denotes the Karhunen–Loève decomposion of $\bar{\alpha}_n$. Finally,

$$\text{CvM} = \sum_{j=1}^{\infty} \langle \bar{\alpha}_n, f_j \rangle^2.$$

Hence, if in a testing situation CvM exceeds a critical value so that H_0 needs to be rejected, the statistician is invited to check in detail the components to get a deeper insight of what may be the reason which led to a rejection. See Durbin and Knott (1972) for a thorough discussion.

Now, the above approach only makes sense if the underlying distribution is described through its d.f. In many applications, in particular in survival analysis, this will not be the case. Rather, a distribution there is often given through its hazard measure Λ defined as

$$\mathrm{d}\Lambda = \frac{\mathrm{d}F}{1 - F^{-}}.$$

Here and in the following, $G^{-}(x) = \lim_{y \uparrow x} G(y)$ denotes the left-hand limit of a function G at x. If G is continuous at x, then $G^{-}(x) = G(x)$. Coming back to empirical d.f.'s, it turns out that the process

$$M_n(x) = n^{1/2} \left[F_n(x) - \int_{-\infty}^{x} [1 - F_n^{-}(y)] \Lambda(\mathrm{d}y) \right]$$

plays an important role in this context. See also Aki (1986). Actually, M_n is the martingale part in the Doob–Meyer decomposition of α_n. It is sometimes referred to as "the basic martingale"; see, e.g., Shorack and Wellner (1986). By now it has become an important tool for model diagnosis in survival analysis, in particular when the data are at risk of being censored from the right.

It is the purpose of the next section to derive the decomposition of M_n into its principal components. A brief discussion of the limit distribution of the associated CvM test statistic is presented in section 3. We only mention that the decomposition with estimated parameters, i.e., when $\Lambda = \Lambda_\theta$ and θ needs to be estimated, is much more lengthy and will be studied elsewhere.

2. Main Results

As for α_n, statistics based on M_n will be distribution-free after a proper transformation only when F is continuous. In this paper we only deal with this case. Actually, as before,

$$\begin{aligned} M_n(x) &= n^{1/2}\left[\bar{F}_n(F(x)) - \int_{-\infty}^{x} \frac{1-\bar{F}_n(F(y))}{1-F(y)} F(\mathrm{d}y)\right] \\ &= n^{1/2}\left[\bar{F}_n(t) - \int_0^t \frac{1-\bar{F}_n(s)}{1-s}\mathrm{d}s\right] \equiv \bar{M}_n(t), \end{aligned}$$

with $t = F(x)$. Therefore, we restrict ourselves to $\bar{M}_n$. First, note that

$$\begin{aligned} \bar{M}_n(t) &= n^{1/2}\left[\bar{F}_n(t) - n^{-1}\sum_{i=1}^{n}\int_0^{t\wedge U_i}\frac{\mathrm{d}s}{1-s}\right] \\ &= n^{1/2}\left[\bar{F}_n(t) + n^{-1}\sum_{i=1}^{n}\ln(1-t\wedge U_i)\right] \qquad (2) \\ &= n^{1/2}\left[\bar{F}_n(t) + n^{-1}\sum_{i=1}^{n}\left(1_{\{U_i\le t\}}\ln(1-U_i) + 1_{\{U_i>t\}}\ln(1-t)\right)\right]. \end{aligned}$$

Secondly, to obtain the eigenfunctions and eigenvalues, note that, as a martingale in standard time, $\bar{M}_n$ has the same covariance K, eigenfunctions f_j and eigenvalues λ_j as the standard Brownian motion, namely

$$\begin{aligned} K(s,t) &= \min(s,t), \quad 0\le s,t\le 1 \\ f_j(t) &= \sqrt{2}\sin\frac{(2j-1)\pi t}{2} \end{aligned}$$

and

$$\lambda_j = \frac{4}{\pi^2(2j-1)^2}, \quad j=1,2,\ldots$$

For simplicity, let $a_j = \frac{(2j-1)\pi}{2}$. We are now in a position to present the principal components of $\bar{M}_n$.

Theorem 1. *The principal components of* $\tilde{M}_n$ *are given by*

$$\langle\bar{M}_n, f_j\rangle = \frac{\sqrt{2}}{a_j\sqrt{n}}\sum_{j=1}^{n}\left[\cos(a_jU_i) - \sum_{k=0}^{\infty}\sum_{m=0}^{\infty}\frac{(-1)^m a_j^{2m}U_i^{2m+k+1}}{(2m)!(2m+k+1)}\right].$$

Proof. From (2), we obtain

$$\begin{aligned}
\langle \bar{M}_n, f_j \rangle &= \int_0^1 \bar{M}_n(t) f_j(t) \mathrm{d}t \\
&= n^{1/2} \int_0^1 \Big[\bar{F}_n(t) + n^{-1} \sum_{i=1}^n \big(1_{\{U_i \le t\}} \ln(1-U_i) \\
&\qquad\qquad + 1_{\{U_i > t\}} \ln(1-t) \big) \Big] f_j(t) \mathrm{d}t \\
&= n^{1/2} \int_0^1 \bar{F}_n(t) f_j(t) \mathrm{d}t + n^{-1/2} \sum_{i=1}^n \Big[\ln(1-U_i) \int_{U_i}^1 f_j(t) \mathrm{d}t \\
&\qquad\qquad + \int_0^{U_i} f_j(t) \ln(1-t) \mathrm{d}t \Big] \\
&= n^{-1/2} \sum_{i=1}^n (1 + \ln(1-U_i)) \int_{U_i}^1 f_j(t) \mathrm{d}t \\
&\qquad\qquad + n^{-1/2} \sum_{i=1}^n \int_0^{U_i} f_j(t) \ln(1-t) \mathrm{d}t.
\end{aligned}$$

Recalling that $f_j(t) = \sqrt{2} \sin(a_j t)$, the first sum becomes

$$\frac{\sqrt{2}}{a_j \sqrt{n}} \sum_{i=1}^n (1 + \ln(1-U_i)) \cos(a_j U_i).$$

As to the second sum, integration by parts yields

$$\begin{aligned}
&\sqrt{\frac{2}{n}} \sum_{i=1}^n \int_0^{U_i} \sin(a_j t) \ln(1-t) \mathrm{d}t \\
&= \frac{\sqrt{2}}{a_j \sqrt{n}} \sum_{i=1}^n \left\{ - \cos(a_j t) \ln(1-t) \Big|_0^{U_i} - \int_0^{U_i} \frac{\cos(a_j t)}{1-t} \mathrm{d}t \right\} \\
&= \frac{\sqrt{2}}{a_j \sqrt{n}} \sum_{i=1}^n \left\{ - \cos(a_j U_i) \ln(1-U_i) - \int_0^{U_i} \frac{\cos(a_j t)}{1-t} \mathrm{d}t \right\}.
\end{aligned}$$

Summarizing we obtain

$$\langle \bar{M}_n, f_j \rangle = \frac{\sqrt{2}}{a_j \sqrt{n}} \sum_{i=1}^n \left\{ \cos(a_j U_i) - \int_0^{U_i} \frac{\cos(a_j t)}{1-t} \mathrm{d}t \right\}.$$

Finally, since

$$\cos(a_j t) = \sum_{m=0}^{\infty} (-1)^m \frac{(a_j t)^{2m}}{(2m)!}$$

and

$$\frac{1}{1-t} = \sum_{k=0}^{\infty} t^k \text{ for } 0 < t < 1,$$

we get

$$\int_0^{U_i} \frac{\cos(a_j t)}{1-t} \mathrm{d}t = \sum_{k=0}^{\infty} \sum_{m=0}^{\infty} \frac{(-1)^m a_j^{2m}}{(2m)!} \int_0^{U_i} t^{2m+k} \mathrm{d}t,$$

whence the conclusion. □

If we expand the cosines into a series, we obtain an alternate representation of the j-th principal component which sheds some new light on what is really compared within $\langle \bar{M}_n, f_j \rangle$.

Corollary 1. *The principal components of* $\tilde{M}_n$ *may be represented as*

$$\langle \bar{M}_n, f_j \rangle = \sqrt{2n} \sum_{m=0}^{\infty} \frac{(-1)^m a_j^{2m-1}}{(2m)!} \left[\int t^{2m} \bar{F}_n(\mathrm{d}t) - \sum_{k=0}^{\infty} \frac{\int t^{2m+k+1} \bar{F}_n(\mathrm{d}t)}{2m+k+1} \right].$$

The terms in brackets incorporate properly weighted empirical moments of all orders larger than or equal to $2m$. Under the null hypothesis, i.e., when the data come from a uniform distribution, the expectation of the bracketed term in Corollary 1 equals

$$\begin{aligned}
&\int_0^1 t^{2m} \mathrm{d}t - \sum_{k=0}^{\infty} \frac{\int_0^1 t^{2m+k+1} \mathrm{d}t}{2m+k+1} \\
&= \frac{1}{2m+1} - \sum_{k=0}^{\infty} \frac{1}{(2m+k+1)(2m+k+2)} \\
&= \frac{1}{2m+1} - \sum_{k=0}^{\infty} \left[\frac{1}{2m+k+1} - \frac{1}{2m+k+2} \right] = \frac{1}{2m+1} - \frac{1}{2m+1} = 0,
\end{aligned}$$

as anticipated. The summation over m in the corollary guarantees, that these internal comparisons are taken into account for each m. The tricky weighting is needed for

$$\mathrm{Var}\langle \bar{M}_n, f_j \rangle = \lambda_j$$

and to achieve that all $\langle \bar{M}_n, f_j \rangle, j = 1, 2, \ldots$ are uncorrelated.

Of course, the statistician may feel free to choose other weightings also. E.g., one might fix one or a few m and take the corresponding bracketed term in Corollary 1 in the expansion of $\langle \bar{M}_n, f_j \rangle$. As seen before, they are centered under the null hypothesis. After multiplication with $n^{1/2}$, they are jointly asymptotically normal. Each variance equals

$$\int_0^1 \left(t^{2m} - \sum_{k=0}^{\infty} \frac{t^{2m+k+1}}{2m+k+1} \right)^2 \mathrm{d}t,$$

while covariances typically no longer vanish.

3. Limit Distribution of the CvM Test Statistic

The CvM test statistic associated with M_n equals

$$\int M_n^2(x) F(\mathrm{d}x) = \int_0^1 \bar{M}_n^2(t)\mathrm{d}t = \sum_{j=1}^{\infty} \langle \bar{M}_n, f_j \rangle^2.$$

In the limit, as $n \to \infty$, the series equals in distribution

$$\sum_{j=1}^{\infty} \langle B, f_j \rangle^2 = \sum_{j=1}^{\infty} \lambda_j Z_j^2, \tag{3}$$

where B is a standard Brownian motion, λ_j is as in the previous section and $Z_1, Z_2, \ldots$ are i.i.d. standard normal random variables. The distribution of (3) has been studied, in a different context, by Rothman and Woodroofe (1972). We briefly review their approach here, since it also nicely applies to our situation. First Rothman and Woodroofe (1972) noticed that by independence and since $Z_i \sim \mathcal{N}(0,1)$, the Laplace transform of (3) equals

$$\begin{aligned}\Psi(t) &= \prod_{i=1}^{\infty} \mathbb{E}(e^{-t\lambda_j Z_i^2}) = \prod_{j=1}^{\infty} (1 + 2\lambda_j t)^{-1/2} \\ &= \prod_{j=1}^{\infty} \left(1 + \frac{4(\sqrt{2t})^2}{\pi^2 (2j-1)^2} \right)^{-1/2} = \left[\cosh(\sqrt{2t}) \right]^{-1/2}.\end{aligned}$$

The last term may be expanded as follows:

$$\left[\cosh(\sqrt{2t}) \right]^{-1/2} = 2^{1/2} \sum_{j=0}^{\infty} \binom{-1/2}{j} \exp\left[-(2j + 1/2)\sqrt{2t} \right].$$

According to Breiman (1968), a random variable X with d.f.

$$H_\alpha(x) = 2\left(1 - \Phi\left(\frac{\alpha}{\sqrt{x}}\right)\right), x > 0, \alpha > 0,$$

and Φ being the d.f. of $\mathcal{N}(0,1)$, has Laplace transform

$$\Psi(t) = e^{-\alpha\sqrt{2t}}.$$

From this we may conclude that (3) has distribution

$$G(x) = 2^{3/2} \sum_{j=0}^{\infty} \binom{-1/2}{j} \left[1 - \Phi\left(\frac{4j+1}{2\sqrt{x}} \right) \right].$$

Orlov (1972) computed some selected numerical values of $G(x)$ which were reprinted in Table 1 on page 748 of Shorack and Wellner (1986).

References

Aki, S. (1986). Some test statistics based on the martingale term of the empirical distribution function, *Ann. Inst. Statist. Math.* **38**, pp. 1–21.

Breiman, L. (1968). *Probability*, Addison-Wesley.

Durbin, J. and Knott, M. (1972). Components of Cramér-von Mises statistics, I, *J. Roy. Statist. Soc. B* **34**, pp. 290–307.

Orlov, A. (1972). On testing the symmetry of distributions, *Theory Prob. Applic.* **17**, pp. 357-361.

Rothman, E. D. and Woodroofe, M. (1972). A Cramér–von Mises type statistic for testing symmetry, *Ann. Math. Statist.* **43**, pp. 2035–2038.

Shorack, G. R. and Wellner, J. A. (1986). *Empirical processes with applications to statistics.* Wiley.

PART 4

Papers on Time Series Analysis

Chapter 16

Fast Scatterplot Smoothing Using Blockwise Least Squares Fitting

Alexander Aue and Thomas C. M. Lee

University of California, Davis

The *blocking* method offers a quick and simple procedure for obtaining a scatterplot smooth from a noisy data set. The idea is to partition the data into blocks of equal size and fit a low order polynomial to each block. An important component of this method is the choice of the number of blocks. This article develops an automatic selection method for choosing the number of blocks. The proposed method is based on the minimum description length principle, and both its theoretical and empirical properties are studied. In the context of local linear regression, the method is also applied to construct pilot estimates for a "plug-in" bandwidth selector.

Contents

1. Introduction

This article concerns quick and simple methods for performing nonparametric curve estimation. The need for such fast methods is apparent. For example, these fast methods are essential for interactive exploratory data analysis, they are the basic ingredient for many dimension reduction techniques, and they are also frequently adopted to obtain fast "pilot" estimates so that more refined curve estimates can be computed.

The particular method that this article focuses on is the so-called "blocking" method described, for example, in Hardle and Marron (1995). To apply this method one first partitions the data into blocks of equal size and then fits a polynomial of low order to each block. Of course, the number of blocks must be pre-specified before the actual partitioning. In this article an automatic method for selecting the number of blocks is developed. This method is based on the Minimum Description Length (MDL) principle (Rissanen (1989); see also Barron, et al. (1998)), which is known to be a special case of the more general technique of complexity regularization (e.g., see Barron (1991) and Liu and Moulin (2001)). Building on the work of Liu and Moulin (2001), bounds on the estimation errors of the complexity regularized blocking estimates are established.

The rest of the article is organized as follows. Section 2 provides background for the blocking method, develops the proposed method for choosing the number of blocks, and presents theoretical results. In Section 3 it is demonstrated that the blocking method can be applied to obtain pilot estimates for smoothing parameter selection in the context of local linear regression. Simulation results are reported in Section 4, and a real data example is given in Section 5. Lastly conclusions are offered in Section 6.

2. Automatic Curve Estimation Using Blocking

2.1. *Background*

Suppose observed is a set of observations $\{x_i, y_i\}_{i=1}^n$ satisfying

$$y_i = m(x_i) + \epsilon_i,$$

where m is the regression function and the ϵ_i's are zero mean independent and identical Gaussian noise with common variance σ^2. The design density for x_i is denoted as $f(x)$, and is assumed to have a compact support $[a, b]$. Suppose we estimate m using k equi–sized blocks, in the sense that all k blocks contain roughly the same number of observations. Denote the jth block as B_j, $j = 1, \ldots, k$. Then an order p blocking estimate for m can be written as

$$\hat{m}_{\mathrm{B}}(x; k) = \sum_{j=1}^{k} \left[\hat{\beta}_{j0} I_{\{x \in B_j\}} + \hat{\beta}_{j1} x I_{\{x \in B_j\}} + \cdots + \hat{\beta}_{jp} x^p I_{\{x \in B_j\}}\right], \quad (1)$$

where the $\hat{\beta}_{jl}$'s are polynomial coefficient estimates and $I_{\{E\}}$ is the indicator function for the event E. The goal is to choose a "best" k. Quite often, the order p is fixed *a priori*. In this article we set $p = 4$, although the proposed method for selecting k can also be extended to select its value. For simplicity, the dependence of p is suppressed in the notation of $\hat{m}_{\mathrm{B}}(x;k)$. Once k and p are specified, the $\hat{\beta}_{jl}$'s can be obtained uniquely by least-squares regression.

Notice that estimates of the derivatives of m can be obtained by differentiating (1). For example, the second derivative $m''(x)$ can be estimated by:

$$\hat{m}''_{\mathrm{B}}(x;k) = \sum_{j=1}^{k} \left[2\hat{\beta}_{j2} I_{\{x\in B_j\}} + 6\hat{\beta}_{j3} x I_{\{x\in B_j\}} + \cdots + (p-1)p\hat{\beta}_{jp} x^{p-2} I_{\{x\in B_j\}}\right]. \tag{2}$$

Such estimates are often used as "pilot" estimates which are required for constructing other more refined "plug-in" type nonparametric regression estimates; see Section 3 for an example in local linear regression.

2.2. *Automatic Selection of Number of Blocks Using MDL*

This subsection presents the proposed method for choosing k. The method is based on the MDL principle developed by Rissanen (1989).

In the current context the MDL principle *defines* the best k, or the best fitting blocking estimate, as the one that produces the shortest code length that completely describes the data y_i's. Here the code length of an object can be taken as the amount of memory space that is required to store the object. One way to store the data y_i's is to split the y_i's into two parts: (i) a fitted model plus (ii) the portion of the data that is unexplained by the fitted model; i.e., the residuals. For the current problem, a fitted model is $\hat{m}_{\mathrm{B}}(x;k)$, which can be completely specified by k and the $\hat{\beta}_{jl}$'s, while the residuals are $y_i - \hat{m}_{\mathrm{B}}(x_i;k)$. If $L(z)$ denotes the code length of object z, one has the following decomposition

$$\begin{aligned} L(\text{"data"}) &= L(\text{"fitted model } \hat{m}_{\mathrm{B}}(x;k)\text{"}) + L(\text{"residuals"}) \\ &= L(k) + L(\hat{\beta}_{jl}; j = 1,\ldots,k, l = 0,\ldots,p) + L(\text{"residuals"}), \end{aligned}$$

and MDL defines the best $\hat{m}_{\mathrm{B}}(x;k)$ (or, equivalently, the best k) as the minimizer of $L(\text{"data"})$.

It is shown in the appendix that $L(\text{"data"})$ admits the following approximation:

$$\mathrm{MDL}(k) = \log_2 k + \frac{p+1}{2} \sum_{j=1}^{k} \log_2 n_j + \frac{1}{2\sigma^2} \sum_{i=1}^{n} \{y_i - \hat{m}_{\mathrm{B}}(x_i;k)\}^2, \quad (3)$$

where n_j is the number of observations in the jth block. We propose to select k as the minimizer of $\mathrm{MDL}(k)$. Notice that one could also simultaneously select k and p as the joint minimizer of $\mathrm{MDL}(k)$. However, such a joint minimization would slow down the procedure and also our numerical experience seems to suggest that the practical gain is minimal. Thus this issue of automatic choice for p is not considered in this article.

For the practical minimization of $\mathrm{MDL}(k)$, one needs to replace the unknown σ^2 with an independent estimate $\hat{\sigma}^2$. Following Mallows (1973), we estimate σ^2 using the following normalized residual sum of squares

$$\hat{\sigma}^2 = \frac{1}{n-(p+1)k_{\mathrm{max}}} \sum_{i=1}^{n} \{y_i - \hat{m}_{\mathrm{B}}(x_i;k_{\mathrm{max}})\}^2,$$

where k_{max} is the maximal value of k that one would attempt when minimizing $\mathrm{MDL}(k)$. In practice we found that the final regression estimate is insensitive to the choice of k_{max}, as long as k_{max} is large enough. In this paper we set $k_{\mathrm{max}} = 10$.

2.3. *Complexity Regularization*

As mentioned before, the MDL principle is a special case of the more general technique complexity regularization (e.g., Barron (1991) and Liu and Moulin (2001)). This subsection provides theoretical bounds for the estimation errors of these complexity regularized blocking estimates.

In complexity regularization, the best fitting model $\hat{m}_{\mathrm{CR}}$ is defined as the minimizer of

$$-\ln p(y_1,\ldots,y_n|\hat{m}_{\mathrm{CR}}) + \mu\Phi(\hat{m}_{\mathrm{CR}}),$$

where the regularization parameter μ controls the tradeoff between the log-likelihood term $\ln p(\cdot|\hat{m}_{\mathrm{CR}})$ and the complexity measure $\Phi(\hat{m}_{\mathrm{CR}})$ for $\hat{m}_{\mathrm{CR}}$. If $\mu = 1$ and $\Phi(\hat{m}_{\mathrm{CR}})$ is taken as the code length $L(\hat{m}_{\mathrm{CR}})$ for $\hat{m}_{\mathrm{CR}}$, then the above complexity regularized solution simply reduces to the MDL solution. For the present blocking problem, the complexity regularized solution $\hat{m}_{\mathrm{CR}}$ is given by

$$\operatorname*{argmin}_{\hat{m}_{\mathrm{B}} \in \mathcal{M}_{\mathrm{B}}} \left[\frac{1}{2\sigma^2} \sum_{i=1}^{n} \{y_i - \hat{m}_{\mathrm{B}}(x_i;k)\}^2 + \mu \left\{ \log_2 k + \frac{p+1}{2} \sum_{j=1}^{k} \log_2 n_j \right\} \right],$$

where $\mathcal{M}_\mathrm{B}$ is the class of all blocking regression estimates that are being considered.

The L_2 loss and risk of $\hat{m}_\mathrm{CR}$ are respectively defined as

$$\mathrm{loss}(\hat{m}_\mathrm{CR}) = \frac{1}{n}\sum_{i=1}^{n}\{m(x_i) - \hat{m}_\mathrm{CR}(x_i)\}^2 \quad \text{and} \quad \mathrm{risk}(\hat{m}_\mathrm{CR}) = E\{\mathrm{loss}(\hat{m}_\mathrm{CR})\}.$$

The error bounds on $\mathrm{loss}(\hat{m}_\mathrm{CR})$ and $\mathrm{risk}(\hat{m}_\mathrm{CR})$ are expressed in terms of an *index of resolvability*, which characterizes the compressibility of the true regression m. This index of resolvability, $R_\mu(m)$, which has been used for example by Barron (1991) and Liu and Moulin (2001), is defined as

$$\begin{aligned} R_\mu(m) &= \frac{1}{n}\min_{\hat{m}_\mathrm{B}\in\mathcal{M}_\mathrm{B}}\left[\|m - \hat{m}_\mathrm{B}\|^2 + 2\sigma^2\mu L(\hat{m}_\mathrm{B})\right] \\ &= \frac{1}{n}\min_{\hat{m}_\mathrm{B}\in\mathcal{M}_\mathrm{B}}\left[\sum_{i=1}^{n}\{m(x_i) - \hat{m}_\mathrm{B}(x_i;k)\}^2\right. \\ &\qquad\qquad \left. + \; 2\sigma^2\mu\left\{\log_2 k + \frac{p+1}{2}\sum_{j=1}^{k}\log_2 n_j\right\}\right]. \end{aligned}$$

With the use of $R_\mu(m)$, we summarize our results in the following theorem.

Theorem 1. *Consider the blocking estimation problem described in Section 2. For $\mu > 1$ and $0 < \delta < 1$, we have*

$$P\left\{\mathit{loss}(\hat{m}_\mathrm{CR}) < \frac{\mu+1}{\mu-1}R_\mu(m) + \frac{4\mu^2\sigma^2}{n(\mu-1)}\ln\frac{1}{\delta}\right\} \geq 1-\delta$$

and

$$\mathit{risk}(\hat{m}_\mathrm{CR}) < \frac{\mu+1}{\mu-1}R_\mu(m) + \frac{4\mu^2\sigma^2}{n(\mu-1)}.$$

This theorem can be proved by a direct application of Theorem 3.1 of Liu and Moulin (2001) using the substitution $\nu = 2(\ln 2)\sigma^2\mu$.

3. Pilot Estimation in Local Linear Regression

As suggested by various authors (e.g., Ruppert, et al. (1995), Müller (2000), and Climov, et al. (2002)), another application of the blocking method is that it can be used to construct fast pilot estimates for other more refined "plug–in" nonparametric smoothing methods. This section focuses on local linear regression (e.g., see Fan and Gijbels (1995)).

The local linear regression estimate for a regression function $m(x)$ is given by $\hat{m}_h(x) = \hat{a}$, where $\hat{a}$, together with $\hat{b}$, are defined as the joint minimizer of

$$\sum_{i=1}^{n} [y_i - \{a + b(x_i - x)\}]^2 \, K_h(x - x_i)$$

(e.g., see Fan and Gijbels (1996)). In the above h is the bandwidth, K is the kernel function, and $K_h(x) = K(x/h)/h$. Throughout this paper a Gaussian kernel is used. It is widely known that the choice of h is crucial.

Ruppert, et al. (1995) propose a "plug-in" type bandwidth selector for selecting h. Their method relies on the following expression for the optimal choice of h

$$h_{\text{opt}} = \frac{\sigma^2(b-a)\int K^2(x)dx}{n\theta\{\int x^2K(x)dx\}^2}, \quad \text{where} \quad \theta = \int \{m''(x)\}^2 f(x)dx. \tag{4}$$

Observe that, in the above expression, the only unknown quantities are θ and σ^2. To obtain an estimate for h_{opt}, the authors suggest replacing these unknown quantities with different pilot estimates. Their simplest such pilot estimate $\hat{\theta}_k$ for θ is a "blocked quartic" estimate (i.e., $p = 4$):

$$\hat{\theta}_k = n^{-1} \sum_{i=1}^{n} \hat{m}''_{\text{B}}(x_i; k), \tag{5}$$

where $\hat{m}''_{\text{B}}(x_i; k)$ is given by (2) and the number of blocks k is chosen as the minimizer of the following C_p type criterion:

$$C_p(k) = \frac{1}{\sigma^2} \sum_{i=1}^{n} \{y_i - \hat{m}_{\text{B}}(x_i; k)\}^2 + 2(p+1)k. \tag{6}$$

The corresponding pilot estimate for σ^2 is

$$\hat{\sigma}_k^2 = \frac{1}{n-5k} \sum_{i=1}^{n} \{y_i - \hat{m}_{\text{B}}(x_i; k)\}^2. \tag{7}$$

Ruppert, et al. (1995) termed the bandwidth obtained by plugging-in the estimates (5) and (7) into (4) the "rule-of-thumb" bandwidth. These authors also develop two other plug–in bandwidth selectors: the "direct plug-in" and the "solve-the-equation" selectors. These two selectors are more sophisticated but also computationally more expensive.

It is shown in the next simulation section that the practical performance of the "rule-of-thumb" bandwidth selector is improved if the C_p criterion (6) is replaced by the MDL criterion (3).

4. Simulation

This section reports the results of a simulation study that was conducted to evaluate the performances of five different smoothing methods. These five methods are

- blockCp: blocking estimate (1) where the number of blocks k is chosen by the C_p criterion (6);
- blockMDL: blocking estimate (1) where k is chosen by the proposed MDL criterion (3);
- llrCp: local linear regression where the bandwidth is chosen by the "rule-of-thumb" method of Ruppert, et al. (1995);
- llrMDL: local linear regression where the bandwidth is chosen by a modified "rule-of-thumb" method, in which the C_p criterion (6) is replaced by the MDL criterion (3); and
- llrDPI: local linear regression where the bandwidth is chosen by the "direct plug-in" method of Ruppert, et al. (1995).

In the above llrDPI is the most computationally expensive procedure, as it adopts a more sophisticated plug-in strategy. It is the procedure recommended by Ruppert, et al. (1995). In this simulation study it is used as a benchmark for comparing the performances of the remaining four procedures.

4.1. *Experimental Setup*

The experimental setup used in this simulation study was designed by Professor Steve Marron. The setup used here was designed to study the effects of varying the (i) noise level, (ii) design density and (iii) degree of spatial variation in an independent and effective fashion. The idea is as follows. Totally three sets of numerical experiments are to be performed. For each set of experiments, only one of the above three experimental factors (e.g., noise level) is changed while the remaining two are being kept unchanged. Within each set of experiments, the factor under examination is changed six times. Thus there are altogether 18 different configurations. In this way it is believed that patterns can be more easily detected. The number of replications for each of the 18 configurations was 200. The setup specification is listed in Table 1, and typical simulated data sets for these configurations are displayed in Figures 1 to 3.

For each simulated data set, the mean-squared-errors $\text{MSE}(\hat{m}) = \frac{1}{n}\|m-$

$\hat{m}\|^2$ was used to evaluate the quality of any curve estimates $\hat{m}$. The averages and standard errors of these MSE($\hat{m}$) values for the 18 different configurations are given in sections (a) and (b) of Table 2 and section (a) of Table 3.

Paired t-tests were also applied to test if the difference between mean MSE values of any two methods is significant or not. To adjust for multiple testing, the significance level used was 1.00%. The methods were also ranked in the following manner. If the mean MSE value of a method is significantly less than the remaining four, it will be assigned a rank 1. If the mean MSE value of a method is significantly larger than one but less than three methods, it will be assigned a rank 2, and similarly for ranks 3 to 5. Methods having non-significantly different mean MSE values will share the same averaged rank. The resulting rankings are tabulated in section (b) of Table 3.

4.2. *Results*

The overall t-test rankings for llrDPI, llrMDL, llrCp, blockMDL and blockCp are, respectively, 2.08, 2.03, 3.14, 3.42 and 4.33. Therefore, when comparing the rankings for llrMDL and llrCp, and for blockMDL and blockCp, it seems to suggest that the MDL criterion (3) is superior to the C_p criterion (6). Also, when comparing the performances of llrMDL to the computationally more demanding llrDPI, it also suggests that llrMDL is a promising method for computing fast and simple pilot estimates for other more refined nonparametric smoothing procedure.

5. Real Data Example

Displayed in Figure 4 is the motorcycle data set analyzed for example by Hardle and Marron (1995). Regression estimates obtained using methods llrDPI, llrMDL and blockMDL are also given in Figure 4. For this data set both the MDL criterion (3) and the C_p criterion (6) selected the same number of $k = 2$ blocks, thus estimates for llrCp and blockCp are not displayed. The two blocks, having roughly the same number of data points, are separated at about $x = 0.39$. This introduced an unwelcome discontinuity point in the curve estimate, which can be treated as a shortcoming of the blocking method. However, if the blocking method is used as a pilot estimate, it is capable of producing reliable regression estimates; compare the estimates obtained by llrDPI and llrMDL.

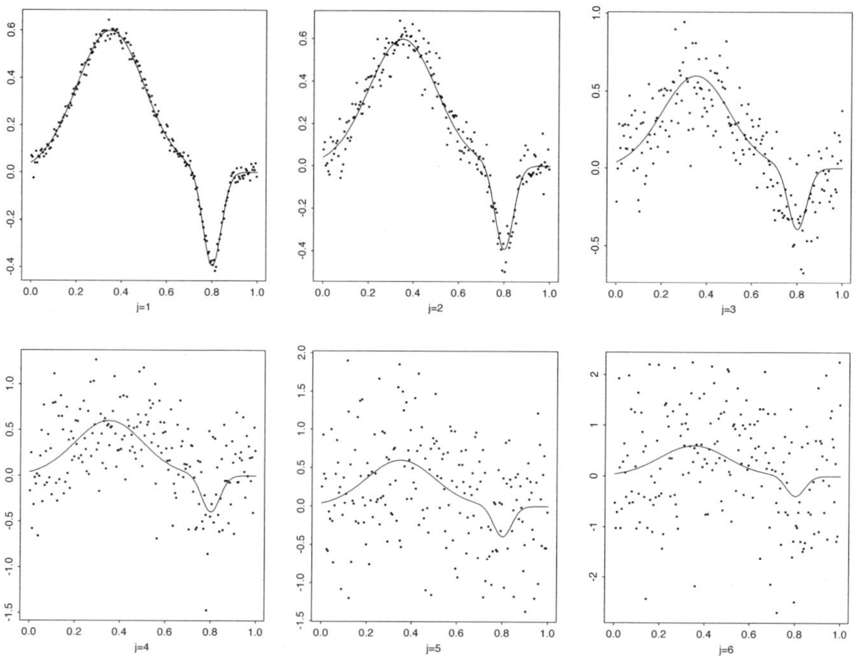

Fig. 1. *Typical simulated noisy data (dots) and the corresponding true regression functions (solid lines) for the "noise level" experimental factor.*

6. Concluding Remarks

This article studies the problem of fast smoothing using the blocking method. In particular an automatic method for choosing the number of blocks is proposed. The proposed method uses the minimum description length principle to define the "best" number of blocks. Theoretical properties of the method are studied via an index of resolvability. Simulation results suggest that the proposed method works really well in practice, and that the blocking estimate can be applied as a quick and reliable method for obtaining pilot estimates for other more refined nonparametric curve estimation methods.

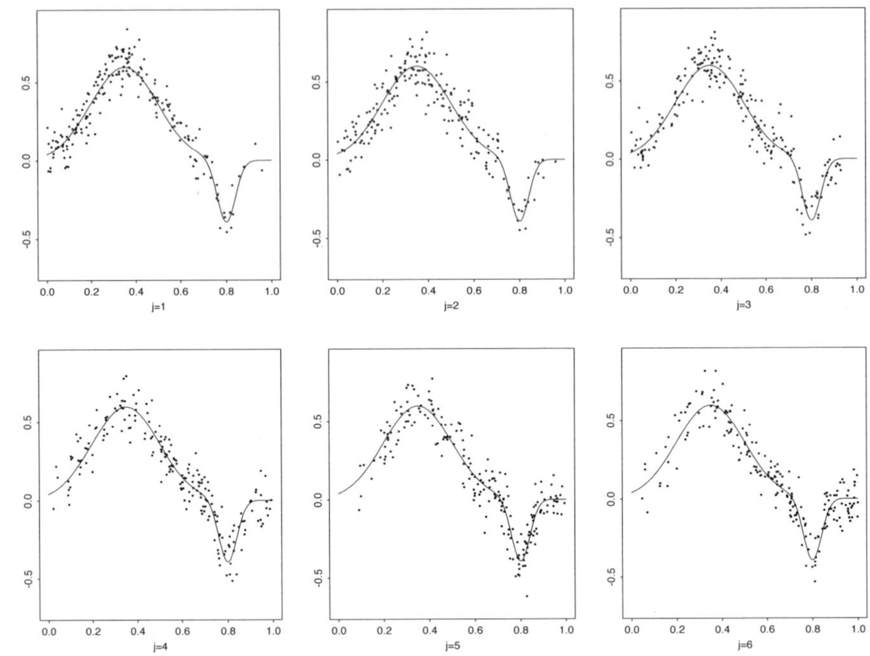

Fig. 2. *Similar to Figure 1 but for the "design density" experimental factor.*

Appendix A. Derivation of MDL(k)

This appendix outlines the derivation of MDL(k), which is an approximation to

$$L(\text{"data"}) = L(k) + L(\hat{\beta}_{jl}; j = 1, \ldots, k, l = 0, \ldots, p) + L(\text{"residuals"}).$$

Since k is an integer, the code length $L(k)$ is $\log_2 k$. To calculate the second term, we use the following result of Rissanen (1989): if a parameter estimate is estimated from N observations, then its code length is asymptotically $\frac{1}{2}\log_2 N$. Since, for any given j, each of the $\hat{\beta}_{jl}$'s is estimated from n_j data points, $L(\hat{\beta}_{j0}, \ldots, \hat{\beta}_{jp}) = \frac{p+1}{2}\log_2 n_j$. Therefore $L(\hat{\beta}_{jl}; j = 1, \ldots, k, l = 0, \ldots, p) = \frac{p+1}{2}\sum_{j=1}^{k}\log_2 n_j$. For the last term, Rissanen demonstrates that it is equal to the negative of the conditional like-

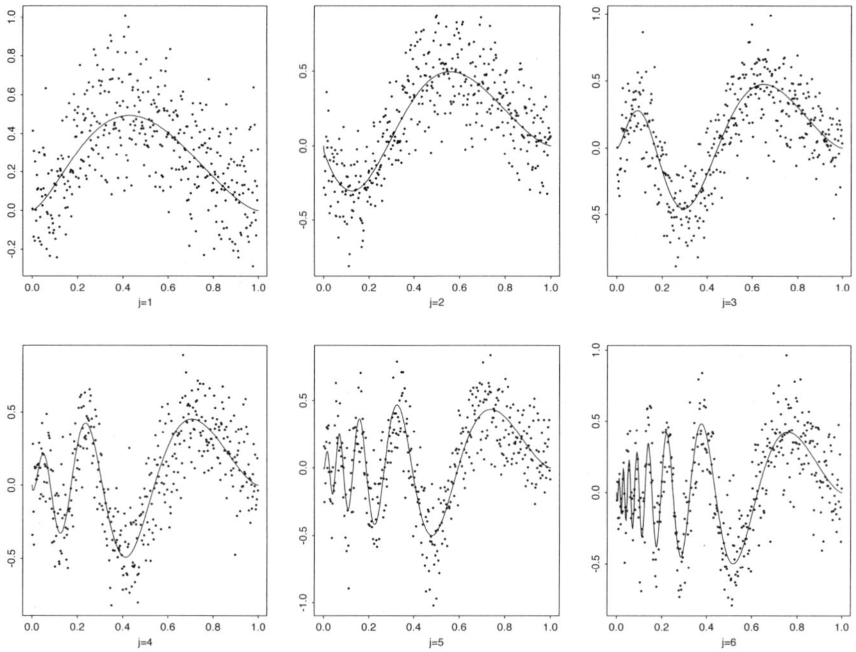

Fig. 3. *Similar to Figure 1 but for the "spatial variation" experimental factor.*

lihood function given the fitted model, which is

$$\log_2 \frac{1}{(2\pi\sigma^2)^{\frac{n}{2}}} \exp\left[-\frac{1}{2\sigma^2}\sum_{i=1}^{n}\{y_i - \hat{m}_{\mathrm{B}}(x_i;k)\}^2\right].$$

Ignoring constant terms, this reduces to $\frac{1}{2\sigma^2}\sum_{i=1}^{n}\{y_i - \hat{m}_{\mathrm{B}}(x_i;k)\}^2$. Now combining these three expressions together we obtain MDL(k).

References

Barron, A. (1991). Complexity Regularization with Application to Artificial Neural Networks, In Roussas, G. (ed.), *Nonparametric Function Estimation and Related Topics* (Kluwer, Dordrecht), 561–576 NATO ASI Series

Barron, A. R., Rissanen, J. and Yu, B. (1998). The Minimum Description Length Principle in Coding and Modeling, *IEEE Transactions on Information Theory* **44**, 6, 2743–2760.

Climov, D., Delecroix, M. and Simar, L. (2002). Semiparametric Estimation in Single Index Poisson Regression: a Practical Approach, *Journal of Applied Statistics* **29**, 7, 1047–1070.

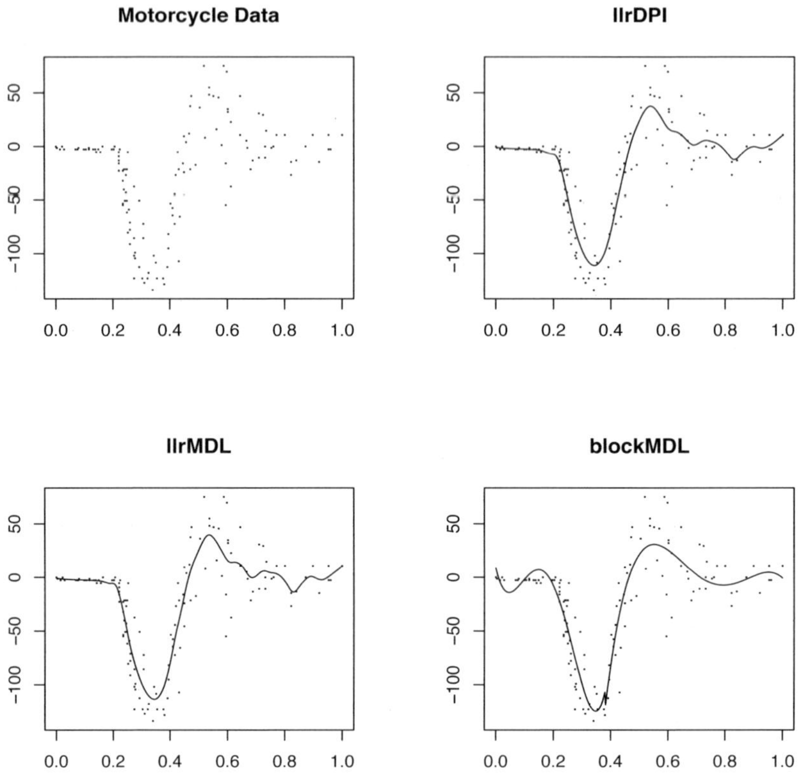

Fig. 4. *The Motorcycle data (dots) and various regression estimates (solid lines).*

Fan, J. and Gijbels, I. (1995). Data–driven bandwidth selection in local polynomial fitting: variable bandwidth and spatial adaptation, *Journal of the Royal Statistical Society Series B* **57**, 2, 371–394.

Fan, J. and Gijbels, I. (1996). *Local Polynomial Modelling and Its Applications* (Chapman and Hall, London).

Härdle, W. and Marron, J. S. (1995). Fast and Simple Scatterplot Smoothing, *Computational Statistics and Data Analysis* **20**, 1–17.

Liu, J. and Moulin, P. (2001). Complexity–Regularized Image Denoising, *IEEE Transactions on Image Processing* **10**, 6, 841–851.

Mallows, C. L. (1973). Some comments on C_p, *Technometrics* **15**, 4, 661–675.

Müller, U. U. (2000). Nonparametric Regression for Threshold Data, *Canadian Journal of Statistics* **28**, 2, 301–310.

Rissanen, J. (1989). *Stochastic Complexity in Statistical Inquiry* (World Scientific, Singapore).

Ruppert, D., Sheather, S. J. and Wand, M. P. (1995). An effective bandwidth selector for local least squares regression, *Journal of the American Statistical Association* **90**, 1257–1270.

Table 1. *Specification of the simulation setup.*

factor	generic form	particular choices
noise level	$y_{ij} = f(x_i) + \sigma_j \epsilon_i$	n=200,$\sigma_j = 0.02 + 0.04(j-1)^2$
design density	$y_{ij} = f(X_{ji}) + \sigma \epsilon_i$	$n = 200, \sigma = 0.1,\ X_{ji} = F_j^{-1}(X_i)$
spatial variation	$y_{ij} = f_j(x_i) + \sigma \epsilon_i$	$n = 400, \sigma = 0.2, f_j(x) = \sqrt{x(1-x)} \sin\left[\frac{2\pi\{1+2^{(9-4j)/5}\}}{x+2^{(9-4j)/5}}\right]$

$j = 1, \ldots, 6;\quad x_i = \frac{i-0.5}{n};\quad \epsilon_i \sim$ iid $N(0,1);\quad f(x) = 1.5\phi\left(\frac{x-0.35}{0.15}\right) - \phi\left(\frac{x-0.8}{0.04}\right)$

$\phi(u) = \frac{1}{\sqrt{2\pi}} \exp(\frac{-u^2}{2});\quad X_i \sim$ iid Uniform$[0,1]$; F_j is the Beta $\left(\frac{j+4}{5}, \frac{11-j}{5}\right)$ c.d.f.

Table 2. (a) Averages and standard errors (in parentheses) of the MSE($\hat{m}$) values for the "noise level" experimental factor. (b) Similar to Section (a) but for the "design density" experimental factor.

Section (a)

j	llrDPI		llrMDL		llrCp		blockMDL		blockCp	
1	0.661	(0.0100)	0.695	(0.00933)	0.706	(0.00958)	0.806	(0.0114)	0.796	(0.0131)
2	3.93	(0.0779)	4.28	(0.0787)	4.36	(0.0895)	4.55	(0.107)	4.99	(0.124)
3	24.0	(0.600)	26.4	(0.659)	28.7	(0.790)	53.1	(0.761)	42.5	(1.02)
4	73.0	(1.83)	67.8	(1.67)	80.8	(2.43)	88.9	(1.51)	112	(2.97)
5	180	(6.30)	158	(5.05)	196	(7.80)	167	(5.23)	238	(11.4)
6	368	(14.2)	322	(10.3)	406	(17.9)	326	(11.1)	477	(26.1)

Section (b)

j	llrDPI		llrMDL		llrCp		blockMDL		blockCp	
1	72.8	(15.4)	60.1	(11.3)	61.9	(10.2)	335	(92.1)	800	(166)
2	37.7	(4.26)	43.6	(9.24)	64.3	(10.7)	127	(13.7)	237	(45.2)
3	24.2	(4.87)	25.8	(2.74)	34.5	(3.69)	57.5	(4.32)	60.9	(6.66)
4	26.0	(7.29)	33.2	(6.10)	41.8	(6.75)	29.9	(2.20)	34.4	(2.95)
5	47.0	(7.15)	59.2	(8.65)	59.5	(8.62)	35.6	(7.10)	51.8	(14.9)
6	140	(24.7)	121	(19.0)	122	(19.7)	79.8	(25.3)	89.8	(25.6)

Table 3. *(a) Similar to Table 2(a) but for the "spatial variation" experimental factor. (b) Paired t-test rankings of the five methods.*

Section (a)

j	llrDPI		llrMDL		llrCp		blockMDL		blockCp	
1	7.26	(0.252)	6.86	(0.218)	8.00	(0.366)	5.72	(0.212)	7.96	(0.533)
2	9.87	(0.243)	9.93	(0.235)	10.9	(0.328)	7.62	(0.197)	11.0	(0.503)
3	13.5	(0.347)	14.3	(0.326)	15.1	(0.386)	14.3	(0.303)	17.3	(0.530)
4	21.6	(0.391)	23.7	(0.394)	24.8	(0.436)	27.9	(0.511)	31.9	(0.626)
5	35.2	(0.489)	36.9	(0.511)	38.7	(0.539)	65.4	(0.770)	62.3	(0.667)
6	60.4	(0.613)	56.1	(0.575)	54.4	(0.544)	87.4	(1.03)	82.5	(0.720)

Section (b)

	noise level					design density					spatial variation				
j	llrDPI	llrMDL	llrCp	blockMDL	blockCp	llrDPI	llrMDL	llrCp	blockMDL	blockCp	llrDPI	llrMDL	llrCp	blockMDL	blockCp
1	1	2	3	4.5	4.5	2	2	2	4	5	4	2	4	1	4
2	1	2	3	4	5	1.5	1.5	3	4.5	4.5	2.5	2.5	4.5	1	4.5
3	1	2	3	5	4	1.5	1.5	3	4.5	4.5	1	2.5	4	2.5	5
4	2	1	3	4	5	3	3	3	3	3	1	2	3	4	5
5	3	1	4	2	5	3	3	3	3	3	1	2	3	5	4
6	3	1.5	4	1.5	5	3	3	3	3	3	3	2	1	5	4
average	1.83	1.58	3.33	3.50	4.75	2.33	2.33	2.83	3.67	3.83	2.08	2.17	3.25	3.08	4.42

Chapter 17

Some Recent Advances in Semiparametric Estimation of the GARCH Model

Jianing Di and Ashis Gangopadhyay

Boston University, Boston, MA, and Johnson&Johnson Pharmaceutical R&D, LLC, San Diego, CA

Semiparametric approaches are often used in estimating the GARCH model parameters. These methods improve the estimation efficiency, as compared to traditional parametric (quasi/pseudo likelihood) approaches, by avoiding the efficiency loss caused by the misspecification of unknown error density. This work summarizes recent developments in semiparametric GARCH estimators. The estimation methodologies, estimator characteristics, advantages and limitations associated with semiparametric estimation of GARCH models are discussed.

Contents

1. The GARCH Model and Its Parameter Estimation

During the last several decades, a significant body of knowledge has been developed in the area of modeling financial time series. In particular, after the introduction of the Autoregressive Conditional Heteroscedasticity

(ARCH) model by Engle (1982) and the Generalized ARCH (GARCH) model by Bollerslev (1986), considerable research has been focused on modeling volatility of observed financial returns. Suppose r_t denotes the observed return and σ_t denotes the volatility. Then the GARCH model assumes the following mathematical form in its original formulation

$$r_t = \sigma_t \epsilon_t \ , \tag{1a}$$

$$sigma_t^2 = \alpha_0 + \sum_{i=1}^{p} \alpha_i r_{t-i}^2 + \sum_{j=1}^{q} \beta_j \sigma_{t-j}^2 \ , \tag{1b}$$

where $\epsilon_t \overset{i.i.d}{\sim} N(0,1)$. The GARCH model captures the well-known empirical characteristics, also known as stylized facts of financial time series (second-order correlation, fat-tail, volatility cluster, etc.) via a simple parametric structure.

Over the years, the GARCH model has been modified in different ways. For example, an additive variance component in the mean equation (1a) can be included to reflect the risk premia (e.g., GARCH-M, Engle, et al. (1987)), a more flexible error distribution f can be considered to allow non-Gaussian innovation (e.g., Baillie and Bollerslev (1989)), an asymmetric parametric form can be assumed to capture the leverage effect (e.g., EGARCH, Nelson (1991) and AGARCH, Engle (1990)), a nonlinear variance modification of the variance equation (1b) can be used to provide extra model flexibility (e.g., NARCH, Higgins and Bera (1992)), and a (fractional) unit root can be incorporated to represent the long memory of the volatility (e.g., IGARCH, Engle and Bollerslev (1986) and FIGARCH, Baillie, et al. (1996)). In addition, typical regression models may be integrated with GARCH model by allowing the innovation to follow a GARCH process. For example, a linear regression with GARCH errors (e.g., Linton (1993)) or an ARMA model with GARCH errors (e.g., Ling and Li (1997), Francq and Zakoïan (2004)). For reviews of various modifications of the GARCH models, see Bollerslev (1994) and Hentschel (1995). The GARCH model has also been generalized to multivariate cases (e.g., MGARCH, Bollerslev, et al. (1988)) and continuous cases (e.g., CGARCH, Drost and Werker (1996), and COGARCH, Klüppelberg, et al. (2004), Brockwell, et al. (2006)).

Besides the extensions of the GARCH model, the problems related to the estimation of GARCH parameters have also attracted significant attention. Different methods have been studied in the literature. Some of the most common approaches include likelihood-based techniques that derive and maximize the likelihood function (e.g., Engle (1982), Bollerslev (1986), Engle and González-Rivera (1991), Linton (1993)), the model-structure-based approaches (e.g., GMM, Rich, et al. (1991), Yule–Walker method, Maercker and Moser (1999)), and the frequency-domain-based methods (e.g., Mikosch and Straumann (2002)). Some of these methods are general in the sense that all model parameters can be estimated together, but other approaches estimate the model parameters in steps or only estimate a certain part of the model. In any case, the core part of the GARCH type models is always the variance equation (1b or similar) and the parameters therein, i.e., $\theta = (\alpha_0, \alpha_1, \ldots, \alpha_p, \beta_1, \ldots, \beta_q)$. For the rest of the paper, we denote by f the true density of the innovations.

In the present work, we will focus on the likelihood-based approaches in estimating the variance equation (1b). The likelihood-based estimation of this model has been studied under various scenarios (Engle (1982), Bollerslev (1986), Weiss (1986), Engle and González-Rivera (1991), Linton (1993), Lumsdaine (1996), Drost and Klaassen (1997), Yang (1998), González-Rivera and Drost (1999), Berkes, et al. (2003), Mukherjee (2006), Hafner and Rombouts (2007), Di and Gangopadhyay (2011), Di and Gangopadhyay (2010)). While the estimation of the GARCH parameters is not particularly difficult, the evaluation of the characteristics of those estimators is in fact non-trivial, even when the true innovation density f is known. Among the challenges, the most commonly studied are the asymptotic distribution and efficiency of the estimators.

The goal of this work is to discuss the estimation methodology and related implications associated with semiparametric estimation of the GARCH parameters, where assumptions on the error distribution is relaxed by taking it to be any absolutely continuous probability density f. This is an important generalization of the GARCH model, since in practice the conditional normality assumption of r_t is rarely satisfied for financial data. In fact, empirical evidences suggest that for financial data the conditional normality assumption is usually violated, and specification of an appropriate parametric error distribution is often difficult.

This work explores various semiparametric estimation techniques that are currently available in the literature. The semiparametric estimators of GARCH models can be broadly classified into two types: first, estimators involving a two-step process that utilizes an initial parameter value to obtain the final estimator; and the second, where no such initial input is necessary. There exist questions in terms of the efficiency of the estimators and the identifiability of the parameters being estimated. This paper provides an overview of the current knowledge of these issues, and points out some of the open questions.

To simplify notations, the discussion is restricted to the GARCH(1,1) model. The estimation and evaluation of the parameters of higher order GARCH models or more general conditional heteroscedastic models can often be accomplished similarly. The innovation distribution is assumed to have zero mean and unit variance. While the innovation mean is sometimes estimated in a location-scale model together with other model parameters (Steigerwald (1992)), the zero mean assumption can always be achieved by appropriate pre-whitening of the data. The unit variance assumption is not required for parameter estimation under some model parametrization, but is added here to avoid the parameter identification problem (Di (2007), Di and Gangopadhyay (2011)). We will also assume the necessary conditions for the existence of ergodic and stationary GARCH process (Nelson (1991), Bougeroland Picard (1992)). Finally, in reference to the established results, the original conditions may not be listed in detail to simplify the discussion.

The paper is organized as follows: the rest of Section 1 provides a review of parametric and semiparametric estimation in the GARCH setup. Section 2 discusses the methodology of typical likelihood-based semiparametric GARCH estimators. Examples based on simulated data are discussed to compare the performance of the estimators. Section 3 discusses the efficiency of the semiparametric estimators. In particular, factors that affect the estimation efficiency are discussed. The last section outlines the future developments of the semiparametric GARCH estimation.

1.1. *Parametric GARCH Estimators*

Consider the GARCH(1,1) model

$$r_t = \sigma_t \epsilon_t \ , \tag{2a}$$

$$\sigma_t^2 = \alpha_0 + \alpha_1 r_{t-1}^2 + \beta_1 \sigma_{t-1}^2 \ , \tag{2b}$$

where $\epsilon_t \overset{i.i.d}{\sim} f(0,1)$. Let θ^* denote the true value of $\theta = (\alpha_0, \alpha_1, \beta_1)$. When f is known, the maximum likelihood (ML) estimator can be derived by maximizing the log-likelihood function

$$L_n(\theta) = \frac{1}{n}\sum_{t=1}^{n} l_t(\theta) = \frac{1}{n}\sum_{t=1}^{n} \log\left[\frac{1}{\sigma_t(\theta)} f\left(\frac{r_t}{\sigma_t(\theta)}\right)\right] \ , \tag{3}$$

where the unconditional log-density of the initial observation is assumed to be asymptotically negligible and thus dropped. The MLE is, under regularity conditions, asymptotically efficient in the sense of Cramér-Rao bound

(González-Rivera and Drost (1999), see also Weiss (1986), Wooldridge (1994), Lumsdaine (1996)):

$$\sqrt{n}(\hat{\theta}_n^{MLE} - \theta^*) \xrightarrow{\mathcal{D}} N(0, \Sigma_{MLE}) , \tag{4}$$

where Σ_{MLE} represents the inverse of the Fisher information.

The ML estimation of θ becomes significantly more challenging when f is unknown, since the true likelihood function cannot be directly formulated. To accommodate this, the quasi-ML approaches that assume an arbitrary functional form to the unknown f have been extensively studied (e.g., Weiss (1986), Lee and Hansen (1994), Lumsdaine (1996), Newey and Steigerwald (1997), Berkes and Horváth (2003), Francq and Zakoïan (2004), Mukherjee (2006)). More specifically, the QML estimator (QMLE) that maximizes the quasi likelihood function

$$L_n^{(g)}(\theta) = \frac{1}{n}\sum_{t=1}^{n} l_t(\theta) = \frac{1}{n}\sum_{t=1}^{n} \log\left[\frac{1}{\sigma_t(\theta)} g\left(\frac{r_t}{\sigma_t(\theta)}\right)\right] \tag{5}$$

can be derived, where g is an arbitrary parametric replacement of the unknown f. By selecting some specific g, the ergodic theorem can be used to show that the corresponding QMLE is consistent, even if $g \neq f$. For example, under various moment conditions (Weiss (1986), Bollerslev and Wooldridge (1992), Lee and Hansen (1994), Lumsdaine (1996)), the most commonly used Gaussian QMLE (by choosing $g = N(0,1)$) is consistent with an asymptotic distribution given by

$$\sqrt{n}(\hat{\theta}_n^{GQMLE} - \theta^*) \xrightarrow{\mathcal{D}} N(0, \Sigma_{GQMLE}) , \tag{6}$$

where $\Sigma_{GQMLE} \geq \Sigma_{MLE}$ in the matrix sense (see also González-Rivera and Drost (1999)). Berkes and Horváth (2004) also discussed a general approach of deriving consistent QMLE.

Different innovation densities can be considered in the QMLE. For example, Student's t distribution has been used to capture the heavy tails that are commonly observed financial returns (Baillie and Bollerslev (1989)). Other distributions can also be applied to meet specific empirical characteristics of the data. Using simulated data, Venter and de Jongh (2004) reported that the QMLE based on normal inverse Gaussian (NIG) distribution appears to be the most efficient among the cases considered in the study.

However, certain limitations are also introduced by the fact that, in the QML approach, a specific (and most of the time incorrect) parametric form is assumed for f. First of all, restrictive conditions on the assumed density g are required in order for the QMLE to have good large sample properties. For example, Berkes and Horváth (2003) showed that, for $1 < v < 2$, the Gaussian QMLE satisfies $\hat{\theta}_n^{GQMLE} - \theta^* = o_P(n^{-(1-1/v)})$ if and only if $E[\epsilon_t^{2v}] < \infty$. In particular, the Gaussian QMLE is $\sqrt{n}$-consistent if $E[\epsilon_t^{2(2+\delta)}] < \infty$ for some $\delta > 0$. It is easy to see that such a condition is always needed for Gaussian QMLE as ϵ_t^4 is involved in Fisher information matrix based on the normal density. Similar results could be found in Mikosch and Straumann (2002) and Hall and Yao (2003), who argued that in the presence of a heavy-tailed error distribution, the QMLE may follow a non-Gaussian asymptotic distribution. Secondly, loss of estimation efficiency occurs for a QMLE if $g \neq f$. For example, it can be shown that the Gaussian QMLE inflates the estimation variability by a factor of $(\kappa - 1)/2$ over the true MLE (Engle and González-Rivera (1991), González-Rivera and Drost (1999), Francq and Zakoïan (2004)), where κ is the kurtosis of f. Thirdly, the QMLE is not generally consistent. For example, Mukherjee (2006) showed that the general QMLE of an ARCH model is asymptotically Gaussian but biased with an asymptotic distribution given by

$$\sqrt{n}(\hat{\theta}_n^{QMLE} - c_{g,f}\theta^*) \xrightarrow{\mathcal{D}} N(0, \Sigma_{QMLE}) , \tag{7}$$

where g is the assumed density, $c_{g,f} > 0$ is the unique constant such that

$$E\left[1 + \frac{\epsilon_t}{c_{g,f}^{1/2}} \frac{g'\left(\epsilon_t/c_{g,f}^{1/2}\right)}{g\left(\epsilon_t/c_{g,f}^{1/2}\right)}\right] = 0 , \tag{8}$$

and $\Sigma_{QMLE} \geq \Sigma_{MLE}$. This constant $c_{g,f}$ can be easily calculated. For example, $c_{g,f} = 1$ if $g = f$ or $g = N(0,1)$, and $c_{g,f} = 0.82$ if $g = t_{15}$ and $f = F_{10,50}$. See Fan, et al. (2010) for a similar discussion.

1.2. *Semiparametric GARCH Estimators*

When a simple parametric form is assumed to the unknown pdf f, it simplifies the estimation procedure. However, the trade-off is the corresponding efficiency loss that occurs if the assumed density g deviates from f, and this kind of inefficiency cannot be improved by enlarging the sample size. A natural modification to this QML approach is to find a better estimate of the unknown f. One may always allow g to be represented by a (finite dimensional) shape parameter η. For example, this η can be the degrees of freedom if g is taken to be a Student's t distribution. If $g(\eta)$ is a correct specification for the underlying density f, then as $n \to \infty$, $g(\hat{\eta}_n)$ is expected to get close to f by utilizing a good estimator $\hat{\eta}_n$ of η. Nevertheless, the corresponding estimation procedure often lacks the necessary orthogonality between the shape parameter and the parameter of interest (Stein (1956), Engle and González-Rivera (1991), Bickel, et al. (1993), Drost and Klaassen (1997)). Thus, the efficiency loss always exists if f is modeled parametrically. A possible remedy is to include the shape parameter into an infinite dimensional nuisance space, and capture f nonparametrically. In this case, the estimation can potentially be largely improved by not making any spurious assumptions about the form of f. In this paper, it is this semiparametric approach that will be discussed in detail.

In the statistics literature, the term *semiparametric* has been used to refer to different estimation scenarios. One commonly studied problem is where certain components of the (regression) model are characterized parametrically, while other components are modeled nonparametrically (Ruppert, et al. (2003)). In GARCH models, this idea is often incorporated by having a nonparametric functional form for the volatility function (2b, see Pagan and Schwert (1990), Härdle and Tsybakov (1997), Yang, et al. (1999), Linton and Mammen (2005), Yang (2006)). However, in the present work we focus on the scenario involving a fully parametrically specified model incorporating an unknown innovation density. This model is a natural generalization of the parametric GARCH model that takes into account the empirical difficulties associated with accurate parametrization of the conditional distribution of the returns r_t. In particular, we are interested in studying the approaches in which the unknown density f is nonparametrically estimated (Engle and González-Rivera (1991), Linton (1993), Steigerwald (1993), Andrews (1994), Drost and Klaassen (1997), Drost, et al. (1997), Sun and Stengos (2006), Hafner and Rombouts (2007), Di (2007), Di and Gangopadhyay (2011; 2010)). Estimation methods that completely avoid specification of the unknown f are also available (Rich, et al. (1991), Davison and Eddelbüttel (1995), Koenker and Zhao (1996), Maercker and Moser (1999)), but will not be discussed here.

2. Likelihood Based Semiparametric GARCH Estimation

Define the semiparametric MLE (SMLE) as the maximizer of the semiparametric likelihood function

$$\hat{L}_n(\theta) = \frac{1}{n}\sum_{t=1}^{n} l_t(\theta) = \frac{1}{n}\sum_{t=1}^{n} \log\left[\frac{1}{\sigma_t(\theta)}\hat{f}_n\Big(\frac{r_t}{\sigma_t(\theta)}\Big)\right], \tag{9}$$

where $\hat{f}_n$ is a nonparametrically derived estimate of the unknown f. Note that the SMLE is really a special case of the QMLE in the sense that the fixed function g assumed in the QMLE is replaced by an estimate $\hat{f}_n$ in the SMLE. The SMLE is intuitively better than the QMLE since a suitably chosen $\hat{f}_n$ may converge to f in some sense, while g doesn't. Therefore with a sufficiently large sample size (which is generally the case in financial time series), the semiparametric likelihood based on $\hat{f}_n$ can better reflect the true likelihood, as compared to the quasi likelihood based on g. To ensure this, it is essential for $\hat{f}_n$ to have the desired asymptotic properties. Hence, the study of the SMLE focuses primarily on the issues related to the construction $\hat{f}_n$, with particular emphasis on the utilization of residual sample in the construction.

2.1. *Two-step Semiparametric Estimators*

Most of the SML approaches available in the literature involve a two-step implementation. In the first step, an initial parameter value is input into the model to create a sample of model residuals recursively via equations (1). To ensure the quality of the residual sample, in the initial step a good plug-in parameter value is necessary. If the initial plug-in is an estimate, ideally this estimator needs to converge fast to the true value, so that for sufficiently large sample size the residual sample approximates a sample realization from f. In the second step, the semiparametric likelihood (9) is derived and maximized. In this step, $\hat{f}_n$ needs to be constructed accurately and good numerical techniques should be applied to find the (local) maximizer of (9). This maximization step can also be replaced by a single order Newton-Raphson influence function, which is usually asymptotically equivalent to direct maximization. In this latter case the SMLE can be viewed as an update of the input parameter value. See Yang (1998) for a finite sample comparison between the two approaches.

Replacing g with $\hat{f}_n$ recaptures some of the efficiency loss associated with the input parameter value, or creates extra orthogonality between θ and the tangent space, and thus this method is often used to derive more efficient (or even adaptive) estimators (Bickel (1982), Linton (1993), Drost and Klaassen (1997), Drost, et al. (1997), Yang (1998), Hafner and Rombouts (2007)). We will discuss this topic further in Section 3.

Engle and González-Rivera (1991) first introduced a semiparametric approach in which the sample residuals are created by plugging in the Gaussian QMLE and $\hat{f}_n$ is calculated via the discrete maximum penalized likelihood (DMPL) approach (Tapia and Thompson (1978), Scott, et al. (1980)). Simulations were carried out under the assumption that f follows the Gamma or t_5 distribution. Clear efficiency gain over the Gaussian QMLE was reported when f is Gamma but not t_5.

A limitation of the method proposed by Engle and González-Rivera (1991) is that the unknown pdf f is estimated via DMPL, which is a finite dimensional linear interpolation of the classical histogram. This estimator is not sufficiently smooth near the tails of the distribution, and does not extend naturally to the estimation of the derivatives of f. This may explain why the asymptotic characteristics of the estimator $\hat{\theta}_n^{SMLE-EG}$ were not derived analytically in their paper.

Still, this work had stimulated significant research activity in the two-step semiparametric estimation of the GARCH model. Several authors had extended the idea to allow more general estimation procedures (Steigerwald (1993), Linton (1993), Drost and Klaassen (1997), Hafner and Rombouts (2007), Di and Gangopadhyay (2011)). These extensions focused on different model specifications, but maintained the same two-step estimation mechanism. The Gaussian QMLE was the usual choice of the initial input parameter due to its consistency and computational simplicity, but the unknown innovation distribution f (or the unknown score/influence function, if the Newton-Raphson update is performed) was often estimated via kernel density estimation approach to provide more flexibility.

Steigerwald (1993) studied a two-step SMLE in which the unknown pdf f is estimated via a trimmed kernel. More specifically, an even trimming function $d(\cdot)$ was used to restrict the estimated density to zero outside a pre-specified compact set. Under regularity conditions, the proposed estimator is consistent and asymptotically Gaussian:

$$\sqrt{n}(\hat{\theta}_n^{SMLE2-ST} - \theta^*) \xrightarrow{\mathcal{D}} N(0, \Sigma_{SMLE2-ST}) , \tag{10}$$

where $\Sigma_{SMLE2-ST}$ depends on $d(\cdot)$ and $\Sigma_{SMLE2-ST} > \Sigma_{MLE}$.

In a later study, Hafner and Rombouts (2007) investigated the semiparametric estimation in multivariate GARCH models with innovations based on a class of spherical distributions. The kernel estimator of the score/influence function was also trimmed. However, unlike the approach of Steigerwald (1993) where the trimming was only applied to extreme residuals, the trimming boundary used in Hafner and Rombouts (2007) also included conditions involving the values of $\hat{f}_n$ and $\hat{f}'_n/\hat{f}_n$. In addition, the trimming boundary was asymptotic in the sense that the boundary expanded as a function of the sample size. By selecting the boundary carefully, they showed that the two-step SMLE with sample size dependent trimming functions for the (multivariate) GARCH parameter θ, denoted as $\hat{\theta}_n^{SMLE2-HR}$, satisfies

$$\sqrt{n}(\hat{\theta}_n^{SMLE2-HR} - \theta^*) \xrightarrow{\mathcal{D}} N(0, \Sigma_{SMLE2-HR}) , \tag{11}$$

where $\Sigma_{SMLE2-HR} \geq \Sigma_{MLE}$, but reaches the efficiency bound for consistent semiparametric estimators (see also González-Rivera and Drost (1999) and Section 3).

This kind of trimming technique is often applied in estimating functionals of a probability density function to eliminate the effect of extreme observations (i.e., the boundary effect), so that $\sqrt{n}$-consistent estimators can be derived (Robinson (1988b), Bickel (1982), Bickel, et al. (1993), Newey and McFadden (1994), Andrews (1995), Lewbel (2000)). However, by retaining only the observations with moderate values, trimming causes loss of information, especially when f is heavy tailed. This explains the large asymptotic variance of $\hat{\theta}_n^{SMLE-ST}$ and $\hat{\theta}_n^{SMLE-HR}$. Even when the boundaries are allowed to expand with n, the expansion is controlled by terms that are stated in orders of magnitude such as $O(\cdot)$ or $o(\cdot)$. Therefore the finite sample performances of these methods are largely unknown. See Section 1.3.3.2 for further discussion on this topic.

Di and Gangopadhyay (2011) showed that the two-step SMLE with a regular kernel density estimator can be consistent and asymptotically Gaussian even if the kernel estimator $\hat{f}_n$ is not trimmed. In addition, since there is no information loss as is usually the case with trimming, the asymptotic variance-covariance matrix of their two-step SMLE matches that of the true MLE. However, the trade-off is that their two-step SMLE involves a slowly-converging bias term:

$$\sqrt{n}(\hat{\theta}_n^{SMLE2-DG} - \theta^* - b_n^{SMLE2-DG}) \xrightarrow{\mathcal{D}} N(0, \Sigma_{SMLE2-DG}) , \tag{12}$$

where $\Sigma_{SMLE2-DG} = \Sigma_{MLE}$, and $b_n^{SMLE2-DG} \to 0$ at a rate slower than $O(n^{-1/2})$. This bias term b_n should be expected due to the use of a non-trimmed kernel estimator, and its nonparametric convergence rate has also been seen in other semiparametric models (Speckman (1988), Härdle (1990)). Equation (12) illustrates several interesting facts:

(1) The term $b_n^{SMLE2-DG}$ in (12) dominates the convergence rate of $\hat{\theta}_n^{SMLE2-DG}$, and therefore $\hat{\theta}_n^{SMLE2-DG}$ is not $\sqrt{n}$-consistent. In addition, the bias and the variance of $\hat{\theta}_n^{SMLE2-DG}$ cannot be balanced by altering the bandwidth used in the kernel estimator – the bias always converges more slowly than the variability. Thus to optimize the

overall rate of convergence, the bandwidth should always be chosen to minimize the bias, subject to the necessary conditions to ensure the asymptotic properties of the SMLE.

(2) By relating (12) to (7) and viewing the SMLE2-DG as a special case of the QMLE, we know $b_n^{SMLE2-DG} - (1 - c_{\hat{f}_n,f})\theta^* = o_P(n^{-1/2})$. Then by (8) and consistency of kernel estimator, we have $c_{\hat{f}_n,f} \to 1$, and thus (7) also implies $b_n^{SMLE2-DG} \to 0$.

(3) Comparing (12) to (10-11), it can be seen that there is a bias-variance trade-off controlled by the trimming function. More specifically, the term $b_n^{SMLE2-DG}$ in (12) appears due to the bias associated with the estimation of f'/f by $\hat{f}_n'/\hat{f}_n$, but the trimming method helps to remove this bias at the expense of larger variance.

2.2. *One-step Semiparametric Estimation*

The two-step SMLE relies on an initial estimate of the parameter θ, thus the two-step semiparametric approach can be viewed as an improvement over this initial estimate. For example, González-Rivera and Drost (1999) showed that the theoretical information bound of the two-step SMLE is lower as compared to the general QMLE. Hafner and Rombouts (2007) established a similar result under a multivariate setup. However, it turns out that the reliance on an initial input is a drawback of the two-step approach, since the performance of the SMLE is highly sensitive to the choice of the input estimate (Di and Gangopadhyay (2010)).

Figure 1 illustrates the sensitivity of the two-step SMLE with respect to the initial input parameter value. It is clear that the characteristics (bias and variance) of the input parameter get carried over to the final two-step SMLE. Interestingly, an increase in the input estimator's variability not only increases the variability of the SMLE2-DG, but also extends the right tail, and therefore creates a positive bias. On the other hand, a negative bias associated with the input estimator not only moves the SMLE2-DG to the left, but also decreases the variability. Other typical two-step SMLEs (SMLE2-EG, SMLE2-ST and SMLE2-HR) exhibit similar patterns in their sampling distributions. In practice, when a pre-selected estimate of θ is used as the initial input, the performance (or even the existence) of the initial estimator varies with the innovation distribution f. This variation is the source of non-robustness of the two-step approach.

A one-step semiparametric approach was briefly mentioned in Engle and González-Rivera (1991), and was subsequently investigated by several

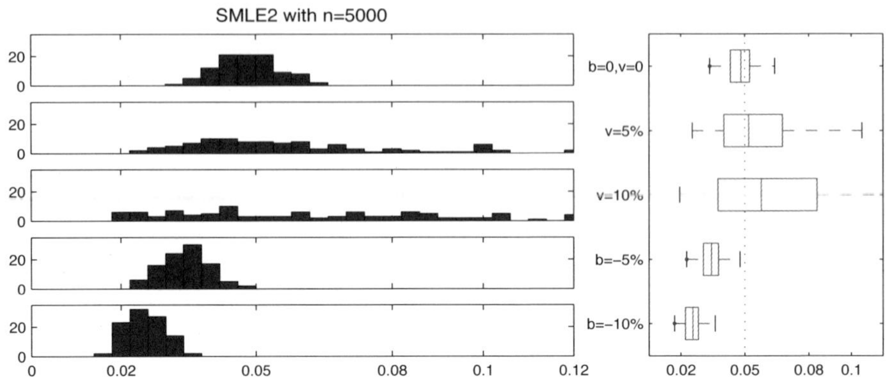

Fig. 1. Sampling distribution (one hundred replications) of the estimates of α_1 with data generated from $\epsilon_t \sim N(0,1)$. The initial estimators are created by adding a random noise with mean of 0 to -0.005 (-10% of the true value $\alpha_1^* = 0.05$) or a standard deviation of 0 to 0.005 to α_1^*. f is estimated using Gaussian kernel. Notation used b: bias, v: std.

authors (Venter and de Jongh (2004), Di and Gangopadhyay (2010)). All these studies considered the maximization of semiparametric likelihoods similar to those in two-step SMLE. However, in an one-step estimation procedure, the residual sample is not created from a pre-defined value or estimate. Instead, the parameter being estimated is utilized directly to generate the residuals. In other words, for a two-step SMLE, the $\hat{f}_n$ in (9) is estimated from $\hat{\epsilon}_1(\tilde{\theta}_n), \ldots, \hat{\epsilon}_n(\tilde{\theta}_n)$, where $\tilde{\theta}_n$ is the initial input and $\hat{\epsilon}_t(\tilde{\theta}_n) = r_t/\hat{\sigma}_t(\tilde{\theta}_n)$ such that $\hat{\sigma}_t(\tilde{\theta}_n)$ is the volatility derived recursively using $\tilde{\theta}_n$ and $r_{i<t}$. On the other hand, in a one-step SMLE, the $\hat{f}_n$ in (9) is estimated from $\hat{\epsilon}_1(\theta), \ldots, \hat{\epsilon}_n(\theta)$. As a result, in the one-step procedure, $\hat{f}$ becomes a function of θ, and consequently a part of the likelihood to be maximized. The concept of the single order Newton-Raphson update remains valid in the one-step approach. However, since the procedure requires an initial parameter value to be updated, in this case it is unclear how exactly the update can be implemented.

Di and Gangopadhyay (2010) showed that the one-step SMLE is consistent and asymptotically Gaussian. In addition, it also has a slowly converging bias term just like its two-step counterpart:

$$\sqrt{n}(\hat{\theta}_n^{SMLE1-DG} - \theta^* - b_n^{SMLE1-DG}) \xrightarrow{\mathcal{D}} N(0, \Sigma_{SMLE1-DG}) , \tag{13}$$

where $b_n^{SMLE1-DG} > b_n^{SMLE2-DG}$ but still $\to 0$ at a rate slower than $O(n^{-1/2})$, and $\Sigma_{SMLE1-DG} > \Sigma_{SMLE2-DG} = \Sigma_{MLE}$. This one-step algorithm does not rely on the initial estimate of θ, and thus more robust to the characteristics of the unknown f. However, as can be seen from the result, the trade-off is that a larger estimation variability is associated with this robustness.

This one-step SML approach introduces a parameter identifiability problem. Recall in (1) and (2) we assume $\text{Var}[\epsilon_t] = 1$. This condition is necessary to ensure the joint identifiability of θ and the overall scale of the GARCH model. In a two-step SMLE that uses the Gaussian QMLE as the initial input, this identifiability problem is naturally resolved as the quasi likelihood function is built-up using $N(0,1)$ as the distribution. In the one-step procedure, on the other hand, an extra step is needed to ensure that the maximization is carried out within the parameter space that reflects the unit variance of ϵ_t, i.e., $\Theta = \{\theta : \text{Var}[\epsilon_t(\theta)] = 1\}$. Alternatively, the model can be reparameterized to ensure the identifiability of the parameters (Linton (1993), Drost and Klaassen (1997)). See Section 1.3.2 for a discussion on this approach.

To ensure parameter identifiability, Di and Gangopadhyay (2010) standardized the residual sample before it is utilized to derive the kernel estimate $\hat{f}_n$. The standardization step guarantees that the model residuals have zero mean and unit variance. In fact, the standardization is an essential step in this context. Fig. 2 illustrates the parameter identifiability issue between α_0 and α_1, and the effect of such standardization step. It

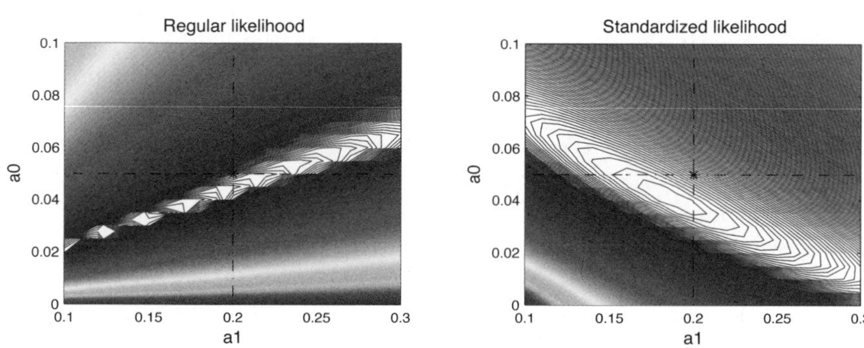

Fig. 2. The log-likelihood contours projected onto the 2-dim space generated by (α_0, α_1) (fix $\beta_1 = 0.9$). Data are generated using a standardized t_4 distribution with $n = 5000$. The left graph illustrates the non-standardized one-step log-likelihood function, and the right graph illustrates the standardized one-step log-likelihood function. The true value of α_0 and α_1 are 0.05 and 0.2, respectively (Di and Gangopadhyay (2010)).

is clear that the semiparametric likelihood is not unimodal without the standardization step.

An alterative standardization approach would be to estimate $\hat{f}_n$ first from the unstandardized residuals, and then standardize the kernel estimate such that it has zero mean and unit variance. The simulations reported in Venter and de Jongh (2004) suggested that both approaches yield similar results.

It is worth pointing out that the two-step estimation algorithm proposed by Engle and González-Rivera (1991) also included a step that standardized the model residuals to have zero mean and unit variance. However, this step does not seem to be necessary since the Gaussian QMLE incorporated in the first step of their procedure is based on the standard Gaussian density, which already utilizes the unit variance assumption. In addition, the effect of this standardization step on the estimation efficiency of $\hat{\theta}_n^{SMLE2-EG}$ is unclear.

2.3. *Two-step SMLE vs. One-step SMLE*

In general, a two-step approach can always be converted into a one-step approach, and vice versa. For example, consider the general QML approach in which a functional form g that includes some shape parameter is used to replace the unknown f. This one-step method can be split into two steps, in the sense that one can always first fit the model using the Gaussian QMLE and then fit the parametric density g using the model residuals, and subsequently derive the quasi likelihood. In the long run, given the consistency of the Gaussian QMLE, the two-step QMLE should be no worse than the original one-step QMLE. Venter and de Jongh (2004) reported similar performances of the two methods using simulations at a relatively small sample size ($n = 500$).

In the semiparametric context, similar results are also expected. Consider a two-step SML approach. If the initial input parameter is consistent, then as sample size increases, the model residual well approximates a sample that follows f. As a result, the residual space on which the unknown density and likelihood are estimated narrows down to a more accurate one. On the other hand, the input value used in the one-step approach varies as one searches freely within the entire parameter space, and a new residual sample is created each time the parameter value changes. In summary, the semiparametric likelihood function of a one-step approach is created from a much larger residual space, and hence is more variable. This explains

Table 1. Comparison of the SMLE1-DG and SMLE2-DG with different innovation densities.

innovation	method	measure	500	2000	5000	8000	10000
$F_{10,50}$	SMLE1	bias	0.0035	-0.0043	-0.0016	-0.0015	-0.0021
		mse	0.0251	0.0091	0.0060	0.0048	0.0040
	SMLE2	bias	0.0015	-0.0040	-0.0016	-0.0019	-0.0024
		mse	0.0284	0.0117	0.0080	0.0049	0.0049
t_4	SMLE1	bias	0.0072	0.0027	0.0001	0.0001	0.0006
		mse	0.0410	0.0229	0.0097	0.0101	0.0071
	SMLE2	bias	0.0150	0.0019	0.0012	0.0001	0.0017
		mse	0.0634	0.0294	0.0151	0.0132	0.0117
Bimodal	SMLE1	bias	0.0024	-0.0030	-0.0015	-0.0015	-0.0014
		mse	0.0336	0.0136	0.0078	0.0058	0.0060
	SMLE2	bias	-0.0016	-0.0046	-0.0044	-0.0031	-0.0030
		mse	0.0258	0.0111	0.0069	0.0059	0.0056
Gaussian	SMLE1	bias	0.0035	-0.0027	-0.0016	-0.0010	-0.0006
		mse	0.0379	0.0142	0.0085	0.0060	0.0062
	SMLE2	bias	0.0024	-0.0034	-0.0022	-0.0007	-0.0009
		mse	0.0328	0.0126	0.0080	0.0061	0.0055

the fact that $\Sigma_{SMLE1-DG} > \Sigma_{SMLE2-DG}$ (Di and Gangopadhyay (2011; 2010)).

However, the larger asymptotic variance matrix of the one-step SML approach may not be a huge disadvantage at moderate sample sizes, since the bias of the semiparametric estimators is in fact the primary factor that affects their performance. For example, since $b_n^{SMLE1-DG} > b_n^{SMLE2-DG}$, the bias of the SMLE1-DG might be 'less negative' for certain choices of f (Di and Gangopadhyay (2010)). Table 1 displays a simulation based comparison between the SMLE1-DG and SMLE2-DG. For brevity, results are given only for the parameter α_1 with true value 0.05. It can be seen that there is no universal preference for either of the two approaches. Although the SMLE1-DG generally illustrates a smaller absolute bias, it only outperforms the SMLE2-DG in terms of the mean square error when f is taken as $F_{10,50}$ or t_4. It is also worth noting that since t_4 does not have finite 4th-moment, the Gaussian QMLE has impaired asymptotic property (Berkes and Horváth (2003), Hall and Yao (2003)), which consequently affects the performance of SMLE2-DG. The simulation confirms the advantage of SMLE1-DG in this case, as it does not use Gaussian QMLE as the input estimator.

3. Efficiency of the Semiparametric GARCH Estimators

Semiparametric GARCH models are introduced to improve the efficiency of GARCH estimation over the commonly used QMLEs. Figure 3 illustrates the efficiency advantage of SMLEs via a simulation example. It can be seen that the SMLEs exhibit smaller variability as compared with the QMLEs. However, it is also clear that the SMLEs are not as optimal as the MLE,

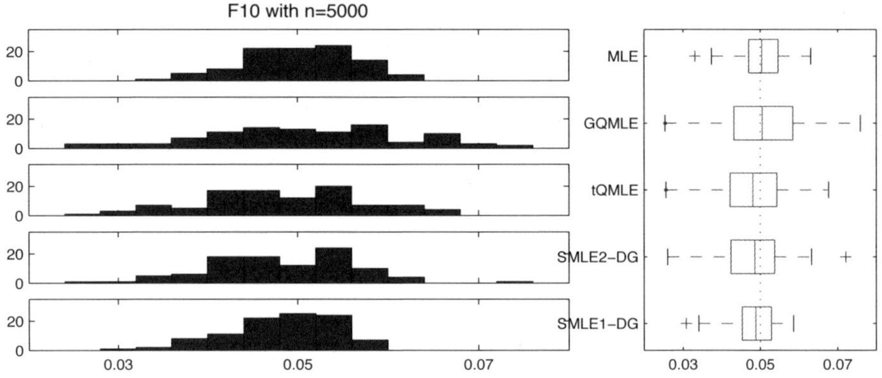

Fig. 3. Sampling distribution (one hundred replications) of the estimates of α_1 with data generated from $\epsilon_t \sim F_{10,50}$ (standardized). GQMLE and tQMLE are based on Gaussian and t_4 distributions, respectively. In SMLE1-DG and SMLE2-DG f is estimated using Gaussian kernel.

which utilizes the full set of information related to f. This raises a natural question – how much can we gain by using the semiparametric approach?

3.1. *The Bias-variance Trade-off of the Semiparametric GARCH Estimators*

Efficiency bounds (Stein (1956), Koshevnik and Levit (1976), Pfanzagl and Wefelmeyer (1982), Bickel, et al. (1993)) reflect the best attainable estimation precision as given by the theoretical limit of the variance of the asymptotic distribution, and therefore is often used to evaluate and compare the efficiencies of estimators. In the semiparametric scenario, these bounds set a standard so that the efficiency loss due to the introduction of a nonparametric component can be measured (Wellner (1985), Bickel, et al. (1993), Newey (1990)).

Consider a parametric model with parameter of interest θ and the unknown nuisance parameter η, and let $\phi = (\theta, \eta)$. Define the inner product $\langle U, V \rangle = \mathrm{E}[UV']$. It is now well-known (Stein (1956)) that the Cramér-Rao bound for the estimation of θ is

$$\langle S_\theta - PS_\eta, S_\theta - PS_\eta \rangle^{-1}, \quad \text{with } P = \langle S_\theta, S_\eta \rangle \langle S_\eta, S_\eta \rangle^{-1}, \tag{14}$$

where S_θ and S_η are the score functions of the parameters. Note that $\langle S_\theta - PS_\eta, S_\theta - PS_\eta \rangle$ represents the variance matrix of the regression residuals of S_θ on S_η, therefore is larger when S_θ and S_η are less linearly correlated. Intuitively, the less related S_θ and S_η are, the less useful is the information regarding η for estimating θ, and thus is more likely that θ can be estimated accurately without the knowledge of η. Geometrically this term is often viewed as the norm of the projection residual of S_θ onto S_η. As a result, the Cramér-Rao bound for θ reaches the minimum when S_θ and S_η are orthogonal. This orthogonality is also represented by a Fisher information matrix that is block-diagonal with respect to θ and η.

In a semiparametric model, η is replaced by the unknown density f. When f is formulated nonparametrically, the nonparametric part can be captured by a family of infinite number of parametric submodels. Correspondingly, the information bound is given by the supremum of the Cramér-Rao bound for all submodels, which can be viewed as a projection of S_θ on the space that is spanned by the scores of the nuisance parameters in all submodels, denoted as S_f. This space is the so-called tangent space, which is commonly formulated as the mean-square closure of the linear combinations of these nuisance scores (Pfanzagl and Wefelmeyer (1982)), given necessary compactness assumptions.

González-Rivera and Drost (1999) studied the efficiency bound of the semiparametric GARCH model. They quantified the semiparametric efficiency loss by

$$V_{MLE}^{-1} - V_{SMLE}^{-1} \propto m - \frac{4}{\kappa - 1 - \zeta^2}, \tag{15}$$

where V_{MLE} and V_{SMLE} are the efficiency bounds for MLE and the SMLE, respectively, $m = \int [1 + \epsilon f'(\epsilon)]^2 / f(\epsilon) d\epsilon$, κ and ζ are the kurtosis and skewness of f, respectively. See Linton (1993) and Drost and Klaassen (1997) for related discussions. See also Hafner and Rombouts (2007) for the similar results in multivariate framework. Although, interestingly, for certain specific choices of f, we have $V_{MLE}^{-1} = V_{SMLE}^{-1}$ (González-Rivera (1997)), in general the difference in (15) is positive definite, i.e., the asymptotic variance of the semiparametric estimators is guaranteed to be larger than that of the MLE, which utilizes the full knowledge of f.

In contrast, equation (12) says that the asymptotic variance can indeed be reduced by allowing a slowly-converging bias. The convergence rate of the bias term $b_n^{SMLE2-DG}$ is affected by several factors, but it is always at a nonparametric level ($\prec n^{-1/2}$). This bias-variance trade-off can be intuitively outlined following Newey (1990, Section 4.4). Consider an estimator $\hat{\theta}_n$ as the solution of the score function, which is a typical m-estimator such that

$$\sqrt{n}(\hat{\theta}_n - \theta^*) = a_n + b_n\sqrt{n}(\hat{\phi}_n - \phi^*) + o_P(1) ,$$

where a_n satisfies the CLT and b_n converges to a constant in probability, and ϕ represents the complete set of parameters that includes θ. In this semiparametric case, we have $\phi = (\theta, f)$, and part of the estimator $\hat{\phi}_n$ is given by a functional of kernel estimate (i.e., $\hat{f}_n$). Obviously, this kernel estimator produces a bias unless appropriate trimming is performed (Härdle (1990), Newey and McFadden (1994), Andrews (1995)).

3.2. *Adaptive Semiparametric GARCH Estimators*

Adaptiveness is an important concept in semiparametric estimation. It refers to the scenario where an estimator is derived using only a subset of information, but performs as precisely as an estimator based on the complete information set. For example, the least square estimator (LSE) in linear regression does not make use of the innovation density, but it becomes equivalent to the MLE when the innovations are Gaussian, and therefore it is adaptive in this case. In our specific case with an unknown f, a semiparametric estimator of the parameter of interest is said to be adaptive if it is asymptotically equivalent to an estimator derived with the full knowledge of f. In other words, an adaptive estimator exists when the unknown nonparametric part does not contain 'useful' information in estimating the parameter of interest. A number of review articles can be found in the context of general i.i.d. case (Bickel (1982), Robinson (1988a), Robinson (1988b), Newey (1990), Bickel, et al. (1993)), time series model (Kreiss (1987a), Kreiss (1987b), Drost, et al. (1997)) and the GARCH model (Steigerwald (1992), Linton (1993), Drost and Klaassen (1997), González-Rivera (1997), González-Rivera and Drost (1999), Ling and McAleer (2003)).

Adaptive estimators are often studied via the score function. More precisely, an adaptive estimator may be found when the score S_θ is orthogonal to the tangent space spanned by S_f. In particular, if a GARCH SMLE is adaptive, then we should have $V_{MLE}^{-1} = V_{SMLE}^{-1}$. González-Rivera (1997) showed that this equality is linked to a set of differential equations such that their solution yields a family of standardized distributions of the symmetric form

$$f(\epsilon) = \frac{\lambda^{\lambda/2}}{2^{\lambda/2}\Gamma(\lambda/2)}|\epsilon|^{\lambda-1}e^{-\lambda\epsilon^2/2} , \tag{16}$$

or of the asymmetric form

$$f(\epsilon) = (c_1 I_{\epsilon\leq 0} + c_2 I_{\epsilon>0})|\epsilon|^{A-1}e^{-A\epsilon^2/2+Bu} , \tag{17}$$

where I is the indicator function, $A = 2/(\kappa - 1 - \zeta^2)$, $B = \zeta A$, c_1 and c_2 are normalizing constants, κ and ζ are the kurtosis and skewness of f.

The small family of densities proposed by González-Rivera (1997) is required to create the necessary orthogonality between the score of parameter of interest θ and the nuisance parameter f in order to derive an adaptive estimator. This approach is elegant; but in practice showing that the unknown f belongs to this class of distributions can be difficult. Linton (1993) and Drost and Klaassen (1997) followed a different path by reparametrizing the GARCH model (2) as

$$r_t = \sigma_t \epsilon_t , \tag{18a}$$

$$\sigma_t^2 = \alpha_0(1 + \dot{\alpha}_1 r_{t-1}^2) + \beta_1\sigma_{t-1}^2 , \tag{18b}$$

where $\dot{\alpha}_1 = \alpha_1/\alpha_0$ reflects the relative effect of the observations on future volatility. Under this parametrization, the parameter of interest becomes $\theta = (\dot{\alpha}_1, \beta_1)$. Although this model setup is incomplete for the practical use of the GARCH model, it introduces several benefits. First, the unit variance assumption for ϵ_t can be removed without compromising the parameter identifiability. This is due to the fact that by taking the ratio, the linear relationship between α_0 and α_1 (see Fig. 2), which is the factor confounded with the variance of ϵ_t, is no longer being estimated. More importantly, orthogonality can be established between the new parameter of interest and the tangent space that now includes the non-identifiable parameter α_0. This largely expands the existence of the adaptive estimation – adaptive estimator of $\theta = (\dot{\alpha}_1, \beta_1)$ can be found for much more flexible f. Linton (1993, based on an ARCH model) and Drost and Klaassen (1997) showed that by assuming f is symmetric, the $\hat{\theta}_n^{ADP}$ defined as

$$\hat{\theta}_n^{ADP} = \tilde{\theta}_n + n^{-1/2}\hat{I}_n^{-1}(\tilde{\theta}_n, \hat{f}_n)\hat{S}_n(\tilde{\theta}_n) , \tag{19}$$

is an adaptive estimator of the GARCH model, where $\tilde{\theta}_n$ is any $\sqrt{n}$-consistent estimator of $\theta = (\dot{\alpha}_1, \beta_1)$, $\hat{I}_n$ and $\hat{S}_n$ are the consistent estimator of the information matrix and the score, respectively, and $\hat{f}_n$ is a kernel estimate of f based on the model residual sample created by $\tilde{\theta}_n$. Similar results have been established in more general GARCH models. See Sun and Stengos (2006) for a discussion on GARCH model with asymmetric volatility structure, and Hafner and Rombouts (2007) for a similar result in multivariate GARCH models.

The approach outlined in (19) is quite typical (see Bickel (1982)). The idea is to update a $\sqrt{n}$-consistent estimator using a consistent estimator of the influence function. The asymptotic normality is usually a result of the local asymptotic normality (LAN, Le Cam (1960), Drost, et al. (1997)). In proving the asymptotic theory, trimming is used in estimating the density or the score. In addition, sample splitting and data discretization are often used to create necessary data independence to simplify the proof (Le Cam (1960), Bickel (1982), Schick (1986), Klaassen (1987), Bickel, et al. (1993), Linton (1993), Drost and Klaassen (1997)). The assumption that f is symmetric in Linton (1993) and Drost and Klaassen (1997) was later removed by Yang (1998). In fact, it has been shown that the symmetry condition is necessary for adaptive estimation only if the parameter of interest includes additional components, e.g., the free intercept in the mean equation of a regression model with GARCH errors.

3.3. *Factors that Impact the Efficiency*

There are still several other factors that can affect the efficiency of the semiparametric GARCH estimators, either asymptotically or at finite sample size. Understanding the roles that these factors play in semiparametric GARCH estimation may improve the performance of the SMLE. Following is a brief list of the factors, which are given inline with the steps by which a semiparametric GARCH estimator is derived:

(1) An initial input is used to derive the model residuals
(2) A nonparametric estimator of the unknown density is derived
(3) The semiparametric likelihood function is derived
(4) The semiparametric likelihood function is maximized.

Item 1 requires a 'good' initial estimate of the parameter. An 'imperfect' initial estimate will produce a 'less than perfect' residual sample, and thus the benefit of the semiparametric approach is impaired. See Di and Gangopadhyay (2011) for a discussion on how the convergence rate of the SMLE relates to the initial input estimate. However, this is a somewhat different discussion, and we will not go into the details here. In addition, if a 'good' initial estimate is difficult to find, the one-step semiparametric approach (Venter and de Jongh (2004), Di and Gangopadhyay (2010)) can always be used. Items 2-4 deserve careful consideration, and we will explore these issues in the following subsections.

3.3.1. *Selections of Kernel and Bandwidth*

Factors related to item 2 have to do with the ways the nonparametric density estimate is derived. In particular, if a kernel method is used (see Engle and González-Rivera (1991) for a different approach), one needs to address the issues related to the selection of the kernel K and the bandwidth h_n. Consider model (2) or (18). Denote $\hat{\epsilon}_t(\theta) = r_t/\sigma_t(\theta)$, and consider the kernel density estimate

$$\hat{f}_n(\cdot) = \frac{1}{nh_n}\sum_{t=1}^{n} K\Big(\frac{\cdot - \hat{\epsilon}_t(\tilde{\theta}_n)}{h_n}\Big) ,$$

where $\tilde{\theta}_n$ is the initial estimate of θ. When f is assumed to be symmetric (Linton (1993), Drost and Klaassen (1997), Yang (1998)), the kernel estimator is given by

$$\hat{f}_n(\cdot) = \frac{1}{2nh_n}\sum_{t=1}^{n} \Big[K\Big(\frac{\cdot - \hat{\epsilon}_t(\tilde{\theta}_n)}{h_n}\Big) + K\Big(\frac{\cdot + \hat{\epsilon}_t(\tilde{\theta}_n)}{h_n}\Big)\Big] .$$

In kernel estimation, to ensure the consistency of the estimator the kernel needs to be smooth enough, and the bandwidth needs to go to zero at particular rates. Table 2 provides a summary of the choices of kernel and bandwidth in typical studies of semiparametric estimators of GARCH model. In deriving the asymptotic properties of the semiparametric estimators, kernel estimates of density derivatives of up to f'' may be necessary. Unlike the usual situations where i.i.d. samples are used to estimate the density, in semiparametric estimation of the GARCH model, the model residuals $\hat{\epsilon}_1(\tilde{\theta}_n)$, ..., $\hat{\epsilon}_n(\tilde{\theta}_n)$ are generally correlated. Therefore, whereas $nh_n^5/\log(1/h_n) \to \infty$ is sufficient for the (a.s.) uniform consistency of $\hat{f}_n''$ (Silverman (1978)), the corresponding semiparametric results often require extra smoothing achieved either by using a smoother kernel, by enlarging the bandwidth, by trimming out the contributions of the extreme residuals, or by utilizing some combinations of these approaches.

In kernel estimation, bandwidth h_n controls the trade-off between the bias and variance. In particular, when the bandwidth h_n increases, the bias

Table 2. The selection of kernel and bandwidth in typical semiparametric GARCH estimators.

Author(s)	Kernel K	Bandwidth h_n	Trim♮?
Linton (1993)	Gaussian	$h_n \to 0$	Yes
Steigerwald (1993)	General	$h_n \to 0$, $nh_n^3 \to \infty$, $nh_n^4 \to 0$	No
Drost and Klaassen (1997)	Logistic	$h_n \to 0$, $nh_n^4 \to \infty$	Yes
Yang (1998)	Gaussian	$h_n \to 0$, $nh_n^5 \to \infty$	Yes
Hafner and Rombouts (2007)	Gaussian product	$h_n \to 0$, subject to data dim.	Yes
	Gamma product	$h_n \to 0$, subject to data dim.	Yes
Di and Gangopadhyay (2011)	General	$h_n \to 0$, $nh_n^6 \to \infty$	No
	Smooth (e.g., Gaussian)	$h_n \to 0$, $nh_n^5 \to \infty$	No

♮ The bandwidth is also involved in the definition of trimming parameters.

increases and the variance decreases. According to Di and Gangopadhyay (2011), this bias-variance trade-off is also observed in the finite sample performance of their estimator SMLE2-DG (Fig. 4). Asymptotically, a

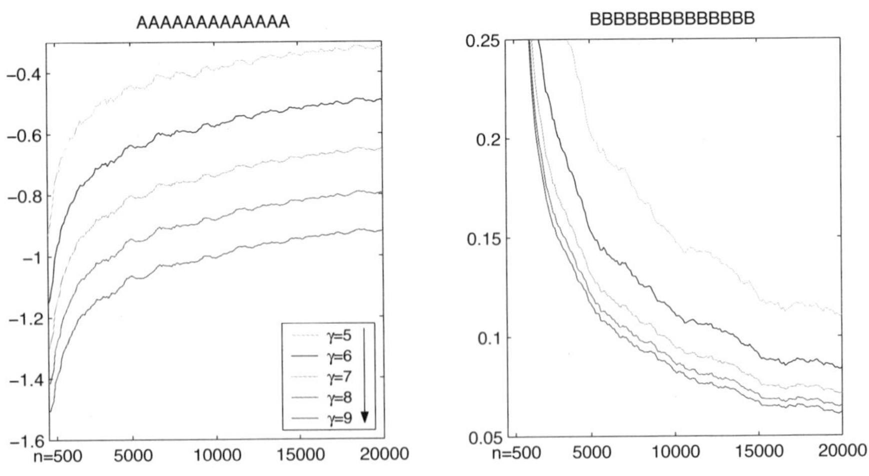

Fig. 4. Convergence of the leading term (multiplied by a factor) of the finite-sample bias and variance of the SMLE2-DG associated with different bandwidths $h_n = n^{-1/\gamma}$. Data simulated from the standardized $F_{10,50}$ distribution. Gaussian kernel is used. Only the α_1 component is displayed.

larger h_n reduces the convergence rate of the the bias $b_n^{SMLE2-DG}$ (Di and Gangopadhyay (2011)). They also argued that the bias term always converges at a slower rate as compared to the variance term, therefore the optimal performance of their SMLE is always achieved by choosing the smallest possible h_n subject to the necessary conditions that ensure the asymptotic characteristics of the estimator.

3.3.2. *Trimming the Semiparametric Likelihood*

In kernel smoothing, trimming functions are often used to remove the effects of extreme observations. In estimating the functionals of a probability density, asymptotic trimming is often applied to eliminate the so-called boundary effect (Robinson (1988b), Bickel (1982), Bickel, et al. (1993), Newey and McFadden (1994), Andrews (1995), Lewbel (2000)). For example, the estimation of $T(1/f)$, where T is some continuous function, is often carried out by plugging-in a kernel estimate defined only at the samples with the absolute value bounded by some $a_n > 0$. The trimming boundary a_n expands to the boundary of the support of f (which is possibly $\pm\infty$) to allow asymptotic inclusion of the full dataset. The expansion is slow as compared to the shrinkage of the bandwidth h_n to eliminate the boundary effect, but also fast enough as compared to $n^{-1/2}$ to ensure that the trimming-induced bias vanishes asymptotically.

In the context of semiparametric GARCH estimation, the asymptotic properties of a semiparametric estimator is related to that of $\sqrt{n}\hat{S}_n$, where $\hat{S}_n$ is the score with respect to the semiparametric likelihood (9). In particular, proving the adaptiveness of an SMLE is essentially the same as proving the asymptotic equivalence of $\sqrt{n}\hat{S}_n$ and $\sqrt{n}S_n$, where S_n is the score with respect to the true likelihood (3), for all θ within a neighborhood of the true value. Due to the fact that $\hat{S}_n$ is a functional of the kernel density estimate $\hat{f}_n$, some form of asymptotic trimming is often required to ensure the $\sqrt{n}$-consistency of the semiparametric score $\hat{S}_n$ (Linton (1993), Steigerwald (1993), Drost and Klaassen (1997), Yang (1998), Hafner and Rombouts (2007)).

By trimming out the effect of the 'shocks' on certain components of the semiparametric likelihood, the requirement on the bandwidth may be relaxed while still providing enough smoothing to ensure the consistency of the estimator. Consider the trimming introduced in Hafner and Rombouts (2007) as an example (see Linton (1993) for a similar formulation). In their approach, the score of the tth observation is defined as

$$\hat{S}_{n,t}(\hat{\epsilon}_t) = \begin{cases} \hat{f}'_n(\epsilon_t)/\hat{f}_n(\epsilon_t), & \text{if } \hat{f}_n(\epsilon_t) \geq d_n,\ |\hat{f}'_n(\epsilon_t)/\hat{f}_n(\epsilon_t)| \leq c_n,\ |\epsilon_t| \leq e_n; \\ 0, & \text{otherwise.} \end{cases}$$

The asymptotic normality of their semiparametric estimator is established under the conditions $h_n \to 0$, $d_n \to 0$, $c_n \to \infty$, $e_n \to \infty$, $c_n h_n \to 0$, and $nh_n^3 c_n^{-2} e_n^{-2} \to \infty$ (for univariate data). For different choices of the trimming boundary, the bandwidth can be adjusted accordingly. For example, if one chooses $c_n = n^{1/6}$ and $e_n^{1/32}$, then the requirement on h_n is roughly $nh_n^5 \to \infty$. On the other hand, if one chooses $c_n = e_n \sim \log(n)$, then the requirement on h_n becomes approximately $nh_n^3 \to \infty$.

However, it is important to note that due to the extra smoothing afforded through trimming, a narrower h_n can be used in establishing the asymptotic distribution of the semiparametric estimator. However, such a narrow bandwidth may not be sufficient in showing the convergence of $\hat{f}_n$, $\hat{f}'_n$ and $\hat{f}''_n$ to f, f' and f''. Without these convergence results, the asymptotic properties of the SMLE could be affected. For example, if $\hat{f}''_n$ converges to a value different from f'', the asymptotic variance matrix of the semiparametric estimator will certainly be different from that of the MLE's (Di and Gangopadhyay (2011)).

3.3.3. *Maximizing the Semiparametric Likelihood*

As discussed in Section 1.2.1, the maximization of the semiparametric likelihood is often replaced by a single-step Newton-Raphson update of the initial input parameter estimate. Although this approach is asymptotically equivalent to direct maximization, which is in fact a multi-step Newton-Raphson update until convergence is reached, their finite sample performances can be quite different. Yang (1997) argued that the single order Newton-Raphson performs poorly with small sample size since it utilizes the outer-product of the gradient to estimate the information matrix. In a later study, Yang (1998) provided further simulation-based evidence in GARCH estimation that the direct maximization of semiparametric likelihood can produce an estimator with less variability.

4. Future Developments in Semiparametric GARCH Estimation

During the last three decades, semiparametric estimation of the GARCH-type models have been extensively studied in the literature. Theoretical results have been derived and compared to the traditional parametric estimators. In various scenarios investigated, the semiparametric estimators have been shown to be at least asymptotically superior to their traditional parametric counterparts. However, there are still several open questions that future research can focus on.

Finite sample performance. While the research on asymptotic theory of semiparametric methods is well developed, very few investigations have been carried out on finite sample properties of the SMLEs. In particular, since the semiparametric estimation process involves the estimation of the unknown density, the level of additional estimation inefficiency introduced by this step for small sample sizes is unknown in almost all cases (see Di (2007) for a related discussion). For example, in several situations fully efficient semiparametric estimators can be derived if the semiparametric likelihood is trimmed before maximization (Linton (1993), Drost and Klaassen (1997), Yang (1998), Ling and McAleer (2003)). Although the impact of trimming can be asymptotically negligible, in practice the trimming inevitably causes certain loss of tail information. This loss of information and its impact on the estimators for finite sample size need to be fully quantified.

Kernel and bandwidth. Selection of the kernel and bandwidth is an old but important question. These factors affect both the small- and large-sample characteristics of the semiparametric GARCH estimators. Among the related questions, the most interesting one is the selection of bandwidth to control the convergence rate of the estimator and the reduction of the estimation bias. Di and Gangopadhyay (2011) showed that in their semiparametric estimation procedure, the bias and the variance cannot be balanced by any choice of h_n without trimming. There does not appear to be any analytical result on the elimination of bias in this framework. Di (2007) reported unsatisfactory evaluation of the bias via bootstrapping.

Selection of the semiparametric methods based on f. With all the available estimation procedures, there has not yet been any systematic comparison between the approaches. Several Monte Carlo studies are available, but the results are usually unclear in the sense that the preference for a specific approach varies when the underlying innovation density f changes. In other words, the performances of these methods have not been linked with the characteristics of f. While González-Rivera and Drost (1999) successfully connected the semiparametric efficiency bound with f, their result is quite general and only reveals the theoretical limit of an SMLE. It is expected that some density-specific criteria, which maybe defined as a 'metric' between f and a benchmark density, could be established and utilized in choosing the estimation procedure.

Continuous GARCH model. Due to the development of modern database technology, financial time series are now available on a virtually continuous basis. Multiple versions of continuous (G)ARCH processes have been studied in the literature (Nelson (1990), Drost and Werker (1996), Klüppelberg, et al. (2004), Brockwell, et al. (2006)). Corresponding GARCH diffusions or GARCH jump-diffusions have been constructed, and strategies for estimating parameters in continuous GARCH models have been developed. For example, Drost and Werker (1996) showed that under their model construction, the knowledge of the discrete time GARCH parameters at any single frequency completely determines the continuous time coefficients of the GARCH process. Therefore the semiparametric estimation of the discrete time GARCH parameters at an available frequency should suffice in estimating the parameters in the original continuous process. However, how the efficiency gain associated with semiparametric estimation in the discrete case translates to the continuous case is still an open question.

Empirical application. Another natural question is that whether any practical task can be accomplished more accurately by using semiparametric estimators of GARCH models. Engle and González-Rivera (1991) had discussed an application of the semiparametric approach for the £/\$ exchange rate returns. Since the return series exhibits clear peak and heavy tail, the validity of the assumption of unconditional Gaussian or unconditional t is questionable. They showed that the semiparametric approach is more appropriate in representing the distribution of the return series. Drost and Klaassen (1997) studied the exchange rates of a total of fifteen currencies w.r.t. the US dollar under their adaptive estimation setup. They compared the semiparametric approach to the QMLE via bootstrapping, and reported that the semiparametric estimators have smaller estimation standard error. These studies illustrate the practical benefits of semiparametric estimation of the GARCH parameters. However, the advantages of the semiparametric methods in estimating the return volatility have not been thoroughly considered. For example, for the models that are used to price the options or to evaluate the value at risk (VAR), the return volatility is a critical input. If the volatility can be estimated more accurately using the semiparametric approach, the performances of those models can be potentially improved as well.

Acknowledgment

Ashis Gangopadhyay wishes to express his deepest gratitude and thanks to Professor P. K. Bhattacharya, who was his Ph.D. thesis advisor, for Professor Bhattacharya's immense support, guidance and generosity over many years.

References

Andrews, D. (1994). Asymptotics for semiparametric econometric models via stochastic equicontinuity, *Econometrica* **62**, pp. 43–72.

Andrews, D. (1995). Nonparametric kernel estimation for semiparametric models, *Econometric Theory,* **11** pp. 560–596.

Baillie, R. and Bollerslev, T. (1989). The message in daily exchange rates: a conditional-variance tale, *Journal of Business and Economic Statistics* **7**, 647–671.

Baillie, R., Bollerslev, T. and Mikkelsen, H. (1996). Fractionally integrated generalized autoregressive conditional heteroskedasticity, *Journal of Econometrics* **74**, 1, pp. 3–30.

Berkes, I. and Horváth, L. (2003). The rate of consistency of the quasi maximum likelihood estimator, *Statistics and Probability Letters* **61**, 2, pp. 133–143.

Berkes, I. and Horváth, L. (2004). The efficiency of the estimators of the parameters in GARCH Procceses, *The Annals of Statistics* **32**, 2, pp. 633–655.

Berkes, I., Horváth, L. and Kokoszka, P. (2003) GARCH processes: structure and estimation, *Bernoulli* **9**, 2, pp. 201–227.

Bickel, P. (1982). On Adaptive Estimation, *The Annals of Statistics* **10**, 3, pp. 647–671.

Bickel, P., Klaassen, C., Ritov, Y. and Wellner, J. (1993). *Efficient and Adaptive Inferences in Semi-parametric Models* (Springer).

Bollerslev, T. (1986). Generalized autoregressive conditional heteroskedasticity, *Journal of Econometrics* **31**, pp. 307–327.

Bollerslev, T., Engle, R. and Nelson, D. (1994). ARCH models, in Engle, R. and McFadden, D. (eds.), *Handbook of Econometrics*, Vol. 4 (Elsevier).

Bollerslev, T., Engle, R. and Wooldridge, J. (1988). A captical asset pricing model with time varying covariances, *Journal of Political Economy* **96**, pp. 116–131.

Bollerslev, T. and Wooldridge, J. (1992). Quasi-maximum likelihood estimation and inference in dynamic models with time-varying covariances, *Econometric Reviews* **11**, 2, pp. 143–172.

Bougerol, P. and Picard, N. (1992). Stationarity of GARCH processes and of some nonnegative time series, *Journal of Econometrics* **52**, pp. 115–127.

Brockwell, P., Chadraa, E. and Lindner, A. (2006). Continuous-time GARCH processes, *The Annals of Applied Probability* **16**, 2, 790–826.

Davison, R. and Eddelbüttel, D. (1995). Semiparametric estimation of ARCH models using nearest neighbour regression: some Monte Carlo results, Working paper series G.R.E.Q.A.M., No. 95a23, Universite Aix-Marseille III.

Di, J. (2007). *New Developments in Time Series Analysis (Part I: On the construction and estimation of GARCH models)*, PhD Dissertation, Department of Mathematics and Statistics, Boston University, Boston, MA.

Di, J. and Gangopadhyay, A. (2010). One-step semiparametric estimation of the GARCH model, Technical Report, Department of Mathematics and Statistics, Boston University, Boston, MA.

Di, J. and Gangopadhyay, A. (2011). On the Efficiency of a Semiparametric GARCH Model, *The Econometrics Journal* **15**, 2 (to appear).

Drost, F. and Klaassen, C. (1997). Efficient estimation in semi-parametric GARCH models, *Journal of Econometrics* **81**, pp. 193–221.

Drost, F., Klaassen, C. and Werker, B. (1997). Adaptive estimation in time-series models, *The Annals of Statistics* **25**, 2, pp. 786–817.

Drost, F. and Werker, B. (1996). Closing the GARCH gap: Continuous time GARCH modeling, *Journal of Econometrics* **74**, 1, pp. 31–57.

Engle, R. (1982). Autoregressive conditional heteroskedasticity with estimates of the variance of united kindom inflation, *Econometrica* **50**, pp. 987–1007.

Engle, R. (1990). Discussion: stock market volatility and the crash of 87, *Review of Financial Studies* **3**, pp. 103–106.

Engle, R. and Bollerslev, T. (1986). Modelling the Persistence of Conditional Variances, *Econometric Reviews* **5**, pp. 1–50.

Engle, R. and González-Rivera, G. (1991). Semiparametric ARCH Models, *Journal of Business & Economic Statistics* **9**, 4, pp. 345–359.

Engle, R., Lilien, D. and Robins, R. (1987). Estimating time varying risk premia in the term structure: the ARCH-M model, *Econometrica* **55**, pp. 391–407.

Fan, J., Qi, L. and Xiu, D. (2010). Non-Gaussian quasi maximum likelihood estimation of GARCH models, Available at SSRN:
http://ssrn.com/abstract=1540363.

Francq, C. and Zakoïan, J. (2004). Maximum likelihood estimation of pure GARCH and ARMA–GARCH processes, *Bernoulli* **10**, 5, pp. 605–637.

González-Rivera, G. (1997). A Note on Adaption in GARCH Models, *Econometric Reviews* **16**, 1, pp. 55–68.

González-Rivera, G. and Drost, F. (1999). Efficiency comparisons of maximum-likelihood-based estimators in GARCH models, *Journal of Econometrics* **93**, 1, pp. 93–111.

Hafner, C. and Rombouts, J. (2007). Semiparametric Multivariate Volatility Models, *Econometric Theory* **65**, 23, pp. 251–280.

Hall, P. and Yao, Q. (2003). Inference in ARCH and GARCH Models with Heavy-Tailed Errors, *Econometrica* **71**, 1, pp. 285–317.

Härdle, W. (1990). *Applied nonparametric regression* (Cambridge University Press, Cambridge, UK).

Härdle, W. and Tsybakov, A. (1997). Local polynomial estimators of the volatility function, *Journal of Econometrics*, **81**, pp. 223–242.

Hentschel, L. (1995). All in the family: nesting symmetric and asymmetric GARCH models, *Journal of Financial Economics* **39**, 1, pp. 71–104.

Higgins, M. and Bera, A. (1992). A class of nonlinear ARCH models, *International Economic Review* **33**, pp. 137–158.

Klaassen, C. (1987). Consistent estimation of the influence function of locally asymptotically linear estimators, *The Annals of Statistics* **15**, pp. 1548–1562.

Klüppelberg, C., Lindner, A. and Maller, R. (2004). A continuous time GARCH process driven by a Lévy process: stationarity and second order behaviour, *Journal of Applied Probability* **41**, pp. 601–622.

Koenker, R. and Zhao, Q. (1996). Conditional quantile estimation and inference for ARCH models, *Econometric Theory*, **12**, pp. 793–813.

Koshevnik, Y. and Levit, B. (1976). On a non-parametric analogue of the information matrix, *Theory of Probability and Applications* **21**, pp. 738–753.

Kreiss, J. (1987). On adaptive estimation in stationary ARMA processes, *The Annals of Statistics* **15**, 1, pp. 112–133.

Kreiss, J. (1987). On adaptive estimation in autoregression processes when there are nuisance functions, *Statistics and Decisions* **5**, pp. 59–76

Le Cam, L. (1960). Local asymptotically normal families of distributions, *University of California Publications in Statistics* **3**, pp. 267–284.

Lee, S. and Hansen, B. (1994). Asymptotic theory for the GARCH(1,1) quasi-maximum likelihood estimator, *Econometric Theory* **10**, pp. 29–52.

Lewbel, A. (2000). Asymptotic trimming for bounded density plug-in estimators, Available at http://citeseerx.ist.psu.edu/viewdoc/ download?doi=10.1.1.71.1350&rep=rep1&type=pdf.

Ling, S. and Li, W. (1997). On fractionally integrated autoregressive moving-average time series models with conditional heteroscedasticity, *JASA* **92**, pp. 1184–1194.

Ling, S. and McAleer, M. (2003). Asymptotic theory for a vector ARMA–GARCH model, *Econometric Theory*, **19**, pp. 280–310.

Linton, O. (1993). Adaptive Estimation in ARCH Models, *Econometric Theory*, **9**, 4, pp. 539–569.

Linton, O. and Mammen, E. (2005). Estimating semiparametric ARCH(∞) models by kernel smoothing methods, *Econometrica* **73**, pp. 771–836.

Lumsdaine, R. (1996). Asymptotic Properties of the Quasi-Maximum Likelihood Estimator in IGARCH(1,1) and Covariance Stationary GARCH(1,1) Models, *Econometrica* **64**, 3, pp. 575–596.

Maercker, G. and Moser, M. (1999). Yule–Walker type estimators in GARCH(1,1) models: asymptotic normality and bootstrap, Available at ftp://statlab.uni-heidelberg.de/pub/reports/by.series/beitrag.58.ps.

Mikosch, T. and Straumann, D. (2002). Whittle estimation in a heavy-tailed GARCH(1,1) model, *Stochastic Processes and their Applications* **100**, 1, pp. 187–222.

K. Mukherjee (2006). Pseudo-likelihood estimation in ARCH models, *The Canadian Journal of Statistics* **34**, 2, pp. 143–172.

Nelson, D. (1990). ARCH models as diffusion approximations, *Journal of Econometrics* **45**, pp. 7–38.

Nelson, D. (1991). Conditional heteroskedasticity in asset returns: a new approach, *Econometrica* **59**, pp. 347–370.

Newey, W. (1990). Semiparametric efficiency bounds, *Journal of Applied Econometrics*, **5**, pp. 99–135.

Newey, W. and McFadden, D. (1994). Large sample estimation and hypothesis testing, in Engle, R. and McFadden, D. (eds.), *Handbook of Econometrics*, Vol. 4 (Elsevier).

Newey, W. and Steigerwald, D. (1997). Asymptotic Bias for Quasi-Maximum-Likelihood Estimators in Conditional Heteroskedasticity Models, *Econometrica* **65**, 3, pp. 587–599.

Pagan, A. and Schwert, G. (1990). Alternative models for conditional stock volatility, *Journal of Econometrics* **45**, pp. 267–290.

Pfanzagl, J. and Wefelmeyer, W. (1982). *Contributions to a Gneral Asymptotic Statistical Theory* (Springer–Verlag).

Rich, R., Raymond, J. and Butler, J. (1991). Generalized instrumental variables estimation of autoregressive conditional heteroscedastic models, *Economics Letters*, **35**, pp. 179–185.

Robinson, P. (1988a). Root-n consistent semiparametric regression, *Econometrica* **56**, pp. 931-954.

Robinson, P. (1988b). Semi-parametric Econometrics: A Survey, *Journal of Applied Econometrics* **3**, pp. 35-51.

Ruppert, D., Wand, M. and Carroll, R. (2003). *Semiparametric regression*, 1st edn. (Cambridge University Press, Cambridge, UK).

Schick, A. (1986). On asymptotically efficient estimation in semiparametric models, *Annals of Statistics* **14**, pp. 1139–1151.

Scott, D., Tapia, R. and Thompson, J. (1980). Nonparametric Probability Density Estimation by Discrete Maximum Penalized-Likelihood Criteria, *The Annals of Statistics* **8**, 4, pp. 820–832.

Silverman, B. (1978). Weak and strong uniform consistency of the kernel estimate of a density and its derivatives, *The Annals of Statistics* **6**, 1, pp. 177–184.

Speckman, P. (1988). Kernel smoothing in partial linear models, *Journal of Royal Statistical Society, Series B*, **50**, pp. 413–446.

Steigerwald, D. (1992). Adaptive Estimation in Time Series Regression Models, *Journal of Econometrics* **54**, pp. 251–275.

Steigerwald, D. (1993). Efficient Estimations of Models with Conditional Heteroscedasticity, Working Paper in Economics #5–93, Department of Economics, University of California at Santa Barbara.

Stein, C. (1956). Efficient nonparametric testing and estimation, in *Proceedings of the Third Berkeley Symposium on Mathematical Statistics and Probability*,

Vol. 1.

Sun, Y. and Stengos, T. (2006). Semi-parametric efficient adaptive estimation of asymmetric GARCH models, *Journal of Econometrics*, **133**, pp. 373–38

Tapia, R. and Thompson, J. (1978). *Nonparametric Probability Density Estimation* (Johns Hopkins University Press, Baltimore, MD).

Venter, J. and de Jongh, P. (2004). Selecting an innovation distribution for GARCH models to improve efficiency of risk and volatility estimation, *Journal of Risk* **6**, 3, pp. 27–52.

Weiss, A. (1986). Asymptotic theory for ARCH models: estimation and testing, *Econometric Theory* **2**, pp. 107–131.

Wellner, J. (1985). Semiparametric models: progress and problems, on *Bulletin of the I.S.I., Proceedings of the 45th Session: Invited Papers*, Vol. 4.

Wooldridge, J. (1994). Estimation and inference for dependent process, in Engle, R. and McFadden, D. (eds.), *Handbook of Econometrics*, Vol. 4 (Elsevier).

Yang, J. (1997). Semiparametric maximum likelihood estimation of nonlinear regression models and Monte Carlo evidence, Technical report, Department of Economics, University of Western Ontario.

Yang, J. (1998). Semiparametric Maximum Likelihood Estimators of GARCH Models, Technical report, Department of Economics, University of Western Ontario.

Yang, L. (2006). A semiparametric GARCH model for foreigh exchange volatility, *Journal of Econometrics*, **130**, 2, pp. 365–384.

Yang, L., Härdle, W. and Nielsen, J., Nonparametric autoregression with multiplicative volatility and additive mean, *Journal of Time Series Analysis* **20**, pp. 579–604.

Chapter 18

Extreme Dependence in Multivariate Time Series: A Review

Rituparna Sen and Zheng Tan

University of California, Davis

This paper gives a brief overview on modeling extreme dependence for stationary time series by using Multivariate Extreme Value Theory. The notions of extremal coefficient and extremal correlation are first introduced in the i.i.d. setting and then the extension to a time series setting is presented. Multivariate Maxima of Moving Maxima and Markov Chains are briefly discussed as special cases of stationary time series. Some directions for future work are suggested.

Contents

1. Introduction

Interest in extremal co-movements in financial and insurance markets has risen dramatically due to significant increases in market volatility and evidence for market linkages. See, e.g., Longin and Solnik (2001), Hartman, et al. (2004), Longstaff (2010) to name a few. Problems concerning environmental extremes are often multivariate in character. An example is wind speed data where maximum hourly gusts, maximum hourly mean speeds and the dependence between them are relevant to building safety. An important issue for risk managers is how to quantify different types of risk such as market risk, credit risk, operational risk, etc. Due to the multivariate nature of risk, a particular concern for a risk manager is how to model the dependence between extreme outcomes, although those extreme outcomes occur rarely. Many of these problems involve a time component. In both of the above situations, we are concerned with the behavior of extremes of a multivariate time series. If the time series is stationary, then we can use the methods described in this review to study this behavior. The multivariate extremal index can be used to measures the degree of clustering of extreme values in the sequence. The extremogram is the analogue of the autocorrelation function based on only the extreme values. Another estimator to quantify this relation is the extremal correlation over a threshold.

The field of multivariate extremes is well studied. But most works concentrate on i.i.d vectors or univariate time series. The joint treatment of the time series and multivariate nature of the problems has been done in only a few instances. There is lot of room for development of theory and empirical work in this area. This would enable us to explain characteristics that are found in the observed extreme events of the time series and also gain insight into the properties of events that are more extreme than those observed. We present a review of the literature in this area with some suggestions for future work. We first present a brief summary of multivariate extreme value theory in the i.i.d. case in Section 2. In Section 3, we explore the present literature on multivariate extremes for stationary time series. We present results for some special time series in Section 4. In Section 5, we discuss some areas where work needs to be done.

2. A brief background on Multivariate Extreme Value Theory (IID case)

Consider a sequence $\mathbf{X}_i, i \geq 1$ of D-dimensional random vectors with (multivariate) distribution $\mathbf{F}_i$ and let

$$\mathbf{M}_n = (\max_{i=1,\ldots,n} X_{i,1}, \ldots, \max_{i=1,\ldots,n} X_{i,D})'. \tag{1}$$

Suppose normalizing constants $\mathbf{a}_n > \mathbf{0}$, $\mathbf{b}_n$ and a D-dimensional distribution function G exist such that the sequence of random variables $\mathbf{M}_n$, after normalization, converges in distribution to a random variable $\mathbf{X}$ that has the cumulative distribution function G. Thus for all x that are points of continuity of G, we have

$$P(\mathbf{a}_n^{-1}(\mathbf{M}_n - \mathbf{b}_n) \leq \mathbf{x}) = P(\mathbf{X}_i \leq \mathbf{a}_n\mathbf{x} + \mathbf{b}_n, \forall i \leq n)$$
$$\Rightarrow G(\mathbf{x}) = P(\mathbf{X} \leq \mathbf{x}) \quad \text{as} \quad n \to \infty. \tag{2}$$

G is called a multivariate extreme value distribution (MEVD) and $\mathbf{F}_i$ is said to belong to the domain of attraction of G. Note that all algebraic operations are componentwise. The univariate case has been treated by many authors; c.f. the textbooks by Leadbetter, et al. (1983), Galambos (1987) and Resnick (1987).

One additional aspect of extreme value theory in the multivariate case is the dependence properties of the limit law G in Equation (2). The study of this problem in the i.i.d. case was initiated by Sibuya (1960) for $D = 2$. A complete discussion can be found in Galambos (1987) and Resnick (1987). For the i.i.d. case ($\mathbf{F}_i = \mathbf{F}$), Equation (2) becomes

$$\mathbf{F}^n(\mathbf{a}_n\mathbf{x} + \mathbf{b}_n) \Rightarrow G(\mathbf{x}). \tag{3}$$

The limit G is characterized by the *max stability property*. We say a distribution $H(\mathbf{x})$ is max-stable if for every $t > 0$ there exist functions $\alpha(t) > 0, \beta(t)$ such that

$$H^t(\mathbf{x}) = H(\alpha(t)\mathbf{x} + \beta(t)) = H(\alpha_1(t)x_1 + \beta_1(t), \cdots, \alpha_D(t)x_D + \beta_D(t)). \tag{4}$$

The following theorem, which is Proposition 5.9 in Resnick (1987), describes the equivalence between multivariate extreme value distributions and max-stable distributions.

Theorem 1. *The class of multivariate extreme value distributions is precisely the class of max-stable distribution functions with non-degenerate marginals.*

Note that the univariate marginals G_j of an extreme value distribution G are obviously *univariate extreme value distributions.* Because of max stability the extreme value distribution G is *max infinitely divisible*, i.e. G^s is a multivariate distribution for every $s > 0$, (cf. Balkema and Resnick (1977)). Hence G is *associated*: $\text{Cov}(\phi(\mathbf{Z}), \psi(\mathbf{Z})) > 0$ for any (componentwise) nondecreasing functions ϕ and ψ where G is the distribution of $\mathbf{Z}$. The result that extreme value distributions are associated, is due to Marshall and Olkin (1983). As noted by Esary, Proschan and Walkup (1967), association is stronger than *positive quadrant dependence*: $G(x_1, x_2, \cdots, x_D) \geq \Pi_{j=1}^D G_j(x_j)$.

The extreme type theorems play a central role in the study of extreme value theory. In the univariate context, the extreme type theorems say that for a sequence of i.i.d. random variables with suitable normalizing constants, the limiting distribution of maximum statistics, if it exists, follows one of three types of extreme value distributions: Gumbel, Fréchet and Weibull. In the multivariate context, there have been many attempts to characterize the possible limits, such as Resnick's (1987) point process approach and Pickands's (1981) representation theorem for multivariate extreme value distribution with unit Fréchet marginals. For statistical analysis, no natural parametric family exists for the dependence between the marginal distributions, but there are considerable restrictions on the dependence structure.

One aspect of the analysis is to model the dependence function, test for independence and discriminate between models. Among the several models that have been developed, it is worth mentioning models such as Bivariate Exponential model by Mardia (1970), Nested Logistic model by McFadden (1978), Bivariate Logistic model by Marshall and Olkin (1983), Negative Asymmetric Logistic model by Joe (1989), Asymmetric Logistic model by Tawn (1990), Bilogistic model by Smith (1990) and Dirichlet model by Coles and Tawn (1991). The last paper considers the point process of observations which are extreme in at least one component. Joe (1994) derives several additional parametric families of multivariate extreme-value distributions. The parametric models are fitted, using numerical maximum likelihood. Some multivariate nonnormal data analysis techniques are proposed to aid in the likelihood analysis. All these models together appear to be adequate for inferences in that they cover a wide range of possible dependence patterns.

Another aspect is to quantify the magnitude of dependence for a multivariate max-stable distribution, see Joe (1997) and Coles, et al. (1999). Cooley, et al. (2006) proposes variograms for spatial max-stable random fields while Schlather and Tawn (2003) proposes the extremal coefficient.

A common notion associated with extreme values is that of regular variation. Excluding some special degenerate cases, a random vector with positive components is in the maximum domain of attraction of a multivariate extreme value distribution with the same Frechét marginals if and only if the vector has a distribution which is regularly varying. We present two equivalent definitions of multivariate regular variation; see e.g. Resnick (1987), Section 5.4 and Resnick (2007), Chapter 6.

Definition 1. *Multivariate regular variation* A vector $\mathbf{X}$ in $\mathbb{R}^D$ is regularly varying with index $\kappa > 0$, if one of the following equivalent conditions hold:
(a) There exists a random vector Θ with values on the unit sphere $\mathbb{S}^{D-1}$ such that for every $x > 0$

$$\frac{\mathrm{P}(|\mathbf{X}| > ux, \mathbf{X}/|\mathbf{X}| \in B)}{\mathrm{P}(|\mathbf{X}| > u)} \overset{v}{\Rightarrow} x^{-\kappa}\mathrm{P}(\Theta \in B) \quad \forall B \in \mathfrak{B}(\mathbb{S}^{D-1}) \quad \text{as} \quad u \to \infty. \tag{5}$$

The distribution of Θ is referred to as the spectral measure of $\mathbf{X}$. B is any Borel set. Note that "$\overset{v}{\Rightarrow}$" means vague convergence.
(b) There exists a Radon measure ρ on $\bar{\mathbb{R}}^D \backslash \{0\}$ with $\rho(E) > 0$ for at least one relatively compact set $E \subset \bar{\mathbb{R}}^D \backslash \{0\}$ and a sequence $a_n \uparrow \infty$ of positive constants such that

$$n\mathrm{P}(a_n^{-1}\mathbf{X} \in B) \overset{v}{\Rightarrow} \rho(B) \quad \text{for} \quad \forall B \in \mathfrak{B}(\bar{\mathbb{R}}^D \backslash \{0\}) \quad \text{as} \quad n \to \infty. \tag{6}$$

The measure ρ satisfies the homogeneity property $\rho(tA) = t^{-\kappa}\rho(A)$ for all $A \in \mathfrak{B}(\bar{\mathbb{R}}^D \backslash \{0\})$ and $t > 0$. Here $\bar{\mathbb{R}} = \mathbb{R} \cup \{\pm\infty\}$.

It can be debated, and indeed has, whether the influence of the marginal distributions (regular variation versus lighter-tailed [a]) blurs the interpretation of estimated extremal dependence.

Ramos and Ledford (2008) argues that it is more reasonable to separate the influence of the marginals and the extreme dependence by standardizing the marginals first to unit Fréchet distributions. Only afterwards they investigate bivariate joint tail distributions through the spectral measure. There are two points to be aware of, when standardizing marginals. Firstly, some statistical uncertainty will be introduced into the model by estimating the parameters of transformation. Secondly, transformation of the data involves the whole model. For instance, a linear exponential model will transform into a product model, resulting in a change of the dependence structure as well.

The currently known methods for fitting multivariate extreme value distribution (MEVD) is under the assumption that the observations are independent. In order to get around this assumption, Nadarajah (2001) uses declustering technique to identify independent events in the data record and

[a] The expression light-tailed stands for distribution functions F for which $\int e^{\alpha|x|}F(dx) < \infty$ holds for an $\alpha > 0$. This assumption implies that the random variable X which has distribution function F possesses finite moments of all orders. This includes all bounded random variables and those which have tails that decrease at least exponentially fast.

filter out extreme values from them. Then they use Multivariate Extreme Value Theory for the declustered data.

3. Extremes of stationary time series

Some effort has been devoted to extending the i.i.d classical results to dependent univariate sequences under some conditions such as stationary, mixing conditions etc. For instance, Leadbetter, et al. (1983), contains an abundant account of the theory of extreme values for dependent sequences, both stationary and non-stationary, as well as for stationary continuous time processes at a rigorous mathematical level. A more recent book is Beirlant, et al. (2004).

In this section we consider stationary sequences of random vectors, with $\mathbf{F}_i = \mathbf{F}$. The assumption of stationarity means that for any $h \geq 0$ and any $n \geq 1$, the distribution of $(\mathbf{X}_h, \cdots, \mathbf{X}_{h+n})$ does not depend on h. Define the vector of component-wise maxima as in Equation (1). In this case the extreme value theory is mainly discussed for Gaussian sequences (Lindgren (1974), Amram (1985), Hüsler and Schüpbach (1988)). The general multivariate stationary problem, however, has only been considered in a few papers. An attempt has been made by Villasenor (1976), for the case of bivariate exchangeable sequences. Hsing (1987) and Hüsler (1987) independently extend univariate results to the multivariate case. Sbihi (1988) also discusses the multivariate stationary case. In addition, Hüsler (1987) focuses more on the dependence properties.

In section 2 we defined the class of multivariate extreme value distributions as the limits of distributions of independent random vectors. We shall see in section 3.1 that for stationary time series, the limits of the distributions belong to this class if the long range dependence at extreme levels is suitably restricted. However, the limit distribution need not be the same as that for the maximum of the associated independent sequence $\tilde{\mathbf{X}}_i$ with the same marginal distribution as $\mathbf{X}_i$. The distinction is due to the extremal index introduced in section 3.2, which measures the tendency of extreme values to occur in clusters. We study two other measures of association in sections 1.2.3 and 1.2.4.

3.1. *The extremal limit theorem*

The following theorems are due to Hsing (1989) and Hüsler (1990).

Theorem 2. *Let $\{\boldsymbol{X}_n\}$ be a stationary sequence for which there exist se-*

quences of constant vectors $\boldsymbol{a}_n > 0$ *and* $\boldsymbol{b}_n$*, and a distribution function* G *with non-degenerate marginals such that*

$$P(\boldsymbol{a}_n^{-1}(\boldsymbol{M}_n - \boldsymbol{b}_n) \leq \boldsymbol{x}) \stackrel{D}{\rightarrow} G(\boldsymbol{x}), \quad n \rightarrow \infty. \tag{7}$$

If Condition $D(\boldsymbol{u}_n)$ *below holds with* $\boldsymbol{u}_n = \boldsymbol{a}_n\boldsymbol{x} + \boldsymbol{b}_n$ *for each* $\boldsymbol{x}$ *such that* $G(\boldsymbol{x}) > 0$*, then* G *is a* D*-variate extreme value distribution function.*

Condition $D(\mathbf{u}_n)$ limits the long-range dependence in the process and is given as follows. For a set J of positive integers, let $M(J) = \max_{i \in J} X_i$. Let

$$\mathcal{I}_{j,k}(u_n) = M(I) \leq u_n : I \subset j, \cdots, k \tag{8}$$

be the set of all intersections of the events $X_i \leq u_n, \quad j \leq i \leq k$.

Condition 1. $D(\mathbf{u}_n)$: For all $A_1 \in \mathcal{I}_{1,l}, \quad A_2 \in \mathcal{I}_{l+s,n}$ and $1 \leq l \leq n - s$

$$\mid P(A_1 \cap A_2) - P(A_1)P(A_2) \mid \leq \alpha(n, s) \tag{9}$$

and $\alpha(n, s_n) \rightarrow 0$ as $n \rightarrow \infty$ for some positive integer sequence s_n such that $s_n = o(n)$.

Theorem 3. *Let* G *be a* D*-variate extreme value distribution and let* $\boldsymbol{a}_n > 0$ *and* $\boldsymbol{b}_n$ *be* D*-dimensional vectors such that* $D(\boldsymbol{u}_n)$ *and* $D'(\boldsymbol{u}_n)$ *given below hold for every* $\boldsymbol{u}_n = \boldsymbol{a}_n\boldsymbol{x} + \boldsymbol{b}_n$ *with* $\boldsymbol{x} \in \mathbb{R}^D$ *such that* $G(\boldsymbol{x}) > 0$*. Then*

$$P(\boldsymbol{a}_n^{-1}(\boldsymbol{M}_n - \boldsymbol{b}_n) \leq \boldsymbol{x}) \stackrel{D}{\rightarrow} G(\boldsymbol{x}), \quad n \rightarrow \infty, \tag{10}$$

if and only if

$$F^n(\boldsymbol{a}_n\boldsymbol{x} + \boldsymbol{b}_n) \stackrel{D}{\rightarrow} G(\boldsymbol{x}), \quad n \rightarrow \infty. \tag{11}$$

Condition $D'(\mathbf{u}_n)$ prevents local clustering of extremes and is given by:

Condition 2. $D'(\mathbf{u}_n)$

$$\lim_{k \rightarrow \infty} \limsup_{n \rightarrow \infty} \sum_{j=2}^{\lfloor n/k \rfloor} P(\mathbf{X}_1 > \mathbf{u}_n, \mathbf{X}_j > \mathbf{u}_n) = 0. \tag{12}$$

Ledford and Tawn (2003) presents a range of new diagnostic tools for assessing whether the assumptions concerning the long- and short-range dependence of the series are appropriate and for identifying structure within extreme events. These tools are based on tail characteristics of joint survivor functions but can be implemented by using existing estimation methods for extremes of univariate i.i.d. variables.

3.2. *The multivariate extremal index*

The extremal index, originated by Cartwright (1958) (c.f. Leadbetter, et al. (1983)), is a quantity which allows one to associate the limiting distribution of a dependent sequence to the extreme value distributions. Due to local dependency the extreme values often tend to appear in clusters and the extremal index measures the degree of clustering in the sequence. In a multivariate analog, Nandagopalan (1994) introduces the multivariate extreme index and derived some elementary properties. Whereas in the univariate case, the effect of local dependence is summarized by a single number, the extremal index, the multivariate setting is more difficult: the analogue of the extremal index turns out to be a function. In effect, the multivariate extremal index summarizes temporal dependence at extreme levels, but the strength of dependence can vary with direction.
Let again $\{\mathbf{X}_n\}$ be a stationary sequence of random vectors in $\mathbb{R}^D$ with distribution function F. Assume that there are vectors $\mathbf{a}_n > 0$ and $\mathbf{b}_n$ and D-variate extreme value distributions G and $\tilde{G}$ such that

$$P(\mathbf{a}_n^{-1}(\mathbf{M}_n - \mathbf{b}_n) \leq \mathbf{x}) \overset{D}{\to} G(\mathbf{x}),$$

$$F^n(\mathbf{a}_n\mathbf{x} + \mathbf{b}_n) \overset{D}{\to} \tilde{G}(\mathbf{x}),$$

as $n \to \infty$.
For $D = 1$ (univariate case), let $G(x) := \tilde{G}^{\theta}(x)$, for some constant $\theta \in [0, 1]$, then $\theta = \log G/\log\tilde{G}$ is called the *extremal index*. The definition of extremal index can be extended to multivariate case as follows.

Definition 2. *Extremal index function* of the sequence $\{\mathbf{X}_n\}$ is

$$\theta(\mathbf{v}) = \frac{\log G(\mathbf{x})}{\log\tilde{G}(\mathbf{x})}, \quad \mathbf{v} \in [\mathbf{0}, \infty)\backslash\{\mathbf{0}\},$$

where, for $\mathbf{v} \in [\mathbf{0}, \infty)\backslash\{\mathbf{0}\}$, let $\mathbf{x} = \mathbf{x}(\mathbf{v})$ be such that $v_j = -\log\tilde{G}_j(x_j)$ for $j = 1, ..., D$.

Smith and Weissman (1996) propose a less direct method for estimating the multivariate extremal index based on Pickands dependence function. They show that under some mixing conditions, if the finite dimensional distributions of the time series converge to those of a max-stable process, then the multivariate extremal index of the time series is the same as that of the limiting max-stable process.

3.3. *The extremal coefficient*

Schlather and Tawn (2003) define the extremal coefficient function for the multivariate random variables. Fasen et al. (2010) study a slightly modified version for univariate time series defined as follows:

Definition 3. *Extreme dependence function, extremal coefficient* For arbitrary $d \in \mathbb{N}$, define the lagged vector $X_d := (X_{t_1}, \cdots, X_{t_d}), \quad t_1 < \cdots < t_d$. The extreme dependence functions of $(X_t)_{t \geq 0}$ are defined for any lagged vector X_d and for all $(x_1, \cdots, x_d) \in \mathbb{R}^d$ by

$$\underline{\chi}_{(t_1,\cdots,t_d)}(x_1,\cdots,x_d) := \lim_{n\to\infty} n\mathrm{P}(X_{t_1} > a_n x_1 + b_n, \cdots, X_{t_d} > a_n x_d + b_n),$$

$$\begin{aligned}&\overline{\chi}_{(t_1,\cdots,t_d)}(x_1,\cdots,x_d)\\ &:= \lim_{n\to\infty} n\mathrm{P}(X_{t_1} > a_n x_1 + b_n \ \text{ or } \ \cdots \ \text{ or } \ X_{t_d} > a_n x_d + b_n).\end{aligned}$$

The extremal coefficient function of $(X_t)_{t\geq 0}$ is defined by

$$\chi(t) := \lim_{x \uparrow x_R} \mathrm{P}(X_t > x \mid X_0 > x),$$

where $x_R \leq \infty$ is the right enpoint of the stationary distribution of X_t.

A multivariate version of this is defined in Davis and Mikosch (2009). They consider a strictly stationary sequence of random vectors whose finite-dimensional distributions are jointly regularly varying with some positive index κ. This means that for any $h \geq 1$, the lagged vector $\mathbf{Y}_h = vec(\mathbf{X}_1, \cdots, \mathbf{X}_h)$ satisfies the relation in definition 1.1b with the measure ρ now indexed by h. This class of processes includes, among others, ARMA processes with regularly varying noise, GARCH processes with normally or Studen t-distributed noise and stochastic volatility models with regularly varying multiplicative noise. They define an analog of the autocorrelation function, the extremogram $\gamma_{AB}(h)$, which depends only on the extreme values in the sequence as follows.

Consider two Borel sets A, B in $\bar{\mathbb{R}}^D$ such that $C = A \times \bar{\mathbb{R}}^{D(h-1)} \times B$ is bounded away from zero and $\rho_{h+1}(\partial C) = 0$. Due to the assumption of regular variation, the following limit exists:

$$n\mathrm{P}(a_n^{-1}\mathbf{X}_0 \in A, a_n^{-1}\mathbf{X}_h \in B) = \rho_{h+1}(A \times \bar{\mathbb{R}}_0^{D(h-1)} \times B) := \gamma_{AB}(h). \quad (13)$$

Note that if both A and B are bounded away from zero, then

$$
\begin{aligned}
& n\mathrm{cov}(I_{a_n^{-1}\mathbf{X}_0\in A}, I_{a_n^{-1}\mathbf{X}_h\in B}) \\
&= n[\mathrm{P}(a_n^{-1}\mathbf{X}_0 \in A, a_n^{-1}\mathbf{X}_h \in B) - \mathrm{P}(a_n^{-1}\mathbf{X} \in A)P(a_n^{-1}\mathbf{X} \in B)] \\
&\sim n\mathrm{P}(a_n^{-1}\mathbf{X}_0 \in A, a_n^{-1}\mathbf{X}_h \in B) \\
&\sim \gamma_{AB}(h). \qquad (14)
\end{aligned}
$$

Davis and Mikosch (2009) also propose a natural estimator for the extremogram and study its asymptotic properties under α-mixing. They show asymptotic normality, calculate the extremogram for various examples and consider spectral analysis related to the extremogram.

3.4. *The extremal correlation*

For $D = 2$, Quintos defines the extremal correlation over a threshold (b_1, b_2) as:

$$
\begin{aligned}
\rho_+ = & \frac{\mathrm{E}(\log X_1 - \log b_1)_+(\log X_2 - \log b_2)_+}{\sqrt{\sigma_{+1}^2}\sqrt{\sigma_{+2}^2}} \\
& - \frac{\mathrm{E}(\log X_1 - \log b_1)_+\mathrm{E}(\log X_2 - \log b_2)_+}{\sqrt{\sigma_{+1}^2}\sqrt{\sigma_{+2}^2}}, \qquad (15)
\end{aligned}
$$

where

$$
\sigma_{+i}^2 = \mathrm{E}(\log X_i - \log b_i)_+^2 - (\mathrm{E}(\log X_i - \log b_i)_+)^2, \quad i = 1, 2. \qquad (16)
$$

He uses a modification of the Hill estimator to obtain an estimate of ρ_+. For a univariate time series $(X_t)_{t=1}^T$ ordered as $X_{(1)} \leq \cdots X_{(T)}$, Hill's estimator takes the means over threshold as:

$$
\hat{\alpha}^{-1} = \frac{1}{m}\sum_{t=1}^{m}\left(\log X_{(T-t+1)} - \log X_{(T-m)}\right), \qquad (17)
$$

where m is the number of observations over the threshold.
For the bivariate setting, Quintos defines $\hat{\alpha_i}^{-j}$ as

$$
\hat{\alpha_i}^{-j} = \frac{1}{j!m}\sum_{t=1}^{m}\left(\log X_{i(T-t+1)} - \log X_{i(T-m)}\right)^j, \quad i = 1, 2; j = 1, 2 \qquad (18)
$$

and $\hat{\alpha_0}^{-1}$ as

$$
\hat{\alpha_0}^{-1} = \frac{1}{m}\sum_{t=1}^{m}\left(\log X_{1(T-t+1)} - \log X_{1(T-m)}\right)\left(\log X_{2(T-t+1)} - \log X_{2(T-m)}\right). \qquad (19)
$$

Then the estimator $\hat{\rho}_+$ is given by

$$\hat{\rho}_+ = \frac{\hat{\alpha}_0^{-1} - \hat{\alpha}_1^{-1}\hat{\alpha}_2^{-1}}{\sqrt{2\hat{\alpha}_1^{-2} - (\hat{\alpha}_1^{-1})^2}\sqrt{2\hat{\alpha}_2^{-2} - (\hat{\alpha}_2^{-1})^2}}. \tag{20}$$

He shows that the estimator is consistent and asymptotically normal under Pareto tails and for GARCH(1,1) processes. Although this paper is not very rigorous or general mathematically, it has simple and implemetable ideas. The dimensions can be generalized to $D > 2$ to get a correlation matrix, though it is not clear if it will be positive definite.

4. Some special stationary time series

4.1. *Multivariate maxima of moving maxima*

The process $\mathbf{X}_i, i = \pm 1, \cdots, \pm\infty$ is said to be a multivariate maxima of moving maxima ($M4$) process if

$$X_{id} = \max_{l=1}^{\infty} \max_{k=-\infty}^{\infty} a_{l,k,d} Z_{l,i-k}, \tag{21}$$

where $Z_{l,i}$ are independent unit Fréchet for all $l, i,\quad a_{l,k,d} \geq 0$ and

$$\sum_{l=1}^{\infty} \sum_{k=-\infty}^{\infty} a_{l,k,d} = 1, \quad d = 1, ..., D. \tag{22}$$

For this process,

$$\begin{aligned} &\mathrm{P}(Y_{id} \leq y_{id}, i = 1, \cdots, n, d = 1, \cdots, D) \\ &= \exp\left(-\sum_{l=1}^{\infty} \sum_{m=-\infty}^{\infty} \max_{k=1-m}^{n-m} \max_{d=1}^{D} \frac{a_{l,k,d}}{y_{m+k,d}} \right). \end{aligned} \tag{23}$$

Smith and Weissman (1996) and Zhang (2002) introduce the $M4$ processes. The $M4$ process is max-stable. Conversely, all max-stable processes can be approximated by $M4$ processes, if we exclude certain degenerate cases. In this sense, the problem of modeling extremes of multivariate stationary processes can be restricted to the study of extremes of $M4$ processes. Zhang (2002) explores estimation of parameters of the M4 process.

4.2. *Markov chains*

Perfekt (1997) determines the asymptotic distribution of $\mathbf{M}_n$ for a class of stationary Markov chains by computing the multivariate extremal index. He uses a characterization of the extremal index, which is tailored to applications to Markov chains. The assumptions include mild local and long-range dependence restrictions, as well as suitable conditions on the marginal distribution and the transition probabilities of the chain. Under these conditions, the extremal characteristics of the chain are determined by another Markov chain, which he calls the tail chain associated with the original Markov sequence. They give a rather complete description of the extremal behavior by proving convergence of point processes associated with exceedances of 'high levels'.

5. Discussion

A vibrant area in multivariate extremes is the estimation of tail dependence. To study the extreme dependence around the boundary point $(1, \ldots, 1)^T$, the tail dependence function of $\mathbf{X}$ is defined as

$$\lambda_{\mathbf{X}}(x_1, \ldots, x_D) = \lim_{t \to 0} t^{-1} \mathrm{P}(1 - F_1(X_1) \le tx_1, \ldots, 1 - F_D(X_D) \le tx_D), \tag{24}$$

where $x_1, \cdots, x_D \ge 0$. A similar definition can be given for studying the extreme dependence around the boundary point $(0, \ldots, 0)^T$. For $D = 2$ the tail dependence function $\lambda_{\mathbf{X}}(x, y)$ or the stable tail dependence function $x + y - \lambda_{\mathbf{X}}(x, y)$ can be estimated nonparametrically, see Schmidt and Stadtmüller (2006). Parametric models for the tail dependence function have been suggested and estimated, see Coles (2001) for examples and further references. Klüppelberg (2008) proposes a semi-parametric model for (asymptotically dependent) tail dependence functions via an elliptical copula. Most applications of both, nonparametric and parametric estimation of tail dependence functions are focused on the case $D = 2$, although theoretically both methods are applicable to the case $D > 2$. Hsing, et al. (2004) estimate the tail dependence function nonparametrically in higher dimensions. There has been no work on extending the concept of tail dependence and developing methods of estimation to the time series setting. One way to go about this is as follows. It is straightforward to extend the concept of tail dependence to lagged sequences for a stationary time series

$$\begin{aligned} &\lambda_{\mathbf{X},\mathbf{L}}(x_1, \cdots, x_D, l_1, \cdots, l_D) \\ &= \lim_{t \to 0} \frac{\mathrm{P}(1 - F_{1,s-l_1}(X_{1,s-l_1}) \le tx_1, \cdots, 1 - F_{D,s-l_D}(X_{D,s-l_D}) \le tx_D)}{t}, \end{aligned}$$

for $x_1, \cdots, x_D \ge 0$, and integers $l_1, \cdots, l_D \ge 0$. Because of stationarity, this does not depend on s. These functions can then be estimated as in the i.i.d case by considering lagged vectors. We need to study the properties of these estimators and how they quantify the relation between extremes of multivariate time series.

Extreme value analysis is most interesting, when extremes happen in a pronounced way. Consequently, models with heavy-tailed marginal distributions are of high interest. Heavy tails in linear models originate simply because of a heavy-tailed noise. In nonlinear models, however, regularly varying tails occur, even though the noise sequence may be light-tailed. Nonlinear models, like the ARCH and GARCH models, are of high interest in many areas of application like insurance and finance. Fasen et al. (2010) give a comprehensive account of extreme dependence in univariate non-linear models with heavy tails. These need to be extended to the multivariate setting.

All methods ignore the dependence in time of the extreme events across multiple variables. During periods of stress, the extremes of all the variables do not necessarily happen exactly at the same time. This is closely related to the concepts of contagion and spill-overs in finance. In general, the transmission of shock from one variable to another takes a finite time. This is analogous to the concept of cross-autocorrelation. One can use recurrence and times to study this dependence. Recurrence times are the waiting times between two successive extreme events. Given that there is an extreme event in one series, the conditional residual recurrence time is the waiting time before the next extreme event of the other series. If the time of occurrence of extremes of the two series are independent, then under many commonly encountered time series models like ARCH and GARCH, the distribution of the recurrence time is the same as that of the conditional residual recurrence time. We can use the distance between these two distributions to test for and estimate the amount of contagion between different series.

References

Amram, F. (1985). Multivariate extreme value distributions for stationary Gaussian sequences. *J. Mult. Anal.* **16**, 237–240.

Balkema, A. A. and Resnick, S. I. (1977). Max-infinite divisibility. *J. Appl. Prob.* **14**, 309–313.

Beirlant, J. S., Goegebeur, Y. and Teugels, J. (2004). Statistics of extremes: theory and applications.

Cartwright, D. E. (1958) On estimating the mean energy of sea waves from the highest waves in a record. *Proc. Roy. Soc. London. Ser. A* **247**, 22–28.

Coles, S. G. and Tawn, J. A. (1991) Modeling Extreme Multivariate Events. *J. R. Statist. Soc. B*, **53**, 377–392.

Coles, S., Heffernan, J. and Tawn, J. (1999). Dependence Measures for Extreme Value Analyses. *Extremes*, **2**, 339–365.

Coles, S. G. (2001). An Introduction to Statistical Modeling of Extreme Values. Springer, London.

Cooley, D., Naveau, P. and Poncet, P. (2006). Variograms for spatial max-stable random fields. Statistics for Dependendent Data: STATDEP2005, Springer Lecture Notes in Statistics. Springer, London.

Davis, R. A. and Mikosch, T. (2009). The extremogram: A correlogram for extreme events. *Bernoulli* **15**, 4, 977–1009.

Esary, J. D., Proschan, F. and Walkup, D. W. (1967). Association of random variables with applications. *The Annals of Mathematical Statistics*, **38**, 5. (Oct., 1967), 1466–1474.

Fasen, V., Klüppelberg, C. K. and Schlather, M. (2010). High-level dependence in time series models, *Extremes*, **13** (1), 138–164.

Galambos, J. (1987). The Asymptotic Theory of Extreme Order Statistics. 2nd. ed. Krieger, Florida.

Hartmann, P., Straetmans, S. and C. G. de Vries, C. G. (2004). Asset Market Linkages in Crisis Periods. *Rev. Econ. Statist.* **86**, 313–326.

Hsing, T. (1987). Extreme value theory for multivariate stationary sequences. Technical Report, Texas A&M University. *J. Mult. Anal.* **29**, 274–291.

Hsing, T. (1989) Extreme value theory for multivariate stationary sequences. *Journal of Multivariate Analysis* **29**, 274–291.

Hsing, T., Klüppelberg, C. and Kuhn, G. (2004). Dependence estimation and visualization in multivariate extremes with application to financial data. *Extremes* **7**, 2, 99–121.

Hüsler, J. (1987). Multivariate extreme values in stationary random sequences. Technical Report, University of Bern. To appear in *Stochastic Processes and Applications.*

Hüsler, J. and Schüpbach, M. (1988). Limit results for maxima in non-stationary multivariate Gaussian sequences. Stoch. Proc. Appli. 28 91-99.

Hüsler, J. (1990) Multivariate extreme values in stationary random sequences. *Stochastic Processes and their Applications* **35** 99–108.

Hüsler, J. (1990). Topics in Statistical Dependence, *Lecture Notes-Monograph Series* **16**, 269–281.

Joe, H. (1989) Families of min-stable multivariate exponential and multivariate extreme value distributions. *Statist. Probab. Lett.* **9**, 75–82.

Joe, H. (1994), Multivariate extreme-value distributions with applications to environmental data. *Canadian Journal of Statistics*, **22**, 47-64.

Joe, H. (1997). Multivariate Models and Dependence Concepts. Chapman & Hall/CRC

Klüppelberg, C., Kuhn, G. and Peng, L. (2008) Semi-parametric models for the multivariate tail dependence function the asymptotically dependent, *Scandinavian Journal of Statistics* **35**, 4, 701–718.

Leadbetter, M.R., Lindgren, G., and Rootzen, H. (1983). Extremes and Related

Properties of Random Sequences and Processes. Springer, Series in Statistics, Berlin.

Ledford, A. W. and Tawn, J. A. (2003). Diagnostics for Dependence within Time Series Extremes. *Journal of the Royal Statistical Society, Series B (Statistical Methodology)*, **65**, 2, 521–543.

Lindgren, G. (1974). A note on the asymptotic independence of high level crossings for dependent Gaussian processes. *Ann. Probab.* **2**, 535–539.

Longin, F. and Solnik, B. (2001). Extreme Correlation of International Equity Markets. *J. Finan.* **56**, 649–676.

Longstaff, F. A. (2010) The subprime credit crisis and contagion in financial markets. *Journal of Financial Economics*, **97**, 3, 436–450.

Mardia, K. V. (1970) Families of bivariate distributions. Grin's Statistical Monographs and Courses, No. 27. Hafner Publishing Co., Darien, Conn.

Marshall, A. W. and Olkin, I. (1983). Domains of attraction of multivariate extreme value distributions. *Ann. Probab.* **11**, 168–177.

McFadden, D. (1978) Modelling the choice of residential location. In Spatial Interaction Theory and Planning Models (eds Karlqvist, A., Lundquist, L., Snickers, F. and Weibull, J.), 75–96. Amsterdam: North-Holland.

Nadarajah, S. (2001). Multivariate declustering techniques, *Environmetrics* **12**, 357–365.

Nandagopalan, S. (1994). On the multivariate extremal index. *Journal of Research of the National Institute of Standards and Technology* **99**, 543–550.

Perfekt, R. (1997). Extreme value theory for a class of markov chains with values in $\mathbb{R}^d$. *Advances in Applied Probability* **29**, 138–164.

Pickands, J. III (1981) Multivariate Extreme Value distributions. *Proc. 43rd Session I.S.I. (Buenos Aires)*, 859–878.

Quintos, C. Estimating tail dependence and testing for contagion using tail indices. Working paper.

Ramos, A. and Ledford, A. (2008). A new class of models for bivariate joint tails. *J. Roy. Statist. Soc. Ser. B* **71**, 219-241.

Resnick, S. I. (1987). Extreme Values, Regular Variation, and Point Processes. Springer, Berlin.

Resnick, S. I. (2007). Heavy-Tail Phenomena. Springer, New York.

Sbihi, A. (1988). Limiting distribution of the maxima of multivariate stationary sequences. Technical Report, McGill University.

Schlather, M. and Tawn, J. (2003). A dependence measure for multivariate and spatial extreme values: Properties and inference. *Biometrika* **90**, 139–156.

Schmidt, R. and Stadtmüller, U. (2006). Nonparametric estimation of tail dependence. *Scand. J. Stat.* **33**, 2, 307–335.

Sibuya, M. (1960). Bivariate extreme statistics. *Ann. Inst. Stat. Math.* **11**, 195–210.

Smith, R. L. (1990) Extreme Value Theory. Handbook of Applicable Mathematics (Supplement), W. Ledermann (ed.) John Wiley, Chichester.

Smith, R. L. and Weissman, I. (1996). Characterization and estimation of the multivariate extremal index.

Tawn, J. A. (1988). Bivariate Extreme Value Theory: Models and Estimation,

Biometrika **75**, 3, 397–415.

Tawn, J. A. (1990) Modeling Multivariate Extreme Value Distributions, *Biometrika*, **77**, 2, 245–253.

Villasenor, J. (1976). On univariate and bivariate extreme value theory. Ph.d. Thesis, Iowa State University.

Zhang, Z. (2002). Multivariate extremes, max-stable process estimation and dynamic financial modeling, PhD Thesis, University of North Carolina at Chapel Hill.

Chapter 19

Dynamic Mixed Models for Irregularly Observed Water Quality Data

Robert H. Shumway

University of California, Davis

A generalization of the conventional state-space model adds fixed covariates to the measurement equation while treating the state-vectors as time-varying random effects, resulting in the time series analogue of the classical mixed model. Incorporation of fixed covariates allows flexible models for trends and incorporation of fixed threshold effects. Estimating the measurement matrix produces the time series analog of factor analysis. Specializing the state-space model to the classical mixed model allows analysis of longitudinal and panel data. We generalize the maximum likelihood estimation and smoothing to the irregularly observed vector case where there are fixed covariates and a possibly unknown measurement matrix. The results are illustrated for several environmental series involving stream-flows and pesticide concentrations.

Contents

1. Introduction: The Dynamic Mixed Model

The manifestations of the conventional linear model are pervasive and various forms of this model can be adapted to many practical applications.

Even nonlinear models can often be expressed in linear form through transformations or by approximation. In the time series case, a simple version of the multivariate linear model, expressing a $q \times 1$ output vector $\mathbf{y}_t = (y_{t1}, y_{t2}, \dots, y_{tq})'$ in terms of some $r \times 1$ input vector of fixed functions $\mathbf{z}_t = (z_{t1}, z_{t2}, \dots, z_{tr})'$, for $t = 1, 2, \dots, n$ time-indexed points covers many situations, i.e.,

$$\mathbf{y}_t = \Gamma \mathbf{z}_t + \mathbf{v}_t, \tag{1}$$

where Γ is a $q \times r$ matrix of regression coefficients relating the input and output vectors and $\mathbf{v}_t, t = 1, 2, \dots, n$ are uncorrelated normal vectors with zero means and common covariance matrix R. Maximum likelihood estimation of the regression matrix Γ and the error covariance matrix R are a part of conventional multivariate regression analysis and the case where parts of the vectors $\mathbf{y}_t$ are missing is covered in Johnson and Wichern (1992). In the time series case, it is natural to allow a correlated random component, say $\mathbf{x}_t = (x_{t1}, x_{t2}, \dots, x_{tp})'$ into (1), obtaining a new mixed model of the form

$$\mathbf{y}_t = \Gamma \mathbf{z}_t + A \mathbf{x}_t + \mathbf{v}_t, \tag{2}$$

where $\mathbf{x}_t$ is a random $p \times 1$ stochastic component and A is a $q \times p$ matrix. The random component $\mathbf{x}_t$ satisfies the state equation

$$\mathbf{x}_t = \Phi \mathbf{x}_{t-1} + \mathbf{w}_t. \tag{3}$$

In the above equations, the random component $\mathbf{x}_t$ in (2) is regarded as the unobserved signal, evolving according to the state transition equation (3). Conceptually, the process starts with $\mathbf{x}_0$ drawn from some population, usually multivariate normal, with mean vector $\boldsymbol{\mu}_0$ and initial covariance matrix Σ_0. State errors $\mathbf{w}_t$, assumed to be independent zero-mean vectors with common covariance matrix Q, are generated for $t = 1, 2, \dots, n$ in (3), leading to the unobserved signal values $\mathbf{x}_t$. Finally, the values of the observed process $\mathbf{y}_t$ in (2) are generated by adding the sequence of independent observation errors $\mathbf{v}_t$, assumed to have mean zero with common covariance matrix R, to the covariance component $\Gamma \mathbf{z}_t$ and the linearly mapped signal $A\mathbf{x}_t$. Note that we will sometimes need to let A and Γ be time varying (e.g. A_t and Γ_t) in order to account for missing data. A notable extension to the replicated case and a conditional autoregressive structure, where the covariates appeared in the state equation (3) has been given by Schmid (1996). We do not need replication for the examples considered here, although it could be easily added by assuming independent, identically distributed vectors $\mathbf{y}_{ti}, i = 1, 2, \ldots, N$ if panel data are of primary interest.

The model for the unobserved components, $\mathbf{x}_t$, is quite general and can easily be specialized to the classical mixed model considered by Laird and Ware (1982). For example, Jones (1993) gives the basic approach and Icaza and Jones (1999), among others, show how to apply the EM algorithm, given for the state-space case by Shumway and Stoffer (1982), to the mixed model with a univariate response. The specialization involves setting the error in (3) to zero and the transition matrix to $\Phi = I$. Then, the distribution of the initial state vector can be taken as normal, with zero mean and covariance matrix $\sigma_0^2 I$ $(x_{ti} = x_{t-1,i}, i = 1, 2, \ldots, p)$, and the sequence of state vectors become conventional random effects that do not change over time. More random effects can be added by stacking the state vectors.

Example 1. Conventional Mixed Model (Laird and Ware (1982), Jones (1993), Icaza and Jones (1999))
As an example of the state-space model with replication, consider the very simple mixed model with r fixed effects $\boldsymbol{\beta} = (\beta_1, \ldots, \beta_r)'$ and N random subject effects $b_i, i = 1, \ldots, N$. Then,

$$y_{ti} = \boldsymbol{\beta}'\mathbf{z}_{ti} + b_i + e_{ti},$$

is a model for subject i observed at time t, where $e_{ti} \sim NI(0, \sigma_e^2)$ and $b_i \sim NI(0, \sigma_b^2)$. Note that the main problems of interest are the estimation of the fixed effect $\boldsymbol{\beta}$ and the variance components σ_e^2 and σ_b^2. Writing the observation equation (2) in the form

$$y_{ti} = \boldsymbol{\beta}'\mathbf{z}_{ti} + b_{ti} + e_{ti},$$

the state equation (3) can be formulated as

$$b_{ti} = b_{t-1,i},$$

with $b_{0i} \sim N(0, \sigma_b^2)$. Besides estimators for the variance components, the Kalman smoothers will produce the usual best linear unbiased predictors (BLUP) for the random effects $b_{0i} = E(b_{0i}|y_{ti}, i = 1, \ldots, N, t = 1, \ldots, n)$.

Like the multivariate linear regression model (1), the state-space model can be adapted to many special circumstances involving linear models. In addition, the Kalman filters and smoothers, used in conjunction with the EM algorithm offer an elegant approach to handling incompletely observed multivariate vectors. The state-equations (3) may also be specialized to produce different kinds of non-stationary signal models (e.g. random walks or splines). The fitting of ARMA models to univariate irregularly observed series in state-space form was started by Jones (1980) and continued by Ansley and Kohn (1984), as well as others. The use of state-space models for fitting fully observed economic series with trends and seasonality has been explored by Kitagawa and Gersch (1984; 1996). Texts that contain examples of fitting various time series models in state-space form are Harvey (1991), Durbin and Koopman (2001) and Shumway and Stoffer (2006).

In order to exploit the extensions covered here, we consider two sets of environmental data; the first involves water flow at three contiguous sites; the second example involves levels of pesticides in water under two different sampling methods. In the first case (see Shumway (1998)), we consider estimating mean and covariance parameters in the state-space models (2) and (3) when there are with missing stream flows and then interpolating the missing observations. In the second case considered in Shumway (2000b), we are primarily interested in extracting a common pesticide signal using one series with daily samples and another benchmark series measured weekly.

Example 2. Estimation and Interpolation: Monthly Stream Flows In the analysis of river and stream flow data, the estimation of flow-frequency statistics, such as monthly means and variances, is of interest for monitoring water resources. Difficulties arise when there are significant portions of one or more stream flows missing in a given set of records. There are a number of papers in the hydrology literature (see, for example, Alley and Burns (1983), Salas, et al. (1980) and Vogel and Stedinger (1985)) proposing somewhat ad-hoc solutions to the problem of estimating the means and variances of bivariate stream flows in various stages of completion. Some of the potential difficulties to overcome when observations are missing can be seen by inspecting Figure 1, which shows 454 measured monthly flows at three stations on the Yakima River, where we assume that the original complete flow record had 504 observations for each series. The incomplete series has 50 observations missing from the beginning of the Keechelus Lake series and 50 observations missing from each of the Cachess Lake and Yakima River series. For this particular data, we happen to know the missing values, enabling a comparison of the interpolation and estimation results. Of interest to hydrologists are the monthly flow rates and the mean monthly flows, expressed as a simple average over the 42 years. Figure 2 shows the 12 month profiles for the 42 years and the

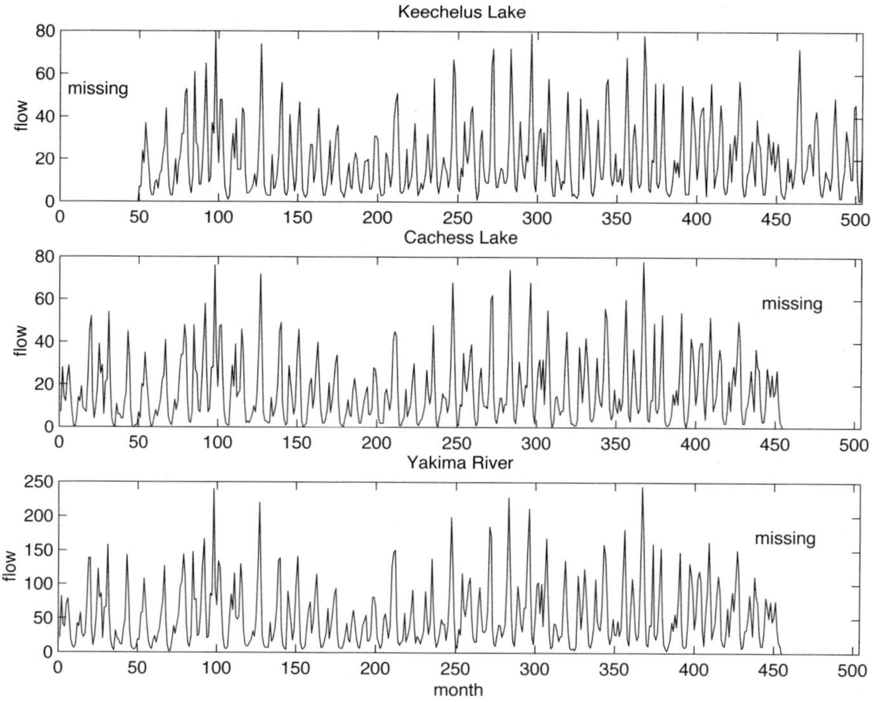

Fig. 1. Monthly flows at three stations on the Yakima River for 42 years. Incomplete data has 50 months missing from the beginning of of Keechelus Lake and 50 months missing from the end of Cachess Lake and Yakima Rivers.

average profile using the fully observed data and we see that the monthly values tend to behave like fixed effects, with the maximum flows occurring generally in the month of April (month 8). Referring to the model defined in (2) and (3), we take

$$\mathbf{y}_t = \Gamma \mathbf{z}_t + \mathbf{a}\mathbf{x}_t + \mathbf{v}_t, \tag{4}$$

where $\mathbf{a} = (a_1, a_2, a_3)'$ is a vector that scales the stochastic components of the three stream flows and where $\Gamma = (\boldsymbol{\mu}_1, \boldsymbol{\mu}_2, \ldots, \boldsymbol{\mu}_{12})$, with $\boldsymbol{\mu}_i = (\mu_{1i}, \mu_{2i}, \mu_{3i})'$ is the 3×1 vector of mean stream flows for month i. Hence, Γ is a 3×12 matrix of means for the three flows over 12 months. The 12×1 vector $\mathbf{z}_t$ has a one in position i when observation t is on month i and zeros elsewhere. To find a sensible model for the signal $\mathbf{x}_t$, Shumway (1998) examined the residuals $\mathbf{y}_t - \hat{\Gamma}\mathbf{z}_t$, based on the part of the record where all series were observed and found that the residuals were highly

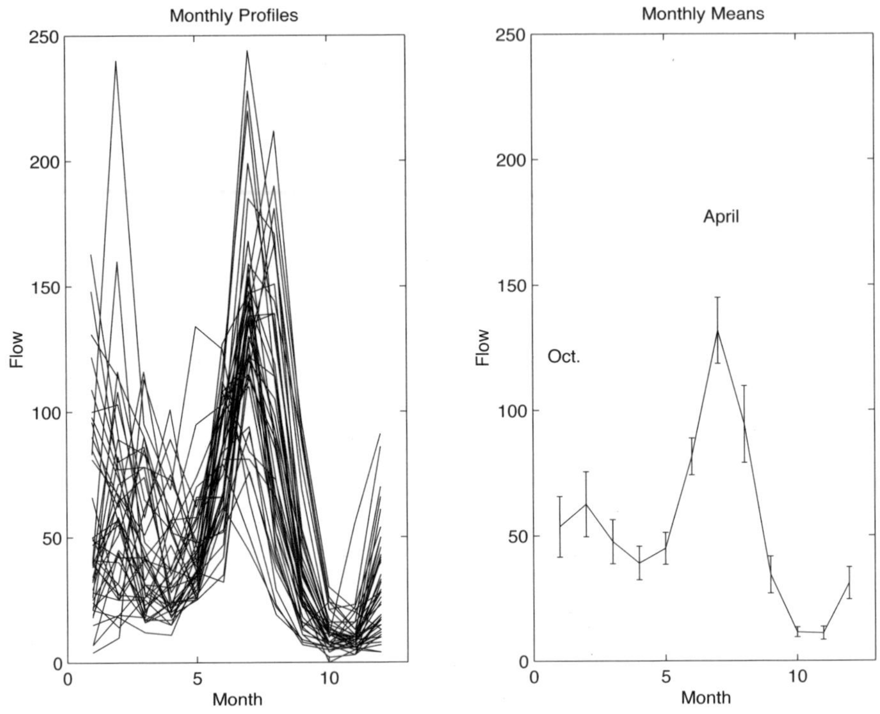

Fig. 2. Monthly profiles for 42 years of flows on Yakima River (left panel) and mean monthly flows (right panel) using complete data, with 95% confidence limits computed from conventional multivariate regresssion.

correlated across streams and that a first-order autoregressive model was reasonable. If we assume that the serially correlated signal is the same, up to a constant multiplier a_i at each stream, we might write the state-space model in the form given above. The state equation expresses the common univariate signal x_t as

$$x_t = \phi x_{t-1} + w_t, \tag{5}$$

where we take var $w_t = q_{11} = 1$ to make the log likelihood identifiable. The parameter set in this case is $\Theta = \{\Gamma, \mathbf{a}, R, \phi\}$, where we set the initial mean, μ_0, to zero and the initial variance at $\sigma_0^2 = 10$. Problems of interest relate to estimating the mean monthly flows, with confidence intervals, and to interpolating the missing observations.

Example 3. Monitoring Pesticide Levels in Water

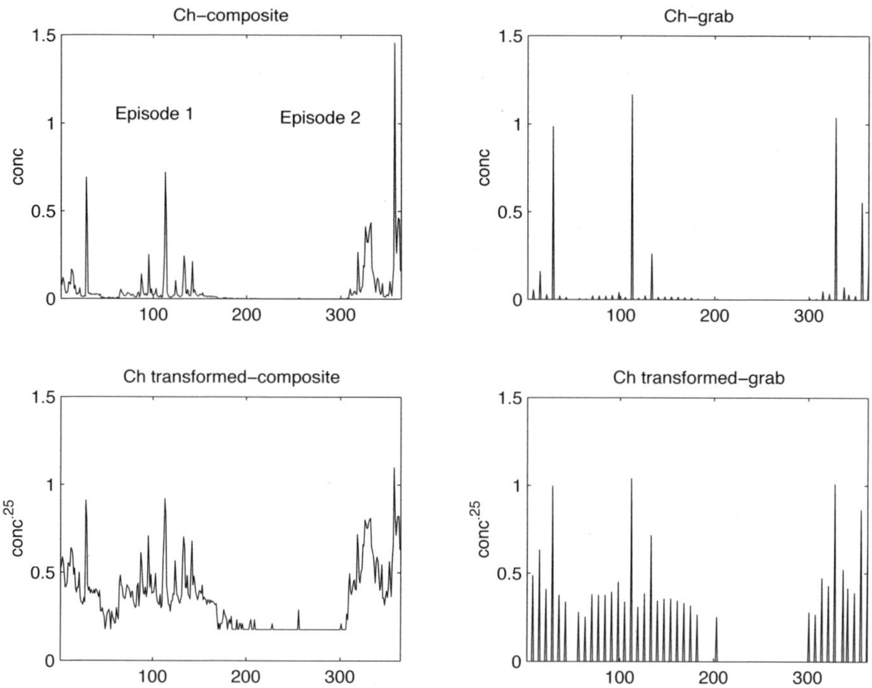

Fig. 3. Observed concentrations of chlorpyrifos in the Orestimba Creek measured daily by an automatic sampler (left panels) and weekly by grab samples (right panel). Transformed (fourth root) concentrations are shown in the bottom panels.

Measurements of pesticide concentrations in water samples tend to be quite volatile and records of daily concentrations measured by an automatic samplers such as those shown in Figure 3 tend to have a number of large peaks and many values that are either zero or below a threshold. Two episodic bursts of activity observed in daily concentrations over the year are marked in the left upper panel of the figure and it is natural to be interested in the predicted profiles of these episodes. The right panel shows pesticide concentrations collected by grab samples on a weekly basis and we note that there are still peaks for the episodes that roughly correspond to those in the left panel. Of interest would be estimated signals corresponding to these episodes, extracted from some sensible model. Because of the large excursions, a transformation might be sensible here and the fourth root, shown in the lower part of Figure 3, seems to produce a more stationary process that might have approximately normally distributed errors. Section 4 shows how to estimate common profile signals for the two episodes using a version of the general model specified in (2) and (3).

In order to fit a reasonable state space model, we noted that the first differences of the transformed daily series, say $y_t^{1/4} - y_{t-1}^{1/4}$ tended to be white so that a sensible model for merging the two series $\mathbf{y}_t = (y_{t1}, y_{t2})'$ into an estimator for unobserved signal, say x_t might be

$$y_{ti}^{1/4} = x_t^{1/4} + v_{ti}, \tag{6}$$

for $i = 1, 2$, where

$$x_t^{1/4} = x_{t-1}^{1/4} + w_t, \tag{7}$$

and the fourth root of the signal is assumed to be a simple random walk. Once the underlying transformed signal and its variance are estimated using the Kalman filters and smoothers, we might assume normality to obtain 95% prediction limits for the smoothed signal. Transforming back to the raw signal will give approximate 95% prediction limits for the raw signal x_t. The predicted signal and its variance, combined with a Bonferroni argument, also will yield various probabilities of interest, for example, the probability that m consecutive values will exceed some pre-specified threshold.

2. Signal Extraction and Interpolation

When data are missing from the observation vector $\mathbf{y}_t$ vector, we will consider the following special notation for the reordered vector of observed and unobserved parts. Let

$$\mathbf{y}_t = \begin{pmatrix} \mathbf{y}_t^{(1)} \\ \mathbf{y}_t^{(2)} \end{pmatrix}$$

denote a partition of $\mathbf{y}_t$, where the $q_1 \times 1$ and $q_2 \times 1$ components contain the observed and unobserved parts respectively for each month t and $q_1 + q_2 = q$. In the case where the vector above is only partially observed, we write the model as

$$\begin{pmatrix} \mathbf{y}_t^{(1)} \\ \mathbf{y}_t^{(2)} \end{pmatrix} = \begin{pmatrix} \Gamma_{1t} \\ \Gamma_{2t} \end{pmatrix} \mathbf{z}_t + \begin{pmatrix} A_{1t} \\ A_{2t} \end{pmatrix} \mathbf{x}_t + \begin{pmatrix} \mathbf{v}_t^{(1)} \\ \mathbf{v}_t^{(2)} \end{pmatrix} \tag{8}$$

with all matrices partitioned into sub-components corresponding to the observed and unobserved parts respectively. Note that, at a particular time point,

$$\operatorname{cov}\begin{pmatrix}\mathbf{v}_t^{(1)}\\ \mathbf{v}_t^{(2)}\end{pmatrix} = \begin{pmatrix}R_{11t} & R_{12t}\\ R_{21t} & R_{22t}\end{pmatrix} \tag{9}$$

defines the partitioned covariance matrix, where the configuration obviously changes as t changes. It will also be convenient to introduce a notation for the complete data and we will use $Y_n = (\mathbf{y}_1, \mathbf{y}_2, \ldots, \mathbf{y}_n)$ to denote the collection of completely observed vectors. Then, we may use $Y_n^{(1)} = (\mathbf{y}_1^{(1)}, \mathbf{y}_2^{(1)}, \ldots, \mathbf{y}_n^{(1)})$ to denote the fully observed part when there are missing components at various times. To continue, we note that the maximum likelihood approach should lead to estimators for the parameter vector $\Theta = (\Gamma, A, R, \Phi, Q)$ under arbitrary missing data configurations. Using standard arguments, we obtain these results in the following sections.

2.1. *Signal Extraction*

The treatment of the general state-space model given by (2) and (3) depends upon being able to compute

$$\mathbf{x}_t^n = E_0\{\mathbf{x}_t | Y_n^{(1)}\}, \tag{10}$$

the optimal smoothed estimators for the process in the presence of missing data and

$$P_t^n = E\{(\mathbf{x}_t - \mathbf{x}_t^n)(\mathbf{x}_t - \mathbf{x}_t^n)' | Y_n^{(1)}\}, \tag{11}$$

the mean square covariance of the smoothed estimators. Because the estimators (10) are for an unobserved process, we will sometimes refer to them as the extracted signals. For the maximum likelihood estimators in the next section, we will also need the covariance between two adjacent smoothed estimators, namely,

$$P_{t,t-1}^n = E\{(\mathbf{x}_t - \mathbf{x}_t^n)(\mathbf{x}_{t-1} - \mathbf{x}_{t-1}^n)' | Y_n^{(1)}\}. \tag{12}$$

These are just the Kalman smoothers, as given in Shumway and Stoffer (2006), Sec. 6.4, for the missing data case (see also Stoffer (1982), Shumway and Stoffer (1982)). In order to compute the values in (10), (11) and (12), we need a powerful recursive procedure based on the Kalman filters and smoothers as covered in the above references.

2.2. *Interpolation*

For solving the problem of interpolating the missing part $\mathbf{y}_t^{(2)}$ of the vector, it is convenient to first take expectations conditionally on $Y_n^{(1)}, \mathbf{x}_t$ and then on $Y_n^{(1)}$, noting that

$$E\{\,*\,|Y_n^{(1)}\} = E\Big\{E\{\,*\,|Y_n^{(1)}, \mathbf{x}_t\}|Y_n^{(1)}\Big\}.$$

For the partition (8) we have $E\{\mathbf{y}_t^{(1)}|Y_n^{(1)}\} = \mathbf{y}_t^{(1)}$ for the observed part and

$$\begin{aligned} E\{\mathbf{y}_t^{(2)}|Y_n^{(1)}\} &= E\{\Gamma_{2t}\mathbf{z}_t + A_{2t}\mathbf{x}_t + \mathcal{B}(\mathbf{y}_t^{(1)} - \Gamma_{1t}\mathbf{z}_t - A_{1t}\mathbf{x}_t)|Y_n^{(1)}\} \\ &= \Gamma_{2t}\mathbf{z}_t + A_{2t}\mathbf{x}_t^n + \mathcal{B}(\mathbf{y}_t^{(1)} - \Gamma_{1t}\mathbf{z}_t - A_{1t}\mathbf{x}_t^n) \end{aligned} \tag{13}$$

for the unobserved part, where

$$\mathcal{B} = R_{21t}R_{11t}^{-1} \tag{14}$$

denotes the regression of the unobserved errors on the observed errors. Note that (13) and (14) provide the appropriate equations for interpolating the missing data vector using the entire sample. In case the vector $\mathbf{y}_t$ is completely unobserved,

$$E\{\mathbf{y}_t|Y_n^{(1)}\} = \Gamma\mathbf{z}_t + A\mathbf{x}_t^n. \tag{15}$$

In the following section, we show how to get maximum likelihood estimators in the general missing data case.

3. Estimation of Parameters

We review the maximum likelihood procedure for estimating parameters using the EM algorithm of Dempster, et al. (1977), as proposed by Shumway and Stoffer (1982). Extensions to that paper covered here are (1) including the covariates $\mathbf{z}_t$ in the model and estimating Γ and (2) estimating the measurement matrix A. The latter of these two procedures can be regarded as the time series version of factor analysis. Hence, the maximum likelihood procedure will be applied to the model given by (2) and (3), assumed to depend on the parameter set $\Theta = (\Gamma, A, R, \Phi, Q)$, with either $\boldsymbol{\mu}_0$ or Σ_0 possibly added; both should not be included at the same time. We may also want to fix A or Γ; setting $\Gamma = 0$ is common when there are no fixed effects. It should be noted that a later application of the EM algorithm to unobserved component systems in economics was by Watson and Engle (1983).

The computation of variances and covariances of the estimated parameters has been an unwieldy proposition because the EM algorithm does not provide an easily computed version of the information matrix. Direct computation of the information matrix via recursions is possible as in Harvey (1991) or Cavanaugh and Shumway (1996). Versions of the information matrix, obtained from outputs arising naturally in the EM algorithm, such as in Meng and Rubin (1991) or Oakes (1999), are either hard to compute, as in the former, or will involve relatively untractable derivatives as in the latter. A compromise that is easy to apply and will be robust towards distributional assumptions is the bootstrap, as derived in Stoffer and Wall (1991) and we focus on their methodology here.

3.1. *Maximum Likelihood Estimation*

The log likelihood of the incomplete data is the classical innovations form

$$\log L(Y_n^{(1)};\Theta) \propto -\frac{1}{2}\sum_{t=1}^{n}\log|\Sigma_t| - \frac{1}{2}\sum_{t=1}^{n}(\mathbf{y}_t - \Gamma\mathbf{z}_t - A\mathbf{x}_t^{t-1})'\Sigma_t^{-1}(\mathbf{y}_t - \Gamma\mathbf{z}_t - A\mathbf{x}_t^{t-1}) \tag{16}$$

where the filtered Kalman filter values, $\mathbf{x}_t^{t-1}$ are as in the previous section and the innovations covariance matrix is defined as

$$\Sigma_t = AP_t^{t-1}A' + R. \tag{17}$$

For the EM algorithm, we need the value of the complete data log likelihood as well, given by

$$\begin{aligned}\log L(Y_n;\Theta) \propto & -\frac{n}{2}\log|R| - \frac{1}{2}\sum_{t=1}^{n}(\mathbf{y}_t - \Gamma\mathbf{z}_t - A\mathbf{x}_t)'R^{-1}(\mathbf{y}_t - \Gamma\mathbf{z}_t - A\mathbf{x}_t)\\ & -\frac{1}{2}\log|\Sigma| - \frac{1}{2}\,(\mathbf{x}_0 - \boldsymbol{\mu}_0)'\Sigma_0^{-1}(\mathbf{x}_0 - \boldsymbol{\mu}_0)\\ & -\frac{n}{2}\log|Q| - \frac{1}{2}\sum_{t=1}^{n}(\mathbf{x}_t - \Phi\mathbf{x}_{t-1})'Q^{-1}(\mathbf{x}_t - \Phi\mathbf{x}_{t-1}).\end{aligned} \tag{18}$$

To apply the EM algorithm, note that successively maximizing

$$Q(\Theta|\Theta_0) = E_0[\log L(Y_n;\Theta)|Y_n^{(1)}] \tag{19}$$

as a function of Θ, with Θ_0 the value of the parameter at the preceding iteration will increase the incomplete data log likelihood and converge, under appropriate regularity conditions (see Wu (1983)), to a unique maximum.

Examining the expression (19), it is clear that we will need the expressions for the conditional means and covariances derived in the previous section. It will be convenient to assume that the covariance matrix of $\mathbf{x}_0$ can be specified and just concentrate on the parameters Γ, A, R in the first term and the parameters Φ, Q in the last term, leading to $\Theta = \{\Gamma, A, R, \Phi, Q\}$ as the set of parameters to be estimated. For the parameters Φ and Q, Shumway and Stoffer (1982) give maximum likelihood estimators for μ, Φ, Q, R under the assumption that R is diagonal. For the more general model with the additional parameters Γ and A,

$$R = \frac{1}{n}\sum_{t=1}^{n} E_0\{(\mathbf{y}_t - \Gamma\mathbf{z}_t - A\mathbf{x}_t)(\mathbf{y}_t - \Gamma\mathbf{z}_t - A\mathbf{x}_t)'|Y_n^{(1)}\}. \tag{20}$$

Taking conditional expectations in (20) and minimizing with respect to Γ and A leads to

$$(\Gamma,\, A) = \left(\sum_{t=1}^{n} E_0\{\mathbf{y}_t|Y_n^{(1)}\}\mathbf{z}_t',\, \sum_{t=1}^{n} E_0\{\mathbf{y}_t\mathbf{x}_t'|Y_n^{(1)}\}\right)\, G^{-1}, \tag{21}$$

where

$$G = \begin{pmatrix} \sum \mathbf{z}_t\mathbf{z}_t' & \sum \mathbf{z}_t\mathbf{x}_t^{n\prime} \\ \sum \mathbf{x}_t^n\mathbf{z}_t' & \sum(x_t^n\mathbf{x}_t^{n\prime} + P_t^n) \end{pmatrix} \tag{22}$$

with $\mathbf{x}_t^n$ and P_t^n given by (10) and (11). The details for computing the components in the matrices given in (21) and (22) and the estimator for R are tedious and can be found in Shumway (2000b).

3.2. *Bootstrap Estimation of Standard Errors*

As we have previously noted, there are a number of approaches to estimating the variances and covariances of the maximum likelihood estimators (see, for example, Meng and Rubin (1991), Oakes (1999)). We propose here a bootstrap estimator as given by Stoffer and Wall (1991). The procedure, summarized in Shumway and Stoffer (2006), Sec. 4.7, does not even require the Gaussian assumption although we do use the maximizer of the innovations log likelihood (16).

Suppose that we obtain maximum likelihood estimators for $\Theta = \{\Gamma, A, R,\ \Phi, Q\}$ as $\hat{\Theta} = \{\hat{\Gamma}, \hat{A}, \hat{R}, \hat{\Phi}, \hat{Q}\}$, where we may stack the components in a vector $\Theta(\hat{\Theta})$. Define the residuals from these estimators as

$\hat{\mathbf{v}}_t = \mathbf{y}_t - \hat{\Gamma} z_t - \hat{A}\hat{\mathbf{x}}_t^{t-1}$ and construct scaled residuals of the form

$$\hat{\boldsymbol{\epsilon}}_t = \hat{\Sigma}_t^{-1/2}\hat{\mathbf{v}}_t. \tag{23}$$

where a hat over a quantity indicates that that it has been evaluated at the maximum likelihood estimator $\hat{\Theta}$. Then, draw a random sample, say $\hat{\boldsymbol{\epsilon}}_t^*, t = 1, \ldots, n$ without replacement from the scaled residuals. Rescale the residuals, i.e., use

$$\mathbf{v}_t^* = \hat{\Sigma}_t^{1/2}\hat{\boldsymbol{\epsilon}}_t^* \tag{24}$$

to obtain residuals with the correct time varying covariance matrix. To reconstruct the data, note that

$$\mathbf{y}_t^* = \hat{\Gamma}\mathbf{z}_t + \hat{A}_t\hat{\mathbf{x}}_t^{t-1} + \mathbf{v}_t^* \tag{25}$$

and compute the values of $\hat{\mathbf{x}}_t^{t-1}$ using (6.19) of Property P6.1 of Shumway and Stoffer (2006) and the maximum likelihood estimators $\hat{\Theta}$ with (6.21) of P6.1 replaced by

$$\hat{\mathbf{x}}_t^t = \hat{\mathbf{x}}_t^{t-1} + \hat{K}_t\mathbf{v}_t^*. \tag{26}$$

Use the reconstructed bootstrap sample to compute maximum likelihood estimators $\hat{\Theta}^* = \{\hat{\Gamma}^*, \hat{A}*, \hat{R}^*, \hat{\Phi}^*, \hat{Q}^*\}$ using the maximum likelihood procedure described in Section 3. Note that the residuals for times when there are missing points are kept as missing or partially missing in the procedure, i.e., the identities of the missing points are kept the same throughout the sampling procedure. This implies that the results will be conditional on the particular missing data pattern observed in the data. An alternate approach might be to use a random mechanism to generate the missing points within each bootstrap replication.

The above bootstrap steps are repeated a large number, B, times, obtaining $\{\hat{\Theta}_b^*,\ b = 1, 2, \ldots, B\}$ The finite sample distributions of $\hat{\Theta} - \Theta_0$ are approximated by the the distribution of $\{\hat{\Theta}_b^* - \hat{\Theta},\ b = 1, 2, \ldots, B\}$, For example, the estimated variance of the estimated parameter $\hat{\theta}_i$ can be computed as

$$\hat{\sigma}_{\hat{\theta}}^2 = \frac{1}{B-1}\sum_{b=1}^{B}(\hat{\theta}_b^* - \bar{\theta}^*)^2. \tag{27}$$

where $\bar{\theta}^*$ denotes the mean of the bootstrap estimators.

4. Examples

We return here to a consideration of the monthly stream flow series in Figure 1 and the observed pesticide concentrations in Figure 2. For the monthly stream flow series, the questions of primary interest relate to estimating monthly mean parameters under serial correlation. Figure 2 makes it plausible to assume the existence of a fixed monthly mean, because the 42 profiles are roughly parallel with fairly consistent peak months. Furthermore, taking residuals from these monthly means yields residuals that are highly correlated over space and time. Modeling these residuals as the sum of an autoregressive component and error allows estimation of the monthly means under serial correlation and the interpolation of the missing stream flows. For the concentrations of the pesticide chloropyrifos, the interest is more in estimating the common signal by combining the weekly benchmark grab samples with the automatic daily values to estimate a common signal. In Figure 3, we see that the common values of the series occur during the short episodic bursts that punctuate the longer periods when one is observing nondetectable levels.

4.1. *Estimation of Monthly Mean Stream Flows*

The analyses discussed above apply only to a bivariate observations, observed according to a very restricted missing data pattern. In general, it is important to be able to handle more than two stream flows at a time with arbitrary missing data patterns rather than the restricted pattern considered above with observations only missing in a block of one of the series. In addition, there is a high probability that there will be serial correlation in the series after the means have been fitted and the estimation procedure should take this into account. Finally, there will always be a preference for full maximum likelihood estimation with incomplete data. This section proposes a full maximum likelihood solution to the parameter estimation problem when there are an arbitrary number of incompletely observed stream flows.

The model given in (4) and (5) was applied to both the incomplete data given in Figure 1 and the complete data (not shown) for comparison purposes. The estimated scale parameters using (21) are shown in Table 1, with the standard errors computed by the bootstrap shown in parentheses.

The estimators of the state transitions $\hat{\phi}$, also computed from (21), were $\hat{\phi} = 0.56$ for both the incomplete and complete data, with estimated

Table 1. Maximum likelihood estimators (se) for scale parameters.

	$\hat{a}_1$(se)	$\hat{a}_2$(se)	$\hat{a}_3$(se)
Incomplete	5.84(1.26)	5.77(1.18)	16.33(3.66)
Complete	5.79(1.18)	5.71(1.14)	16.25(3.56)

standard deviations of .11 in both cases. The estimated measurement error covariances

$$\hat{R} = \begin{pmatrix} 74(76) & 60(61) & 192(194) \\ 60(61) & 52(53) & 167(167) \\ 192(194) & 167(167) & 551(548) \end{pmatrix}$$

differed little for the complete and incomplete data, where the format of the matrix is incomplete(complete). It should be noted that the variances of the scale estimators in the complete data case were sometimes slightly larger than those computed for the incomplete data. This can happen due to slight random fluctuations in the bootstrap runs.

The estimated mean profiles were of primary interest to the hydrologists and we note from Figure 4 that they are essentially the same, with standard errors that are about 30% larger in the incomplete case and about twice as large for the stream 3 means. The estimated mean profiles are basically identical.

As a final task, there will be interest in interpolating the missing stream flows in Figure 1. For this, we can use (13) and (14) from Section 3. Since the values were nearly the same, a scatter plot, as shown in Figure 5 for two of the streams gives more detailed information. This plot shows the true known flows on the horizontal axis and the interpolated values on the vertical axis. Note that the points are distributed quite evenly about the line, with no particular bias, which indicates that the interpolation does very well.

4.2. *Signal Extraction for Monitoring Pesticide Concentrations in Water*

Monitoring levels of pesticide concentrations in water often involves measurements at contiguous locations made by different sampling techniques. A simple example involves the concentrations of the pesticide chlorpyrifos taken from Orestimba Creek in California between May, 1996 and May 1997 (364 days). Two different sampling methods were used, namely (1) automatic collection hourly and composited over a 24 hour period every

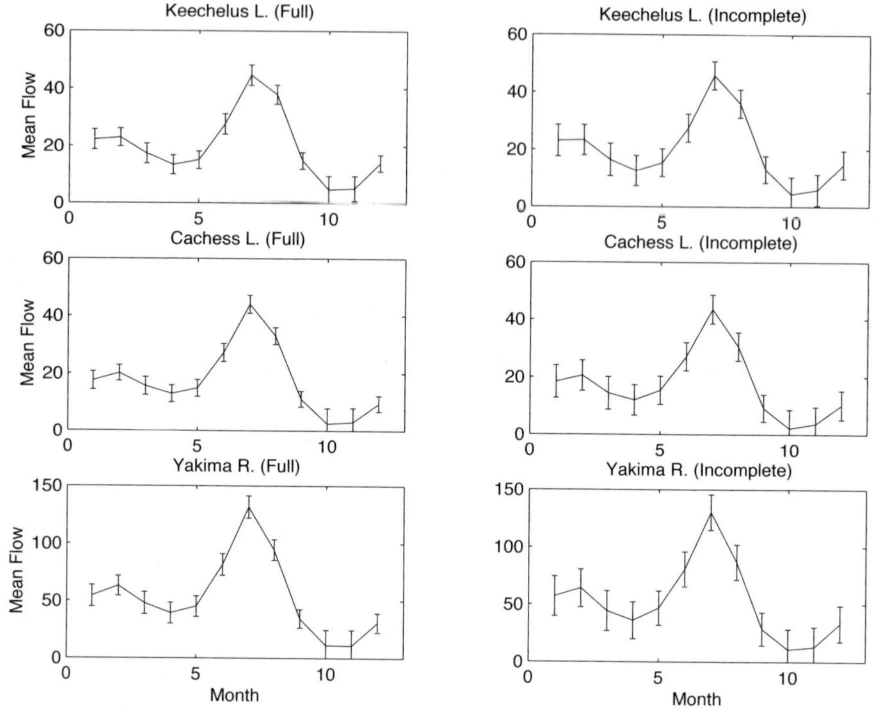

Fig. 4. Estimated monthly flow profiles and 95% confidence intervals in the dependent case for three series, with standard errors estimated by the bootstrap. Left panel shows means estimated from complete data whereas right panel gives comparable results for the incomplete data in Figure 1.

day and (2) grab samples collected weekly. Both series are shown in Figure 3, with the samples that failed to measure below a detection limit replaced by the detection limit. Independent samples which have a number of non-detections can be treated as in Shumway, et al. (1989). Because the non-detections for pesticide concentrations are far below the conventional values, the more sophisticated treatment in this case was not applied. The original daily values of the composite sampler and the weekly grab sample series have been previously shown in Figure 3.

The relative large sporadic excursions observed in the concentrations indicates a nonlinear behavior which could be treated in a number of ways. One might try various transformations to achieve a series that is more linear and approximately Gaussian. General power transformations are possible and the result of applying a fourth root transformation is shown in the

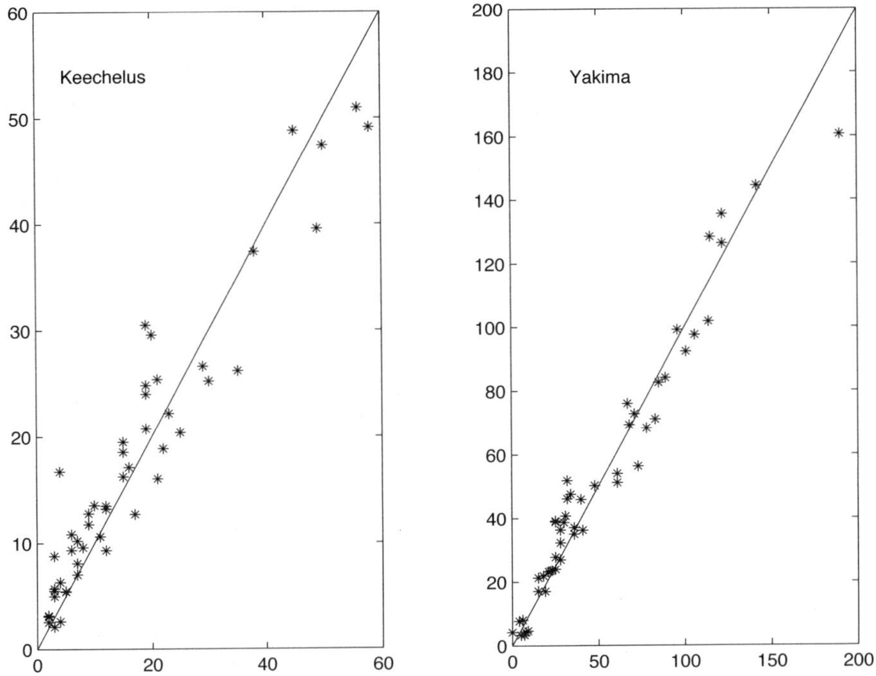

Fig. 5. Scatterplot showing true and predicted flows for unobserved portions of Keechelus and Yakima Rivers.

bottom two panels of Figure 3. This transformation produces histograms that are approximately normal and the transformed series exhibit a more stationary behavior. The treatment of the process in terms of stochastic volatility models such as ARCH or GARCH (see Engle (1982), Shepard (1996), Shumway and Stoffer (2006)) is also possible but we do not consider those models here.

Again, the EM algorithm of Section 3 was applied, leading to the measurement covariances $r_{11} = 0.0011, r_{12} = 0.0004, r_{22} = 0.0080$ and state variance $q_{11} = 0.0050$. Interpreting these, the full series has measurement standard error 0.033, whereas the weekly series has a standard error of 0.089, as might be expected. The correlation between the two measurement errors was small ($r = 0.14$). The standard deviation of the state error 0.071. The signal to noise ratios of the two measurements are about 5 to $1(q_{11}/r_{11})$ for the daily samples and less than 1 to 1 (q_{22}/r_{22}) for the incomplete weekly data.

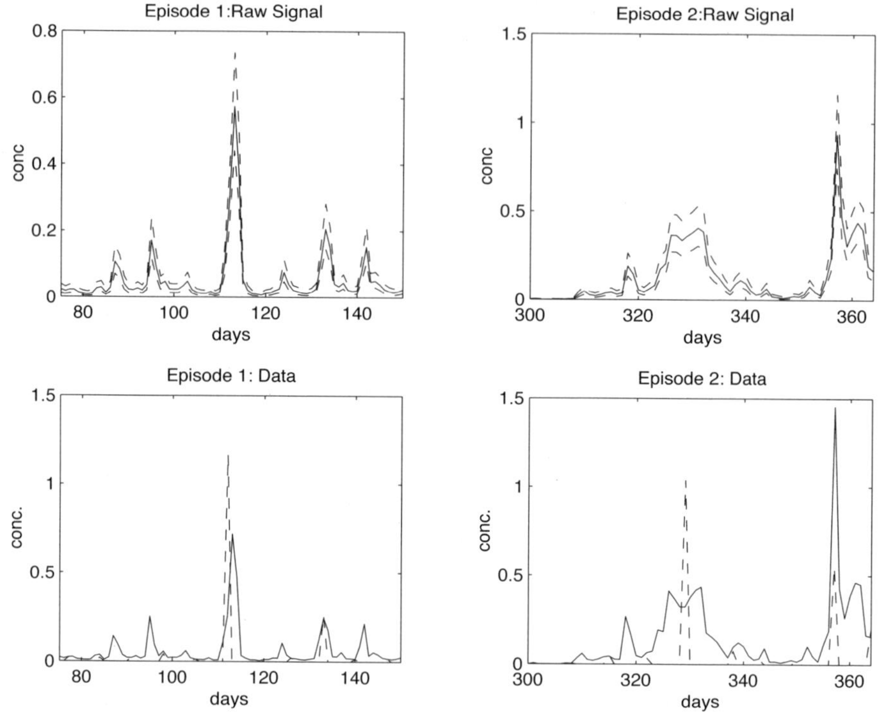

Fig. 6. Estimated signals focused on two high concentration episodes. Dashed lines are 95% prediction limits in upper panels. Bottom panels show daily series as solid lines and weekly benchmarks as dashed lines.

Figure 6 summarizes the estimated chlorpyrifos signal and its 95% prediction limits for several episodes. It is clear that high concentration episodes will be of more interest than the majority of values tending to lie at or below the detection limits and we plot two of these episodes and their 95% limits in Figure 7. These signal profiles, shown in the upper panels with their 95% prediction intervals, can be taken as the signature of the two merged observed series shown in the lower panels.

5. Discussion and Extensions

The intent here has been to adapt the state-space model to situations where there can be a mixture of time varying fixed and random effects that are serially correlated over time. Additionally, we have given a complete characterization of the EM algorithm that works for very general missing data patterns and enables interpolating missing values with a correlated measurement structure. The estimation of the scaling matrix A for the random component $\mathbf{x}_t$ provides a time series version of the factor analysis model.

While the theory can work for any longitudinal mixed model by specializing the measurement and state equations, it is best for data with quite severe missing data patterns. In this context, the method provides maximum likelihood estimators for all parameters and estimated variances via the bootstrap of Stoffer and Wall (1991). As an additional output, one gets an extracted signal vector $\mathbf{x}_t^n$ and its covariance matrix P_t^n, enabling one to fuse the common random features of a number of series, possibly observed at different intervals, into a consistent estimator for the signal. Finally, there is an optimal method for interpolating missing observations.

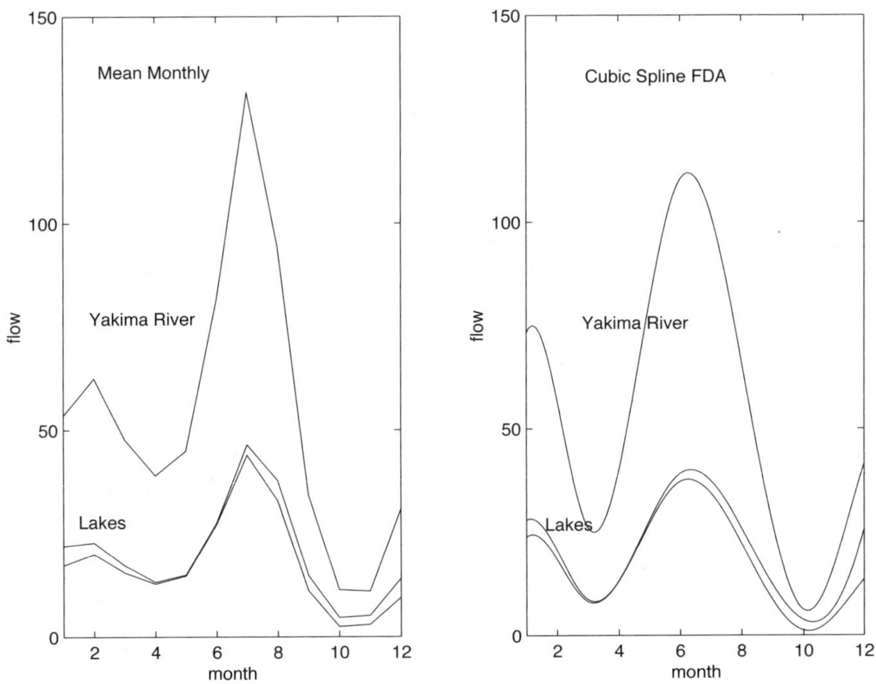

Fig. 7. Estimated monthly flow profiles using means in the left panel and a cubic spline with three knots in the right panel.

It should be noted that extended versions of the observation equation can handle more general situations with only modest modifications. For example, for a functional data analysis model, the observation equation might be written in the form

$$\mathbf{y}_t = \Gamma\Psi\mathbf{z}_t + v_t \tag{28}$$

where the new fixed covariate vector is of the form $\Psi\mathbf{z}_t$. In the case of the monthly profiles in Figure 2, we might consider taking a cubic spline of the form

$$\sum_{k=0}^{3} \gamma_{is}\left(\frac{2s-1}{24}\right)^k + \sum_{k=4}^{k_n+4} \gamma_{ik}\left(\frac{2s-1}{24} - s_k\right)_+^3$$

for river i at month s for the matrix product. Note that $(x - s_k)_+^3 = 0$ for $x < s_k$, where s_k defines the location of knot k Taking $k_n = 3$ knots for the flow reduces the number of parameters in the matrix Γ from $12 \times 3 = 36$ to $7 \times 3 = 21$. Taking the knots at $k_n = 3, 6, 10$ months gives the estimated curves in Figure 7 for the Yakima River and the two lakes. Note that the smooth functional data form seems to miss at the endpoints and at the peaks relative to the mean values although the squared error in R is lower for the functional data model.

Dedication and Acknowledgments

This paper is dedicated to my highly esteemed colleague, Professor P. K. Bhattacharya on his eightieth birthday, in appreciation for guiding me through many theoretical thickets during our years together in the University of California, Davis, Department of Statistics

This research was supported in part by the Hydrologic Engineering Center, U.S. Army Corps of Engineers under Contract DACW05-97-0982 and the California Department of Pesticide Regulation under Agreement 99-0248 with the University of California.

I am grateful to an anonymous referee who corrected a number of misprints and pointed out an apparent inconsistency in the bootstrap variances which led to replicating the original bootstrap sampling results a number of times.

References

Alley, W. M. and Burns, A. W. (1983). Mixed-station extension of monthly streamflow records. *J. Hydraulic Eng.*, **109**, 1272–1284.

Anderson, T. W. (1984). *An Introduction to Multivariate Analysis, 2nd ed.*. New York: Wiley.

Ansley, C. F. and Kohn, R. (1984). On the estimation of ARIMA models with missing values. In Parzen, E. (ed.) *Time Series analysis of Irregularly Spaced Observations.* New York: Springer-Verlag.

Cavanaugh, J. E. and Shumway, R. H. (1996). On computing the expected Fisher Information matrix for state-space model parameters. *Statistics and Probability Letters*, **26**, 347–355.

Dempster, A. P., Laird, N. M. and Rubin, D. B. (1977). Maximum likelihood from incomplete data via the EM algorithm. *J. Roy. Statist. Soc., Ser. B*, **39**, 1–38.

Durbin, J. and Koopman, S. J. (2001). *Time Series Analysis by State Space Methods*, Oxford: Oxford University Press.

Engle, R. (1982). Autoregressive conditional heteroscedasticity with estimates of the variance of United Kingdom inflation. *Econometrica*, **50**, 987–1007.

Harvey, A. C., (1991). *Forecasting Structural Time Series Models and the Kalman Filter.* Cambridge: Cambridge University Press.

Icaza, G. and Jones, R. H. (1999). A state-space EM algorithm for longitudinal data. *J. Time Series Analysis*, **20**, 537–550.

Johnson, R. A. and Wichern, D. W. (1992). *Applied Multivariate Analysis, 3rd ed.*. Englewood Cliffs, NJ: Prentice-Hall.

Jones, R. H. (1980). Maximum likelihood fitting of ARMA models to time series with missing observations. *Technometrics*, **22**, 389–395.

Jones, R. H. (1993). *Longitudinal Data With Serial Correlation: A State Space Approach.* London: Chapman & Hall.

Kitagawa, G. and Gersch, W. (1984). A smoothness priors modeling of time series with trend and seasonality. *J. Amer. Statist. Assoc.*, **79**, 378–389.

Kitagawa, G. and Gersch, W. (1996). *Smoothness Priors Analysis of Time Series.* New York: Springer-Verlag.

Laird, N. M. and Ware, J. H. (1982). Random effects models for longitudinal data. *Biometrics*, **73**, 13–22.

Matalas, N. C. and Langbein, W. B. (1962). Information content of the mean. *J. Geophys. Research*, **67**, 3441–3448.

Meng, X.-L. and Rubin, D.B. (1991). Using EM to obtain asymptotic variance-covariance matrices: The SEM Algorithm. *J. Amer. Statist. Assoc.*, **86**, 899–909.

Meng, X.-L. and Rubin, D. B. (1993). Maximum likelihood estmation via the ECM algorithm: A general framework. *Biometrika 80*, 267–278.

Oakes, D. (1999). Direct calculation of the information matrix via the EM algorithm. *J. Royal Statist. Soc. B*, **61**, 479–482.

Salas, J. D., Delleur, J. W., Yenevica, V. and Lane, N. (1980). Water Resource Publications. P.O. Box 2841, Littleton, Colorado.

Schmid, C.H. (1996). Fitting first-order conditional autoregressive models to longitudinal data. *J. Amer. Statist. Assoc.*, **91**, 1322–1330.

Shepard, N. (1996). Statistical aspects of ARCH and stochastic volatility. In D.R. Cox, D.V. Hinkley and O.E. Barndorff-Nielson (eds.), *Time Series Models in Econometrics, Finance and Other Fields*, 1–100. London: Chapman and Hall.

Shumway, R. H. (1998). Flow frequency estimation in the presence of missing data. Final Scientific Report, Contract DACS05-0982, Hydrological Engineering Center, Army Corps of Engineers, 609 E. 2nd. St., Davis, CA 95616.

Shumway, R.H. (2000a). Statistical approaches to assessing pesticide concentrations in the DPR Surface Water Database Final Report, Agreement No. 99-0248, California Environmental Protection Agency, Department of Pesticide Regulation, 830 K. St., Sacramento, CA 95814-3510.

Shumway, R.H. (2000b). Dynamic mixed models for irregularly observed time series. *Resenhas- Reviews of the Institute of Mathematics and Statistics*,(4), 433–456, University of Sao Paulo, Brazil: USP Press.

Shumway, R. H. and Stoffer, D. S. (1982). An approach to time series smoothing and forecasting using the EM algorithm. *J. Time Series Analysis*, **3**, 253–264.

Shumway, R. H., Azari, A. S. and Johnson, P. (1989). Estimating mean concentrations under transformation for environmental data with detection limits. *Technometrics*, **31**, 347–356.

Shumway, R. H. and Stoffer, D. S. (2006, 2nd. ed.) *Time Series Analysis and Its Applications With R Examples.* New York: Springer-Verlag.

Stoffer, D. S. (1982). Estimation of parameters in linear dynamic models with missing observations. *Dissertation, University of California, Davis*, One Shields Ave., Davis, CA 95616.

Stoffer, D. S. and Wall, K. (1991). Bootstrapping state space models: Gaussian maximum likelihood estimation and the Kalman filter. *J. Amer. Statist. Assoc.*, **86**, 1024–1033.

Vogel, R. M. and Stedinger, J. R. (1985). Minimum variance streamflow record augmentation procedures. *Water Resources Res.*, **21**, 715–723.

Watson, M. W. and Engle, R. F. (1983). Alternative algorithms for the estimation of dynamic factor, MIMIC and varying coefficient regression models. *J. Econometrics*, **23**, 385–400

Wu, C.F. (1983). On the convergence properties of the EM algorithm. *Ann. Statist.*, **11**, 95–103.

PART 5

Papers on Asymptotic Theory

Chapter 20

Asymptotic Behavior of the Kernel Density Estimators for Nonstationary Dependent Random Variables with Binned Data

Jean-François Lenain, Michel Harel and Madan L. Puri

Faculté des Sciences et Techniques, France; IUFM du Limousin, France; IMT, Universite Paul Sabatier, France; and Indiana University, USA

We study the performance of the Kernel Density Estimators when the discretization, the nonstationarity and the dependence of the random variables are all present. We establish the results only on the bias and the mean square error of the estimators under two different forms. We prove that the precision in the expression of the bias depends on the form of the estimator.

Contents

1. Introduction

1.1. *Kernel density estimators*

The simplest and the most popular nonparametric estimators of an unknown density function f are the histogram and the kernel density estimators (KDE). The former provides one of the earliest statistical tools. The latter is a product of the more recent explosion of interest in methods for the nonparametric density estimation. The KDE can provide pleasingly smooth curves in contrast to the piecewise constant and piecewise linear forms of the histogram and frequency polygon respectively.

Let $\{X_1, \ldots, X_n\}$ be a sequence of random variables from a univariate distribution with density f. The KDE $\widehat{f}$ of f of Rosenblatt (1971) based on these data is defined by

$$\widehat{f}(x) = \frac{1}{nh} \sum_{i=1}^{n} K\left(\frac{x - X_i}{h}\right),$$

where $K(\cdot)$ is the kernel function and $h = h(n)$ is a scalar smoothing parameter (the window width).

The most classic kernel functions are:

triangular	:	$K(t) = (1 - \lvert t \rvert)$
Epanechnikov	:	$K(t) = \frac{3}{4}\left(1 - t^2\right)$
quartic biweight	:	$K(t) = \frac{15}{16}\left(1 - t^2\right)^2$
triweight	:	$K(t) = \frac{35}{32}\left(1 - t^2\right)^3$
cosines	:	$K(t) = \frac{\pi}{4}\cos\left(\frac{\pi}{2}t\right)$
Gaussian	:	$K(t) = \frac{1}{\sqrt{2\pi}}\exp\left(-\frac{1}{2}t^2\right).$

The kernel estimator of a probability density function inherits its smoothness properties from the kernel density chosen by the investigator. Nevertheless, for computational (and especially graphical) reasons, exact kernel density estimators are often presented in (piecewise constant) discretized or (piecewise linear) interpolated form, using the exact estimate only at some grids of points. The problems of rounding errors or discretization of the data imply errors in the values of the variables. In practice, we generally use discretized data because we have only the information on these data (rounded data) or we wish to reduce the number of calculations for the estimation. The properties of the kernel estimators depend both on the type of discretization and the nature of the kernel (Hall and Wand (1996), Jones (1989)). But the results for the discretization on the kernel estimator were established only when the variables were supposed to be stationary and independent. The hypothesis of stationarity is difficult to accept when we study natural phenomena with so many conditions of heterogeneousness. We can assume that the heterogeneousness is non local and comes from local nonstationarity: This means that the effects are decreasing with the distance and disappear at infinity. For the kernel density, the known results are from Harel and Puri when the random variables are absolutely regular with an initial measure which is not invariant (Harel and Puri (1996), Harel and Puri (1999)). One of the simple cases is an ARMA process where the initial measure is very far from the invariant measure: the difference between the nonstationary process and the stationary process decreases geometrically. In this paper, we generalize the results of Hall and Wand (1996) on the bias of the binned kernel density estimators from the independent and stationary case to the strong mixing and nonstationary case. The paper of Hall and Wand seems to be the most recent work on this field.

1.2. *Discretized kernel density estimators*

With the discretized data, let $\{a_j\} = \{a_0 + j\delta\}_{j\in\mathbb{Z}}$ be a grid of points with an arbitrary origin a_0, the estimator $\widehat{f}$ is replaced by the estimator $\widehat{f_G}$ for which the calculations are divided in two steps: discretizing before applying the kernel-smoothing prescription; that is, base the KDE not directly on the underlying continuous data, but on some binned version of it by using a kernel T with the window width δ. We define, as in Hall and Wand (1996), the estimator $\widehat{f_G}$ of f by

$$\widehat{f_G}(x) = \frac{1}{nh}\sum_{j\in\mathbb{Z}} K\left(\frac{x-a_j}{h}\right)\sum_{i=1}^{n} T\left(\frac{X_i - a_j}{\delta}\right)$$

which is known as the discretized form of the estimator $\widehat{f_G}$.

However, this estimate has also an other equivalent form called "missing random variables form" which is a function of the initial randoms and the missing data called the rounding errors. This form depends on the kernel T.

If we denote by a the rounding function, $\lfloor . \rfloor$ the integer closest to x (if x is a half integer, we shall assume for definiteness that $\lfloor x \rfloor = x + \frac{1}{2}$) and $[x]$ is the largest integer less than or equal to x,

$$y \to a(y) = \delta\left\lfloor \frac{y-a_0}{\delta} \right\rfloor + a_0 \quad \text{if the values are rounded to the nearest value}$$

$$y \to a(y) = \delta\left[\frac{y-a_0}{\delta} \right] + a_0 \quad \text{if the values are rounded down.}$$

If we define a sequence $\{Z_i\}$ of non observed data distributed on the interval $[0,1]$ or $[-\frac{1}{2},\frac{1}{2}]$ by

$$Z_i = \frac{X_i - a(X_i)}{\delta}$$

then we get another form for $\widehat{f_G}(x)$:

If T is the rounded kernel:

$$\widehat{f_G}(x) = \frac{1}{nh}\sum_{i1}^{n} K\left(\frac{x-X_i}{h} + \frac{\delta}{h}Z_i\right), \tag{1}$$

$$Z_i \in [0,1] \quad \text{(rounded down) or}$$

$$Z_i \in [-1/2,1/2] \quad \text{(rounded to the nearest value).}$$

If T is the triangular kernel:

$$\widehat{f_G}(x) = \frac{1}{nh}\sum_{i=1}^{n}\left[(1-Z_i)\,K\left(\frac{x-X_i}{h} + \frac{\delta}{h}Z_i\right)\right.$$
$$\left. + Z_i K\left(\frac{x-X_i}{h} + \frac{\delta}{h}(Z_i-1)\right)\right], \tag{2}$$

$Z_i \in [0,1]$,

or equivalently

$$\widehat{f_G}(x) = \frac{1}{nh}\sum_{i=1}^{n}\left(\frac{1}{2}-Z_i\right)K\left(\frac{x-X_i}{h} + \frac{\delta}{h}\left(Z_i+\frac{1}{2}\right)\right)$$
$$+\frac{1}{nh}\sum_{i=1}^{n}\left(\frac{1}{2}+Z_i\right)K\left(\frac{x-X_i}{h} + \frac{\delta}{h}\left(Z_i-\frac{1}{2}\right)\right),$$

$Z_i \in [-1/2,1/2]$.

This form is called "missing random variables form" of $\widehat{f_G}$. From the properties of the convergence of the sequence $\{Z_i\}$, we obtain a different approach on the discretization (see e.g. Hall (1982) and Hall and Wand (1996)).

The papers cited in Section 1.1 and most of the relevant existing ones assume stationarity and the independence of the observed random variables. Many sequences of random variables encountered in practice are neither stationary nor independent. This justifies the need for studying some existing work in a nonstationary and dependent framework. Our main objective in this paper is to extend the results of Hall and Wand (1996), Hall (1982) and

Jones (1989) to the case when the sequence F_i of the distribution functions of the $X_i's$ converges in a sense to be made precise later in Section 1.3 to a limiting distribution function F^* and the sequence X_i is strongly mixing with some rate. In Section 2, we state the sequence of hypotheses which we consider in Section 3 where we also show that the missing random variables form of the estimator $\widehat{f_G}$ is an asymptotically unbiased estimator of f^*, the density function associated with the distribution function F^*. We also give a detailed expression of the bias. Finally, we show that the estimator $\widehat{f_G}$ converges to f^* in the mean square. In Section 4, we show that the discretized form of $\widehat{f_G}$ defined in (1) gives more precise results in the expression of the bias. The proofs of the theorems, lemmas and propositions are deferred to Section 5.

1.3. *Hypotheses of the convergence of the nonstationary process*

For a stationary sequence of random variables $(X_i^*)_{i\geq 1}$, the covariance between two variables X_i^* and X_j^* $(i<j)$ depends only on the distance $j-i$. The joint distribution of X_i^* and X_j^* is denoted by F_{j-i}^* with marginal F^*. In the nonstationary case, this covariance depends also on the distance between the random variables X_i and X_j. The joint distribution of X_i and X_j is denoted by $F_{i,j}$ with marginals F_i and F_j. Harel and Puri (1996) assume that the rates of convergence are geometric with respect to the norm of total variations (NTV) between $F_{i,j}$ and F_{j-i}^*:

$$\left\|F_{i,j}-F_{j-i}^*\right\|_{\mathsf{V}} = O(\tau^i), \quad 0<\tau<1\ ,\ 1\leq i<j.$$

(In $\mathbb{R}^d$, the NTV is proportional to the distance L_1 between the density functions.)

In the introduction of the discretized estimators, we noticed that we can write the same estimator in two forms, one from the discretized variables and the other from the continuous and missing variables. From a generalization to the nonstationary case of Lemma 3 of Hall (1983), the missing variables converge to uniformly distributed random variables. After the statement of our hypotheses in Section 2, we will use them to get the mean square error (MSE) between the estimators $\widehat{f_G}$ and f^* (the limiting stationary density) and also to calculate the bias. The method of "missing variables" consists first in writing a relation between $\widehat{f_G}$ and $\widehat{f}$ (the Rosenblatt estimator) and then in studying the convergence of $\widehat{f_G}$ to $\widehat{f}$. In practice, the missing variables form is easier to manipulate than the discretized form. We get better results for the expression of the bias if we use the discretized form when the kernels and the densities have at least third derivatives.

2. Hypotheses and notations

In this section, we state a set of hypotheses which we will use in some of the theorems, lemmas and propositions stated and proved in the subsequent sections.

Hypothesis 2.1:

Let $\{X_i\}_{i\in\mathbb{N}}$ be a sequence of real random variables (r.r.v.) for which the initial measure is not invariant.

Let $\{X_i^*\}_{i\in\mathbb{N}}$ be a sequence of strictly stationary r.r.v. and independent of the sequence $\{X_i\}_{i\in\mathbb{N}}$ (the asterisk denotes the stationarity).

Let $\{Z_i\}_{i\in\mathbb{N}}$ be a sequence of missing variables coming from a discretization with values in an interval $\Delta \subset R$ with length equal to 1.

We assume the existence and the continuity of the functions which are defined below (d.f.= distribution function, p.d.f.=probability density function, m.d.f.=marginal distribution function):

r.v.		*d.f.*	p.*d.f.*	*m.d.f.*
X_i	$i = 1,\ldots,n$	F_i	f_i	
X_1^*		F^*	f^*	
(X_i, X_j)	$1 \le i < j \le n$	$F_{i,j}$	$f_{i,j}$	
(X_i^*, X_j^*)	$1 \le i < j \le n$	F_{j-i}^*	f_{j-i}^*	F^*
(X_i, Z_i)	$i = 1,\ldots,n$	G_i	g_i	
(X_i, X_j, Z_i, Z_j)	$1 \le i < j \le n$	$G_{i,j}$	$g_{i,j}$	

We suppose also that:

$$\sup_{i\in\mathbb{N}}\sup_{x\in\mathbb{R}} f_i(x) < \infty \qquad \sup_{\substack{i,j\in\mathbb{N}\\ i<j}}\ \sup_{\substack{x\in\mathbb{R}\\ u\in\Delta}} g_{i,j}(x,u) < \infty$$

$$\sup_{x\in\mathbb{R}} |f^{*\prime}(x)| < \infty \qquad \sup_{\substack{i,j\in\mathbb{N}\\ i<j}}\ \sup_{\substack{x,y\in\mathbb{R}\\ z=x,y}} \left|\frac{\partial f_{j-i}^*(x,y)}{\partial z}\right| < \infty$$

$$\text{For} \quad j > i\ , \quad \left\|F_{i,j} - F_{j-i}^*\right\|_{\mathsf{V}} = O(\eta(i)) \ \rightarrow\ 0 \quad \text{if} \quad i \rightarrow \infty.$$

Hypothesis 2.2:

K is a nonnegative bounded symmetric function which is absolutely continuous and differentiable piecewise such that:

$$\int K(x)dx = 1 \ , \qquad \int x^2 K(x)dx < \infty \ ,$$

$$\int xK'(x)dx < \infty \ , \qquad \sup_{x\in\mathbb{R}} |K'(x)| < \infty \ , \qquad \int |K'(x)|\, dx < \infty,$$

$h = h(n)$ and $\delta = \delta(n)$ are two sequences of positive constants such that:

$$h \to 0 \ , \quad \delta \to 0 \qquad \text{if } \ n \to \infty$$
$$\frac{\delta}{h} \to 0 \ , \quad nh \to \infty \quad \text{if } \ n \to \infty.$$

The nonstationarity rates η satisfy:

$$\sum_{i=1}^{\infty} \eta\,(i) < \infty.$$

We note that under these conditions, $\|K\|_p$ and $\|K'\|_p$ are finite for any $p \geq 1$: This follows from the inequality

$$\int |K'(t)|^p \, dt \leq \sup_{x\in\mathbb{R}} \left(|K'(x)|^{p-1} \right) \int |K'(t)|\, dt.$$

Remark: *The condition $\sum_{i=1}^{\infty} \eta\,(i) < \infty$ is particularly satisfied for a geometric rate:*

$$\eta\,(i) = O\left(\tau^i\right) \ , \qquad 0 < \tau < 1$$

or an arithmetic rate:

$$\eta\,(i) = O\left(i^{-\tau}\right) \ , \qquad \tau > 1.$$

Hypothesis 2.3:

The sequence $\{X_i\}$ is strongly mixing with rate $\alpha(n)$ satisfying:

$$\alpha(n) \downarrow 0 \qquad \text{if } \ n \to \infty \ .$$

For the sequence $h = h(n)$, there exists a nondecreasing sequence $m = m(n)$ satisfying the following two conditions with $1 < \lambda < \infty$:

$$h(n)\ m(n) \to 0 \qquad \text{if } \ n \to \infty$$
$$h(n)^{-1/\lambda} \sum_{i=m}^{n} \alpha\,(i)^{1/\lambda} \to 0 \qquad \text{if } \ n \to \infty. \tag{3}$$

Hypothesis 2.4:

α and $h(n)$ satisfy the following conditions if $n \to \infty$ with $1 < \lambda < \infty$:

$$\sum_{i=1}^{n} i\,\alpha(i)^{1/\lambda} < \infty \quad , \qquad n^{\lambda/3}\, h(n) \to \infty. \tag{4}$$

Remark*: $\sum_{i=1}^{n} \alpha(i)^{1/(2\lambda)} < \infty$ is a sufficient condition for which $\sum_{i=1}^{n} i\,\alpha(i)^{1/\lambda} < \infty$; in fact we have:*

$$\sum_{i=1}^{n} i\,\alpha(i)^{1/\lambda} \leq n \sum_{i=1}^{n} \alpha(i)^{1/\lambda} = \sum_{k=1}^{n}\sum_{i=1}^{n} \alpha(i)^{1/\lambda}$$
$$\leq \sum_{k=1}^{n} \sqrt{\alpha(k)^{1/\lambda}} \sum_{i=1}^{n} \sqrt{\alpha(i)^{1/\lambda}} = \left(\sum_{i=1}^{n} \alpha(i)^{1/(2\lambda)}\right)^2$$

because α is non-increasing.

3. Missing random variables form of the estimator

In this section, we use the expressions of the estimator $\widehat{f_G}(x)$ described in Section 1. For each type of discretization, we have a different estimator. We will consider mainly the rounded functions: the discretization kernel is an indicator function which may be symmetric or not.

3.1. *Bias*

The estimator for rounded data is given by:

$$\widehat{f_G}(x) = \frac{1}{nh} \sum_{i=1}^{n} K\left(\frac{x - X_i}{h} + \frac{\delta}{h} Z_i\right), \; Z_i \in \Delta$$

where $\Delta = [0, 1[$ if the values are rounded down or $= [-1/2, 1/2[$ if the values are rounded to the nearest value. From a Taylor development of the kernel K, the expression of $\widehat{f_G}(x)$ is a function of $\widehat{f}(x)$, the Rosenblatt estimate.

The expression of the bias is given by the following Proposition:

Proposition 1 *Under Hypotheses 2.1 and 2.2, the estimators $\widehat{f_G}$ are asymptotically unbiased estimators of f^*, the density function of the in-*

variant measure. More, we have the following expression for the bias:

$$E\widehat{f_G}(x) - f^*(x) = \left[E\widehat{f^*}(x) - f^*(x)\right] + O\left(\frac{1}{nh}\right) + O\left(\frac{\delta^2}{h^2} + \delta \int_\Delta u du + \frac{\delta^2}{h}\right) \tag{5}$$

where $\left[E\widehat{f^*}(x) - f^*(x)\right]$ *is the bias of the Rosenblatt estimate. We have* $\frac{\delta^2}{h^2} = 0$ *if the support of* K is *infinite.*

Proof: See Section 5.

The following theorem evaluates more precisely the expression of the bias.

Theorem 1. *Assume that the assumptions of Proposition 1 hold. Assume that* f^* *and* K *have a third derivative and* $K''' \in L_1$*. Then the expectation of* $\widehat{f_G}(x)$ *can be written as:*

$$\begin{aligned} E\widehat{f_G}(x) &= f^*(x) + \frac{1}{2}h^2\mu_2(K)f^{*\prime\prime}(x) + o(h^2) - \delta\ \mu_1(K')f^{*\prime}(x) \int_\Delta u du \\ &+ \left(\frac{1}{2}\frac{\delta^2}{h^2}\int K''(t)dt\ f^*(x) + \frac{1}{4}\delta^2\mu_2(K'')f^{*\prime\prime}(x)\right)\int_\Delta u^2 du \\ &+ O\left(\frac{\delta^3}{h^3}\right) + O\left(\frac{1}{nh} + \frac{\delta^2}{h^2}\right) \end{aligned} \tag{6}$$

where $\mu_1(f) = \int tf(t)dt$ *and* $\mu_2(f) = \int t^2 f(t)dt$.

Proof: See Section 5.

The first line of (6) corresponds to the expectation of the Rosenblatt estimator, the following lines to the effects of the discretization and the last line to the nonstationarity. The expression of the bias depends on the value of $\int_\Delta udu$ (equal to zero for a symmetric discretisazion) or on the value of $\int K''(t)dt$, but our interest is limited to the precise expression of the bias because it is in the term $O\left(\frac{\delta^2}{h^2}\right)$ which comes from the nonstationarity and its value is not known. It follows that δ has to be a $o(h^2)$ to imply a negligeable effect on the discretization on the bias.

Corollary 1. For the triangular kernel defined in (1), we get:

$$
\begin{aligned}
E\widehat{f_G}(x) - f^*(x) &= \frac{1}{2}h^2\mu_2(K)\ f^{*\prime\prime}(x) + o(h^2) + O\left(\frac{1}{nh}\right) \\
&+\frac{1}{12}\frac{\delta^2}{h^2}\int K''(t)dt\ f^*(x) + \frac{1}{24}\delta^2\mu_2(K'')\ f^{*\prime\prime}(x) \\
&+O\left(\frac{\delta^3}{h^3}\right) \qquad (7)
\end{aligned}
$$

Proof: Section 5.

3.2. *Estimation of the Mean Square Error (MSE) between $\widehat{f_G}$ and f^**

We now establish the MSE between the estimator $\widehat{f_G}(x)$ and the Rosenblatt estimator $\widehat{f}$.

Theorem 2. *Under the Hypotheses 2.1 and 2.2, if the sequence $\{X_i\}$ is strongly mixing with rate $\alpha(n) \underset{n\to\infty}{\to} 0$ and if $m = o(n)$ such that*

$$
\frac{1}{nh}\ h^{-1/\lambda}\sum_{i=m}^{n}\alpha\left(i\right)^{1/\lambda} \to 0 \qquad \text{as} \ \ n\to\infty\ ,
$$

then the estimator $\widehat{f_G}$, defined previously, converges to f^* in the mean square.

Proof : See Section 5

4. Discretized form of the estimator

This Section gives more precise results in the expression of the bias. We do not suppose the existence of the joint distribution functions (the distribution functions G_i and g_i defined in Hypothesis 2.1).

4.1. *Hypotheses*

Hypotheses 4.1 and 4.3 below replace Hypotheses 2.1 and 2.3 of Section 2.

Hypothesis 4.1:

Let $\{X_i\}_{i\in\mathbb{N}}$ be a sequence of real random variables (r.r.v.) for which the initial measure is not invariant.

Let $\{X_i^*\}_{i\in\mathbb{N}}$ be a sequence of strictly stationary r.r.v. and independent of the sequence $\{X_i\}_{i\in\mathbb{N}}$ (the asterisk denotes the stationarity)

We suppose the existence and the continuity of the fonctions which are defined below (d.f.= distribution function, p.d.f.=probability density function, m.d.f.=marginal distribution function):

v.a.r.		d.f.	p.d.f.	m.d.f.
X_i	$i = 1, ..., n$	F_i	f_i	
X_1^*		F^*	f^*	
(X_i, X_j)	$1 \leq i < j \leq n$	$F_{i,j}$	$f_{i,j}$	
$\left(X_i^*, X_j^*\right)$	$1 \leq i < j \leq n$	F_{j-i}^*	f_{j-i}^*	F^*

We suppose that f^* is absolutely continuous and differentiable and that

$$\sup_{i\in\mathbb{N}}\sup_{x\in\mathbb{R}} f_i(x) < \infty, \qquad \sup_{i\in\mathbb{N}}\ \sup_{x,y\in\mathbb{R}} f_{i,j}(x,y) < \infty$$

$$\sup_{x\in\mathbb{R}} f^{*\prime}(x) < \infty, \qquad \int |f^{*\prime}(x)|\, dx < \infty, \qquad \sup_{\substack{i,j\in\mathbb{N}\\ k=i,j}}\ \sup_{x\in\mathbb{R}} \left|\frac{\partial f_{j-i}^*(x_i,x_j)}{\partial x_k}\right| < \infty,$$

$$\left\|F_{i,j} - F_{j-i}^*\right\|_{\mathsf{V}} = O\left(\eta\left(i\right)\right) \to 0 \quad \text{if } i \to \infty.$$

Hypothesis 4.2 = Hypothesis 2.2

Hypothesis 4.3:

The sequence $\{X_i\}$ is strongly mixing with rate:

$$\alpha(n) \to 0 \qquad \text{if } n \to \infty$$

For the sequence $h = h(n)$, there exists a nondecreasing sequence $m = m(n)$ satisfying the following two conditions with $1 < \lambda < \infty$:

$$h(n)\, m(n) \to 0 \qquad \text{if } n \to \infty$$

$$h(n)^{-1/\lambda - 1} \sum_{i=m(n)}^{n} \alpha\left(i\right)^{1/\lambda} \to 0 \qquad \text{if } n \to \infty. \tag{8}$$

4.2. *Bias*

We show how the evaluation of the bias can be improved by using the discretized form of the estimates $\widehat{f_G}$. This improvement comes from the conditions on the second and third derivatives. If we use the same conditions on differentiability as in Section 3, the results are weaker because the error is bounded by $\frac{\delta}{h}$ instead of $\frac{\delta^2}{h^2}$.

Proposition 2. *Under Hypotheses 4.1 and 4.2, the estimators* $\widehat{f_G}$ *with their discretized forms are asymptotically unbiased estimators of* f^*, *the density function of the invariant measure. Moreover, we have the following expression for the bias:*

$$E\widehat{f_G}(x) - f^*(x) = \left[E\widehat{f^*}(x) - f^*(x)\right] + O(\delta) + O\left(\frac{\delta}{h}\right) + O\left(\frac{1}{nh}\right) \tag{9}$$

where $\left[E\widehat{f^*}(x) - f^*(x)\right]$ *is the bias of the Rosenblatt estimator.*

Proof: See Section 5.

To evaluate more precisely the bias, if we consider the discretized form of the estimator associated to the sequence of strictly stationary r.r.v. defined by

$$\widehat{f_G^*}(x) = \frac{\delta}{h} \sum_{j \in \mathbb{Z}} K\left(\frac{x - a_j}{h}\right) \frac{1}{n\delta} \sum_{i=1}^{n} T\left(\frac{X_i^* - a_j}{\delta}\right), \tag{10}$$

we deduce the following result.

Theorem 3. *Assume that the assumptions of Proposition 2 hold. Asume that* K *and* f^* *have a third derivative where the two first derivatives are at least absolutely continuous,* K' *to* $K''' \in L_1$, f'^* *to* $f^{*'''}$ *are bounded. Then, the difference between the expectations of the two estimators can generally be written as:*

$$\begin{aligned}
&E\widehat{f_G^*}(x) - E\widehat{f^*}(x) \\
&= \delta\left(\mu_1(T) - \int_\Delta u du\right) f^{*\prime}(x) + \frac{1}{2}\delta^2\left(\mu_2(T) - \int_\Delta u^2 du\right) f^{*\prime\prime}(x) \\
&+O\left(\delta^3\right) - \left[\delta\mu_1(K') f^{*\prime}(x) + O\left(\frac{\delta^2}{h^2}\right)\right]\int_\Delta u du \\
&+\left[\frac{1}{2}\frac{\delta^2}{h^2}\int K''(t)\, dt f^*(x) + \delta^2\left(\frac{1}{4}\mu_2(K'') - \mu_1(K')\right) f^{*\prime\prime}(x)\right. \\
&\left.+O\left(\frac{\delta^3}{h^3}\right)\right]\int_\Delta u^2 du + O\left(\frac{\delta^3}{h^3}\right) \qquad (11)
\end{aligned}$$

Proof: See Section 5.

For a rounded discretization, the expression can be simplified. In that case, $\int_\Delta u du$ and $\int_\Delta u^2 du$ are respectively equal to $\mu_1(T)$ and $\mu_2(T)$ and we deduce easily the following corollary.

Corollary 2. Under the conditions of Theorem 3, for a rounded discretization, we get

$$\begin{aligned}
&E\widehat{f_G}(x) \\
&= f^*(x) + \frac{1}{2}h^2\mu_2(K) f^{*\prime\prime}(x) + o(h^2) + O\left(\delta^3\right) + O\left(\frac{1}{nh}\right) \\
&- \left[\delta\mu_1(K') f^{*\prime}(x) + O\left(\frac{\delta^2}{h^2}\right)\right]\int_\Delta u du \\
&+\left[\frac{1}{2}\frac{\delta^2}{h^2}\int K''(t)\, dt f^*(x) + \delta^2\left(\frac{1}{4}\mu_2(K'') - \mu_1(K')\right) f^{*\prime\prime}(x)\right]\int_\Delta u^2 du \\
&+O\left(\frac{\delta^3}{h^3}\right). \qquad (12)
\end{aligned}$$

We have to compare this expression with (6): the main terms are identical but (12) gives more precise information on the order of the size of the errors, particularly for the symmetric discretizations. In fact the term $O\left(\frac{\delta^2}{h^2}\right)$ is contained in the integral $\int_\Delta u du$, which allows us to use the terms of order 2 in the case of a symmetric discretization. If we use the missing random variables form of the estimator defined in Section 1.2, it is impossible to use this information.

If we want that the discretization has negligible effect on the bias, it is necessary in some cases that δ is $o\left(h^2\right)$; that is the case:

- for an asymmetric discretization where $\int_\Delta udu \neq 0$: In essence, it concerns the missing random variables form of the estimator where the values are rounded down as defined in Section 1.2;
- for kernels such that $\int K''(t)dt \neq 0$: among the most known kernels, the Epanechnikov kernel $K = \frac{3}{4}\left(1 - x^2\right)$ and the cosinus kernel $K = \frac{\pi}{4}\cos\left(\frac{\pi}{2}x\right)$ have to be avoided for discretized data.

Remarks:

Hall and Wand (1996) obtain an expression similar to that in (6), for a symmetric discretization with stationary variables under very similar conditions:

$$
\begin{aligned}
& E\widehat{f_G^*}(x) - E\widehat{f^*}(x) \\
&= \delta^s \mu_s(T) f^{*(s)}(x) + o(\delta^s) + \left(\frac{\delta}{h}\right)^{2t}(-2t!)^{2t} B_{2t} J\left(K^{2t-1}\right) \\
&\quad + o\left(\left(\frac{\delta}{h}\right)^{2t}\right) \qquad (13)
\end{aligned}
$$

where $\mu_s(T)$ is the first non-null moment of T, $(2t-1)$ the order of the derivative of K which is continuous by pieces with a finite number of jumps, B a Bernoulli number, and J is the function representing theses jumps. We are surprised on the absence of terms depending of δ/h in the expressions of the bias given by Jones (1989).

5. Proofs of the results

This section contains the proofs of the main results stated in this paper, as well as some lemmas for proving them.

Proof of Proposition 1:

First, we prove the following lemma:

Lemma 1. *Let K be a positive and symmetric density function (typically a kernel) such that K and K' are bounded and belong to L_1, $\int x^2K(x)dx < \infty$, $\int xK'(x)dx < \infty$.*

Let a be a positive real number in the neighbor of 0, *then*

1 : $\int K'(t-a)\,dt = O(a)$ if the support of K *is compact*

2 : $\int |tK'(t-a)|\,dt = \int tK'(t)\,dt + O(a)$.

Proof. 1- If the support of K is compact ($K = 0$ outside of $[-1,1]$), we have

$$\begin{aligned}\int_{-1}^{1} K'(t-a)\,dt &= \int_{-1-a}^{1-a} K'(v)\,dv \\ &= \int_{-1-a}^{-1} K'(v)\,dv + \int_{-1}^{1} K'(v)\,dv - \int_{1-a}^{1} K'(v)\,dv,\end{aligned}$$

$$\begin{aligned}\left|\int_{-1}^{1} K'(t-a)\,dt\right| &\le \left|\int_{-1-a}^{-1} K'(v)\,dv\right| + \left|\int_{-1}^{1} K'(v)\,dv\right| + \left|\int_{1-a}^{1} K'(v)\,dv\right| \\ &\le 0 + 0 + a\sup_t |K'(t)|.\end{aligned}$$

2- We have

$$\begin{aligned}\int_{-\infty}^{\infty} |tK'(t-a)|\,dt &= \int_{-\infty}^{\infty} |(t+a)K'(t)|\,dt \\ &= \int_{-\infty}^{\infty} tK'(t)\,dt + a\int_{-\infty}^{\infty} |K'(t)|\,dt.\end{aligned}$$

□

From the Taylor expansion of $\widehat{f_G}(x)$, we get, when δ/h goes to 0, with $\epsilon(.)$ a bounded function with values between 0 and 1 :

$$\begin{aligned}&E\widehat{f_G}(x) = \frac{1}{nh}\sum_{i=1}^{n} EK\left(\frac{x-X_i}{h} + \frac{\delta}{h}Z_i\right) \\ &= \frac{1}{nh}\sum_{i=1}^{n} EK\left(\frac{x-X_i}{h}\right) + \frac{1}{nh}\sum_{i=1}^{n} E\left[\frac{\delta}{h}Z_iK'\left(\frac{x-X_i}{h} + \frac{\delta}{h}Z_i\epsilon(Z_i)\right)\right] \\ &= E_A + E_B.\end{aligned} \tag{14}$$

Here we use a finite expansion which allows us to verify the conditions for which the last term is bounded. We calculate, for each term, the difference between the nonstationary expression, E_A or E_B, and its stationary equivalent expression, denoted respectively by E_A^* and E_B^*. The first term deals only with the nonstationarity while the remaining terms deal both the nonstationarity and the discretization:

$$\begin{aligned}E_A &= \frac{1}{nh}\sum_{i=1}^{n}\int K\left(\frac{x-y_i}{h}\right)[f_i(y_i)-f^*(y_i)]\,dy_i\\ &\qquad +\frac{1}{nh}\sum_{i=1}^{n}\int K\left(\frac{x-y_i}{h}\right)f^*(y_i)dy_i\\ &= (E_A-E_A^*)+E_A^*,\end{aligned}$$

$$\begin{aligned}E_B &= \frac{1}{nh}\sum_{i=1}^{n}\int_\Delta\int \frac{\delta}{h}u_iK'\left(\frac{x-y_i}{h}+\frac{\delta}{h}u_i\epsilon\left(u_i\right)\right)[g_i(y_i,u_i)-f^*(y_i)]\,dy_idu_i\\ &+\frac{1}{nh}\sum_{i=1}^{n}\int_\Delta\int \frac{\delta}{h}u_iK'\left(\frac{x-y_i}{h}+\frac{\delta}{h}u_i\epsilon\left(u_i\right)\right)f^*(y_i)dy_idu_i\\ &= (E_B-E_B^*)+E_B^*.\end{aligned}$$

From the results of convergence with respect to the norm of the total variation, we get for each difference, with M a generic constant:

$$\begin{aligned}|E_A-E_A^*| &\le \frac{1}{nh}\sup_t\left|K\left(\frac{x-t}{h}\right)\right|\sum_{i=1}^{n}\int|f_i(y_i)-f^*(y_i)|\,dy_i\\ &= \frac{1}{nh}\,M\,O\left(\sum_{i=1}^{n}\|F_i-F^*\|_{\mathsf{V}}\right),\\ |E_B-E_B^*| &\le \frac{1}{nh}\frac{\delta}{h}\sup_{t,u}\left|uK'\left(\frac{x-t}{h}+\frac{\delta}{h}u\epsilon\left(u\right)\right)\right|\\ &\qquad\times\sum_{i=1}^{n}\int_\Delta\int|g_i(y_i,u_i)-f^*(y_i)|\,dy_idu_i\\ &= \frac{1}{nh}\frac{\delta}{h}\,M\,O\left(\sum_{i=1}^{n}\|G_i-F^*\times U\|_{\mathsf{V}}\right).\end{aligned}$$

These allow us to write, from Hypothesis 2.1 and a generalization of

Lemma 3 of Hall (1983) from the stationary case to the nonstationary case

$$
\begin{aligned}
&E\widehat{f_G}(x)\\
&= E_A^* + E_B^* + O\left(\frac{1}{nh}\right) O\left(\sum_{i=1}^{n} \|F_i - F^*\|_{\mathsf{V}} + \frac{\delta}{h}\sum_{i=1}^{n} \|G_i - F^* \times U\|_{\mathsf{V}}\right)\\
&= E_A^* + E_B^* + O\left(\frac{1}{nh}\right) O\left(\sum_{i=1}^{n} O(\eta(i)) + \frac{\delta}{h}\sum_{i=1}^{n} (O(\delta) + O(\eta(i)))\right)\\
&= E_A^* + E_B^* + O\left(\frac{1}{nh} + \frac{\delta^2}{h^2}\right)
\end{aligned}
$$

E_A^* is the Rosenblatt estimator for stationary variables. On the other hand, we have by stationarity and the change of variables: $y = x - ht$,

$$
\begin{aligned}
E_B^* &= \frac{1}{nh} n \int_\Delta h \int \frac{\delta}{h} u K'\left(-t + \frac{\delta}{h} u\epsilon(u)\right) f^*(x+ht)dtdu\\
&= -\frac{\delta}{h}\int_\Delta \int uK'\left(t - \frac{\delta}{h}u\epsilon(u)\right)\left[f^*(x) + htf^{*\prime}(x + ht\,\epsilon(t))\right]dtdu\\
&= -\frac{\delta}{h}\int_\Delta u \left\{\int K'\left(t - \frac{\delta}{h}u\epsilon(u)\right)dt\right\} f^*(x)\; du\\
&\quad -\delta\int_\Delta u\left\{\int t\,K'\left(t - \frac{\delta}{h}u\epsilon(u)\right) f^{*\prime}(x + ht\,\epsilon(t))\,dt\right\} du \qquad (15)
\end{aligned}
$$

From an integration with respect to t, we can see that the first term of E_B^* is equal to zero only for kernels with infinite supports and equal to $\dfrac{\delta^2}{h^2}$ for kernels with compact support. Using Lemma 1, we have $\left|\int_{-1}^{1} K'\left(t - O\left(\frac{\delta}{h}\right)\right)dt\right| = O\left(\frac{\delta}{h}\right)$. The absolute value of the second term is bounded and we have

$$
|E_B^*| \le O\left(\frac{\delta^2}{h^2}\right) + \delta\; \left|\int_\Delta u\left\{\sup_z |f^{*\prime}(z)| \int \left|t\,K'\left(t - \frac{\delta}{h}u\epsilon(u)\right)\right| dt\right\} du\right|
$$

Also, from Lemma 1, we have: $\int \left|tK'\left(t + O\left(\frac{\delta}{h}\right)\right)\right| dt = \int tK'(t)dt + O\left(\frac{\delta}{h}\right)$ because $f^{*\prime}$ is bounded, $K' \in L_1$, and $\int tK'(t)\,dt < \infty$ by the hypothesis of Lemma 1. E_B^* is $O\left(\dfrac{\delta^2}{h^2} + \delta\int_\Delta udu + \dfrac{\delta^2}{h}\right)$. We get the expression (6) which goes to zero 0 when $n \to \infty$ from Hypothesis 2.2. This proves Proposition 1. □

Proof of Theorem 1. At last, we get a more precise expression of the bias which is given in (6) if we evaluate an additional term E_C in the

Taylor expression (14):

$$\begin{aligned} E_C &= \frac{1}{nh}\sum_{i=1}^{n}\int_{\Delta}\int \frac{\delta^2}{h^2}u_i^2K''\left(\frac{x-y_i}{h}\right)[g_i(y_i,u_i)-f^*(y_i)]\,dy_idu_i \\ &\quad +\frac{1}{nh}\sum_{i=1}^{n}\int_{\Delta}\int \frac{\delta^2}{h^2}u_i^2K''\left(\frac{x-y_i}{h}\right)f^*(y_i)dy_idu_i \\ &= (E_C-E_C^*)+E_C^*. \end{aligned}$$

From

$$\begin{aligned} |E_C-E_C^*| &\leq \frac{1}{nh}\frac{\delta^2}{h^2}\sup_{t,u}\left|u^2K''\left(\frac{x-t}{h}\right)\right| \\ &\qquad \times\sum_{i=1}^{n}\int_{\Delta}\int |g_i(y_i,u_i)-f^*(y_i)|\,dy_idu_i \\ &= \frac{1}{nh}\frac{\delta^2}{h^2}\,M\,O\left(\sum_{i=1}^{n}\|G_i-F^*\times U\|_{\mathsf{V}}\right) \end{aligned}$$

and

$$\begin{aligned} &E_C^* \\ &= \frac{1}{2}\frac{\delta^2}{h^2}\int K''(t)\left[f^*(x)+htf^{*\prime}(x)+\frac{1}{2}h^2t^2f^{*\prime\prime}(x)+O\left(h^3\right)\right]dt\int_{\Delta}u^2du. \end{aligned}$$

If $\mu_1(K''')<\infty$, the last term is equal to

$$\frac{\delta^3}{h^3}\int_{\Delta}\int u^3K'''\left(t-\frac{\delta}{h}u\epsilon(u)\right)htf^{*\prime}(x)\,dtdu = O\left(\frac{\delta^3}{h^2}\right)$$

instead of $O\left(\frac{\delta^3}{h^3}\right)$. Finally, we get the expression (6) from the preceeding results on the expectation. □

Proof of Corollary 1. For the triangular kernel, the second term E_B disappears by symmetry and we get (7). □

Proof of Theorem 2. It is an easy consequence of the triangular inequality:

$$\sqrt{E\left[\widehat{f_G}(x)-f^*(x)\right]^2} \leq \sqrt{E\left[\widehat{f_G}(x)-\widehat{f}(x)\right]^2}+\sqrt{E\left[\widehat{f}(x)-f^*(x)\right]^2} \tag{16}$$

and the following two Lemmas:

Lemma 2. *Under Hypotheses 2.1 and 2.2, the estimator $\widehat{f_G}$, defined previously, converges to the Rosenblatt estimator $\widehat{f}$ in the mean square.*

Lemma 3. *Under Hypotheses 2.1 and 2.2, if the sequence $\{X_i\}$ is strongly mixing with rate $\alpha(n) \underset{n\to\infty}{\to} 0$ and if $m = o(n)$ such that*

$$\frac{1}{nh}\, h^{-1/\lambda} \sum_{i=m}^{n} \alpha\,(i)^{1/\lambda} \to 0 \qquad \text{if } n \to \infty\,,$$

then the Rosenblatt estimator $\widehat{f}$ converges to f^* in the mean square. Moreover, if Hypothesis 2.3 is satisfed, then

$$E\left(\widehat{f}(x) - f^*(x)\right)^2 dx = \frac{1}{nh}\,\|K\|_2^2\, f^*\,(x) + O\left(\left(\frac{1}{nh} + h\right)^2\right). \tag{17}$$

Proof of Lemma 2: Consider the rounded case: from the development identical to (6) we have:

$$\begin{aligned}\widehat{f_G}(x) - \widehat{f}(x) &= \frac{1}{nh}\sum_{i=1}^{n}\frac{\delta}{h} Z_i K'\left(\frac{x - X_i}{h} + \frac{\delta}{h} Z_i \epsilon\,(Z_i)\right), \qquad Z_i \in \Delta \\ &= \frac{1}{nh}\,\frac{\delta}{h}\,\sum_{i=1}^{n} A_i.\end{aligned}$$

The mean square convergence from $\widehat{f_G}$ to $\widehat{f}$ is given under minimal hypotheses: MSE can be divided into the sum of square terms and the sum of product terms:

$$\begin{aligned}MSE &= E[\widehat{f_G}(x) - \widehat{f}(x)]^2 \\ &= \frac{1}{n^2h^2}\frac{\delta^2}{h^2}\; E\left[\sum_{i=1}^{n} A_i\right]^2 \\ &= \frac{1}{n^2h^2}\frac{\delta^2}{h^2}\left[\sum_{i=1}^{n} E\left(A_i^2\right) + 2\sum_{i=1}^{n-1}\sum_{j=i+1}^{n} E\left(A_i A_j\right)\right].\end{aligned}$$

We have:

$$
\begin{aligned}
& MSE \\
& \leq \frac{1}{n^2h^2}\frac{\delta^2}{h^2}\; n\; h\; \max_i \sup_{t,u} \left(g_i\left(x+ht,u\right)\right) \\
& \qquad \times \int_{\Delta} u_i^2 \left\{ \int K'\left(-t_i + \frac{\delta}{h} u_i \epsilon\left(u_i\right)\right)^2 dt_i \right\} du_i \\
& \quad + \frac{1}{n^2h^2}\frac{\delta^2}{h^2}\; n^2\; h^2\; \max_{i,j} \sup_{u,v,s,t} g_{i,j}\left(x+hs, x+ht, u, v\right) \\
& \quad \times \int_{\Delta} \left|u_j\right| \int_{\Delta} \left|u_i\right| \left\{ \iint \left| K'\left(-t_i + \frac{\delta}{h} u_i \epsilon\left(u_i\right)\right)\right| \right. \\
& \quad \times \left. \left| K'\left(-t_j + \frac{\delta}{h} u_j \epsilon\left(u_j\right)\right)\right| dt_i dt_j \right\} du_i du_j \\
& = O\left(\frac{1}{nh}\frac{\delta^2}{h^2}\right) + O\left(\frac{\delta^2}{h^2}\right).
\end{aligned}
$$

Thus the $MSE = O\left(\frac{\delta^2}{h^2}\right)$ which converges to zero. □

Proof of Lemma 3: We have:

$$
\begin{aligned}
E\left[\widehat{f}(x) - f^*(x)\right]^2 &= E\left[\widehat{f}(x) - E\widehat{f}(x)\right]^2 + \left[E\widehat{f}(x) - f^*(x)\right]^2 \\
&= Var\left(\widehat{f}(x)\right) + O\left(\left(\frac{1}{nh} + h\right)^2\right).
\end{aligned}
$$

If we denote $K_i = K\left(\frac{x - X_i}{h}\right)$, $Var\left(\widehat{f}(x)\right)$ can be divided into the sum of square terms and the sum of product terms. The last terms are divided in two blocks such that $|j-i| <$ or $\geq m = o\left(n\right)$. With $j > i$,

$$
\begin{aligned}
Var\left(\widehat{f}(x)\right) &= \frac{1}{n^2h^2}\sum_{i=1}^{n} E\left(K_i - EK_i\right)^2 \\
&\quad + \frac{2}{n^2h^2}\left(\sum\sum_{1<j-i\leq m} + \sum\sum_{j-i>m}\right) Cov\left(K_i, K_j\right) \\
&= VAR + COV_1 + COV_2.
\end{aligned}
$$

Similarly, if we denote

$$
K_i^* = K\left(\frac{x - X_i^*}{h}\right), \text{ and } VAR^* = \frac{1}{n^2h^2}\sum_{i=1}^{n} E\left(K_i^* - EK_i^*\right)^2,
$$

we have

$$\begin{aligned}
&\left|VAR - VAR^*\right| \\
&\leq \frac{1}{n^2h^2}\sum_{i=1}^{n}\int\left(K\left(\frac{x-y_i}{h}\right)+O\left(h\right)\right)^2 \left|f_i\left(y_i\right)-f^*\left(y_i\right)\right| dy_i \\
&\leq \frac{1}{n^2h^2}\ \sup\left|K\left(\frac{x-y_i}{h}\right)+O\left(h\right)\right|^2 \sum_{i=1}^{n}\int\left|f_i\left(y_i\right)-f^*\left(y_i\right)\right| dy_i \\
&= O\left(\frac{1}{n^2h^2}\right). \qquad (18)
\end{aligned}$$

By stationarity, VAR^* satisfies:

$$\begin{aligned}
VAR^* &= \frac{1}{n^2h^2}nh\int K\left(-t\right)^2\ f^*\left(x+ht\right)dt - \frac{1}{n^2h^2}nh^2\left(f^*\left(x\right)^2+o\left(1\right)\right) \\
&= \frac{1}{nh}\left\|K\right\|_2^2 f^*\left(x\right)+o\left(n^{-1}\right). \qquad (19)
\end{aligned}$$

As $EK_i = O\left(h\right)$, we get:

$$\begin{aligned}
COV_1 &= \frac{2}{n^2h^2}\sum\sum_{1<j-i\leq m} E\left(K_i-EK_i\right)\left(K_j-EK_j\right) \\
&= \frac{2}{n^2h^2}\sum\sum_{1<j-i\leq m}\left(E\left(K_iK_j\right)-O\left(h^2\right)\right) \\
&= \frac{2}{n^2h^2}\sum\sum_{1<j-i\leq m} E\left(K_iK_j\right)+O\left(\frac{m}{n}\right).
\end{aligned}$$

We have:

$$\begin{aligned}
E\left(K_iK_j\right) &\leq h^2\ \max_{i,j}\sup_{s,t}\left(f_{i,j}(x+hs,x+ht)\right)\iint K(t_i)K(t_j)\ dt_idt_j \\
&= O\left(h^2\right).
\end{aligned}$$

and then

$$COV_1 = O\left(\frac{m}{n}\right). \qquad (20)$$

From the well known inequality on the covariances, we get :

$$
\begin{aligned}
|COV_2| &\leq \frac{C}{n^2h^2} \sum_{j-i>m}\sum \alpha(j-i)^{1/\lambda} \left| \int \left| K\left(\frac{x-y_i}{h}\right)\right|^p f_i(y_i)dy_i \right|^{1/p} \\
&\quad \times \left| \int \left| K\left(\frac{x-y_j}{h}\right)\right|^p f_j(y_j)dy_j \right|^{1/p} \\
&\leq \frac{C}{n^2h^2} \sum_{i=m}^{n} (n-i)\,\alpha(i)^{1/\lambda}\ h^{2/p} \\
&\qquad \times \left(\max_k \left| \int |K(t_k)|^p f_k(x+ht_k)dt_k \right|^{1/p} \right)^2 \\
&\leq C\, n^{-1}h^{2/p-2} \sum_{i=m}^{n} \alpha(i)^{1/\lambda}\ \max_k \sup_y \left(f_k(y)^{2/p} \right)\ \|K\|_p^2 \\
&= O\left(\frac{1}{nh}\ h^{-1/\lambda} \sum_{i=m}^{n} \alpha(i)^{1/\lambda} \right). \qquad (21)
\end{aligned}
$$

From the hypotheses, we deduce that COV_1 and COV_2 converge to zero and the convergence in mean square is proved.

On the other hand, under Hypothesis 2.3, COV_1 are COV_2 are $o\left(\frac{1}{nh}\right)$, and of course, negligible with respect to VAR^*, which implies the second part of Lemma 3. □

Theorem 2 follows.

Proof of Proposition 2. Let Λ_j be the domain of definition of $T\left((y-a_j)/\delta\right)$. Recall the definition of the discretized forms of $\widehat{f_G}(x)$ and $\widehat{f_G^*}(x)$:

$$
\widehat{f_G}(x) = \frac{\delta}{h} \sum_{j\in\mathbb{Z}} K\left(\frac{x-a_j}{h}\right) \frac{1}{n\delta} \sum_{i=1}^{n} T\left(\frac{X_i-a_j}{\delta}\right), \qquad (22)
$$

$$
\widehat{f_G^*}(x) = \frac{\delta}{h} \sum_{j\in\mathbb{Z}} K\left(\frac{x-a_j}{h}\right) \frac{1}{n\delta} \sum_{i=1}^{n} T\left(\frac{X_i^*-a_j}{\delta}\right). \qquad (23)
$$

We now show that the nonstationarity implies the order of approximation $n^{-1}h^{-1}$ for the expectation. In fact, for M a generic constant

$$\begin{aligned}
&\left|E\widehat{f_G}(x) - E\widehat{f^*_G}(x)\right| \\
&\le \frac{1}{nh}\sum_{j\in\mathbb{Z}} K\left(\frac{x-a_j}{h}\right)\sum_{i=1}^{n}\int_{\Lambda_j}\left|T\left(\frac{y_i-a_j}{\delta}\right)[f_i(y_i)-f^*(y_i)]\right|dy_i \\
&\le \frac{1}{nh}\ \max_j \sup_t \left|K\left(\frac{x-a_j}{h}\right)T\left(\frac{t-a_j}{\delta}\right)\right| \sum_{i=1}^{n}\sum_{j\in\mathbb{Z}}\int_{\Lambda_j}|f_i(y_i)-f^*(y_i)|\,dy_i \\
&= \frac{1}{nh}\ M\sum_{i=1}^{n}\sum_{j\in\mathbb{Z}}\int_{\Lambda_j}|f_i(y_i)-f^*(y_i)|\,dy_i.
\end{aligned}$$

Λ_j is always in a finite number of intervals B_{j+k} (with $k=\{0\}$ or $\{0,1\}$ for the usual cases, and $B_{j+k}=[a_{j+k}-\delta/2, a_{j+k}+\delta/2[$ or $B_{j+k}=[a_{j+k}, a_{j+k}+\delta[$. We can write:

$$\left|E\widehat{f_G}(x) - E\widehat{f^*_G}(x)\right| \le \frac{1}{nh}\ \sum_{i=1}^{n}\ \sum_{j\in\mathbb{Z}}\sum_{k}\int_{B_{j+k}}|f_i(y_i)-f^*(y_i)|\,dy_i.$$

As $\cup_{j\in\mathbb{Z}}B_{j+k}=\mathbb{R}$, we get by a summation of the integrals with respect to the index j :

$$\begin{aligned}
\left|E\widehat{f_G}(x) - E\widehat{f^*_G}(x)\right| &\le \frac{1}{nh}\ \sum_{k}\ \sum_{i=1}^{n}\int|f_i(y_i)-f^*(y_i)|\,dy_i \\
&= O\left(\frac{1}{nh}\ \sum_{i=1}^{n} O\left(\eta\left(i\right)\right)\right) \\
&= O\left(\frac{1}{nh}\right). \qquad (24)
\end{aligned}$$

The expression of the expectation of $\widehat{f^*_G}(x)$ can be obtained directly from the expectation of $\widetilde{f^*}(a_j) = \dfrac{1}{n\delta}\sum_{i=1}^{n} T\left(\frac{X^*_i - a_j}{\delta}\right)$:

$$E\widehat{f^*_G}(x) = \frac{\delta}{h}\sum_{j\in\mathbb{Z}} K\left(\frac{x-a_j}{h}\right)[f^*(a_j)+O(\delta)].$$

From Lemma 3.3 of Carbon, et al. (1997), we get

$$\begin{aligned}
&\frac{\delta}{h}\sum_{j\in\mathbb{Z}} K\left(\frac{x-a_j}{h}\right) f^*(a_j) \\
&= \frac{1}{h}\int K\left(\frac{x-y}{h}\right) f^*(y)dy + O\left(\frac{\delta}{h}\int\left|\left(K\left(\frac{x-y}{h}\right) f^*(y)\right)'\right| dy\right) \\
&= E\widehat{f^*}(x) + O\left(\frac{\delta}{h}\ \|K'\|_1\ \sup_t f^*(t) + \delta\ \sup_t |f^{*\prime}(t)|\right) \qquad (25)
\end{aligned}$$

by using for the last term the Hölder inequality. We get the expression of the bias in Proposition 2. □

Proof of Theorem 3. The expression of $E\widehat{f_G^*}(x)$ is obtained from the expansion of $f^*(a_j)$:

$$\begin{aligned}
E\widehat{f_G^*}(x) = \frac{\delta}{h}\sum_{j\in\mathbb{Z}} K\left(\frac{x-a_j}{h}\right) &[f^*(a_j) + \delta\mu_1(T) f^{*\prime}(a_j) \\
&+ \frac{1}{2}\delta^2\mu_2(T) f^{*\prime\prime}(a_j) + O\left(\delta^3\right)\Big]. \qquad (26)
\end{aligned}$$

We denote $\psi(y) = K\left(\frac{x-y}{h}\right) f^*(y)$ and consider the following approximation of the expectation of the Rosenblatt kernel obtained by Taylor expansion with $y_j = a_j$:

$$\begin{aligned}
E\widehat{f^*}(x) &= \frac{1}{h}\int K\left(\frac{x-y}{h}\right) f^*(y)dy \\
&= \frac{1}{h}\sum_{j\in\mathbb{Z}}\int_{B_j}\Big[\psi(a_j) + (y_j - a_j)\psi'(a_j) + \frac{1}{2}(y_j-a_j)^2\psi''(a_j) \\
&\qquad + O\left((y_j-a_j)^3\psi'''(a_j)\right)\Big] dy_j.
\end{aligned}$$

By the change of variables: $y_j = a_j + \delta u_j$, and noting that $K_j =$

$K\left(\frac{x-a_j}{h}\right)$, we get:

$$
\begin{aligned}
&E\widehat{f^*}(x)\\
&=\frac{1}{h}\sum_{j\in\mathbb{Z}}K_j f^*(a_j)\delta\int_\Delta du_j\\
&+\frac{1}{h}\sum_{j\in\mathbb{Z}}\left[K_j f^{*\prime}(a_j)-\frac{1}{h}K_j' f^*(a_j)\right]\delta^2\int_\Delta u_j du_j\\
&+\frac{1}{2}\frac{1}{h}\sum_{j\in\mathbb{Z}}\left[K_j f^{*\prime\prime}(a_j)-\frac{2}{h}K_j' f^{*\prime}(a_j)+\frac{1}{h^2}K_j'' f^*(a_j)\right]\delta^3\int_\Delta u_j^2 du_j\\
&+O\left(\frac{1}{h}\,\delta^4\sum_{j\in\mathbb{Z}}(K_j f^*(a_j))'''\int_\Delta u_j^3 du_j\right).
\end{aligned}
\tag{27}
$$

The first line of $E\widehat{f^*}(x)$ corresponds to the first term on the right of (26). Each term of the following lines of $E\widehat{f^*}(x)$ is an approximation of the corresponding integral. From Lemma 3.3 of Carbon, et al. (1997), we have for example:

$$
\begin{aligned}
&\delta\sum_{j\in\mathbb{Z}}K'\left(\frac{x-a_j}{h}\right)f^{*\prime}(a_j)\\
&=\int K'\left(\frac{x-y}{h}\right)f^{*\prime}(y)dy+O\left(\delta\int\left|\left(K'\left(\frac{x-y}{h}\right)f^{*\prime}(y)\right)'\right|dy\right)\\
&=-\,h^2\,\mu_1\left(K'\right)f^{*\prime\prime}(x)+O\left(h^3\right)\\
&\quad+O\left(\delta\ \|K''\|_1\ \sup_x|f^{*\prime}(x)|+h\delta\ \|K'\|_1\ \sup_x|f^{*\prime\prime}(x)|\right).
\end{aligned}
$$

For the successive derivatives of f in (12), we get in a similar way:

$$
\begin{aligned}
&\frac{\delta}{h}\sum_{j\in\mathbb{Z}}K\left(\frac{x-a_j}{h}\right)f^{*(k)}(a_j)\\
&=f^{*(k)}(x)+O\left(\frac{\delta}{h}\ \|K'\|_1\ \sup_x\left|f^{*k}(x)\right|+\delta\ \sup_x\left|f^{*(k+1)}(x)\right|\right).
\end{aligned}
$$

□

Acknowledgment

The authors would like to express their sincere thanks to Professor Francisco Samaniego and the referee for their valuable suggestions which improved the presentation of the results considerably.

References

Carbon, M., Garel, B. and Tran L. T. (1997). Frequency polygons for weakly dependant processes. *Stat. & Prob. Letters* **33**, 1–13.

Hall, P. and Wand, M. P. (1996). On the accuracy of binned kernel density estimators. *J. Multivar. Anal.* **56**, 165–184.

Rosenblatt, M. (1971). Curve estimates. *Ann. Math. Stat.* **42**, 6, 1815–1842.

Hall, P. (1982). The influence of rounding errors on some nonparametric estimators of a density and its derivatives. *SIAM J. Appl. Math.* **42**, 390–399.

Hall, P. (1983). Edgeworth expansions of the distribution of Stein's statistic. *Math. Proc. Cambridge Philos. Soc.* **93**, 163–175.

Harel, M. and Puri, M. L. (1996). Nonparametric density estimators based on nonstationnary absolutely regular random sequences. *J. Appl. Math. Stoch. Anal.* **9**, 3, 233–254.

Harel, M. and Puri, M. L. (1999). Conditional empirical processes defined by nonstationary absolutely regular sequences. *J. Multivar. Anal.* **70**, 250–285.

Jones, M. C. (1989): Discretized and interpolated kernel density estimates. *J. Amer. Statist. Assoc.* **84**, 733–741.

Chapter 21

Convergence Rates of an Improved Isotonic Regression Estimator

Hari Mukerjee

Wichita State University

Isotonic regression is a nonparametric uniform kernel regression with a random bandwidth that depends on the observations themselves. The random bandwidth is typically too small to yield a satisfactory convergence rate or asymptotic distribution theory. Here we consider an estimator that adds a fixed bandwidth to the random "skin" resulting from the isotonization process. The width of this skin goes down monotonically to an asymptotic value as the bandwidth and the order of the conditional moments increase so that the convergence rates are the same as those of an unrestricted uniform kernel estimator that is not necessarily monotonic.

Contents

1. Introduction

Let $(X_1, Y_1), \ldots, (X_n, Y_n)$ be IID bivariate random vectors, distributed as (X, Y), with finite conditional means, $m(x) = E[Y|X = x]$. When m is nondecreasing, Brunk (1958) derived the isotonic least squares estimator of m on the set, $\{X_i\}$, by

$$m_n^*(x) = \max_{s \le x} \min_{t \ge x} Av_n([s,t]), \quad \text{where } Av_n(A) = \frac{\sum_{i=1}^n Y_i I(X_i \in A)}{\sum_{i=1}^n I(X_i \in A)}, \ A \subset S. \tag{1}$$

This estimator, with monotone extensions, has been studied extensively. Hanson, et al. (1973) have established strong uniform consistency over an interval with only a first moment type assumption on the conditional distributions of Y. Rao (1969), Wright (1981), and others have shown that $|m_n^*(x) - m(x)| = O_p(n^{-1/3})$ when $m'(x) > 0$, with $n^{1/3}[m_n(x) - m(x)]$ having a nondegenerate (unfamiliar) asymptotic distribution. This may be contrasted with the kernel (or other nonparametric regression) estimators that are asymptotically normal, with the norming of $n^{2/5}$, for the optimal bandwidths.

The estimator in (1) is essentially a uniform kernel estimator with a random bandwidth that depends on the observations, $\{Y_i\}$, in contrast with nonparametric regression. Mukerjee (1988) introduced a two-stage estimator that first isotonizes the data and then smooths the isotonized data with a log-concave kernel that retains the monotonicity, achieving the same convergence rate as the kernel estimator while guaranteeing monotonicity. To obtain a Bahadur-type representation of conditional quantiles as in Bhattacharya and Gangopadhyay (1990), but using isotonic estimators when the quantiles are assumed to be monotone, Mukerjee (1993a) introduced a single-stage estimator that adds a fixed bandwidth, h_n, to the random "skin" resulting from isotonization:

$$\hat{m}_n(x) = \max_{s \le x - h_n} \min_{t \ge x + h_n} Av_n([s,t]), \tag{2}$$

the monotonicity being guaranteed by the fact that as x increases, the set of maximization increases and the set of minimization decreases. Isotonization introduces a first order (conditional) bias term of the same order as the width of the random band. This random skin shrinks monotonically as the bandwidth increases, down to the asymptotic value of $O_{a.s.}((\log\log n/n)^{1/2})$ even when h_n is as small as const. $\times$ $(\log n)^{1/2}/(n \log\log n)^{1/4}$. Thus, for the optimal bandwidth of $h_n =$ const. $\times\, n^{-1/5}$, we get essentially the same asymptotic properties for the restricted and the unrestricted estimators. The key to obtaining these results is the fact that the conditional quantile function may be viewed as the mean regression function of Bernoulli variables, that are uniformly bounded, to which the Hoeffding (1963) and Bernstein (see Bennett (1962)) inequalities could be applied. These inequalities are also applicable if the conditional distributions have a uniformly subnormal tail (Bennett (1962)), but these are severe restrictions. In this paper we consider the convergence rates of $\hat{m}_n(x)$ as a function of h_n and r, the order of the conditional moment, for $r > 1$ at a point where $m'(x) > 0$. The random skin from isotonization decreases monotonically as both h_n and r increase, to the asymptotic value of $O_{a.s.}((\log\log n/n)^{1/2})$, as in the quantile case. The basic tool is an exponential probability inequality for truncated random variables (RVs) developed in Mukerjee (1993b). The purpose of this investigation is to determine almost sure upper bounds of the random bandwidth and of $|\hat{m}_n(x) - \tilde{m}_n(x)|$ as a function of r and h_n, where

$$\tilde{m}_n(x) = Av_n([x - h_n, x + h_n]), \tag{3}$$

the uniform kernel estimator of $m(x)$.

2. Convergence rates of $\hat{m}_n$

Assume that (X, Y) has a joint density. Let f be the marginal density of X, S the support of X, and let

$$H(t) = \sup_{x \in S} P(|Y| \geq t | X = x), \ t \geq 0.$$

Fix x_0 in the interior of S and consider the following assumptions:

$$H(t) \to 0 \text{ as } t \to \infty \text{ and } \sigma_r = \int_0^\infty t^r |dH(t)| < \infty \text{ for some } r > 1; \tag{4}$$

$$m' \text{ exists continuously in a neighborhood of } x_0 \text{ and } m'(x_0) > 0; \tag{5}$$

$$f \text{ is continuous at } x_0 \text{ and } f(x_0) > 0. \tag{6}$$

WLOG we assume that $x_0 = 0$ and we omit the dependence of functions on x for $x = 0$. Let $b_n \to 0$ be a positive sequence and define

$$\bar{m} = \max_s \min_t \{Av_n([s, t]) : -h_n - b_n \leq s \leq -h_n, \ h_n \leq t \leq h_n + b_n\}. \tag{7}$$

Let $\rho = r \wedge 2$. The following theorem gives the almost sure bounds of the random bandwidth as a function of r and h_n. It should be noted that the constants, c_i's and M_i's, in the proof do not depend on n.

Theorem 1. *Under assumptions* (4)–(6), $P(\bar{m} \neq \hat{m} \ i.o.) = 0$
if $b_n \geq M_1(\log\log n/n)^{1/2}$, $h_n + b_n \geq M_2 n^{-1+1/r}$, *and*
(I) $h_n + b_n = O\left(n^{(1-\rho)/r}\right)$ *and* $b_n(h_n + b_n)^2 \geq M_3 n^{-1+(2-\rho)/r} \log n$, *or,*
(II) $n^{(1-\rho)/r} = o(h_n + b_n)$ *and* $b_n(h_n + b_n) \geq M_4 n^{-1+1/r} \log n$
for sufficiently large positive constants, M_1, M_2, M_3 *and* M_4.

Proof. Suppose that, for all $s \leq -h_n$ and $t \geq h_n + b_n$, we have $Av_n([s, h_n + b_n/2)) \leq Av_n([h_n + b_n/2, t])$. Then, by the Cauchy mean value property of averages (Robertson and Wright (1974)), $Av_n([s,t]) \le Av_n([h_n + b_n/2, t])$, so that

$$\hat{m} = \max_s \min_t \{Av_n([s,t]) : s \le -h_n,\ h_n \le t \le h_n + b_n\}.$$

By a similar argument, we note that

$$\hat{m} = \max_s \min_t \{Av_n([s,t]) : -h_n - b_n \le s \le -h_n,\ t \ge h_n\}$$

if, for all $s \le -h_n - b_n$ and $t \ge h_n$, we have $Av_n([s, -h_n - b_n/2)) \le Av_n([-h_n - b_n/2, t])$. Putting these two observations together and taking complements of the events, we see that $P(\bar{m} \ne \hat{m})$ is bounded above by the sum of

$$P(\max_{s \le -h_n} Av_n([s_n, h_n + b_n/2)) > \min_{t \ge h_n + b_n} Av_n([h_n + b_n/2, t])) \quad \text{and} \tag{8}$$

$$P(\max_{s \le -h_n - b_n} Av_n([s_n, -h_n - b_n/2)) > \min_{t \ge h_n} Av_n([-h_n - b_n/2, t])). \tag{9}$$

To prove the theorem, we show that the probabilities of the events in (8) and (9) occurring infinitely often is zero under the conditions of the theorem. Since the analyses are similar, we consider (8) only.

The expression (8) is bounded above by the sum of

$$P(\min_{t \ge h_n + b_n} Av_n([h_n + b_n/2, t]) \le m(h_n/2 + b_n/2)) \quad \text{and} \tag{10}$$

$$P(\max_{s \le -h_n} Av_n([s, h_n + b_n/2)) \ge m(h_n/2 + b_n/2)). \tag{11}$$

Again, since the proofs are similar, we consider an upper bound of (10) only.

For $t \ge h_n + b_n$, let

$$N(t) = \sum_{i=1}^{n} I(h_n + b_n/2 \le X_i \le t).$$

By assumptions (5) and (6), for all n large enough,

$$E[Av_n([h_n + b_n/2, t])] - m(h_n/2 + b_n/2) \ge c_1(h_n + b_n) \tag{12}$$

for some $c_1 > 0$, whenever $t \ge h_n + b_n$. Letting

$$\Sigma'(\cdot) = \Sigma_{i=1}^{n}(\cdot) I(h_n + b_n/2 \le X_i \le t)$$

and using (12), we note that (10) is bounded above by

$$\begin{aligned} &P(\max_{t \ge h_n + b_n} \{N^{-1}(t)|\Sigma'(Y_i - EY_i)| \ge c_1(h_n + b_n)) \\ &\le n \max_{t \ge h_n + b_n} P(N^{-1}(t)|\Sigma'(Y_i - EY_i)| \ge c_1(h_n + b_n)). \end{aligned} \tag{13}$$

For each $n \geq 1$, let $Y_{in} = Y_i I(|Y_i| \leq n^{1/r}),\ 1 \leq i \leq n$. Now,

$$\frac{1}{N(t)}|\Sigma'(Y_i - EY_i)| \leq \frac{1}{N(t)}[|\Sigma'(Y_i - Y_{in})| + |\Sigma'(Y_{in} - EY_{in})| + |\Sigma'(EY_{in} - EY_i)|]. \tag{14}$$

Bounds for each of the terms on the *rhs* were given in a series of lemmas in Mukerjee (1993b), — here we only give some indications of the proofs. The numberings of the lemmas are the same as in the quoted paper. [For ease of comparisons, we make the following observations about the notation and results in Mukerjee (1993b). The paper is about uniform convergence of nearest neighbor regression estimators on a compact subset of $\mathbb{R}^d$ when σ_r is assumed to be finite only for an arbitrary $r > 1$. The results we need here are simplified because we consider pointwise convergence only. The weights c_{in} in the paper is the random $1/N(t)$ in (14); $f(n)$ is $N(t)$; the constant C in the paper is 1 in our case; and the multiplier ${}_N C_{\leq d}$ in Lemma 2.7, denoting the upper bound of the number of possible values of the regression function estimator in $\mathbb{R}^d$, is 1 in our case of pointwise convergence.] By Lemma 2.2, using $P(|Y_n| > n^{1/r}\ i.o.) = 0$ from assumption (5),

$$N^{-1}(t)|\Sigma'(Y_i - Y_{in})| = O_{a.s.}(N^{-1}(t)). \tag{15}$$

By Lemma 2.3, using $E[|Y|I(|Y| > n^{1/r})] \leq E[|Y|^r |n^{1/r}|^{1-r} I(|Y| > n^{1/r})]$ for $r > 1$,

$$\Sigma' N^{-1}(t)(EY_i - EY_{in})| \leq \sigma_r n^{-1+1/r}. \tag{16}$$

By Lemma 2.7, for every $\epsilon > 0$,

$$\begin{aligned} &P(N^{-1}(t)|\Sigma'(Y_{in} - EY_{in})| \geq \epsilon|\mathcal{A}) \\ &\leq 2\exp[-N(t)\epsilon \min\{c_2\epsilon(2n)^{(\rho-2)/r},\ n^{-1/r}/8\}] \quad a.s., \end{aligned} \tag{17}$$

where $c_2 = 2^{\rho+3}\sigma_\rho$, $\sigma_\rho = \sup_{x\in S} E[|Y|^\rho|X = x]$, and $\mathcal{A}$ is the σ-field generated by $\{X_i\}$. Derivation of (17) uses the fact that $|Y_{in} - EY_{in}| \leq 2n^{1/r},\ 1 \leq i \leq n$, Lemma 2.4 that generalizes a classical result (when $\rho = r = 2$) to show that $|Ee^{t(Y_{in} - EY_{in})} - 1| \leq \sigma_\rho (2n)^{(2-\rho)/r} t^2$ for $|t| \leq n^{-1/r}/2$, and Lemma 2.5 that gives an exponential probability inequality for weighted sums of independent mean-zero RVs with a uniformity condition on their moment generating functions near the origin that is due to Hanson and Koopmans (1965). Note that the negative of the exponent in (17) may be written as

$$c_3 N(t) n^{(\rho-2)/r} \epsilon \min\{\epsilon, c_4 n^{(1-\rho)/r}\} \tag{18}$$

for some $c_3 > 0$ and $c_4 > 0$.

By assumption (6) and the law of the iterated logarithm for the empirical distribution function (EDF), there exists $c_5 > 0$ and $c_6 > 0$ such that

$$P(N(h_n + b_n) - c_5 n b_n < -c_6 (n \log\log n)^{1/2} \; i.o.) = 0.$$

If $b_n \geq M_1 (\log\log n/n)^{1/2}$ for an M_1 sufficiently large,

$$P(N(h_n + b_n) < c_7 n b_n \; i.o.) = 0 \quad \text{for some } c_7 > 0. \tag{19}$$

On the set, $\{N(h_n + b_n) \geq c_7 n b_n\}$, if $h_n + b_n \geq M_2 n^{-1+1/r}$ for an M_2 sufficiently large, from (15)–(17) and (19) we have, for some $c_8 > 0$, (recall that $\mathcal{A}$ is the σ-field generated by the X_i's)

$$\begin{aligned}
&P(N^{-1}(t)|\Sigma'(Y_i - EY_i))| \geq c_1(h_n + b_n)|\mathcal{A}) \\
&\leq P((c_7 n b_n)^{-1}|\Sigma'(Y_{in} - EY_{in}))| \geq c_1(h_n + b_n) - \sigma_r n^{-1+1/r}|\mathcal{A}) \\
&\leq P((c_7 n b_n)^{-1}|\Sigma'(Y_{in} - EY_{in}))| \geq c_8(h_n + b_n)|\mathcal{A}) \\
&\leq e^{-d_n},
\end{aligned} \tag{20}$$

where

$$\begin{aligned}
d_n &= c_3(c_7 n b_n) n^{(\rho-2)/r} c_8(h_n + b_n) \min\{c_8(h_n + b_n),\; c_4 n^{(1-\rho)/r}\} \\
&= c_9 n^{1-(2-\rho)/r} b_n (h_n + b_n) \min\{(h_n + b_n),\; c_{10} n^{(1-\rho)/r}\}
\end{aligned} \tag{21}$$

for some $c_9 > 0$ and $c_{10} > 0$. In case (I) of the theorem, $h_n + b_n = O(n^{(1-\rho)/r})$ and $b_n(h_n + b_n)^2 \geq M_3 n^{-1+(2-\rho)/r} \log n$. Thus (21) indicates that for some $c_{11} > 0$,

$$d_n \geq c_{11} n^{1-(2-\rho)/r} b_n (h_n + b_n)^2 \geq c_{11} M_3 \log n$$

so that ne^{-d_n} is summable if M_3 is large enough. In case (II) of the theorem, $n^{(1-\rho)/r} = o(h_n + b_n)$ and $b_n(h_n + b_n) \geq M_4 n^{-1+1/r} \log n$. Then, for some $c_{12} > 0$,

$$d_n \geq c_{12} n^{1-(2-\rho)/r} M_4 n^{-1+1/r} \log n \, n^{(1-\rho)/r} = c_{12} M_4 \log n$$

so that ne^{-d_n} is summable if M_4 is large enough. An application of the Borel–Cantelli Lemma completes the proof. □

2.1. *The random bandwidth*

Given r, the maximal conditional moment for which σ_r is finite, Theorem 1 provides the ranges of fixed (h_n, b_n)'s necessary for $\bar{m} = \hat{m}$ eventually almost surely. Thus, given this maximal r (an assumption), we may choose h_n and make conservative estimates of the random bandwidth introduced by the isotonization process using the minimum of these b_n's for which

$P(\bar{m} = \hat{m}\ i.o.) = 0$. From now on, these b_n's will represent these conservative estimates, and they will be given in almost sure large or small order notation without the subscript, *a.s.* From the theorem, we always have

$$b_n \geq M_1(\log\log n/n)^{1/2} \quad \text{and} \quad h_n + b_n \geq M_2 n^{-1+1/r}$$

for some $M_1, M_2 > 0$. For the sake of completeness, we consider the case when $h_n = O(b_n)$, but our main interest is to reduce b_n by increasing h_n for a given r. However, from consideration of the symmetric uniform kernel estimator, $\tilde{m}_n$ in (3), we will restrict h_n to be no larger than const.$\times n^{-1/5}$, the optimal bandwidth under sufficiently large r, when the square of the bias of $\tilde{m}_n$ is of the same order as its variance.

In case (I) of the theorem,

$$\begin{aligned} &n^{-1+(2-\rho)/r}\log n = O((h_n+b_n)^3) = O(n^{3(1-\rho)/r}) \\ &\Longrightarrow -1+(2-\rho)/r < 3(1-\rho)/r \Longrightarrow 2\rho - r < 1. \end{aligned}$$

Since $r > 1$ and $\rho = r \wedge 2$, the last inequality can be satisfied only if

$$r > 3 \Longrightarrow \rho = 2,\ h_n + b_n = O(n^{-1/r}) \text{ and } b_n(b_n+h_n)^2 \geq M_3 n^{-1}\log n \quad (22)$$

for some $M_3 > 0$.

From the last inequality above, if $h_n = O(b_n)$ we have $b_n = O(\log n/n)^{1/3}$ since this also satisfies $h_n + b_n = O(n^{-1/r})$ for $r > 3$. Note that this covers the case of $h_n = 0$, corresponding to the standard isotonic estimator, m_n^*, in (1).

If $b_n = o(h_n)$, then $h_n = O(n^{-1/r})$ and $b_n h_n^2 \geq M_5 n^{-1}\log n$ for some $M_5 > 0$ so that $b_n = O(\log n/(nh_n^2))$. Since $h_n = O(n^{-1/r})$, depending on the value of r we could have b_n ranging from $o(\log n/n)^{1/3}$, if r is just barely more than 3 and $h_n = n^{-1/r}$, to its asymptotic value of $O((\log\log n/n)^{1/2})$ if $r > 4$ and $h_n = \sqrt{\log n}/(n\log\log n)^{1/4}$. Further increase of r or h_n does not shrink the random band any further from the law of the iterated logarithm for the EDF.

In case (II) of the theorem, we note that $n^{-1+1/r} = n^{(1-r)/r} = O(n^{(1-\rho)/r}) = o(h_n + b_n)$.

If $h_n = O(b_n)$ then

$$b_n((h_n+b_n) \geq M_4 n^{-1+1/r}\log n \Longrightarrow b_n \geq M_6(n^{-1+1/r}\log n)^{1/2} \quad \text{for some } M_4, M_6 > 0,$$

and

$$n^{(1-\rho)/r} = o(h_n+b_n) = o(b_n) \Longrightarrow (1-\rho)/r \leq (-1+1/r)/2 \Longrightarrow 2\rho - r \geq 1 \Longrightarrow r \leq 3.$$

This means $b_n = O((\log n/n)^{1/3})$ when $r = 3$ and it goes up to very large values as r goes down.

If $b_n = o(h_n)$ then $b_n h_n \geq M_7 n^{-1+1/r} \log n$ for some $M_7 > 0$. This implies $(n^{-1+1/r} \log n)^{1/2} = o(h_n)$ and $b_n = O(n^{-1+1/r} h_n^{-1} \log n)$. We may choose any h_n as long as $(n^{-1+1/r} \log n)^{1/2} = o(h_n)$ and $h_n = O(n^{-1/5})$ for any $r > 1$. If we choose $h_n = n^{(-1+1/r)/2} \log n$, then, since $(1-\rho)/r = (1 - r \wedge 2)/r < (-1+1/r)/2 = (1-r)/2r$ for $r > 1$, we have $n^{(1-\rho)/2} = o(h_n + b_n)$, and $b_n = O(n^{(-1+1/r)/2}) + O((\log\log n/n)^{1/2})$, the second term arising from the fact that b_n cannot go down below this by the law of the iterated logarithm.

2.2. *Comparison with the uniform kernel estimator*

The fixed bandwidth in the unrestricted uniform kernel regression is of the form $h_n = \text{const.} \times n^{-\delta}$ for some $1/5 \leq \delta < 1/3$. The optimal bandwidth corresponds to $\delta = 1/5$ when the square of the bias and the variance of the estimators are of the same order and we have asymptotic normality with a norming of $n^{2/5}$ for $r > 5/2$. If $r > 5/2$ and $h_n = n^{-1/5}$, then $n^{(1-\rho)/r} = n^{(1-r)/r} = o(h_n)$ and case (II) of the theorem applies. Thus,

$$b_n = O(n^{-1+1/r} h_n^{-1} \log n) = O(n^{-4/5+1/r} \log n) = o(n^{-2/5}).$$

This implies $|\hat{m}_n(x_0) - \tilde{m}_n(x_0)| = o(n^{-2/5})$ so that $n^{2/5}[\hat{m}_n(x_0) - m(x_0)]$ and $n^{2/5}[\tilde{m}_n(x_0) - m(x_0)]$ have the same asymptotic normal distributions. Thus, we get the monotonicity of our estimator "free of charge."

3. Concluding remarks

The standard isotonic regression estimator, m_n^* in (1), is a uniform kernel estimator with a random bandwidth. This bandwidth is very narrow, resulting in poor convergence rates due to large variance. We have considered an improved isotonic estimator that adds a fixed band of larger width to the random one. This results in shrinking the width of the random band so that the restricted estimator is convergence equivalent to the unrestricted uniform kernel estimator for all statistical purposes.

Although only pointwise convergence was considered, if assumptions (5) and (6) were assumed to hold uniformly in a closed interval contained in the interior of the support, S of X, then the same convergence rates could be shown to hold uniformly in the interval as in Mukerjee (1993b).

The estimator, $\hat{m}$, is not smooth. Mukerjee (1988) employed a 2-stage estimator that smooths the standard isotonic estimator, m_n^*, using a smooth kernel of bandwidth $n^{-1/5}$. The resulting estimator is isotonic, smooth, and yields an asymptotic normal distribution with the norming of $n^{2/5}$. Friedman and Tibshirani (1984) suggested another 2-stage estimator that smooths the data first and then isotonizes this smoothed estimator. Since isotonization can be carried out only on finite sets, this 2-stage estimator is not smooth. However, Mammen (1991) showed that it is better than that of Mukerjee (1988) in the second order. As a future project, it would be of interest to see if smoothing $\hat{m}$ of this paper with a smooth kernel will match the second order properties of the Friedman–Tibshirani estimator while providing a smooth monotone estimator.

We conclude with a note on an algorithm for computing $\hat{m}_n(x)$. The well known pool adjacent violator algorithm (PAVA) computes m_n^* very efficiently. The resulting estimator produces the so-called level sets,— sets of constancy of m_n^*. For a given x, if $x - h_n$ lies in a level set, L_1 and $x + h_n$ lies in a level set, L_2, then, using the arguments to justify PAVA, it is easy to show that

$$\hat{m}(x) = \max_s \min_t \{Av_n([s,t]) : s \le x - h_n, s \in L_1, \ t \ge x + h_n, t \in L_2\}.$$

References

Bennett, G. (1962). *J. Amer. Statist. Assoc.* **57**, 33.
Bhattacharya, P. K. and Gangopadhyay, A. K.(1990). *Ann. Statist.* **18**, 1400.
Brunk, H. D.(1958). *Ann. Math. Statist.* **29**, 437.
Friedman, J. and R. Tibshirani, R. (1984). *Technometrics* **26**, 242.
Hanson, D. L. and Koopmans, L. H.(1965). *Ann. Math. Statist.* **36**, 559.
Hanson, D. L., Pledger, G. and Wright, F. T.(1973). *Ann. Statist.* **1**, 401.
Hoeffding, W.(1963). *J. Amer. Statist. Assoc.* **58**, 13.
Mammen, E.(1991). *Ann. Statist.* **19**, 724.
Mukerjee, H.(1988). *Ann. Statist.* **16**, 741.
Mukerjee, H.(1993a). *Ann. Statist.* **21**, 924.
Mukerjee, H.(1993b). *Ann. Statist.* **21**, 681.
Rao, B. L. S. Prakasa(1969). *Technical Report No. 9, Department of Statistics and Probability, Michigan State University.*
Robertson, T. and Wright, F. T.(1974). *Ann. Statist.* **2** 528.
Wright, F. T.(1981) *Ann. Statist.* **10**, 443.

Chapter 22

Asymptotic Distribution of the Smallest Eigenvalue of Wishart(N, n) When $N, n \to \infty$ Such That $N/n \to 0$

Debashis Paul

University of California, Davis

Suppose that X is an $N \times n$ matrix with i.i.d. standard Gaussian (real or complex) entries. We show that the distribution of the smallest eigenvalue of XX^* (where $*$ denotes the Hermitian transpose of X), after appropriate centering and scaling, converges in distribution to a "reflected" Tracy–Widom distribution, when $N, n \to \infty$ such that $N/n \to 0$.

Contents

1. Introduction

In recent years, several important contributions have been made to the study of asymptotic distribution of the extreme eigenvalues of a random matrix of growing dimension. Pioneering works by Tracy and Widom (1994a; 1994b) led to the discovery of a new class of distributions generally referred to as the Tracy–Widom laws. Tracy and Widom (1994a) derived the distributional limits of the largest eigenvalue of a Complex Gaussian Wigner matrix, as N, the dimension of the matrix, grows without bound. They later derived (Tracy and Widom (1996)) the limiting distribution for the largest eigenvalue of the Symmetric Gaussian Wigner matrix. Johnstone (2001) proved that, when $n, N \to \infty$ such that $N/n \to \gamma \in (0, 1]$, similar distributional limits can be derived for the largest eigenvalue of the complex and real Wishart(N, n) matrices. The approach by Johnstone has been refined further by El Karoui (2005) and Ma (2008) to prove results about the rates of convergence of the appropriately scaled and centered largest eigenvalues of *white* Wishart matrices to the respective Tracy–Widom distributions. El Karoui (2003) extended the result of Johnstone (2001) to show that the same limiting distributions hold even when $N/n \to 0$ (or ∞). The limiting distribution of the smallest eigenvalue of random matrices has also received attention in the literature. Baker, et al. (1998) derived the asymptotic limit of the smallest eigenvalue of complex Wishart(N, n) matrices when $N/n \to (0, 1)$. This article is devoted to deriving the asymptotic distribution of the smallest eigenvalue of the (unscaled) sample covariance matrix $V := XX^*$, where X is an $n \times N$ matrix with i.i.d. (real or complex) $N(0, 1)$ entries, when $N, n \to \infty$ in such a manner that $N/n \to 0$. The derivation makes use of the techniques introduced in Johnstone (2001) and El Karoui (2003). By interchanging the role of N and n, versions of the results can be applied to the smallest nonzero eigenvalue of the matrix $V := XX^*$ when $N/n \to \infty$.

The asymptotic results derived here can be used to test whether the population covariance matrix Σ is the identity matrix when the observations are independent samples from a multivariate Gaussian distribution with covariance matrix Σ and the dimension is much larger than the sample size. Under $H_0 : \Sigma = I$, the limiting distribution of the smallest nonzero eigenvalue of the sample covariance matrix, after appropriate normalization, is a Tracy–Widom law. If $\Sigma \neq I$, the behavior of the distribution of the smallest nonzero eigenvalue will be different and this fact can be utilized to propose an appropriate test.

1.1. *Results on the largest eigenvalue*

Denote the eigenvalues of V in decreasing order by $l_1^N \geq \ldots \geq l_N^N$. Johnstone (2001) proved the following result about the distributional convergence of the centered and scaled largest eigenvalue of the Wishart matrices when the entries of the data matrix are i.i.d. Gaussian (real or complex) with mean zero and variance 1.

Theorem 1. *Suppose that the entries of X are i.i.d. complex $N(0,1)$. If $N/n \to \gamma \in (0,1]$, as $n \to \infty$, then*

$$\frac{l_1^N - \widetilde{\mu}_{nN}}{\widetilde{\sigma}_{np}} \Longrightarrow W_2, \tag{1}$$

where

$$\widetilde{\mu}_{nN} = (\sqrt{n} + \sqrt{N})^2, \qquad \widetilde{\sigma}_{nN} = (\sqrt{n} + \sqrt{N}) \left(\frac{1}{\sqrt{n}} + \frac{1}{\sqrt{N}} \right)^{1/3}.$$

If the entries of X are i.i.d. real $N(0,1)$, then as $n \to \infty$,

$$\frac{l_1^N - \widetilde{\mu}'_{nN}}{\widetilde{\sigma}'_{nN}} \Longrightarrow W_1, \tag{2}$$

where

$$\widetilde{\mu}'_{nN} = (\sqrt{n-1} + \sqrt{N})^2, \qquad \widetilde{\sigma}'_{nN} = (\sqrt{n-1} + \sqrt{N}) \left(\frac{1}{\sqrt{n-1}} + \frac{1}{\sqrt{N}} \right)^{1/3}.$$

Here the random variables W_1 and W_2 follow the Tracy–Widom(1) and Tracy–Widom(2) laws, respectively.

For the case $\gamma = 0$, proved by El Karoui (2003), $\widetilde{\sigma}_{nN}$ can be replaced by $n^{1/2}N^{-1/6}$.

Remark 1. Tracy and Widom (1994a) showed that W_2 appears as the limiting law of the largest eigenvalue of the GUE (Gaussian Unitary Ensemble) after an appropriate normalization. Tracy and Widom (1996) proved that W_1 is the limiting law of the (appropriately normalized) largest eigenvalue of the GOE (Gaussian Orthogonal Ensemble). Let F_1 and F_2 denote the cumulative distribution function (c.d.f.) of W_1 and W_2, respectively. Then F_1 and F_2 are determined through a function $q(x)$ satisfying the *Painlevé II* differential equation:

$$q''(x) = xq(x) + 2q^3(x), \tag{3}$$

with the feature that $q(x) - \mathrm{Ai}(x) \to 0$ as $x \to \infty$, where $\mathrm{Ai}(x)$ denotes the Airy function. A brief introduction to the Airy function is given in Section 5.5. Then F_2 is given by

$$F_2(s) = \exp\left(-\int_s^\infty (x-s)q^2(x)dx\right), \quad s \in \mathbb{R}, \tag{4}$$

and F_1 is given by

$$F_1(s) = \exp\left(-\frac{1}{2}\int_s^\infty (q(x) + (x-s)q^2(x))dx\right), \quad s \in \mathbb{R}. \tag{5}$$

1.2. *Results on the smallest eigenvalue*

Main contribution of this paper are the following theorems.

Theorem 2. *Suppose that, entries of X are i.i.d. complex $N(0,1)$ random variables. Let l_N^N be the smallest eigenvalue of $V = XX^*$. If $N = N(n) \to \infty$, as $n \to \infty$ such that $N/n \to 0$, then*

$$\frac{l_N^N - \mu_{nN}}{\sigma_{nN}} \Longrightarrow W_2', \tag{6}$$

where

$$\mu_{nN} = (\sqrt{n_+} - \sqrt{N_+})^2, \qquad \sigma_{nN} = n_+^{1/2} N_+^{-1/6}, \tag{7}$$

with $n_+ = n + \frac{1}{2}$, $N_+ = N + \frac{1}{2}$, and the c.d.f. G_2 of W_2' satisfies

$$G_2(x) = 1 - F_2(-x), \qquad x \in \mathbb{R}. \tag{8}$$

Theorem 3. *Suppose that, entries of X are i.i.d. real $N(0,1)$ random variables. Let l_N^N be the smallest eigenvalue of $V = XX^T$. If $N = N(n) \to \infty$, as $n \to \infty$ such that $N/n \to 0$, then*

$$\frac{l_N^N - \mu_{n-1,N}}{\sigma_{n-1,N}} \Longrightarrow W_1', \tag{9}$$

where $\mu_{n-1,N}$, $\sigma_{n-1,N}$ are as in (7), with n replaced by $n-1$. The c.d.f. G_1 of W_1' satisfies

$$G_1(x) = 1 - F_1(-x), \qquad x \in \mathbb{R}. \tag{10}$$

In other words, the limiting distribution of the smallest eigenvalue, after normalization, follows the "reflected" Tracy–Widom law, where the reflection is about the point 0. Note that, even though in the statement of Theorem 2 we use n_+ and N_+ to define μ_{nN} and σ_{nN}, we can replace these by n and N, respectively, and the result will still hold. However, the specification of the quantities μ_{nN} and σ_{nN} in terms of n and N can affect the quality of approximation by the limiting Tracy–Widom distribution. Similar considerations apply to Theorem 3.

The organization of this paper is as follows. Section 2 contains the representation of the probability distribution of l_N^N in terms of the Fredholm determinant of a certain operator, and has outlines of the major steps necessary in proving Theorem 2. Section 3 is devoted to proving the results stated in Section 2. In Section 4, the asymptotic analysis is extended to the case of real Gaussians, and Theorem 3 is proved. Section 5 contains some of the details of the derivations.

Remark 2. Henceforth, it is assumed that $N = N(n) \to \infty$ as $n \to \infty$ such that $N/n \to 0$. Also, without loss of generality, it is assumed that $N < n$. Moreover, μ_N and σ_N are used to denote μ_{nN} and σ_{nN}, or $\mu_{n-1,N}$ and $\sigma_{n-1,N}$, as is appropriate. The former is applicable to the complex Gaussian (Theorem 2) setting, and the latter to the real Gaussian (Theorem 3) case. The meaning of μ_N and σ_N will be clear from the context. We shall also drop the superscript N from the quantities l_i^N.

2. Operator kernels and determinantal representation

In this section, the different steps that lead to the proof of (6) are indicated. We primarily follow the approach taken in Johnstone (2001) for establishing (1). This involves two steps, the first of which is to show that the quantity $(l_N - \mu_N)/\sigma_N$ converges in distribution to W_2' when the matrix X has complex $N(0,1)$ entries (i.e., the real and imaginary parts of the random variables are independently distributed as $N(0,1/2)$). The second step uses the tools developed by Tracy and Widom (1996) and Widom (1999) to extend the result to the case when X has i.i.d. real $N(0,1)$ entries. We first give an outline of the first step.

In many of the standard problems in random matrix theory the joint distribution of the sample eigenvalues is explicitly known in terms of the determinant of a certain matrix. From this one can derive the useful representation

$$\mathbb{E}_N\left(\prod_{i=1}^{N}(1+f(l_i))\right) = \det(I + S_N f) \tag{11}$$

where S_N is an explicit operator kernel and det refers to the Fredholm determinant of an operator (see Reed and Simon (1972), or Gohberg and

Krein (1969) for relevant definitions and results on operator theory). When $f(x) = -\chi_{(t,\infty)}(x) = -1_{(t,\infty)}(x)$, one gets

$$P^N(t) = \mathbb{P}_N(l_1 \leq t) = \det(I - S_N\chi_{(t,\infty)}).$$

Similarly, when $f(x) = -\chi_{(0,t)}(x)$, one gets

$$G^N(t) = \mathbb{P}_N(l_N \geq t) = \det(I - S_N\chi_{(0,t)}). \tag{12}$$

Thus, the task of proving (1) and (6) is now reduced to showing that, if one chooses $t = \widetilde{\mu}_{nN} + \widetilde{\sigma}_{nN}s$ for the former, and $t = \mu_{nN} - \sigma_{nN}s$ for the latter, for any $s \in \mathbb{R}$, one gets $F_2(s)$ as the limit for both $P^N(t)$ and $G^N(t)$.

2.1. *Complex Gaussian entries*

Tracy and Widom (1994a) demonstrated that in the case of complex Gaussian entries, it is possible to get a relatively simple expression for the operator kernel S_N appearing in (11). Through the use of some transformations of variables (see Lemma 1 below), it turns out that the proof of convergence of distribution for l_N (or l_1) can be obtained by showing that the *Fredholm determinant* of a certain operator, related to S_N, converges. This is done in Johnstone (2001) by proving the convergence in trace class norm of this operator. Building on Widom (1999), Johnstone gave an integral formula for this operator. The problem is then reduced to showing that this integral converges, uniformly, over certain sets. The proof of Theorem 2 will follow the same pattern, and will rely heavily on El Karoui (2003) and Olver (1974; 1975; 1980) for certain approximations.

As a precursor, we introduce some notations. Let $\alpha := n - N$, and for later use, we define $\kappa_m := m + (\alpha + 1)/2$. When, $m = N$, we drop the subscript, and so $\kappa \equiv \kappa_N = N + (\alpha + 1)/2$. Further, we define

$$\lambda := \frac{\alpha}{2}, \qquad l := \frac{\kappa}{\lambda}, \quad \text{and} \quad \beta := \sqrt{2(l-1)}. \tag{13}$$

Let L_k^α denote the k-th degree *Laguerre polynomial* with parameter α (see Szegö (1939)). We define a weight function $\omega(x)$ by

$$\omega(x) = \begin{cases} x^{\alpha/2}e^{-x/2} & \text{if} \quad x > 0 \\ 0 & \text{if} \quad x \leq 0, \end{cases}$$

and define $\Phi_k(x)$ for $k \geq 0$ by

$$\Phi_k(x) = \sqrt{\frac{k!}{(k+\alpha)!}}\, \omega(x) L_k^\alpha(x). \tag{14}$$

Remark 3. An important fact is, if $\alpha \geq 6$ then the functions $\Phi_k(x)/x$ are twice continuously differentiable, and these functions and their derivatives vanish on the negative real axis. This fact will be crucial in the asymptotic analysis that follows.

The kernel S_N in (11) is given by Mehta (1991):

$$S_N(x, y) = \sum_{k=0}^{N-1} \Phi_k(x)\Phi_k(y). \tag{15}$$

We define $\xi_k(x) := \Phi_k(x)/x$ (defined through limits as $x \to 0$) and $a_N := \sqrt{Nn}$, and

$$\phi(x) := (-1)^N \sqrt{\frac{a_N}{2}} \left(\sqrt{n}\xi_N(x) - \sqrt{N}\xi_{N-1}(x)\right), \tag{16}$$

$$\psi(x) := (-1)^N \sqrt{\frac{a_N}{2}} \left(\sqrt{N}\xi_N(x) - \sqrt{n}\xi_{N-1}(x)\right). \tag{17}$$

From now on, unless otherwise specified, χ will denote the indicator function of the interval $(0, t)$, and $\overline{\chi}$ will denote the indicator function of (s, ∞) (here t and s are generic, with $t > 0$, and $s \in \mathbb{R}$). We also define,

$$\tau(s) := \mu_N - \sigma_N s \quad \text{and} \quad \chi_\tau := 1_{(s, \mu_N/\sigma_N)}. \tag{18}$$

With these notations we have the following lemma whose proof is by direct verification and the observation that S_N is a compact operator on $L^2(0, \infty)$.

Lemma 1. *If $t = \tau(s)$, then the eigenvalues of $S_N\chi$, as an integral operator on $L^2(0, t)$, are the same as the eigenvalues of $S_\tau\chi_\tau$, where S_τ is an integral operator with kernel*

$$S_\tau(x, y) = \sigma_N S_N(\tau(x), \tau(y)) \tag{19}$$

and acting on $L^2(s, \infty)$.

A consequence of Lemma 1 is that, $G^N(\tau(s)) = \det(I - S_\tau\chi_\tau)$, where G^N is defined through (12). The following lemma gives an integral representation of the kernel S_τ which will be used for subsequent analyses. Note that this is similar to the representation for the corresponding kernel used by Johnstone (2001).

Lemma 2. *Let* $\phi_\tau(x) = \sigma_N\phi(\tau(x)) = \sigma_N\phi(\mu_N - \sigma_N x)$ *and* $\psi_\tau(x) = \sigma_N\psi(\tau(x)) = \sigma_N\psi(\mu_N - \sigma_N x)$. *Then for* $\alpha \geq 6$,

$$S_\tau(x,y) = -\int_0^\infty [\phi_\tau(x+z)\psi_\tau(y+z) + \psi_\tau(x+z)\phi_\tau(y+z)]\,dz. \quad (20)$$

Notice that the integral is really over the interval $[0, \mu_N/\sigma_N - (x \vee y)]$ and it is well defined.

Proof of Lemma 2. Widom (1999) gives the following representation for the kernel of the commutator $[D, S_N]$ of S_N,

$$\left(\frac{\partial}{\partial x} + \frac{\partial}{\partial y}\right) S_N(x,y) = -\phi(x)\psi(y) - \psi(x)\phi(y).$$

Even though Widom's formula is for $x, y \in \mathbb{R}_+$, by Remark 2.1, it can be extended to the whole real line. Therefore, since $d\tau(x)/dx = -\sigma_N$,

$$\left(\frac{\partial}{\partial x} + \frac{\partial}{\partial y}\right) S_\tau(x,y) = \phi_\tau(x)\psi_\tau(y) + \psi_\tau(x)\phi_\tau(y). \quad (21)$$

Now we argue as in Tracy and Widom (1996). Construct a kernel $T_\tau(x,y)$ by defining it as the right hand side of (20). Since

$$\begin{aligned}&\left(\frac{\partial}{\partial x} + \frac{\partial}{\partial y}\right) [\phi_\tau(x+z)\psi_\tau(y+z) + \psi_\tau(x+z)\phi_\tau(y+z)]\\ &= \frac{\partial}{\partial z} [\phi_\tau(x+z)\psi_\tau(y+z) + \psi_\tau(x+z)\phi_\tau(y+z)],\end{aligned}$$

it follows that

$$\begin{aligned}&\left(\frac{\partial}{\partial x} + \frac{\partial}{\partial y}\right) T_\tau(x,y)\\ &= -\int_0^\infty \frac{\partial}{\partial z} [\phi_\tau(x+z)\psi_\tau(y+z) + \psi_\tau(x+z)\phi_\tau(y+z)]\,dz\\ &= \phi_\tau(x)\psi_\tau(y) + \psi_\tau(x)\phi_\tau(y),\end{aligned}$$

where the last step follows from the fact that $\phi_\tau, \phi'_\tau, \psi_\tau$ and ψ'_τ all vanish at ∞. Therefore, S_τ and T_τ both satisfy (21), and so they differ only by a function of the form $g(x-y)$. Since both S_τ and T_τ converge to zero as $x, y \to \infty$ independently, the function g must be identically zero, thus concluding the proof. □

2.2. *Representation in terms of Airy operator*

Let $\overline{S}$ denote the Airy operator with kernel

$$\overline{S}(x,y) = \frac{\mathrm{Ai}(x)\mathrm{Ai}'(y) - \mathrm{Ai}(y)\mathrm{Ai}'(x)}{x-y} = \int_0^\infty \mathrm{Ai}(x+u)\mathrm{Ai}(y+u)du, \quad (22)$$

where Ai is the Airy function. It is known (Tracy and Widom, 1994a) that viewing $\overline{S}$ as an operator on $L^2(s,\infty)$, one has,

$$\det(I - \overline{S\chi}) = F_2(s),$$

where W_2 is the Tracy–Widom(2) law. So, in order to prove Theorem 2, it suffices to show that, for all $s \in \mathbb{R}$,

$$\det(I - S_\tau\chi_\tau) \to \det(I - \overline{S\chi}). \quad (23)$$

The key step is to show that, the operator $S_\tau\chi_\tau$ converges to $\overline{S\chi}$ in trace class norm. This is because, as Johnstone (2001) has shown, this implies (23). Following Johnstone (2001), the task can be reduced to showing that the operators H_τ^ϕ and H_τ^ψ on $L^2(s,\infty)$, with kernels

$$H_\tau^\phi(x,y) = \phi_\tau(x+y-s) \qquad \text{and} \qquad H_\tau^\psi(x,y) = -\psi_\tau(x+y-s),$$

respectively, converge in Hilbert-Schmidt norm on $L^2(s,\infty)$ to the operator H, with kernel

$$H(x,y) = \frac{1}{\sqrt{2}}\mathrm{Ai}(x+y-s).$$

This is because, from (22) and (20), as operators on $L^2(s,\infty)$, $\overline{S}$ and S_τ can be represented as

$$S_\tau = H_\tau^\phi H_\tau^\psi + H_\tau^\psi H_\tau^\phi, \qquad \overline{S} = 2H^2. \quad (24)$$

Remark 4. From this point on, for simplicity of derivations, it is crucial to follow the convention that N takes even integer values only. The result for arbitrary sequences $N = N(n)$ can be easily obtained from this by using the interlacing inequality

$$l_{N+1}^{N+1} \leq l_N^N \leq l_{N-1}^{N-1}. \quad (25)$$

To see the importance of this requirement, we observe that (as shown later) the asymptotic behavior of the kernel S_τ is connected to the asymptotic behavior of the approximation (49) of the *Whittaker functions*. If we choose N to be even, then the value of the term $\sin(\pi\lambda\beta^2/2)$ in the aforementioned equation becomes 1 or -1, according to the choice of λ and β, (in (13) and (77), respectively) thereby considerably simplifying the steps in proving the key result Proposition 2. Thus, henceforth, we assume that N is an even integer.

The following lemmas are the building blocks for showing the convergence of operator kernels in various function space norms.

Lemma 3. *As $n \to \infty$, for every $s \in \mathbb{R}$,*

$$\phi_\tau(s) \to \frac{1}{\sqrt{2}}\mathrm{Ai}(s), \qquad \psi_\tau(s) \to -\frac{1}{\sqrt{2}}\mathrm{Ai}(s). \tag{26}$$

Lemma 4. *For every $M > 0$ (fixed), for some positive n_0, uniformly in $n > n_0$, and for every $s_0 \in \mathbb{R}$,*

$$|\phi_\tau(s)|, |\psi_\tau(s)| \leq C_1(s_0)e^{-\frac{1}{3}s^{3/2}} \qquad \text{for } s \in [s_0, \mu_N/\sigma_N - M] \tag{27}$$

$$\leq C_2(M)n2^{-n/4} \qquad \text{for } s \in [\mu_N/\sigma_N - M, \mu_N/\sigma_N] \tag{28}$$

where $C_1(s_0), C_2(M) > 0$ are constants independent of N, n and s.

Remark 5. To see how Lemma 3 and Lemma 4 will suffice to conclude the argument, recall that $\phi_\tau(s)$ and $\psi_\tau(s)$ both vanish on $[\mu_N/\sigma_N, \infty)$. On the other hand, $\mathrm{Ai}(s)$ converges to zero very fast as $s \to \infty$ (see Section 3). Thus, to show that the Hilbert-Schmidt norm of operators $H_\tau^\phi - H$ and $H_\tau^\psi - H$ converge to zero, it is enough to limit attention to the (double) integral of the squares of these kernels over intervals of the form $[s_0, \mu_N/\sigma_N]^2$. Then by (26) and (27), it is possible to apply the Dominated Convergence Theorem to the kernels restricted to $[s_0, \mu_N/\sigma_N - M]^2$. To show that the remaining part is negligible, we simply use the finite length of the interval $[\mu_N/\sigma_N - M, \mu_N/\sigma_N]$, the fact that the interval $[s_0, \mu_N/\sigma_N]$ has length of the order $\mu_N/\sigma_N \sim n^{1/2}N^{1/6} = o(n)$, and the bound (28).

The following proposition allows us to establish convergence of S_τ to $\overline{S}$ in various function space norms.

Proposition 1. As $n \to \infty$, for every $x, y \in \mathbb{R}$,

$$S_\tau(x, y) \to \overline{S}(x, y). \tag{29}$$

Moreover, for every $M > 0$, for every $s_0 \in \mathbb{R}$, there is $n_0 \geq 7$ such that, for every $n \geq n_0$, uniformly on $(x, y) \in [s_0, \infty) \times [s_0, \infty)$, the following bound holds.

$$|S_\tau(x,y)| \le C_3(s_0)e^{-\frac{1}{3}(x^{3/2}+y^{3/2})} + C_4(M)h_{up}(x)h_{up}(y), \tag{30}$$

where

$$h_{up}(x) = \begin{cases} x2^{-x/4} & if\ x \ge \frac{1}{4}\log 2 \\ h_{up}(\frac{1}{4}\log 2) & if\ x < \frac{1}{4}\log 2. \end{cases} \tag{31}$$

Proof. The pointwise convergence (29) follows easily from Lemma 3 and Lemma 4 by the Dominated Convergence Theorem. Since $(x+y)^{3/2} \ge x^{3/2} + y^{3/2}$, for $x, z > 0$, the representation (20) can be used to check that uniformly over $(x,y) \in [s_0, \mu_N/\sigma_N] \times [s_0, \mu_N/\sigma_N]$,

$$|S_\tau(x,y)| \le C_3(s_0)e^{-\frac{1}{3}(x^{3/2}+y^{3/2})} + C_4(M)n^2 2^{-n/2} \tag{32}$$

for some constants $C_3(s_0)$ and $C_4(M)$. (30) now follows by noting that, $S_\tau(x,y) = 0$ if $\min\{x,y\} > \mu_N/\sigma_N$, the fact that h_{up} is a non-increasing function with $h_{up}(n) = n2^{-n/4}$, and that $n \ge \mu_N/\sigma_N$ for $n \ge n_0$. □

3. Asymptotic analysis: Complex case

In this section, we prove Lemmas 3 and 4. The proofs use the approach of Section 4 of El Karoui (2003) and Section 5 of Johnstone (2001). As a first step, we define

$$f_N(z) = \sigma_N^{-1/2}\frac{1}{\sqrt{n!N!}}W_{N+\frac{\alpha+1}{2},\frac{\alpha}{2}}(z), \tag{33}$$

$$h_N(z) = \sigma_N^{-1/2}\frac{1}{\sqrt{(n-1)!(N-1)!}}W_{N-1+\frac{\alpha+1}{2},\frac{\alpha}{2}}(z), \tag{34}$$

where $W_{a,b}$ are the *Whittaker functions*. Here dependence on n is suppressed, but it must be remembered that $f_N(z) = f_{n,N}(z)$ and $h_N(z) = h_{n,N}(z)$ and $n = n_N$ satisfies $N/n \to 0$. The following result (Temme (1990), formula (3.1)) provides a connection between Whittaker functions and Laguerre polynomials. Recall that $\kappa_m = m + (\alpha+1)/2$ and $\lambda = \alpha/2$, and define

$$w_m(z) \stackrel{def}{=} z^{(\alpha+1)/2}e^{-z/2}L_m^\alpha(z) = \frac{(-1)^m}{m!}W_{\kappa_m,\lambda}(z), \qquad z \in \mathbb{R}_+ \tag{35}$$

for m and α integers. Note that $w_m(z) = \sqrt{(m+\alpha)!/m!}\ z^{1/2}\Phi_m(z)$ for $z > 0$. The primary aim is to establish the following result.

Proposition 2. Let $z = \tau(s) = \mu_N - \sigma_N s$ (refer to (18)). As $n \to \infty$, for every $s \in \mathbb{R}$,

$$f_N(z) \to \mathrm{Ai}(s) \tag{36}$$

$$h_N(z) \to -\mathrm{Ai}(s) \tag{37}$$

Proposition 2, together with (14), (16) and (17), will be used to establish Lemma 3. In this direction, we begin with a differential equation for $w_m(z)$. Dropping the subscript m for convenience, we have the *Whittaker differential equation* for w:

$$\frac{d^2 w}{dz^2} = \left(\frac{1}{4} - \frac{\kappa}{z} + \frac{\lambda^2 - 1/4}{z^2}\right) w. \tag{38}$$

Following Olver (1980), we observe that the two *turning points* of (38), (i.e., the zeros of the ODE, thus named due to the fact that each of them separates an interval on which the solution decays exponentially from another where the solution oscillates), are

$$\xi_1 = 2l - 2\sqrt{l^2 - 1} \qquad \text{and} \qquad \xi_2 = 2l + 2\sqrt{l^2 - 1},$$

after the substitution $\xi = z/\lambda$ since $l = \kappa/\lambda$ (cf. (13)). Of interest is an asymptotic expansion of the functions $w_N(z)$ and $w_{N-1}(z)$ in a neighborhood of the point $\lambda\xi_1$. From (13), it follows that, $\mu_N = \lambda\xi_1$. Furthermore, as $n, N \to \infty$ such that $N/n \to 0$, we have

$$\alpha \sim n, \qquad \lambda = \frac{n-N}{2} \sim \frac{n}{2}, \qquad \kappa = \frac{n+N}{2} \sim \frac{n}{2},$$
$$l \sim 1 + \frac{2N}{n}, \qquad \sqrt{l^2 - 1} \sim 2\sqrt{\frac{N}{n}}, \tag{39}$$

where the notation $\sim$ means that the ratio of the LHS and RHS converges to 1. Notice that the two turning points ξ_1 and ξ_2 coalesce at 2 under the stated assumptions of Theorem 2, and this is the point where the current analysis technically differs from that of Johnstone (2001).

Let $W(\xi) := w(\lambda\xi)$. Then the differential equation (38) is converted into

$$\frac{d^2 W}{d\xi^2} = \left(\lambda^2 g(\xi) - \frac{1}{4\xi^2}\right) W, \tag{40}$$

where

$$g(\xi) = \frac{(\xi - \xi_1)(\xi - \xi_2)}{4\xi^2}. \tag{41}$$

We define an auxiliary variable ζ through the equations

$$\int_{\beta}^{\zeta}(x^2-\beta^2)^{1/2}dx = \int_{\xi_2}^{\xi} g^{1/2}(x)dx, \qquad \text{if } \xi_2 \le \xi < \infty, \tag{42}$$

$$\int_{-\beta}^{\zeta}(x^2-\beta^2)^{1/2}dx = \int_{\xi_1}^{\xi} (-g)^{1/2}(x)dx, \qquad \text{if } \xi_1 \le \xi \le \xi_2, \tag{43}$$

$$\int_{\zeta}^{-\beta}(x^2-\beta^2)^{1/2}dx = \int_{\xi}^{\xi_1} g^{1/2}(x)dx, \qquad \text{if } 0 < \xi \le \xi_1. \tag{44}$$

Olver (1980) uses the *Liouville transform* (treating ξ as a function of ζ):

$$W = \left(\frac{d\xi}{d\zeta}\right)^{1/2} V, \qquad \left(\frac{d\zeta}{d\xi}\right)^2 = \frac{g(\xi)}{\zeta^2-\beta^2} \tag{45}$$

to transform the differential equation (40) into

$$\frac{d^2V}{d\zeta^2} = \{\lambda^2(\zeta^2-\beta^2) + \Psi(\beta^2,\zeta)\}V. \tag{46}$$

Here

$$\Psi(\beta^2,\zeta) = -\frac{(\xi')^2}{4\xi^2} + (\xi')^{1/2}\frac{d^2}{d\zeta^2}\left((\xi')^{-1/2}\right),$$

where ξ' denotes derivative of ξ with respect to ζ. The advantage of this transformation is that, useful bounds on the solution of (46) are available.

For the derivation of the results here, we are concerned with

$$z = \tau(s) = \mu_N - \sigma_N s, \qquad \text{or equivalently,} \qquad \xi = \xi_1 - \frac{\sigma_N}{\lambda}s$$

when $s \in [s_0, \mu_N/\sigma_N)$, so that $z > 0$. Therefore, among the equations defining ζ, the attention can be restricted to (43) and (44) only. The reason is, for every fixed $s_0 \in \mathbb{R}$, for sufficiently large values of n, $\xi \in (0, (\xi_1+\xi_2)/2)$ uniformly over $s \in [s_0, \mu_N/\sigma_N)$. Important facts about this transformation are listed below.

(1) ζ is a monotonically increasing function of ξ;
(2) ζ ranges from $-\infty$ to ∞ as ξ ranges from 0 to ∞;
(3) $\zeta < 0$ or > 0 according as ξ is less than or greater than $(\xi_1+\xi_2)/2$;
(4) $|\zeta| \ge 1$ if either $\xi \le \xi_1$ or $\xi \ge \xi_2$.

Olver (1980) provides an asymptotic expansion of the solution of (46), where he expands the Whittaker functions in terms of *parabolic cylinder functions.*

He shows that, for $\zeta < 0$, the two linearly independent solutions of the equation (46) are of the form,

$$\begin{aligned} v_2(\lambda^2, \beta^2, \zeta) &= U(-\frac{1}{2}\lambda\beta^2, -\zeta\sqrt{2\lambda}) + \varepsilon_2(\lambda^2, \beta^2, \zeta) \\ \text{and} \quad \overline{v}_2(\lambda^2, \beta^2, \zeta) &= \overline{U}(-\frac{1}{2}\lambda\beta^2, -\zeta\sqrt{2\lambda}) + \overline{\varepsilon}_2(\lambda^2, \beta^2, \zeta), \end{aligned} \tag{47}$$

(cf. equations (3.13) and (3.14) of Olver (1980)), where $U(b, x)$ and $\overline{U}(b, x)$ are parabolic cylinder functions; and ε_2 and $\overline{\varepsilon}_2$ are errors in approximation with bounds given by

$$\begin{aligned} \varepsilon_2(\lambda^2, \beta^2, \zeta) &= \mathbf{E}^{-1}(-\frac{1}{2}\lambda\beta^2, -\zeta\sqrt{2\lambda})\mathbf{M}(-\frac{1}{2}\lambda\beta^2, -\zeta\sqrt{2\lambda})O(\lambda^{-2/3}); \quad (48) \\ \overline{\varepsilon}_2(\lambda^2, \beta^2, \zeta) &= \mathbf{E}(-\frac{1}{2}\lambda\beta^2, -\zeta\sqrt{2\lambda})\mathbf{M}(-\frac{1}{2}\lambda\beta^2, -\zeta\sqrt{2\lambda})O(\lambda^{-2/3}); \end{aligned}$$

(cf. (3.15) and (3.16) of Olver (1980)); for large values of λ, and uniformly with respect to $\beta \in [0, A]$ $(A > 0)$ and $\zeta \in (-\infty, 0]$. Here $\mathbf{E}$ and $\mathbf{M}$ are *weight and modulus functions* associated with U (see Olver (1975), p. 156). It is shown in Section 5.3 that only one of these solutions is recessive, namely, v_2. A crucial point of the analysis here is that, for every fixed value of α and N, $z^{1/2}\Phi_N(z) \to 0$ and $z^{1/2}\Phi_{N-1}(z) \to 0$ as $\xi \to 0+$. This means that the solution we are looking for is recessive. This ensures that the function $W_{\kappa,\lambda}(\lambda\xi)$ is a multiple of $v_2(\lambda^2, \beta^2, \zeta)(d\xi/d\zeta)^{1/2}$. Equation (3.18) of Olver (1980) gives an exact form for this multiplicative constant. Accordingly,

$$\begin{aligned} W_{\kappa,\lambda}(\lambda\xi) &= [\sin(\frac{1}{2}\pi\lambda\beta^2) + O(\lambda^{-2/3})](2\lambda)^{1/4}\{\lambda(2 + \frac{1}{2}\beta^2)/e\}^{\lambda(1+\beta^2/4)} \\ &\quad \times \left(\frac{\zeta^2 - \beta^2}{\xi^2 - 4l\xi + 4}\right)^{1/4} \xi^{1/2}\{U(-\frac{1}{2}\lambda\beta^2, -\zeta\sqrt{2\lambda}) + \varepsilon_2(\lambda^2, \beta^2, \zeta)\}, \end{aligned} \tag{49}$$

the O-terms here being independent of ζ and uniform with respect to $\beta \in [0, A]$.

Remark 6. Since the focus is on asymptotic solutions to (46) when $\zeta \to -\infty$, or equivalently, solutions to (40) as $\xi \to 0+$, one could use a different Whittaker function $M_{\kappa,\lambda}$ that satisfies the relationship

$$z^{(\alpha+1)/2}e^{-z/2}L_m^{\alpha}(z) = \frac{(m+\alpha)!}{m!\alpha!}M_{\kappa_m,\alpha}$$

(Temme (1990), equation (3.1)). As a standard solution to (40) (recessive as $\xi \to 0+$), it has the expansion (cf. equation (3.25) of Olver (1980))

$$M_{\kappa,\lambda}(\lambda\xi) = (2\lambda)^{3/4}(2\lambda/e)^{\lambda(1-\frac{1}{4}\beta^2)}(1+\frac{1}{4}\beta^2)^{-\lambda(1+\frac{1}{4}\beta^2)} \\ \times\left(\frac{\zeta^2-\beta^2}{\xi^2-4l\xi+4}\right)^{1/4}\xi^{1/2}\{U(-\frac{1}{2}\lambda\beta^2,-\zeta\sqrt{2\lambda})+\varepsilon_2(\lambda^2,\beta^2,\zeta)\}, \tag{50}$$

valid when $\beta \in [0,\infty)$ and $\zeta \in (-\infty, 0]$. However, from an asymptotic analysis point of view, the expansion (49) is equivalent to (50), except for a sign change. Moreover, (49) allows us to use some of the results from El Karoui (2003).

The final step of the asymptotic analysis of f_N and h_N is to study the asymptotic behavior of parabolic cylinder functions in terms of Airy functions. Recall that, since s is restricted to the interval $[s_0, \mu_N/\sigma_N]$ for some $s_0 \in \mathbb{R}$, for sufficiently large n, $\zeta < 0$, or $\xi < (\xi_1+\xi_2)/2$. Then, equation (5.12) of Olver (1975) gives the following approximation for the function $U(-\frac{1}{2}\lambda\beta^2, -\zeta\sqrt{2\lambda})$.

$$U(-\frac{1}{2}\lambda\beta^2,-\zeta\sqrt{2\lambda}) = \Xi\ [\mathrm{Ai}(\theta)+\mathcal{E}^{-1}(\theta)\mathcal{M}(\theta)O((\lambda\beta^2)^{-1})], \tag{51}$$

where

$$\Xi = \frac{(4\pi)^{1/4}}{(\lambda\beta^2)^{1/12}}[\Gamma(\frac{1}{2}(1+\lambda\beta^2))]^{1/2}\beta^{1/2}\left(\frac{\eta}{\zeta^2-\beta^2}\right)^{1/4}, \tag{52}$$

and the functions $\mathcal{E}$ and $\mathcal{M}$ satisfy the relations (cf. equations (5.15) and (5.16) of Olver (1975)),

$$\mathcal{E}(x) = \left(\frac{\mathrm{Bi}(x)}{\mathrm{Ai}(x)}\right)^{1/2}, \qquad \mathcal{M}(x) = \{2\mathrm{Ai}(x)\mathrm{Bi}(x)\}^{1/2}, \qquad (x \geq c), \tag{53}$$

$$\mathcal{E}(x) = 1, \qquad \mathcal{M}(x) = \{\mathrm{Ai}^2(x)+\mathrm{Bi}^2(x)\}^{1/2}, \qquad (x \leq c), \tag{54}$$

where $c(= -0.36605...)$ is the smallest root, in absolute value, of the equation $\mathrm{Ai}(x) = \mathrm{Bi}(x)$ where $Bi(x)$ is defined in (125).

In order to carry out this analysis, and obtain bounds on the errors, Olver (1975) introduced a new auxiliary variable η which is defined in terms of ξ such that,

$$\frac{2}{3}\eta^{3/2}\beta^2 = \int_{\xi}^{\xi_1} g^{1/2}(t)dt, \qquad \text{for } \ 0 < \xi \leq \xi_1 \tag{55}$$

$$\frac{2}{3}\eta^{3/2}\beta^2 = -\int_{\xi_1}^{\xi} (-g(t))^{1/2}dt, \qquad \text{for } \ \xi_1 \leq \xi \leq \xi_2, \tag{56}$$

where g is defined through (41). We also introduce a new variable θ by

$$\theta = (\lambda\beta^2)^{2/3}\eta. \tag{57}$$

The significance of this transformation will become clear in the next few sections. A first goal is to study the asymptotic behavior of η by establishing appropriate bounds for θ.

3.1. *Asymptotics for* η

3.1.1. *Case :* $s \in [0, \mu_N/\sigma_N)$

In this case $\xi \leq \xi_1$ and η is defined by equation (55).

Lemma 5. *For every* $M > 0$*, and* $0 < \delta < 1$*, there is* $n_0 \geq 1$ *such that for* $n \geq n_0$*, uniformly over* $s \in [0, \mu_N/\sigma_N - M]$*,*

$$\theta^{3/2} = \lambda\beta^2\eta^{3/2} \geq \sqrt{1-\delta}s^{3/2}. \tag{58}$$

Proof. Set $x(u) = \xi_1 - (\sigma_N/\lambda)u$ for $u \geq 0$. Then,

$$\begin{aligned} \sigma_N^2 g(x(u)) = \sigma_N^2 \frac{(x(u)-\xi_1)(x(u)-\xi_2)}{4\xi^2} &= \frac{u\sigma_N^3}{\lambda}\frac{(\xi_2-\xi_1)+\frac{\sigma_N}{\lambda}u}{4\left(\xi_1-\frac{\sigma_N}{\lambda}u\right)^2} \\ &= \frac{u\sigma_N^3}{\lambda}\frac{4\sqrt{l^2-1}+\frac{\sigma_N}{\lambda}u}{4\left(\frac{\mu_N}{\lambda}\right)^2\left(1-\frac{\sigma_N}{\mu_N}u\right)^2}. \end{aligned} \tag{59}$$

Further, by (13) and (39),

$$2\sqrt{l^2-1} = \sqrt{2}\beta(l+1)^{1/2} \sim 2\beta \ \text{ and } \ \lambda\beta^2 = 2N+1,$$

implying that

$$\beta = \sqrt{\frac{2N+1}{\lambda}} \sim 2\sqrt{\frac{N}{n}}\,.$$

If $u \in [0, \mu_N/\sigma_N - M]$, then for any $\delta > 0$,

$$\left(\frac{\mu_N}{\lambda}\right)^2\left(1-\frac{\sigma_N}{\mu_N}u\right)^2 \leq \left(\frac{\mu_N}{\lambda}\right)^2 \leq 4(1-\delta/2)^{-1},$$

for sufficiently large n, since $\mu_N/\lambda = \xi_1 \sim 2$. Thus, from (59), for n sufficiently large, uniformly over $u \in [0, \mu_N/\sigma_N - M]$,

$$\sigma_N^2 g(x(u)) \geq \frac{\sigma_N^3\sqrt{l^2-1}}{\lambda} u \left(\frac{\mu_N}{\lambda}\right)^{-2} \left(1 - \frac{\sigma_N}{\mu_N}u\right)^{-2} > (1-\delta/2)\frac{\sigma_N^3\beta}{4\lambda}u. \tag{60}$$

Now, since

$$\frac{\sigma_N^3\beta}{4\lambda} \sim \frac{(n^{1/2}N^{-1/6})^3(2N^{1/2}n^{-1/2})}{4\left(\frac{n-N}{2}\right)} \sim \frac{n^{3/2}N^{-1/2}N^{1/2}n^{-1/2}}{n} \to 1, \tag{61}$$

for large enough n, it follows that

$$\sigma_N^2 g(x(u)) \geq (1-\delta)u \qquad \text{uniformly in } \; u \in [0, \mu_N/\sigma_N - M]. \tag{62}$$

Since $\xi = x(s) = \xi_1 - (\sigma_N/\lambda)s$, from (62), we get (for large enough n)

$$\begin{aligned}\frac{2}{3}\lambda\eta^{3/2}\beta^2 &= \frac{\lambda}{\sigma_N}\int_\xi^{\xi_1}\sqrt{\sigma_N^2 g(t)}dt \\ &= \int_0^s \sqrt{\sigma_N^2 g\left(\xi_1 - \frac{\sigma_N}{\lambda}u\right)}\,du \geq \sqrt{1-\delta}\int_0^s \sqrt{u}\,du,\end{aligned}$$

which establishes (58). □

Case: $s \in [\mu_N/\sigma_N - M, \mu_N/\sigma_N)$
In this situation (58) is not adequate to give a usable lower bound for θ. Instead, we have the following.

Lemma 6. *For every $M > 0$, and $\delta \in (0,1)$, there is an $n_0 \geq 1$, such that for $n \geq n_0$, uniformly over $s \in [\mu_N/\sigma_N - M, \mu_N/\sigma_N)$,*

$$\frac{2}{3}\theta^{3/2} = \frac{2}{3}\lambda\beta^2\eta^{3/2} \geq \sqrt{1-\delta}\;\frac{n}{2}\;\left[-\log\left(\frac{\mu_N/\sigma_N - s}{2M}\right)\right]. \tag{63}$$

Proof. As before, let $\xi = x(u) = \xi_1 - \frac{\sigma_N}{\lambda}u$, with $u \in [\mu_N/\sigma_N - 2M, \mu_N/\sigma_N)$. Then, from (59), for sufficiently large n,

$$\sigma_N^2 g(x(u)) \geq \frac{u}{\beta}\,\frac{\sigma_N^3\beta}{4\lambda}\,\frac{\frac{\sigma_N}{\lambda}u}{\left(\xi_1 - \frac{\sigma_N}{\lambda}u\right)^2} \geq (1-\delta/2)\frac{\mu_N}{\sigma_N}\sqrt{\frac{n}{N}}\left(\xi_1 - \frac{\sigma_N}{\lambda}u\right)^{-2}, \tag{64}$$

where the last step follows from (61), the fact that $\beta \sim 2\sqrt{\frac{N}{n}}$, and that,

$$\frac{\sigma_N}{\lambda}u \in \left[\xi_1 - 2M\frac{\sigma_N}{\lambda}, \xi_1\right) \sim \xi_1 = \frac{\mu_N}{\lambda} \sim 2,$$

uniformly over $u \in [\mu_N/\sigma_N - 2M, \mu_N/\sigma_N)$. Consequently,

$$\begin{aligned}
&\frac{2}{3}\lambda\eta^{3/2}\beta^2 \\
&= \frac{\lambda}{\sigma_N}\int_{\xi}^{\xi_1}\sqrt{\sigma_N^2 g(x)}dx \;=\; \int_0^s \sqrt{\sigma_N^2 g\left(\xi_1 - \frac{\sigma_N}{\lambda}t\right)}dt \\
&\geq \sqrt{1-\delta/2}\sqrt{\frac{\mu_N}{\sigma_N}}\left(\frac{n}{N}\right)^{1/4}\int_{\frac{\mu_N}{\sigma_N}-2M}^{s}\left(\xi_1 - \frac{\sigma_N}{\lambda}t\right)^{-1} dt \\
&= \sqrt{1-\delta/2}\sqrt{\frac{\mu_N}{\sigma_N}}\left(\frac{n}{N}\right)^{1/4}\frac{\lambda}{\sigma_N}\int_{\xi_1-\frac{\sigma_N}{\lambda}2M}^{\frac{\sigma_N}{\lambda}s}\frac{1}{\xi_1 - u}du \\
&= \sqrt{1-\delta/2}\sqrt{\frac{\mu_N}{\sigma_N}}\left(\frac{n}{N}\right)^{1/4}\frac{\lambda}{\sigma_N}\left[\log\left(\frac{\sigma_N}{\lambda}2M\right) - \log\left(\xi_1 - \frac{\sigma_N}{\lambda}s\right)\right] \\
&\geq \sqrt{1-\delta/2}\,\frac{\lambda}{\mu_N}\left(\frac{\mu_N}{\sigma_N}\right)^{3/2}\left(\frac{n}{N}\right)^{1/4}\left[-\log\left(\frac{\mu_N/\sigma_N - s}{2M}\right)\right].
\end{aligned}$$

Since $\lambda/\mu_N \to 1/2$, and $(\mu_N/\sigma_N)^{3/2}(n/N)^{1/4} \sim n$, (63) follows. □

3.1.2. **Case:** $s \in [s_0, s_1]$

This situation is relatively simple. Only concern is with the sign of s. So, if s_0 and s_1 are of opposite signs, just split the interval as $[s_0, 0] \cup [0, s_1]$. Note that on the interval $[0, s_1]$, the inequality (58) still holds. In fact, it is trivial to check that in this case, uniformly on $s \in [0, s_1]$, as $n \to \infty$,

$$\theta^{3/2} = \lambda\beta^2\eta^{3/2} = s^{3/2}(1 + o(1)). \tag{65}$$

To verify this, recall (59) and (61), and observe that

$$\xi_1 \gg (\xi_2 - \xi_1) \sim 4\beta \gg \frac{\sigma_N}{\lambda}s \qquad \text{for} \quad s \in [0, s_1].$$

The situation where $s \in [s_0, 0]$ is different only in that, now

$$\xi_1 \leq \xi = x(s) = \xi_1 - \frac{\sigma_N}{\lambda}s \leq \frac{\xi_1 + \xi_2}{2}$$

for sufficiently large n (so that $-\zeta \in (0, 1)$). Therefore, η satisfies the equation (56), where $\eta^{3/2}$ must be interpreted as $-|\eta|^{3/2}$ since $\eta < 0$. Now, by following essentially the same arguments as in case of the interval $[0, s_1]$ with $s_1 > 0$, it can be shown that (65) holds uniformly over $s \in [s_0, 0]$

3.2. *Asymptotics for f_N*

In this subsection, as in the following one, we rely heavily on the results derived in the appendices of El Karoui (2003). The first step is to show that the function f_N can be bounded uniformly by some function over different range of values for s. Then (36) would be proved by using the relevant information that we gather in the process. For the rest of this section Δ is used as a generic constant not dependent on n,N or s and may change from one line to the next.

3.2.1. *Uniform bounds on f_N*

The strategy here is to consider the cases $s \in [s_1, \mu_N/\sigma_N)$ (where $s_1 > 0$), and $s \in [s_0, s_1]$ for (potentially) $s_0 < 0$ separately, and then use Lemma 5 and Lemma 6 for the first case, and (65) in the second scenario.

Case: $s \in [s_1, \mu_N/\sigma_N)$, with $s_1 > 0$

Lemma 7. *Set*

$$\xi = \frac{\tau(s)}{\lambda} = \frac{\mu_N}{\lambda} - \frac{\sigma_N}{\lambda}s = \xi_1 - \frac{\sigma_N}{\lambda}s.$$

For every $M > 0$ and $s_1 > 0$, for all sufficiently large n, we have

$$|f_N(\lambda\xi)| \leq \Delta(s_1)\exp\left(-\frac{1}{2}s^{3/2}\right), \qquad s \in [s_1, \mu_N/\sigma_N - M) \tag{66}$$

$$\leq \Delta(M)\left(1 - \frac{\sigma_N}{\mu_N}s\right)^{1/2} \exp\{-\frac{n}{3}\left[-\log\left(\frac{\mu_N/\sigma_N - s}{2M}\right)\right]\},$$
$$s \in [\mu_N/\sigma_N - M, \mu_N/\sigma_N). \tag{67}$$

Proof. From the definition of $U(b, x)$ on p. 156 of Olver (1975), one has $|U(b, x)| \leq \mathbf{E}^{-1}(b, x)M(b, x)$. Hence, for n large enough so that $\zeta < 0$, from (49), it follows that

$$|W_{\kappa,\lambda}(\lambda x)| \leq (2\lambda)^{1/4}\{\lambda(2 + \frac{1}{2}\beta^2)/e\}^{\lambda(1+\beta^2/4)}\left(\frac{\zeta^2 - \beta^2}{\xi^2 - 4l\xi + 4}\right)^{1/4}\xi^{1/2}$$
$$\times \mathbf{ME}^{-1}(1 + O(\lambda^{-2/3})). \tag{68}$$

From (48), (68), (98) and (99),

$$|f_N(z)| \leq K_{n,N}\xi^{1/2}\left(\frac{\eta}{\xi^2 - 4l\xi + 4}\right)^{1/4}\mathcal{E}^{-1}(\theta)\mathcal{M}(\theta)(1+O((\lambda\beta^2)^{-1}\vee\lambda^{-2/3}))$$

where

$$K_{n,N} = \frac{(2\lambda)^{1/4}\{\lambda(2+\frac{1}{2}\beta^2)/e\}^{\lambda(1+\beta^2/4)}(4\pi)^{1/4}[\Gamma(\frac{1}{2}(1+\lambda\beta^2))]^{1/2}\beta^{1/2}}{(\lambda\beta^2)^{1/12}\sigma_N^{1/2}\sqrt{n!N!}}. \tag{69}$$

In Section 5.2 it is shown that $K_{n,N} \sim 2^{2/3}(N/n)^{1/4}$. Also, since $s \in [0, \mu_N/\sigma_N)$, which implies, $0 < \xi \le \xi_1$, which in turn implies that, $\theta \ge 0$, it follows from (53), (54), and properties of functions Ai and Bi that, $\mathcal{M}(\theta) \le \Delta\theta^{-1/4}$ and $\mathcal{E}^{-1}(\theta) \le \Delta\exp(-(2/3)\theta^{3/2})$. We combine all these to get,

$$\begin{aligned}|f_N(\lambda\xi)| &\le \Delta\left(\frac{N}{n}\right)^{1/4}\xi^{1/2}\left(\frac{\eta}{\xi^2-4l\xi+4}\right)^{1/4}\theta^{-1/4}\exp\left(-\frac{2}{3}\theta^{3/2}\right)\\ &\le \Delta\left(\frac{N}{n}\right)^{1/4}N^{-1/6}\left(\frac{\xi^2}{\xi^2-4l\xi+4}\right)^{1/4}\exp\left(-\frac{2}{3}\theta^{3/2}\right)\\ &= \Delta\left(\frac{N}{n}\right)^{1/4}N^{-1/6}(4g(\xi))^{-1/4}\exp\left(-\frac{2}{3}\theta^{3/2}\right),\end{aligned} \tag{70}$$

where the second inequality follows from the identity $\theta = \lambda^{2/3}\beta^{4/3}\eta$, and the fact that $\lambda\beta^2 = 2N+1$, and for the last step we just use the definition of g in (41).

From the first inequality in (60), (39) and the fact that $\sigma_N \sim n^{1/2}N^{-1/6}$, it follows that for n sufficiently large, uniformly over $s \in [s_1, \mu_N/\sigma_N)$,

$$(4g(\xi))^{-1/4} \le \Delta\sigma_N^{1/2}s^{-1/4}\left(1-\frac{\sigma_N}{\mu_N}s\right)^{1/2}.$$

This also implies that

$$\left(\frac{N}{n}\right)^{1/4}N^{-1/6}(4g(\xi))^{-1/4} \le \Delta s^{-1/4}\left(1-\frac{\sigma_N}{\mu_N}s\right)^{1/2}. \tag{71}$$

On the other hand, by (58), for sufficiently large n,

$$\text{uniformly on } s \in [0, \mu_N/\sigma_N - M), \quad \frac{2}{3}\theta^{3/2} \ge \frac{1}{2}s^{3/2}.$$

Also, by (63), for sufficiently large n,

$$\text{uniformly on } s \in [\mu_N/\sigma_N - M, \mu_N/\sigma_N), \quad \frac{2}{3}\theta^{3/2} \ge -\frac{n}{3}\log\left(\frac{\mu_N/\sigma_N - s}{2M}\right).$$

Therefore, (66) and (67) follow from (70) and (71). □

Case: $s \in [s_0, s_1]$

In this case, it is enough show that $f_N(\lambda\xi)$ as a function of s is uniformly bounded. We combine this with (66) to get a bound for f_N which is of the same form as (66) but now with $s_1 \in \mathbb{R}$ arbitrary. First, we observe that $\mathbf{E}^{-1} \le 1$ (by eq. (5.22) of Olver (1975)). Recalling that, if $s \in [s_0, s_1]$, then $\zeta \le 0$ for sufficiently large n, we use the display between equations (6.12) and (6.13) of Olver (1975), together with $\lambda\beta^2 = 2N+1 \ge 1$, to conclude that

$$\frac{\mathbf{M}(-\frac{1}{2}\lambda\beta^2, -\zeta\sqrt{2\lambda})}{[\Gamma(\frac{1}{2}(1+\lambda\beta^2))]^{1/2}} \le \Delta\frac{\beta^{1/2}}{(\lambda\beta^2)^{1/12}}\left(\frac{\eta}{\zeta^2-\beta^2}\right)^{1/4}.$$

Combining these bounds for $\mathbf{E}^{-1}$ and $\mathbf{M}$ with (68), we get

$$|f_N(\lambda\xi)| \le K_{n,N}\Delta\left(\frac{\eta}{\xi^2-4l\xi+4}\right)^{1/4}\xi^{1/2}. \tag{72}$$

But, on the interval $s \in [s_0, s_1]$, $\xi \to 2$, and

$$\xi^2 - 4l\xi + 4 = \frac{s\sigma_N}{\lambda}\{(\xi_2-\xi_1) + \frac{\sigma_N}{\lambda}s\} = 4s\sigma_N\beta\lambda^{-1}(1+o(1))$$

(since $\xi_2 - \xi_1 = 4\sqrt{l^2-1} = 4\beta(1+o(1))$). Furthermore, by (65), $\eta = (\lambda\beta^2)^{-2/3}s(1+o(1))$. Therefore,

$$\frac{\eta}{\xi^2-4l\xi+4} \sim \frac{1}{4}\sigma_N^{-1}\lambda\beta^{-1}(\lambda\beta^2)^{2/3} \sim 2^{-2/3}2^{-4}\frac{n}{N} \tag{73}$$

on $[s_0, s_1]$. From (72), (73) and (100) it follows that $f_N(\lambda\xi)$ is uniformly bounded for $s \in [s_0, s_1]$.

3.2.2. *Pointwise convergence: Proof of (36)*

Now, we are in a position to prove (36). Observe that, since $\lambda\beta^2 = 2N+1$, and N is taken to be even, $\sin(\pi\lambda\beta^2/2) = 1$ in (49). In order to compute the limit, we break the interval $[s_0, \mu_N/\sigma_N)$ into two parts: $[s_0, s_1]$ and $[s_1, \mu_N/\sigma_N)$, where $s_1 > 0$. First, we restrict to the case $[s_1, \mu_N/\sigma_N)$. Here, from (48), (98) and (99), for sufficiently large n,

$$|\varepsilon_2(\lambda^2, \beta^2, \zeta)| \le \Xi\mathcal{E}^{-1}(\theta)\mathcal{M}(\theta)O(\lambda^{-2/3}).$$

This, together with (51), implies that

$$f_N(\lambda\xi) = (1+O(\lambda^{-2/3}))K_{n,N}\xi^{1/2}\left(\frac{\eta}{\xi^2-4l\xi+4}\right)^{1/4} \times [\mathrm{Ai}(\theta) + \mathcal{E}^{-1}(\theta)\mathcal{M}(\theta)O((\lambda\beta^2)^{-1} \vee \lambda^{-2/3})] \tag{74}$$

Now, by (65), $\theta \to s$ for every fixed $s \in [s_1, \mu_N/\sigma_N)$, and the terms $\mathcal{E}^{-1}(\theta)$ and $\mathcal{M}(\theta)$ remain bounded on $\mathbb{R}$ (as shown in Section 11.2, p.394–397 of

Olver (1974)). From (73), (100) and the fact $\xi \to 2$, it follows that, as $n \to \infty$,

$$K_{n,N}\xi^{1/2}\left(\frac{\eta}{\xi^2 - 4l\xi + 4}\right)^{1/4} \to 1.$$

Therefore, (36) follows from (74).

Only a minor modification of this argument is needed for the situation when $s \in [s_0, s_1]$. Here, again by (48), (98) and (99),

$$|\varepsilon_2(\lambda^2, \beta^2, \zeta)| \leq \Xi\, O(\lambda^{-2/3}).$$

This leads to,

$$f_N(\lambda\xi) = (1 + O(\lambda^{-2/3}))K_{n,N}\xi^{1/2}\left(\frac{\eta}{\xi^2 - 4l\xi + 4}\right)^{1/4} \times [\mathrm{Ai}(\theta) + \mathcal{E}^{-1}(\theta)\mathcal{M}(\theta)O((\lambda\beta^2)^{-1}) + O(\lambda^{-2/3})]. \quad (75)$$

Hence, again (36) follows.

3.3. *Asymptotics for h_N*

This section contains asymptotic results for h_N defined in 34, which are analogous to the asymptotic behavior of f_N. The basic arguments used in Section 3.3 hold true for h_N as well. For brevity, some tedious details, that can be verified quite easily, are omitted. The fact connecting the asymptotic analysis of f_N and h_N is this observation.

Define $\tilde{s}$ by $\mu_N - \sigma_N s = \mu_{N-1} - \sigma_{N-1}\tilde{s}$. Then

$$\tilde{s} = -\frac{\mu_N - \mu_{N-1}}{\sigma_{N-1}} + \frac{\sigma_N}{\sigma_{N-1}} s.$$

Here, in the definition for μ_{N-1} and σ_{N-1}, we use $n_{N-1} = n_N - 1$. Then

$$\sqrt{\frac{N}{n}}(\mu_N - \mu_{N-1}) \to 1, \qquad \sigma_N \sim n^{1/2}N^{-1/6} \qquad \text{and} \qquad \frac{\sigma_N}{\sigma_{N-1}} \to 1.$$

Notice that, for any fixed s, $\tilde{s}$ is actually a function of s, N and n. Also, due to the relationship stated above,

$$|\tilde{s} - s| \leq \frac{\mu_N - \mu_{N-1}}{\sigma_N} + |s|\left|\frac{\sigma_N - \sigma_{N-1}}{\sigma_N}\right| = O(N^{-1/3}) + |s|O(N^{-1}), \quad \text{uniformly on} \quad s \in \mathbb{R}\,. \quad (76)$$

Moreover,

$$s \in [s_1, \mu_N/\sigma_N) \quad \text{if and only if} \quad \tilde{s} \in \left[-\frac{\mu_N - \mu_{N-1}}{\sigma_{N-1}} + \frac{\sigma_N}{\sigma_{N-1}}s_1, \frac{\mu_{N-1}}{\sigma_{N-1}}\right).$$

This shows that, in particular, if $s > s_1 > 0$ then $\exists\ \tilde{s}_1 > 0$ such that for sufficiently large n, $\tilde{s} > \tilde{s}_1$.

One can define quantities analogous to ζ, η and θ in terms of

$$\kappa = \kappa_{N-1} = N - 1 + \frac{\alpha+1}{2}, \quad \alpha = n - N, \quad \lambda = \frac{\alpha}{2}, \quad \tilde{l} = \frac{\kappa}{\lambda}, \quad \beta = \sqrt{2(\tilde{l}-1)} \tag{77}$$

and $\tilde{s}$. Just to distinguish from the earlier situation, one may denote the corresponding quantities by $\tilde{\zeta}$, $\tilde{\eta}$ and $\tilde{\theta}$, respectively. The lower turning point of (40) will now be denoted by $\tilde{\xi}_1 = \mu_{N-1}/\lambda$, and

$$\tilde{\xi} = \tilde{\xi}_1 - \frac{\sigma_{N-1}}{\lambda}\tilde{s} = \frac{\mu_{N-1} - \sigma_{N-1}\tilde{s}}{\lambda} = \frac{\mu_N - \sigma_N s}{\lambda} = \xi.$$

Inequalities similar to (58) and (63) are available, with s replaced by $\tilde{s}$, in the cases where $s \in [s_1, \mu_N/\sigma_N - M)$ and $s \in [\mu_N/\sigma_N - M, \mu_N/\sigma_N)$. Here $s_1, M > 0$ and $\delta \in (0,1)$ arbitrary. Indeed, since (see (76)) $\tilde{s} \geq (1-\delta')s$ for sufficiently large n, for any $\delta' \in (0,1)$, one can actually get the same inequalities involving s instead of $\tilde{s}$, but with possibly different δ and s_1. These observations lead to the inequality,

$$\begin{aligned}|h_N(\lambda)| \leq \sqrt{\sigma_{N-1}/\sigma_N} K_{n-1,N-1}\xi^{1/2}\left(\frac{\tilde{\eta}}{\xi^2 - 4\tilde{l}\xi + 4}\right)^{1/4} \\ \times\, \mathcal{E}^{-1}(\tilde{\theta})\mathcal{M}(\tilde{\theta})(1 + O(\max\{(\lambda\beta^2)^{-1}, \lambda^{-2/3}\})).\end{aligned}$$

By the preceding discussion, and the analysis in Section 3.2, the following analogue of Lemma 7 is obtained.

Lemma 8. *For every $M > 0$, and $s_1 > 0$, for all sufficiently large n,*

$$|h_N(\lambda\xi)| \leq \Delta s^{-1/4}\exp\left(-\frac{1}{2}s^{3/2}\right), \qquad s \in [s_1, \mu_N/\sigma_N - M) \tag{78}$$

$$\begin{aligned}\leq \Delta\left(1 - \frac{\sigma_N}{\mu_N}s\right)^{1/2} \exp\left(\frac{n}{3}\log\left(\frac{\mu_N/\sigma_N - s}{2M}\right)\right), \\ s \in [\mu_N/\sigma_N - M, \mu_n/\sigma_N).\end{aligned} \tag{79}$$

The case $s \in [s_0, s_1]$ can again be treated the same way, where one gets

$$|h_N(\lambda\xi)| \leq \Delta\sqrt{\frac{\sigma_{N-1}}{\sigma_N}} K_{n-1,N-1}\left(\frac{\tilde{\eta}}{\xi^2 - 4\tilde{l}\xi + 4}\right)^{1/4} \xi^{1/2}.$$

Eventually, we use the asymptotic behavior of $\tilde{\eta}$ and $\xi^2 - 4\tilde{l}\xi + 4$ to show that $h_N(\lambda\xi)$ is bounded on this interval.

The most important difference from the case of f_N is that, now $\lambda\beta^2 = 2N-1$. This implies that, in (49), $\sin(\pi\lambda\beta^2/2) = -1$, since N is chosen to be even. The pointwise convergence (37) now follows from arguments similar to those given in Section 3.2, coupled with the fact that, for every fixed $s \in \mathbb{R}$, $\tilde{\theta} \to s$. This last statement is a result of (76) and (65).

3.4. *Asymptotics for ϕ_τ and ψ_τ*

Following Johnstone (2001), write

$$\phi_\tau = \phi_{I,N} + \phi_{II,N} \qquad \text{and} \qquad \psi_\tau = \psi_{I,N} + \psi_{II,N}, \tag{80}$$

where, with $\alpha = n - N$,

$$\begin{aligned}
\phi_{I,N}(z) &= (-1)^N \sigma_N \sqrt{\frac{a_N}{2}} \sqrt{n}\xi_N(z) \\
&= \frac{1}{\sqrt{2}} \sqrt{n}(Nn)^{1/4} \left(\frac{\sigma_N}{z}\right)^{3/2} \frac{\sigma_N^{-1/2}}{\sqrt{n!N!}} W_{N+\frac{\alpha+1}{2},\frac{\alpha}{2}}(z) \\
&= \frac{1}{\sqrt{2}} \sqrt{n}(Nn)^{1/4} \left(\frac{\sigma_N}{z}\right)^{3/2} f_N(z),
\end{aligned} \tag{81}$$

$$\begin{aligned}
\phi_{II,N}(z) &= -(-1)^N \sigma_N \sqrt{\frac{a_N}{2}} \sqrt{N}\xi_{N-1}(z) \\
&= \frac{1}{\sqrt{2}} \sqrt{N}(Nn)^{1/4} \left(\frac{\sigma_N}{z}\right)^{3/2} \frac{\sigma_N^{-1/2}}{\sqrt{(n-1)!(N-1)!}} W_{N-1+\frac{\alpha+1}{2},\frac{\alpha}{2}}(z), \\
&= \frac{1}{\sqrt{2}} \sqrt{N}(Nn)^{1/4} \left(\frac{\sigma_N}{z}\right)^{3/2} h_N(z),
\end{aligned} \tag{82}$$

and

$$\psi_{I,N}(z) = \sqrt{\frac{N}{n}} \phi_{I,N}(z) \qquad \psi_{II,N}(z) = \sqrt{\frac{n}{N}} \phi_{II,N}(z). \tag{83}$$

These relations follow from the definition of ξ_N and ξ_{N-1} and (33), (34) and (35). We are now in a position to complete the proofs of Lemmas 3 and 4.

Proof of Lemma 3. Let $z = \mu_N - \sigma_N s$, for $s \in \mathbb{R}$. Then, as $n \to \infty$,

$$\begin{aligned}
\sqrt{n}(Nn)^{1/4} \left(\frac{\sigma_N}{z}\right)^{3/2} &= n^{3/4} N^{1/4} \left(\frac{\sigma_N}{\mu_N}\right)^{3/2} (1 + o(1)) \\
&= n^{3/4} N^{1/4} (n^{-1/2} N^{-1/6})^{3/2} (1 + o(1)) \to 1.
\end{aligned}$$

The proof is completed by combining (81), (82) and (83) with the pointwise limits (36) and (37). □

Proof of Lemma 4. First consider $\phi_{I,N}$. The multiplier of $f_N(z)$ (with $z \equiv \lambda\xi$) in (81) can be bounded as:

$$\frac{1}{\sqrt{2}}\sqrt{n}(Nn)^{1/4}\left(\frac{\sigma_N}{z}\right)^{3/2} \tag{84}$$

$$\leq \Delta n^{3/4}N^{1/4}(n^{1/2}N^{-1/6})^{3/2}n^{-3/2}\left(1-\frac{\sigma_N}{\mu_N}s\right)^{-3/2}$$

$$\leq \Delta\left(1-\frac{\sigma_N}{\mu_N}s\right)^{-3/2}. \tag{85}$$

Therefore, by (67), on the interval $s \in [\mu_N/\sigma_N - M, \mu_N/\sigma_N)$,

$$\begin{aligned}
|\phi_{I,N}(\lambda\xi)| &\leq \Delta\left(1-\frac{\sigma_N}{\mu_N}s\right)^{-1}\exp\left[\frac{n}{3}\log\left(\frac{\mu_N/\sigma_N - s}{2M}\right)\right]\\
&= \Delta\frac{\mu_N}{\sigma_N}\exp\left[\left(\frac{n}{3}-1\right)\log(\mu_N/\sigma_N - s) - \frac{n}{3}\log(2M)\right]\\
&\leq \Delta n^{1/2}N^{1/6}\frac{1}{2M}\exp\left[\left(\frac{n}{3}-1\right)\log\left(\frac{\mu_N/\sigma_N - s}{2M}\right)\right]\\
&\leq \Delta n\exp\left[-\frac{n}{4}\log 2\right]\\
&= \Delta n 2^{-n/4},
\end{aligned}$$

where the last inequality follows from the fact that on $[\mu_N/\sigma_N - M, \mu_N/\sigma_N)$, (and assuming $n \geq 12$), $(\mu_N/\sigma_N - s)/2M \leq 1/2$.

Now we focus on the interval $s \in [s_0, \mu_N/\sigma_N - M]$. By (66), (84) and the comment in the paragraph following "**Case:** $s \in [s_0, s_1]$" in Section 3.2,

$$|\phi_{I,N}(\lambda\xi)| \leq \Delta\left(1-\frac{\sigma_N}{\mu_N}s\right)^{-3/2}\exp\left(-\frac{1}{2}s^{3/2}\right). \tag{86}$$

Suppose that, $1-(\sigma_N/\mu_N)s \geq 1/2$. Then the bound of the form (28), for $\phi_{I,N}$, follows trivially from (86). So, consider the case $1-(\sigma_N/\mu_N)s \leq 1/2$. This implies that,

$$\frac{\mu_N}{\sigma_N} - M \geq s \geq \frac{\mu_N}{2\sigma_N}.$$

Hence, for sufficiently large n,

$$s^{3/2}\left(1-\frac{\sigma_N}{\mu_N}s\right) \geq 2^{-3/2}\left(\frac{\mu_N}{\sigma_N}\right)^{3/2}\left(1-\frac{\sigma_N}{\mu_N}s\right) \geq \frac{1}{2}\sqrt{\frac{\mu_N}{\sigma_N}}M,$$

which yields a bound of the form (27) for $\phi_{I,N}$, since the upper bound in (86) can be expressed as

$$\Delta\left[\left(1-\frac{\sigma_N}{\mu_N}s\right)^{-1}\exp\left(-\frac{1}{2}\left(1-\frac{\sigma_N}{\mu_N}s\right)^{-1}\right)\right]$$
$$\times\exp\left[-\frac{1}{2}\left(s^{3/2}-\left(1-\frac{\sigma_N}{\mu_N}s\right)^{-1}\right)\right].$$

We use (78) and (79) to derive similar bounds for $\psi_{II,N}$. Finally, we use (81), (82) and (83) to conclude the proof. □

4. Real Gaussian entries

In this section we suppose that X is an $n \times N$ matrix with real valued random variables X_{ij} distributed independently as $N(0,1)$, for all i,j. The first step in deriving the asymptotic distribution of l_N, the smallest eigenvalue of $V = XX^T$, is a determinantal representation due to Tracy and Widom (1998). Let us denote by $\mathbb{E}_N$ the expectation with respect to the joint distribution of the eigenvalues $l_1 > \cdots > l_N$. Then

$$\mathbb{E}_N\left(\prod_{j=1}^{N}[1+f(l_j)]\right) = \sqrt{\det(I+K_N f)}, \tag{87}$$

where K_N is a matrix kernel defined below. Choose $f = -\chi_{(0,t)}$ to get

$$G^N(t) = \mathbb{P}_N(l_N \geq t) = \sqrt{\det(I - K_N\chi_{(0,t)})}. \tag{88}$$

So, as before, the goal is to show that, if $t = \tau(s) = \mu_N - \sigma_N s$, then as $n \to \infty$, $G^N(t) \to F_1(s)$, where F_1 is the c.d.f. of Tracy–Widom(1) law.

K_N appearing in (87) is an operator with kernel a 2×2 matrix. Following Tracy and Widom (1996) and Widom (1999), $K_N : L_2^2(0,\infty) \to L_2^2(0,\infty)$ has the representation:

$$K_N = \begin{bmatrix} S + \psi\otimes\epsilon\phi & SD - \psi\otimes\phi \\ \epsilon S - \epsilon + \epsilon\psi\otimes\epsilon\phi & S + \epsilon\phi\otimes\psi \end{bmatrix} \tag{89}$$

Here $\epsilon(x) = \mathrm{sign}(x)/2$ and the operator ϵ denotes convolution with the kernel $\epsilon(x-y)$. $\alpha\otimes\beta$ denotes the rank one operator defined by $(\alpha\otimes\beta)f = \langle\beta, f\rangle\alpha$. D denotes the first order differential operator.

Remark 7. A minor but important details is that, the kernel $S = S_N$ appearing in (89), is really $S_{N,n-1}$, i.e., the Laguerre ensemble has parameter

$\alpha = n-1-N$, instead of $n-N$. Therefore, in the expressions for ϕ and ψ appearing in (16) and (17), one needs to change n to $n-1$ and the functions Φ_k appearing in the expression (15), have parameter $\alpha = n-1-N$. Clearly, this does not have any effect in the asymptotic limits.

Formula (89) has exactly the same form as the one derived for Gaussian Orthogonal Ensemble (GOE) in Section V of Tracy and Widom (1996). Much of the limiting arguments here will follow the pattern of Tracy and Widom (1996), so we give only a brief summary of the steps involved. In particular, we derive expressions similar to the ones given in Section 4 of Johnstone (2001), and merely point to the differences involved. The notations are patterned after Tracy and Widom (1996) and Johnstone (2001).

From now on, in order to distinguish the real and complex cases, $G_1^N(t)$ will denote the function appearing in (88), and $G_2^N(t)$ will denote the function appearing in (12). The asymptotic behavior of $G_2^N(t)$, when $t = \mu_N - \sigma_N s$, has already been studied. For the asymptotic analysis of $G_1^N(t)$, we introduce certain quantities associated with the operator $(I - S\chi)^{-1}$.

Let $R(x,y)$ denote the kernel of the resolvent operator of $S\chi$. That is, $(I-S\chi)^{-1} = I + R$. We define

$$\begin{aligned} Q_\epsilon &:= (I-S\chi)^{-1}\epsilon\phi, & \tilde{v}_\epsilon &:= \langle Q_\epsilon, \chi\psi\rangle \\ P &:= (I-S\chi)^{-1}\psi, & q_\epsilon &:= Q_\epsilon(t), \\ \mathcal{P} &:= \int_t^\infty P(x)dx, & \mathcal{R} &:= \int_t^\infty R(x,t)dx \\ & & c_\phi &:= \frac{1}{2}\int_0^\infty \phi(x)dx \end{aligned} \tag{90}$$

Next, define

$$\begin{aligned} &\alpha_1 = Q_\epsilon, \qquad \beta_1 = \chi\psi, \qquad \beta_2 = \delta_t - \delta_0 \\ &\alpha_2 = \frac{1}{2}[(S+RS)(1-\chi) + \langle\psi, 1-\chi\rangle Q_\epsilon] \end{aligned} \tag{91}$$

We can use the same manipulations as in Tracy and Widom (1996) to obtain

$$\frac{(G_1^N(t))^2}{(G_2^N(t))^2} = \det(I - \alpha_1\otimes\beta_1 - \alpha_2\otimes\beta_2) = \det_{1\le j,k\le 2}(\delta_{jk} - \langle\alpha_j,\beta_k\rangle). \tag{92}$$

Details of this derivation are given in Section 5.4. From (90) and (91), it follows that (see Section 5.5), the ratio appearing in (92) can be represented as

$$\frac{(G_1^N(t))^2}{(G_2^N(t))^2} = (1-\tilde{v}_\epsilon)(1-\frac{1}{2}\mathcal{R}) - \frac{1}{2}(q_\epsilon + c_\phi)\mathcal{P} \tag{93}$$

Now, set $t = \tau(s) = \mu_N - \sigma_N s$, and use Lemmas 3 and 4 to evaluate the limits of the different quantities involved in the last equation. Define

$$a(x) = \frac{1}{\sqrt{2}} - \frac{1}{\sqrt{2}}\int_x^\infty \mathrm{Ai}(y)dy, \qquad \overline{Q} = (I - \overline{\chi}\overline{S}\overline{\chi})^{-1}a.$$

El Karoui (2003) shows that,

$$\lim_{n\to\infty} c_\phi = \frac{1}{\sqrt{2}}\,. \tag{94}$$

Then, it can be checked that, as $n \to \infty$,

$$\begin{aligned} q_\epsilon \to -\overline{q} = -\overline{Q}(s), \qquad & \tilde{v}_\epsilon \to \overline{u} = \frac{1}{\sqrt{2}}\langle \overline{Q}, \overline{\chi}\mathrm{Ai}\rangle \\ \mathcal{R} \to \overline{\mathcal{R}}, \qquad & \mathcal{P} \to -\overline{\mathcal{P}}, \end{aligned} \tag{95}$$

where $\overline{\mathcal{R}}$ and $\overline{\mathcal{P}}$ are exactly those occurring in the GOE case, as given in Tracy and Widom (1996). For example, with $t = \mu_N - \sigma_N s$,

$$\begin{aligned} \epsilon\phi(t) &= \epsilon\phi(\mu_N - \sigma_N s) = \frac{1}{2}\int_0^{\mu_N - \sigma_N s}\phi(x)dx - \frac{1}{2}\int_{\mu_N-\sigma_N s}^\infty \phi(x)dx \\ &= \int_0^{\mu_N-\sigma_N s}\phi(x)dx - \frac{1}{2}\int_0^\infty \phi(x)dx \\ &= \int_s^{\mu_N/\sigma_N}\phi_\tau(y)dy - c_\phi \qquad \left(\text{set } \; y = \frac{\mu_N - x}{\sigma_N}\right) \\ &\to \frac{1}{\sqrt{2}}\int_s^\infty \mathrm{Ai}(y)dy - \frac{1}{\sqrt{2}} \;=\; -\,a(s), \end{aligned} \tag{96}$$

where, in the last line we used (94). Again, by Proposition 1, S_τ converges in $L^2(s,\infty)\otimes L^2(s,\infty)$ to $\overline{S}$. This implies that the operator $\chi_\tau S_\tau \chi_\tau$ converges in operator norm on $L^2(\mathbb{R})$ to the operator $\overline{\chi}\overline{S}\overline{\chi}$. Since, $I - \overline{\chi}\overline{S}\overline{\chi}$ is invertible (Tracy and Widom, 1996), it follows that

$$(I - \chi_\tau S_\tau \chi_\tau)^{-1} \to (I - \overline{\chi}\overline{S}\overline{\chi})^{-1} \tag{97}$$

in operator norm. From (96) and (97), the convergence $\tilde{v}_\epsilon \to \overline{u} = \langle \overline{Q}, \overline{\chi}\mathrm{Ai}\rangle/\sqrt{2}$ follows, since $(I - S\chi)^{-1} = I + S\chi(I - \chi S\chi)^{-1}$. The proofs of the other limits appearing in (95) require slightly more work, and can be accomplished with the help of Proposition 1 by following the approach taken in Section VII of Tracy and Widom (1996). It then follows that,

$$\lim_{n\to\infty}\frac{(G_1^N(t))^2}{(G_2^N(t))^2}=\frac{\widetilde{G}_1^2(s)}{\widetilde{G}_2(s)}=(1-\overline{u})(1-\frac{1}{2}\overline{\mathcal{R}})-\frac{1}{2}(\overline{q}-\frac{1}{\sqrt{2}})\overline{\mathcal{P}},$$

where $\widetilde{G}_i(s)=1-G_i(-s)$, $i=1,2$. Set $\mu=\int_s^\infty q(x)dx$. Tracy and Widom (1996) identify that,

$$2\overline{u}=\overline{\mathcal{R}}=-\sqrt{2}\overline{\mathcal{P}}=1-\sqrt{2}\overline{q}=1-e^{-\mu}.$$

Hence,

$$\frac{\widetilde{G}_1^2(s)}{\widetilde{G}_2(s)}=e^{-\mu}=\exp\left(-\int_s^\infty q(x)dx\right)=\frac{F_1^2(s)}{F_2(s)},$$

from which Theorem 3 follows, since $\widetilde{G}_2(s)=F_2(s)$.

5. Some proof details

5.1. *Relationships among* $\mathbf{E}$, $\boldsymbol{\mathcal{E}}$, $\mathbf{M}$ *and* $\boldsymbol{\mathcal{M}}$

If $s\geq 0$ and $\theta=(\lambda\beta^2)^{2/3}\eta$, the following inequalities hold (El Karoui (2003), Section **A2**).

$$\mathbf{E}^{-1}(-\frac{1}{2}\lambda\beta^2,-\zeta\sqrt{2\lambda})\leq\mathcal{E}^{-1}(\theta)(1+O((\lambda\beta^2)^{-1})), \tag{98}$$

$$\mathbf{M}(-\frac{1}{2}\lambda\beta^2,-\zeta\sqrt{2\lambda})\leq\Xi\,\mathcal{M}(\theta)(1+O((\lambda\beta^2)^{-1})), \tag{99}$$

where Ξ is as in (52).

Remark 8. With $s\geq 0$, $\xi\leq\xi_1$, and so (57) holds, implying that $\theta\geq 0$. The definitions of $\mathbf{E}$ and $\mathbf{M}$ depend on the position of the argument with respect to the largest real root, $\rho(b)$ say, of the equation $\overline{U}(b,x)=U(b,x)$. Equation (5.21) of Olver (1975) states that

$$\rho(b)=2(-b)^{1/2}+c(-b)^{-1/6}+O(b^{-5/6})\qquad\text{as}\quad b\to-\infty.$$

where $c(=-0.36605...)$ is the smallest root, in absolute value, of the equation $\mathrm{Ai}(x)=\mathrm{Bi}(x)$. In the present scenario, $b=-\lambda\beta^2/2$. If $s\geq 0$, then $-\zeta\geq\beta$ and so,

$$-\zeta\sqrt{2\lambda}\geq\sqrt{2\lambda\beta^2}\geq\rho^2(-\frac{1}{2}\lambda\beta^2).$$

On the other hand, since $\theta \geq 0 \geq c$, it is also true that, $\mathcal{M} = \{2\text{AiBi}\}^{1/2}$ and $\text{Ai} = 2^{-1/2}\mathcal{E}^{-1}\mathcal{M}$, and hence $\text{Bi} = 2^{-1/2}\mathcal{E}\mathcal{M}$.

Consequence of the position of ξ and ζ

As an important consequence of Remark 8, it follows that (cf. eq. (5.22) of Olver (1975)), for large enough n, and all $s \geq 0$, one can write

$$\mathbf{E}^{-2}(-\frac{1}{2}\lambda\beta^2, -\zeta\sqrt{2\lambda}) = \frac{U}{\overline{U}} = \left(\frac{\mathcal{M}(\theta)\mathcal{E}^{-1}(\theta)}{\mathcal{M}(\theta)\mathcal{E}(\theta)}\right)\frac{2^{-1/2} + O((\lambda\beta^2)^{-1})}{2^{-1/2} + O((\lambda\beta^2)^{-1})}.$$

This implies that, for some n_0, and all $n \geq n_0$, for all $s \geq 0$, (98) holds. (99) is proved by a similar argument.

5.2. *Asymptotic behavior of $K_{n,N}$*

This part is the same as section **A4** of El Karoui (2003). The aim is to show that, as $N, n \to \infty$ such that $N/n \to 0$, asymptotically

$$K_{n,N} \sim 2^{2/3}\left(\frac{N}{n}\right)^{1/4}. \tag{100}$$

To see this, first note that $K_{n,N}$ has the expression (69). Since $\lambda\beta^2 = 2N+1$, it follows that, $\Gamma((1+\lambda\beta^2)/2) = N!$. Also,

$$2\lambda + \frac{1}{2}\lambda\beta^2 = n - N + \frac{2N+1}{2} = n + \frac{1}{2} = n_+.$$

Consequently,

$$\begin{aligned} A_n := \frac{1}{\sqrt{n!}}\left(\lambda(2+\frac{1}{2}\lambda\beta^2)/e\right)^{\lambda(1+\beta^2/4)} &= \frac{(n_+/e)^{n_+/2}}{\sqrt{n!}} \\ \sim \left(\frac{n_+}{n}\right)^{n/2}\left(\frac{n_+}{n}\right)^{1/4}(2\pi e)^{-1/4} & \qquad \text{(Stirling's approximation)} \\ \sim (2\pi)^{-1/4}. & \end{aligned}$$

Now, we rewrite

$$K_{n,N} = \frac{A_n(\lambda\beta^2)^{1/4}(8\pi)^{1/4}}{(\lambda\beta^2)^{1/12}\sqrt{\sigma_N}},$$

from which we obtain (100) by using $A_n(8\pi)^{1/4} \sim \sqrt{2}$ and the fact that,

$$\frac{\lambda\beta^2}{\sigma_N^3} \sim \frac{2N+1}{(n^{1/2}N^{-1/6})^3} \sim 2\left(\frac{N}{n}\right)^{3/2}.$$

5.3. *Recessive solution of (40)*

From the discussions in Section 3.2, and the expression of $\overline{v}_2$ in (47), it is enough to study the principal part of $\overline{U}(-\lambda\beta^2/2, -\zeta\sqrt{2\lambda})$. By equation (5.13) of Olver (1980), this is just Ξ Bi(θ) with Ξ and θ as before. From Olver (1974),

$$\mathrm{Bi}(x) \sim \pi^{-1/2} x^{-1/4} e^{\frac{2}{3}x^{3/2}}. \tag{101}$$

On the other hand, for $0 < \xi < \xi_1$, $\theta = (\lambda\beta^2)^{2/3}\eta > 0 > c$, where c is as in Section 5.1. As a result, the principal part of the function $\overline{v}_2(d\xi/d\zeta)^{1/2}$ is a constant multiple of

$$\left(\frac{\eta}{g(\xi)}\right)^{1/4} \mathrm{Bi}(\theta) \sim C_{n,N} g(\xi)^{-1/4} \exp\left(\frac{2}{3}\theta^{3/2}\right), \tag{102}$$

as $\xi \to 0+$. Here $C_{n,N}$ is an explicit constant depending on n, N which are kept fixed. Since

$$g(\xi) = \frac{\xi^2 - 4l\xi + 4}{4\xi^2} \sim \frac{1}{\xi^2}, \qquad \text{as } \xi \to 0+,$$

it follows that,

$$\frac{2}{3}\beta^2\eta^{3/2} = \int_{\xi}^{\xi_1} g^{1/2} t dt \sim \log\left(\frac{1}{\xi}\right), \qquad \text{as } \xi \to 0+ \ .$$

Substitute in (102) to conclude that, as $\xi \to 0+$, the function $\overline{v}_2(d\xi/d\zeta)^{1/2}$ behaves like

$$C_{n,N}\xi^{1/2} e^{-\lambda \log \xi (1+o(1))} = C_{n,N}\xi^{(1/2-\lambda)(1+o(1))} \to \infty.$$

Hence this is not a recessive solution of (40). Furthermore, similar calculations show that, any constant multiple of $v_2(d\xi/d\zeta)^{1/2}$ is a recessive solution of (40).

5.4. *Representation of the ratio* $(G_1^N(t))^2/(G_2^N(t))^2$

Following the notations in Tracy and Widom (1998), we write $E_1(0, J)$ and $E_2(0, J)$ to denote the probability in the real and the complex case, respectively, of the event that an interval $J \subset \mathbb{R}$ does not contain any eigenvalue. Then we use the matrix kernel of K_N given by (89), and follow the derivation of Section V of Tracy and Widom (1996), to obtain

$$\frac{E_1(0, J)^2}{E_2(0, J)} = \det(I - (S + RS)(I - \chi)\epsilon\chi D - Q_\epsilon \otimes \chi\psi - Q_\epsilon \otimes \psi(1 - \chi)\epsilon\chi D). \tag{103}$$

In their papers Tracy and Widom deal with general finite unions of intervals, but that is not necessary for the present work. Let the interval J be (a_1, a_2) where $a_1 = 0$ and $a_2 = t$. We set

$$\epsilon_k(x) := \epsilon(x-a_k), \quad \delta_k(x) := \delta(x-a_k), \quad R_k(x) := R(x, a_k) = \lim_{\substack{y\to a_k\\ y\in J}} R(x,y),$$

for $k = 1, 2$. To be consistent with the notations used in the earlier sections, observe that we have the identification: $\delta_1 \equiv \delta_0$, $\delta_2 \equiv \delta_t$, $\epsilon_1 \equiv \epsilon_0$ and $\epsilon_2 \equiv \epsilon_t$. Using these, we obtain the following representations for the commutator $[\chi, D] = \chi D - D\chi$.

$$[\chi, D] = \sum_{k=1}^{2} (-1)^k \delta_k \otimes \delta_k \tag{104}$$

$$\epsilon[\chi, D] = \sum_{k=1}^{2} (-1)^k \epsilon_k \otimes \delta_k. \tag{105}$$

Since $\epsilon D = 1$, and $\chi(1-\chi) = 0$,

$$(1-\chi)\epsilon\chi D = (1-\chi)\epsilon[\chi, D]. \tag{106}$$

Hence, the representation

$$(S+RS)(1-\chi)\epsilon\chi D = \sum_{k=1}^{2} (-1)^k (S+RS)(1-\chi)\epsilon_k \otimes \delta_k \tag{107}$$

follows. Combining (104), (105), (106) with (103), we get that the operator appearing in (103) has the representation

$$I - \sum_{k=1}^{2}(-1)^k(S+RS)(1-\chi)\epsilon_k\otimes\delta_k - Q_\epsilon\otimes\chi\psi - \sum_{k=1}^{2}(-1)^k\langle\psi,(1-\chi)\epsilon_k\rangle Q_\epsilon\otimes\delta_k. \tag{108}$$

Observe that

$$\epsilon_1(x) = \frac{1}{2} \quad \text{and} \quad (1-\chi)\epsilon_1(x) = (1-\chi)\epsilon_2(x) = \frac{1}{2}(1-\chi)(x) \quad \text{for } x \in \mathbb{R}_+ \,. \tag{109}$$

Also, over $\mathbb{R}$,

$$\epsilon_1 = \epsilon, \qquad (1-\chi)\epsilon_1 = (1-\chi)\epsilon_2 = (1-\chi)\epsilon.$$

On the other hand, $S(x, y) = 0$ for either $x < 0$ or $y < 0$, and $\psi(x) = 0$ for $x < 0$. Hence the operator in (108) becomes, as an operator on $L^2(0,\infty)$, the operator,

$$I - Q_\epsilon \otimes \chi\psi - \frac{1}{2}[(S+RS)(1-\chi) + \langle\psi,(1-\chi)\rangle Q_\epsilon] \otimes (\delta_2 - \delta_1), \tag{110}$$

from which (91) follows on comparison with (92).

5.5. *Representation of the inner products*

First observe that R is a finite rank operator. Indeed, one can use the representation

$$R = S\chi(I - \chi S\chi)^{-1} S\chi + S\chi \tag{111}$$

to conclude that

$$R(x, y) = \sum_{j=0}^{N-1} \sum_{k=0}^{N-1} c_{jk}(t)\Phi_j(x)\Phi_k(y)\chi(y), \tag{112}$$

for some constants $\{c_{jk}(t) : 0 \le j, k \le N-1\}$. Hence, the kernel $R(x, y)$ is smooth in x and has a simple discontinuity in y. Also, $R(x, y)$ is bounded on $\mathbb{R}^+ \times \mathbb{R}^+$. Furthermore, since all the Φ_k's vanish on $(-\infty, 0]$ (without loss of generality, assuming $\alpha \ge 6$), it follows that

$$R(x, 0) = \lim_{y \to 0+} R(x, y) = 0. \tag{113}$$

Notice that, by (92) and (91),

$$\langle \alpha_1, \beta_1 \rangle = \langle Q_\epsilon, \chi\psi \rangle = \tilde{v}_\epsilon. \tag{114}$$

Again, since $R = S\chi(I - S\chi)^{-1}$,

$$R\epsilon\phi = S\chi(I - S\chi)^{-1}\epsilon\phi = S\chi Q_\epsilon.$$

This implies that

$$R\epsilon\phi(0) = \lim_{x \to 0+} \int_0^t S(x, y)Q_\epsilon(y)dy = 0, \tag{115}$$

since $S(0, y) = 0$ for all y (and S is bounded, and continuous in both x and y, and $R(y, y')$ is bounded on $(0, t)$). Also,

$$Q_\epsilon(0) = \epsilon\phi(0) + R\epsilon\phi(0) = -\frac{1}{2}\int_0^\infty \phi(x)dx = -c_\phi, \tag{116}$$

where the second equality is a consequence of (115). Therefore

$$\langle \alpha_1, \beta_2 \rangle = \langle Q_\epsilon, \delta_t - \delta_0 \rangle = Q_\epsilon(t) - Q_\epsilon(0) = q_\epsilon + c_\phi. \tag{117}$$

An important fact for computing the other inner products is that

$$(S + SR^t)\chi = R, \tag{118}$$

where R^t is the adjoint of the operator R. Further, define

$$b := \langle \psi, (1-\chi)\epsilon \rangle = \frac{1}{2}\langle \psi, (1-\chi) \rangle, \tag{119}$$

where the second equality is due to (109). Then

$$\begin{aligned} \langle \alpha_2, \beta_1 \rangle &= \frac{1}{2}[\langle (S+RS)(1-\chi), \chi\psi \rangle + \langle \psi, (1-\chi) \rangle \langle Q_\epsilon, \chi\psi \rangle] \\ &= \frac{1}{2}\langle (1-\chi), (S+SR^t)\chi\psi \rangle + b\tilde{v}_\epsilon, \end{aligned}$$

where the second equality is due to the fact that $S = S^t$. By (118), the first term is just

$$\frac{1}{2}[\langle (1-\chi), (I+R)\psi \rangle - \langle (1-\chi), \psi \rangle].$$

The latter, following (90), equals $(1/2)\mathcal{P} - b$. Therefore,

$$\langle \alpha_2, \beta_1 \rangle = \frac{1}{2}\mathcal{P} - b + b\tilde{v}_\epsilon. \tag{120}$$

From (113) and (117), similar calculations imply that,

$$\begin{aligned} \langle \alpha_2, \beta_2 \rangle &= \frac{1}{2}\langle (S+RS)(1-\chi), \delta_2 - \delta_1 \rangle + \frac{1}{2}\langle \psi, (1-\chi) \rangle \langle Q_\epsilon, \delta_2 - \delta_1 \rangle \\ &= \frac{1}{2}\mathcal{R} + bq_\epsilon + bc_\phi. \end{aligned} \tag{121}$$

Now, we combine (92) with (114), (117), (120) and (121) to get (93).

Appendix A.

A brief introduction to Airy functions

We quote some basic facts about *Airy functions* from Olver (1974). For $x \in \mathbb{R}$, the Airy function $\mathrm{Ai}(x)$ is defined by the formula

$$\mathrm{Ai}(x) = \frac{1}{\pi}\int_0^\infty \cos\left(\frac{t^3}{3} + xt\right) dt. \tag{122}$$

Although the integral does not die away as $t \to \infty$, the increasingly rapid oscillations induce convergence of the integral. It can be shown by contour integration that an analytic continuation of the Airy function on the complex plane can be obtained and is given by

$$\mathrm{Ai}(u) = \frac{1}{2\pi i}\int_\Gamma e^{u^3/3 - xu} du, \tag{123}$$

where Γ is the contour in the complex plane consisting of the ray joining $e^{-\pi i/3}\infty$ to the origin plus the ray joining the origin to $e^{\pi i/3}\infty$. $\mathrm{Ai}(x)$ is an entire analytic function of x and satisfies the *Airy equation*

$$\frac{d^2y}{dx^2} - xy = 0. \tag{124}$$

The function $\mathrm{Ai}(x)$, up to a constant factor, is the only solution of the *Airy equation* with a rapid decay along the positive real axis. The function $\mathrm{Bi}(x)$ defined by

$$\mathrm{Bi}(x) = \frac{1}{\pi}\int_0^\infty \left[\exp\left(-\frac{t^3}{3} + xt\right) + \sin\left(\frac{t^3}{3} + xt\right)\right] dt \tag{125}$$

is the *standard solution of the second kind* for (124). For large negative x, $\mathrm{Bi}(x)$ has the same amplitude of oscillation as $\mathrm{Ai}(x)$ but differs in phase by $\pi/2$.

References

Baik, J., Ben Arous, G. and Péché, S. (2005). Phase transition of the largest eigenvalue for non-null complex covariance matrices, *Annals of Probability*, **33**, 1643–1697.

Baker, T. H., Forrester, P. J. and Pearce, P. A. (1998). Random matrix ensembles with an effective external charge, *Journal of Physics A*, **31**, 6087–6101.

El Karoui, N. (2003). On the largest eigenvalue of Wishart matrices with identity covariance when n, p and p/n tend to infinity, arXiv.math.ST/0309355.

El Karoui, N. (2005). A rate of convergence result for the largest eigenvalue of complex white Wishart matrices, *Annals of Probability*, **34**, 2077–2117.

Gohberg, I. and Krein, M. (1969). *Introduction to the theory of linear nonselfadjoint operators*, American Mathematical Society, Providence.

Johnstone, I. M. (2001). On the distribution of the largest principal component, *Annals of Statistics*, **29**, 295–327.

Ma, Z. (2008). Accuracy of the Tracy–Widom limit for the largest eigenvalue in white Wishart matrices, *Technical Report*, Stanford University.

Mehta, M. L. (1991). *Random Matrices*, Academic Press.

Olver, F. W. J. (1974). *Asymptotics and Special Functions*, Academic Press.

Olver, F. W. J. (1975). Second-order linear differential equations with two turning points, *Philosophical Transactions of Royal Society of London, Series A*, **278**, 137–174.

Olver, F. W. J. (1980). Whittaker functions with both parameter large : uniform asymptotic expansions in terms of parabolic cylinder functions, *Proceedings of Royal Society of Edinburgh*, **86A**, 213–234.

Péché, S. (2003). Universality of local eigenvalue statistics for random sample covariance matrices, *Ph. D. Thesis*, Ecole Polytechnique Fédérale de Lausanne.

Reed, M. and Simon, B. (1972). *Methods of Modern Mathematical Physics*, Academic Press.

Soshnikov, A. (2002). A note on universality of the distribution of the largest eigenvalues in certain sample covariance matrices, *Journal of Statistical Physics*, **108**, 1033–1056.

Szegö, G. (1939). *Orthogonal Polynomials*, American Mathematical Society.

Temme, N. (1990). Asymptotic estimates for Laguerre polynomials, *Journal of Applied Mathematics and Physics*, **41**, 114–126.

Tracy, C. and Widom, H. (1994a). Level-spacing distribution and Airy kernel, *Communications in Mathematical Physics*, **159**, 151–174.

Tracy, C. and Widom, H. (1994b). Fredholm determinants, differential equations and matrix models, *Communications in Mathematical Physics*, **163**, 33–72.

Tracy, C. and Widom, H. (1996). On orthogonal and symplectic matrix ensembles, *Communications in Mathematical Physics*, **177**, 727–754.

Tracy, C. and Widom, H. (1998). Correlation functions, cluster functions and spacing distribution of random matrices, *Journal of Statistical Physics*, **92**, 809–835.

Widom, H. (1999). On the relation between orthogonal, symplectic and unitary ensembles, *Journal of Statistical Physics*, **94**, 347–363.

PART 6
Curriculum Vitae

Curriculum Vitae of Prodyot K. Bhattacharya

Name: Prodyot K. Bhattacharya

Office Address: Department of Statistics
University of California
Davis, CA 95616
e-mail: pkbhattacharya@ucdavis.edu

Home Address: 2244 Alameda Avenue
Davis, CA 95616
(530) 758-1480

Biography:

Born, September 30, 1930, Calcutta, India
Citizenship: United States
B. Sc., 1951, Calcutta University, India
M.Sc., 1953, Calcutta University, India
Govt. of India Research Scholar — Calcutta University, India, 1954-1957
Research Scholar - Indian Statistical Institute, Calcutta, India, 1957–60
Ph.D., 1960, Calcutta University, India
Research Associate - University of North Carolina at Chapel Hill, 1960–62
Research Associate- Stanford University, 1962–63
Reader - Indian Statistical Institute, Calcutta, India, 1963–65
Associate Professor - University of Arizona at Tucson, 1965–68
Professor - University of Arizona at Tucson, 1968–80
Visiting Professor - University of Minnesota, 1971–72
Visiting Scientist - Indian Statistical Institute, Calcutta, Fall 1978
Visiting Professor - MIT, Spring 1979
Professor - University of California, Davis, 1981–June 1994
Professor Emeritus - University of California, Davis, July 1994 –

PRODYOT K. BHATTACHARYA PUBLICATIONS

1955 Bhattacharya, P. K. Joint test for the mean and variance of a normal population. *Calcutta Statistical Association Bulletin* 6: 1–18.

1956 Bhattacharya, P. K. Comparison of the means of K normal populations. *Calcutta Statistical Association Bulletin* 7:1–16.

1958 Bhattacharya, P. K. On the large sample behavior of a multiple-decision procedure. *Calcutta Statistical Association Bulletin* 8: 43–47

1961 Parthasarathy, K. R. and Bhattacharya, P. K. Some limit theorems in regression theory. *Sankhya: The Indian Journal of Statistics, Series A* 23:91–102.

1962 Bhattacharya, P. K. Some properties of the least squares estimator in regression analysis when the predictor variables are stochastic. *Annals of Mathematical Statistics* 33: 1365–1374.

1963 Bhattacharya, P. K. On an analog of regression analysis. *Annals of Mathematical Statistics* 34: 1459–1473.

1964 Bhattacharya, P. K. and Das Gupta, S. Classification between univariate exponential populations. *Sankhya: The Indian Journal of Statistics, Series A* 26:17–24.

1966 Bhattacharya, P. K. Estimating the mean of multivariate normal population with general quadratic loss function. *Annals of Mathematical Statistics* 37:1819–1824.

1967 Bhattacharya, P. K. Efficient estimation of a shift parameter from grouped data. *Annals of Mathematical Statistics* 38:1770–1787.

1967 Bhattacharya, P. K. Estimation of the probability density function and its derivatives. *Sankhya: The Indian Journal of Statistics, Series A* 29:373–382.

1971 Bhattacharya, P. K. Transformations of Gaussian processes to the

Brownian motion. *Annals of Mathematical Statistics* 42:2008–2017.

1971 Bhattacharya, P. K., Bartels, P. H., Bahr, G. F. and Wied, G. L. A test statistic for detecting the presence of abnormal cells in a sample. *Acta Cytologica* 15:533–544.

1971 Bhattacharya, P. K., Bartels, P. H., Bahr, G. F., Bellamy, J., Bibbo, M. and Wied, G. L. Automated Cytodiagnosis. In *Tutorial Proceedings, Section on Automated Cytology*, University of Chicago Printing Dept., Chicago, Illinois.

1972 Bhattacharya, P. K. Probabilities of large deviations of sums of random variables. *Sankhya: The Indian Journal of Statistics, Series A* 34:9–16.

1972 Bhattacharya, P. K. Order of dependence in a stationary normally distributed two-way series. *Annals of Mathematical Statistics* 43: 1792–1807.

1972 Bhattacharya, P. K. and Smith, R. P. Sequential probability ratio test for the mean value function of a Gaussian process. *Annals of Mathematical Statistics* 43:1861–1873.

1973 Bhattacharya, P. K. and Mallik, A. Asymptotic normality of the stopping times of some sequential procedures. *Annals of Statistics* 1: 1203–1211.

1973 Bhattacharya, P. K., Bartels, P. H., Taylor, J. and Wied, G. L. A decision procedure for automated cytology: test statistic for detecting sample abnormality, and inadequacy. *Acta Cytologica* 17:538–552.

1974 Bartels, P. H., Bhattacharya, P. K., Bellamy, J. C., Bahr, G. F., Bibbo, M. and Wied, G. L. Computer generated, synthetic cell images. *Acta Cytologica* 18:155–164.

1974 Bhattacharya, P. K. Convergence of sample paths of normalized sums of induced order statistics. *Annals of Statistics* 2: 1034–1039.

1975 Bhattacharya, P. K., Bartels, P. H., Bibbo, M., Taylor, J. and Wied, G. L. Estimation procedure for the cellular composition of cervical smears. *Acta Cytologica* 19:366–373.

1976 Bhattacharya P. K. An invariance principle in regression analysis. *Annals of Statistics* 4:621–624.

1976 Bhattacharya, P. K. and Brockwell, P. J. The minimum of an additive process with applications to signal estimation and storage theory. *Z. Wahrscheinlichkeitstheorie verw. Gebiete* 37:51–75.

1981 Bhattacharya, P. K. and Frierson, Jr., D. A nonparametric control chart for detecting small disorders. *Annals of Statistics* 9(3):544–554.

1981 Bhattacharya, P. K. Posterior distribution of a Dirichlet process from quantal Response data. *Annals of Statistics* 9(4):803–811.

1981 Bhattacharya, P. K., Dietrich, J. K., Heckerman, D. G. and Meyers, D. E. Variable wages and prices and the demand for capital with discrete and continuous adjustments. *International Economics Review* 22(2):295–307.

1982 Bhattacharya, P. K., Gastwirth, J. L. and Wright, A. L. Two modified Wilcoxon tests for symmetry about an unknown location parameter. *Biometrika* 69(2):377–382.

1983 Bhattacharya, P. K., Chernoff, H. and Yang, S. S. Nonparametric estimation of the slope of a truncated regression. *Annals of Statistics* 11(2):505514.

1983 Bhattacharya, P. K. Justification of a $K--S$ type test for the slope of a truncated regression. *Annals of Statistics* 11(2):697–701.

1983 Bhattacharya, P. K. and Gastwirth, J. L. A non-homogeneous Markov model of a chain-letter scheme. *Recent Advances in Statistics: A Festschrift in Honor of Hermann Chernoff's Sixtieth Birthday*, 143–174, Rizvi, M. H., Rustagi, I. S. and Siegmund, D. (Eds.). Academic Press.

1984 Bhattacharya, P. K. Induced order statistics: Theory and applications. *Handbook of Statistics*, 4:383–403, P. R. Krishnaiah and P. K. Sell

(Eds.). North Holland, Elsevier Science Publishers.

1984 Gastwirth, J. L. and Bhattacharya, P. K. Two probability models of pyramid or chain letter schemes demonstrating that their promotional claims are unreliable. *Operations Research* 32(3):527–536.

1985 Bhattacharya, P. K. and Mack, Y. P. A two-stage procedure for nonparametric estimation. *Statistics & Decisions, Supplemental Issue* No. 2, 143–153.

1987 Bhattacharya, P. K. and Mack, Y. P. Weak convergence of $k - NN$ density and regression estimators with varying k and applications. *Annals of Statistics* 15(2):976–994.

1987 Bhattacharya, P. K. Maximum likelihood estimation of a change-point in the distribution of independent random variables: General Multiparameter Case. *Journal of Multivariate Analysis* 23(2): 183–208.

1990 Bhattacharya, P. K. and Gangopadhyay, A. Kernel and nearest neighbor estimation of a conditional quantile. *Annals of Statistics* 18(3):1400–1415.

1990 Bhattacharya, P. K. and Mack, Y. P. Multivariate data-driven k-NN function estimation. *Journal of Multivariate Analysis* 35(1):1–11.

1990 Bhattacharya, P. K. Weak convergence of the log likelihood process in the two-phase linear regression problem. *Proceedings of the R. C. Bose Symposium on Probability. Statistics and Design of Experiments.* 145–156, Bahadur, R. R. (Ed.), Wiley Eastern.

1991 Bhattacharya, P. K. Analysis of observational studies from the point of view of nonparametric regression. *Proceedings of the NATO Advanced Studies Institute on Nonparametric Functional Estimation and Related Topics* 587–602, Roussas, G. G. (Ed.), Kluwer

1993 Bhattacharya, P. K. and Zhao, P. Almost sure uniform convergence rates for M-smoothers with non-monotone score functions. *Statistics and Probability Letters* 16: 189–196.

1994 Bhattacharya, P. K. Some aspects of change-point analysis. In *Change-point Problems*, Carlstein, E., Müller, H. G. and Siegmund, D. (Eds.), IMS Lecture Notes — Monograph Series 23: 28–56.

1994 Bhattacharya, P. K. and Zhao, P. A rank-CUSUM procedure for detecting small changes in a symmetric distribution. In *Change-point Problems*, Carlstein, E. Müller, H. G. and Siegmund, D. (Eds.), IMS Lecture Notes—Monograph Series. 23: 57–65

1994 Bhattacharya, P. K. and Müller, H. G. Asymptotics for nonparametric regression. *Sankhya, Series A*, P. C. Mahalanobis Centennial Issue. (Invited Paper)

1996 Bhattacharya, P. K. and Zhao, P. A generalized CUSUM procedure for sequential detection of change-point in a parametric family when the initial parameter is unknown. *Sequential Analysis* 15(4): 311–325.

1997 Bhattacharya, P. K. and Zhao, P. Semiparametric inference in a partial linear mode. *Annals of Statistics* 25(1): 244–262.

1998 Bhattacharya, P. K. and Burman, P. Semiparametric estimation in the multivariate Louisville model. *Journal of Multivariate Analysis* 65(1): 1–18.

1999 Bhattacharya, P. K. and Gastwirth, J. L. Estimation of the odds-ratio in an observational study using bandwidth-matching. *Nonparametric Statistics* 11: 1–12.

2005 Bhattacharya, P. K. A maximal inequality for nonnegative submartingales. *Statistics and Probability Letters* 72: 11–12